Obstructive Airway Diseases

Role of Lipid Mediators

Obstructive Airway Diseases

Role of Lipid Mediators

Edited by

Abhijit Ray
Punit Kumar Srivastava

CRC Press
Taylor & Francis Group
Boca Raton London New York

CRC Press is an imprint of the
Taylor & Francis Group, an **informa** business

CRC Press
Taylor & Francis Group
6000 Broken Sound Parkway NW, Suite 300
Boca Raton, FL 33487-2742

First issued in paperback 2019

© 2012 by Taylor & Francis Group, LLC
CRC Press is an imprint of Taylor & Francis Group, an Informa business

No claim to original U.S. Government works

ISBN-13: 978-1-4398-5140-1 (hbk)
ISBN-13: 978-1-138-37445-4 (pbk)

Visit the Taylor & Francis Web site at
http://www.taylorandfrancis.com

and the CRC Press Web site at
http://www.crcpress.com

Contents

Foreword

Incidences of both asthma and chronic obstructive pulmonary disease (COPD) are increasing throughout the world, especially in the developing countries. There has been a major investment by the pharmaceutical industry in the search for new and more effective treatments of these diseases. Although treatments for mild and moderate asthma are very effective, severe asthma is poorly controlled by existing therapies, and there are no very effective anti-inflammatory treatments for COPD. In both COPD and severe asthma, there is a relative resistance to the anti-inflammatory effects of glucocorticoids; therefore, alternative anti-inflammatory strategies are needed. The development of new anti-inflammatory treatments is dependent on a better understanding of the inflammatory process, particularly in severe asthma and COPD. There has been an enormous increase in the research of protein mediators, particularly cytokines, chemokines, and growth factors, and this has led to several potential therapies, although at the moment these approaches have proved to be very disappointing in terms of clinical benefit. Although less attention has been paid to lipid mediators of inflammation, they play a key role in inflammatory cascades involved in asthma and COPD. Moreover, they signal through G-protein-coupled receptors, which are more amenable to the development of small-molecule antagonists than the more complex cytokine and growth factor receptors. The enzymes involved in the synthesis of lipid mediators have also been discussed in detail and these are important targets for inhibition, particularly as these enzymes tend to generate several lipids.

This book is very timely as it brings together a wealth of information on lipid mediators in obstructive airway diseases. There are comprehensive discussions of the major prostaglandins, leukotrienes, platelet-activating factors, and the more recently described oxo-ETE and sphingolipids, many of which are attractive targets for inhibition. The book also covers the interesting emerging area of anti-inflammatory lipid mediators, such as lipoxins, resolvins, and protectins, which are involved in the resolution of inflammation and might be exploited for therapeutic benefit in the future. The contributors are international authorities on lipid mediators and this makes the book an invaluable resource for anyone interested in the mechanism of airway diseases or in finding new therapeutic approaches for very common and troublesome diseases. Dr. Abhijit Ray and Dr. Punit Srivastava have done an excellent job in putting all this together and I think this book will prove to be a very valuable resource to all those involved in researching, teaching, and studying airway inflammation as well as to those involved in drug development.

Professor Peter J. Barnes, FRS
Head of Respiratory Medicine
Imperial College London

Preface

Incidences of inflammatory airway diseases are on the rise. This book is an attempt to understand inflammatory airway diseases, their prevalence, pathophysiology and existing therapeutic options. In this book we highlight the fact that despite a lot of effort, meaningful therapeutic options for inflammatory airway diseases are far from satisfactory. Thus, need for drug discovery research in this therapeutic segment remains essential. Improved understanding of lipid biology has broadened the scope of looking at different lipid molecules and their signaling mechanisms critically as sources of potential drug discovery targets.

This book explains what constitutes a lipid and how a lipid is broken down to generate biologically active mediators. We look at important lipid mediators, such as arachidonic acid, platelet-activating factor, and lysophosphatidic acid. How these molecules act and what role they play in airway inflammation are discussed. Some chapters of the book discuss the products of arachidonic acid metabolism, namely, leukotrienes, prostaglandins, epieicosatrienoic acid, and oxoeicosatetraenoic acid. There is evidence that different arachidonic acid metabolites play a role in airway inflammation. Drug discovery effort around these lipid metabolites remains an attractive proposition because many of these molecules act on cell surface G-protein-coupled receptors.

Most lipid mediators promote inflammation. However, several lipid-derived mediators actually resolve inflammation and act as anti-inflammatory molecules. In the coming years, it may be possible to address inflammation using the paradigm of resolution. We have devoted a chapter on anti-inflammatory lipid mediators such as lipoxin and on drug discovery efforts to mimic lipoxin action in inflammatory airway disease conditions.

Although intended to describe lipid mediators predominantly, this book also discusses enzymes that play an important role in the process of lipid mediator synthesis. We have included phospholipase A_2 family of enzymes that initiate phospholipid breakdown and 5-lipoxygenase enzyme and 5-lipoxygenase-activating protein that contribute toward the generation of leukotrienes and lipoxins. A chapter each has been devoted to discuss the biology of these proteins and the drug discovery effort around them.

Sphingolipids such as ceramide and sphingosine have been implicated in experimental models of asthma and COPD. Distinct receptors for sphingosine-1-phosphate have been identified in inflammatory cells. We have examined the evidence in support of the role played by sphingosine and ceramide in inflammatory airway disease.

Kinases that are activated by lipid mediators as well as kinases that trigger the generation of lipid messengers have also been discussed. There is evidence emerging in support of these proteins as targets for drug discovery.

We hope this book will be helpful to researchers, teachers, and students of inflammation and lipid biology. The take-home message from this book should include how lipid metabolism starts, important lipid mediators, and how we can tap them to ameliorate inflammatory diseases of the airway.

We take this opportunity to thank the contributors for making this book possible and our publisher for having faith in us. We also thank the management of our organization for allowing us to work on this book. Last but not the least, we acknowledge the patience of our families during the preparation of this book.

Editors

Abhijit Ray is a pharmacologist by training, having pursued graduate studies in pharmacology at the University of British Columbia, Canada, from 1985 to 1992. He completed a three-year postdoctoral training at the University of Ottawa, Canada, in the area of molecular pharmacology from 1992 to 1995. As a member of the drug discovery research team at New Drug Discovery Research, Ranbaxy Laboratories Ltd., India, since 1995, Dr. Ray has played an important role in progressing four molecules for human trial. Two of these molecules were designed to treat airway inflammation associated with asthma and COPD. Dr. Ray has published more than 60 articles in peer-reviewed journals, which include 17 reviews, 22 research papers, and 29 abstracts. Since July 2011, Dr. Ray has been the vice president and head of biology research at Daiichi Sankyo Life Science Research Centre India, Gurgaon, India. As a member of the drug discovery team at Daiichi Sankyo Life Science Research Centre India, he manages a group of 60 scientists engaged in discovery research for drugs in the areas of infectious and inflammatory diseases.

Punit Kumar Srivastava has been the assistant director of biology at Daiichi Sankyo Life Science Research Centre India, Gurgaon, India, since July 2010. Dr. Srivastava was a member of the drug discovery group of Ranbaxy Laboratories Ltd. from 2004 to June 2010. He is a biochemist by training, having pursued his graduate studies in the area of enzymology at the School of Biotechnology, Banaras Hindu University, Varanasi, India, from 1997 to 2001. He completed a three-year postdoctoral training at the Medical College of Wisconsin, in the area of molecular pharmacology from 2001 to 2002 and at the University of Connecticut, in the area of arachidonic acid metabolic pathway from 2002 to 2004. Dr. Srivastava was awarded a gold medal in his postgraduate degree in biochemistry and graduate degree in chemistry from Banaras Hindu University in 1997 and 1995, respectively.

Dr. Srivastava has worked for more than six years in drug discovery research, which has resulted in one molecule entering human trial for COPD. He has published more than 15 articles in peer-reviewed journals and filed at least six patents. As a member of the drug discovery team at Daiichi Sankyo Life Science Research Centre India, Dr. Srivastava is leading projects in the area of inflammation.

Contributors

Puneet Chopra
Daiichi Sankyo Life Science
    Research Centre India
Gurgaon, India

Sunanda Ghosh Dastidar
Daiichi Sankyo Life Science
    Research Centre India
Gurgaon, India

Rishabh Dev
Daiichi Sankyo Life Science
    Research Centre India
Gurgaon, India

Manish Diwan
Daiichi Sankyo Life Science
    Research Centre India
Gurgaon, India

Hua Dong
Department of Entomology
and
UCD Cancer Center
University of California
Davis, California

Suman Gupta
Daiichi Sankyo Life Science
    Research Centre India
Gurgaon, India

Bruce D. Hammock
Department of Entomology
and
UCD Cancer Center
University of California
Davis, California

Kazuhiro Ito
National Heart and
    Lung Institute
Imperial College
London, United Kingdom

Jitesh P. Iyer
Daiichi Sankyo Life Science
    Research Centre India
Gurgaon, India

Bruce D. Levy
Harvard Medical Research
Brigham and Women's Hospital
Boston, Massachusetts

Sanjay Malhotra
Daiichi Sankyo Life Science
    Research Centre India
Gurgaon, India

Nicolas Mercado
National Heart and Lung Institute
Imperial College
London, United Kingdom

Makoto Murakami
Tokyo Metropolitan Institute of
    Medical Science
Tokyo, Japan

Viswanathan Natarajan
Department of Pharmacology and
    Medicine
University of Illinois at Chicago
Chicago, Illinois

Carole A. Oskeritzian
Department of Biochemistry and
    Molecular Biology
Virginia Commonwealth University
Richmond, Virginia

Carlo Pergola
Institute of Pharmacy
Friedrich-Schiller-University Jena
Jena, Germany

Kent E. Pinkerton
Center for Health and the Environment
University of California
Davis, California

William S. Powell
Meakins-Christie Laboratories
McGill University
Montreal, Canada

Abhijit Ray
Daiichi Sankyo Life Science
    Research Centre India
Gurgaon, India

Joshua Rokach
Claude Pepper Institute
and
Department of Chemistry
Florida Institute of Technology
Melbourne, Florida

Jitendra Anant Sattigeri
Daiichi Sankyo Life Science
    Research Centre India
Gurgaon, India

V. Senthil
Daiichi Sankyo Life Science
    Research Centre India
Gurgaon, India

Charles N. Serhan
Harvard Medical Research
Brigham and Women's Hospital
Boston, Massachusetts

Sameer Sharma
Daiichi Sankyo Life Science
    Research Centre India
Gurgaon, India

Punit Kumar Srivastava
Daiichi Sankyo Life Science
    Research Centre India
Gurgaon, India

Yoshitaka Taketomi
Tokyo Metropolitan Institute of
    Medical Science
Tokyo, Japan

Oliver Werz
Friedrich-Schiller
    University Jena
Jena, Germany

Jun Yang
Department of Entomology
and
UCD Cancer Center
University of California
Davis, California

Yutong Zhao
Department of Medicine
University of Pittsburgh School of
    Medicine
Pittsburgh, Pennsylvania

1 Obstructive Airway Diseases
Epidemiology, Etiology, and Therapeutic Options

Abhijit Ray

CONTENTS

1.1 INTRODUCTION

All over the world, a very large number of people suffer from significant morbidity and mortality as a result of diseases affecting the respiratory system. The respiratory system functions to bring oxygen to and remove carbon dioxide from living organisms. Blockage of the respiratory passage as a result of underlying inflammation

1

results in airflow obstruction. Asthma and chronic obstructive pulmonary disease (COPD) are respiratory ailments that exhibit a propensity for airflow blockage. Underlying factors that determine individual susceptibility to asthma and COPD are not very clear. It is believed that an ageing population, changing lifestyles, and the environment, as well as genetics may contribute toward initiation and propagation of these conditions. Although guidelines to treat asthma and COPD have been framed, therapeutic options for both diseases remain limited. In the following sections, we will introduce the respiratory system and diseases that affect the system with special reference to asthma and COPD. Our effort will be to revisit epidemiology and etiology at the cellular and molecular levels for asthma and COPD. We will also look at existing therapeutic options and key unmet needs in this area of obstructive airway diseases.

1.2 RESPIRATORY SYSTEM

In this section, we will briefly acquaint readers with important components of the respiratory system, which is relevant to the discussion in this chapter and in the context of this book. No attempt will be made to describe in detail the anatomy and physiology of the system at the molecular and cellular levels. The respiratory system brings oxygen into our bodies, and helps us get rid of carbon dioxide. The respiratory system consists of all the organs involved in breathing (Figure 1.1), including the

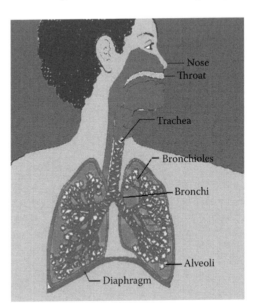

FIGURE 1.1 A schematic representation of the respiratory system. The respiratory system consists of the nasal passage, trachea, bronchi, and alveoli. Air moves in through the nasal passage and passes through the trachea. The trachea divides into two bronchi. Each bronchus again divides into smaller bronchioles that end in the alveolar sac. In inflammatory airway diseases, there is an inflammation of the airway as well as a constriction of the smooth muscle of the tracheal air passage.

nose, pharynx, larynx, trachea, bronchi, and lungs. All the structures mentioned work like a system of pipes through which the air is funneled down into our lungs. There, in very small air sacs called alveoli, oxygen is brought into the bloodstream and carbon dioxide is pushed from the blood out into the air.

Passageways of the respiratory system are lined with mucous membrane made of columnar epithelium. These cells are ciliated from nose to bronchioles. Simple squamous epithelial cells line the alveoli and alveolar ducts. Olfactory cells and mucus cells are embedded in the epithelium. The bronchioles are lined with a simple cuboidal epithelium, which is mainly composed of Clara cells. Alveoli are composed of type I and type II pneumocytes. Type I cells secrete extracellular matrix like glycosaminoglycan and form blood–gas barrier. Type II cells secrete pulmonary surfactant. Type II cells can replicate and give rise to type I and II cell types.

Bordering the alveolar lumen are wandering cells called alveolar phagocytes, or macrophages. These cells engulf dust, bacteria, and other inhaled particles that are trapped in the pulmonary surfactant. After they become filled with debris, macrophages migrate to the bronchioles from where they get carried by ciliary action to the pharynx and swallowed. Alternatively, they may also migrate into the interstitium where they are removed via the lymphatic vessels.

From the perspective of the present chapter, it is the air tubes like trachea and bronchi that are affected in asthma. In COPD, it is the terminal bronchioles and air sacs that are affected.

1.3 EPIDEMIOLOGY OF RESPIRATORY DISEASES

1.3.1 ASTHMA

Asthma and COPD are the two diseases of the respiratory system with a very high rate of prevalence. According to the Global Initiative for Asthma (GINA) guidelines (Bateman et al., 2008; GINA, 2010), in 2006, nearly 46 million people had asthma, of which 9.5 million were children and 27.5 million were adults, including 8.5 million senior citizens (Bateman et al., 2008; GINA, 2010; DATA Monitor, 2006). Asthma prevalence varied from 1% to 18% in countries all over the world. The United States had the largest number of asthma patients, in terms of both absolute number and percentage of the population, followed closely by the United Kingdom, France, and Germany (Table 1.1). High prevalence of asthma has also been reported from Australia and New Zealand (Masoli et al., 2004).

In India, the prevalence of asthma was less than 5% of the population (Masoli et al., 2004). India had, according to various estimates, between 15 and 28 million people with asthma in 2006 (Aggarwal et al., 2006). In a document prepared for the National Commission for Macroeconomics and Health (NCMH), Government of India, Murthy and Sastry (2005) predicted that the number of asthma patients is likely to go up nearly twofold by 2016 in urban and rural areas of India. The number of chronic asthma cases of mild, moderate, and severe categories may increase from 28 million in 2006 to 35 million in 2016 (Murthy and Sastry, 2005). All over the world, the number of people suffering from asthma is likely to rise because of the

TABLE 1.1
Asthma Prevalence in 2006

	Million	Percentage
U.S.	22.5	7.9
U.K.	5.6	10
France	3.1	5.6
Germany	4.1	5.8
Spain	1.8	4.8
Italy	2.5	4.9
Japan	5.4	4.8
Australia		25–30
New Zealand		>20
India	15	<5
China		<5

Source: Adapted from DATA Monitor. 2006. Pipeline insight asthma/
COPD: Targeted therapies on the horizon; Masoli, M. et al.
2004. *Allergy* 59:469–78.

increase in population growth in the United States as well as the increase in prevalence in Africa, Latin America, and Asia (DATA Monitor, 2006).

Asthma can be divided into controlled, partly controlled, and uncontrolled,
depending on the degree and frequency of attack and its response to anti-inflammatory
therapy (Table 1.2). Patients with partly controlled and uncontrolled asthma experience exacerbation that may be fatal unless hospitalization is undertaken. According

TABLE 1.2
Asthma Subtypes

Controlled	Daytime symptom—none
	Limitation of activity—none
	Nocturnal awakening—none
	Rescue medicine—none (<twice per week)
	Exacerbation—none
Partly controlled	Daytime symptom—more than twice/week
	Limitation of activity—any
	Nocturnal awakening—any
	Rescue medicine—more than twice per week
	Exacerbation—one or more/year
Uncontrolled	Three or more features of partly controlled asthma present in any week
	Exacerbation—one in any week

Source: Reprinted from Global Initiative of Asthma (GINA) 2010. Global Strategy for asthma
management and prevention. Available at www.ginaasthma.org, last updated 2010.
With permission.

to the World Health Organization, nearly 250,000 people die of asthma every year (Bateman et al., 2008; GINA, 2010). Asthma is also responsible for the absence from work and is a burden on healthcare because of the cost of hospitalization and treatment. The incidence of severe asthma needing hospital admission is increasing, especially among children. Usually, the cost of treating severe asthma is very high compared to mild and moderate persistent asthma. According to the estimate, severe steroid-resistant or steroid-dependent asthma constitutes 10% of the total asthma population but consumes 50–80% of the healthcare cost (DATA Monitor, 2006; Holgate and Polosa, 2006; Barnes, 2008). In India, the number of acute cases of asthma needing hospitalization is predicted to go up from 0.4 million incidences in 2006 to 0.5 million in 2016 (Table 1.3). The per-patient treatment cost is likely to increase by 1.5-fold from Rs. 9000 per year in 2006 to 13,500 per year in 2016 (Murthy and Sastry, 2005).

1.3.2 COPD

According to the Global Initiative for Chronic Obstructive Lung Disease (GOLD), by 2020, COPD is likely to be the third leading cause of mortality and the fifth leading cause of morbidity. This has been attributed to smoking habits in developed countries, where nearly a quarter of the population still smoke, increase in the ageing population of developed countries, and increase in the population of developing countries like China and India (Barnes, 2004; DATA Monitor, 2006; Rabe et al., 2007; GOLD, 2010).

As shown in Table 1.4, the GOLD has classified COPD patients into mild, moderate, severe, and very severe based on their lung function (Rabe et al., 2007; GOLD, 2010). In 2006, according to DATA Monitor survey, the total COPD population in seven countries was around 28.8 million, of which 8.8 million were mild, 9.4 million were moderate, 6.8 million were severe, and 3.4 million constituted very severe COPD patients (Table 1.5). The United States reported the highest

TABLE 1.3
Patient Number and Cost of Treatment: Chronic and Acute Asthma

	Patient Number (Rs. Million)		Cost of Treatment (Rs./Patient/Year)	
	Chronic[a]	Acute	Chronic	Acute[b]
1996	22	0.34	1825	3909
2006	28	0.42	3447	14,879
2016	35	0.50	5060	21,848

Source: Adapted from Murthy, K. J. R. and J. G. Sastry 2005. Economic burden of asthma. http://www.whoindia.org/LinkFiles/Commision_on_Macroeconomic_and_Health_Bg_P2_Economic_burden_of_asthma.pdf.

[a] Chronic cases include moderate and severe asthma patients.

[b] Acute cases include hospitalization cost per event.

TABLE 1.4
COPD Type

At risk	Chronic symptoms
	Normal spirometry
Mild	FEVI ≥ 80%
	FEV1/FVC < 70%
	With or without chronic symptom
Moderate	50% ≤ FEVI ≤ 80%
	FEV1/FVC < 70%
	With or without chronic symptom
Severe	30% ≤ FEVI ≤ 50%
	FEV1/FVC < 70%
	With or without chronic symptom
Very severe	FEVI ≤ 30% or
	FEVI ≤ 50%
	With chronic respiratory failure
	FEV1/FVC < 70%

Source: Reprinted from Global Initiative of Lung Disease (GOLD) 2010. Global strategy for the diagnosis, management and prevention of chronic obstructive pulmonary disease, available at www.goldcopd. org, last updated 2010. With permission.

number of total COPD patients, including patients in very severe category. The incidence of COPD increases with age as lung function declines. Nearly 9–10% of adults at or beyond 40 years of age exhibit symptoms of COPD. A document detailing the prevalence of COPD in India was compiled by Jindal et al. (2004). COPD prevalence was between 2% and 9% in different areas of the country. According to Murthy and Sastry (2005) the incidence of asthma is likely to go up by 50% in the next 10 years.

TABLE 1.5
COPD Prevalence

	Million	Mild	Moderate	Severe	Very Severe
U.S.	10.6	2.9	3.4	2.8	1.5
U.K.	2.4	0.7	0.9	0.6	0.2
France	2.4	0.7	0.7	0.6	0.4
Germany	3.6	1.2	1	0.8	0.5
Spain	1.6	0.4	0.6	0.4	0.2
Italy	2.5	0.8	0.7	0.6	0.4
Japan	5.7	2.3	1.8	1.1	0.5
Total	28.8				

Source: Adapted from DATA Monitor. 2006. Pipeline insight asthma/COPD: Targeted therapies on the horizon.

1.4 ASTHMA AND COPD: CELLULAR AND MOLECULAR LEVEL

Asthma and COPD affect the respiratory system and lead to breathing difficulty due to the obstruction of the airway caused by bronchoconstriction, inflammation, and mucus secretion. Both diseases have different causative factors and different cellular and molecular mediators, and affect different parts of the airway (Barnes, 2008).

1.4.1 ASTHMA

Asthma is a chronic inflammatory disease of conducting airway. Asthma attack is accompanied by a widespread reversible obstruction of the airway. An initial acute bout of bronchoconstriction is followed by a late phase response. Late phase response is believed to be the consequence of the migration of inflammatory cells into the airway and the activation and release of different inflammatory mediators. Another feature of asthma is airway hyperresponsiveness, where the airway becomes sensitive to noxious agents in the environment, such as allergens, gases, pollutants, etc. An asthma attack is manifested by wheezing, chest tightness, and difficulty in breathing at night or early morning. Many asthma patients experience exacerbation (Table 1.3). Exacerbation is progressive increase in wheezing, chest tightness, and difficulty in breathing. If not controlled, exacerbation can be fatal (GINA, 2010).

Several factors can precipitate an attack of asthma, most important being allergens like pollen, house dust mite, pet dander, cockroach antigen, fungus, and mold. Although 60% of asthma patients are found to be atopic, in severe asthma, atopy may not be the dominant trigger. Airway infection (viral infection), environmental pollutants, occupational sensitizers, cigarette smoke, certain drugs (β-adrenoceptor antagonist, aspirin), diet, psychological status, endocrine factors (obesity, menstrual cycle), etc. play an important role in precipitating bronchoconstriction and asthma attack. In addition, the genetic makeup of a person can make him more susceptible to asthma attack. Obesity is also considered an important determinant of asthma attack.

The cellular and molecular basis of asthma have been researched extensively (Barnes et al., 1998; Wills-Karp, 1999; Busse and Lemansk, 2001; Barnes, 2008). It is well accepted that mast cell degranulation plays an important role in allergen-induced bronchoconstriction of asthma. Cross-linking of high-affinity immunoglobulin receptors on mast cells by immunoglobulin E (IgE) leads to degranulation of mast cells. Spasmogens thus released cause bronchoconstriction. Chemotactic agents attract inflammatory cells like eosinophils that contribute toward local inflammation. However, it has been understood that apart from mast cells and eosinophils, there are many other cell types—circulating inflammatory cells like T-lymphocytes antigen-presenting cells like macrophages and dendritic cells, along with airway structural cells like smooth muscle, fibroblast, epithelial, endothelial cells, goblet cells and nerve cells may also contribute to asthma symptom and severity. Hyperplasia of the smooth muscle can enhance bronchoconstriction. Whereas fibroblast may give rise to the generation of the extracellular matrix component like collagen and produce fibrosis. Goblet cell hyperplasia enhances mucus secretion that acts as a block to airflow. Sensitized nerve ending may exaggerate bronchoconstrictor response in the airway (Figure 1.2).

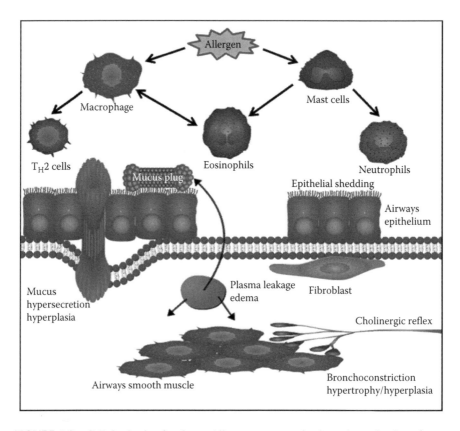

FIGURE 1.2 Cellular basis of asthma. Allergen exposure leads to the activation of mast cells and macrophages. Mast cells release bronchoconstrictor and proinflammatory mediators. Eosinophils, a key inflammatory cell type, are recruited into the airway, which release mediators that can cause localized tissue damage as well as act as chemotactic factor and sensitized cholinergic nerve terminals. In severe asthma, neutrophils are recruited and subepithelial fibrosis is initiated. (Reviewed in Barnes, P. J. 2004. *Nat Rev Drug Discov.* 3:831–44.)

At the molecular level, it is accepted that IgE is central to allergic asthma. Inhaled antigen is exposed to antigen-presenting cells lining the airway—mainly dendritic cells and to some extent epithelial cells and macrophages. Antigens are processed and presented to uncommitted T-lymphocytes (T_H0). T_H0 cells are committed to Type 2 T-lymphocyte (T_H2) phenotype under the influence of cytokines like IL4 and antigen-presenting cells. CD4$^+$ T_H2 cells interact with B lymphocytes to secrete IgE. T_H2 cells also release different cytokines like IL4, IL5, IL3, interleukin 6 (IL6), interleukin 10 (IL10), granulocyte macrophage colony-stimulating factor (GMCSF), etc. These cytokines play an important role in the recruitment of mast cells and eosinophils.

Mast cells express receptors for IgE, the central molecule in allergic asthma. IgE binds to its receptors on mast cells. Subsequent exposure to the allergen leads to the cross-linking of IgE-bound receptors to mast cells, leading to degranulation. Mast cells release a variety of different mediators. Some of these agents like histamine, leukotrienes, platelet-activating factor, prostaglandin D_2, etc. bring about broncho-constriction, increase the permeability of the blood vessel, and cause edema and inflammation. Cytokines like IL5, IL6, interleukin 8 (IL8), and GMCSF help in the recruitment, maturation, and prolongation of the life of mast cells and eosinophils and increase mucus production. Chemotactic proteins like regulated on activation normal T expressed and secreted (RANTES), monocyte chemotactic protein 1 (MCP-1), macrophage inhibitory protein 1B (MIP1B), and macrophage inhibitory protein 1A (MIP1A) contribute toward inflammatory cell recruitment.

Eosinophils are dominant cell types in allergic inflammation. These cells migrate as a result of cytokine and chemokine generation at the site of inflammation. Eosinophils release their granule contents at the site of inflammation, which include eosinophil-derived neurotoxin, eosinophil chemotactic protein, several cytokines [interleukin 2 (IL2), interleukin 3 (IL3), interleukin 4 (IL4), interleukin 5 (IL5), GMCSF, transforming growth factor-β (TGFβ), etc.], and lipid mediators (leukotri-enes, prostaglandins, platelet-activating factor, etc.). All these cause significant damage to the site and attract more inflammatory cells, and the process is perpetu-ated. Although eosinophils have been reported to be the most important inflamma-tory cells in asthma, it is not clear if eosinophils are the cause or the effect. A monoclonal antibody to IL5, a key cytokine that promotes eosinopoiesis, did not have any effect on airway hyperresponsiveness and late asthmatic response in humans (Leckie et al., 2000).

1.4.1.1 Severe Asthma

Corticosteroids are very effective in controlling symptoms of asthma of mild to moderate severity. However, not all asthma patients respond to corticosteroids. Considered as severe asthma patients, steroid-resistant asthmatics constitute 5–10% of the asthma population. These patients show a fixed airway obstruction. Persistent untreated eosinophilic inflammation leads to tissue remodeling, characterized by fixed airflow obstruction due to the increase in amount and size of the smooth muscle or the deposition of extracellular matrix proteins like collagen under the epithelium. This remodeling results in thicker and stiffer airway with progressive decline of airway function. It has been proposed that abnormality in the glucocorticoid signal transduction pathway due to the abnormal expression of receptor subtype or altered binding to glucocorticoid receptor or abnormality in histone deacetylase pathway may contribute toward resistant asthma (Wenzel, 2006; Holgate and Polosa, 2006; Barnes, 2008).

At the cellular level, inflammation in steroid-resistant asthma is marked by the presence of neutrophils. Lymphocytes driving the inflammation are a mixture of Type 1 T-lymphocyte (T_H1) and T_H2 cells. The presence of CD8$^+$ lymphocytes has been reported. At the molecular level, tumor necrosis factor α (TNFα) and IL8 are known to drive inflammation. There is involvement of proteolytic enzymes, such as neutrophil elastase, and matrix metalloproteinase family member, like matrix metallo proteinase 9 (MMP9). Tissue remodeling due to the breakdown of extracellular matrix by metalloproteinases and the deposition of collagen and proteoglycans in subepithelial lamina leads to thicker and stiffer airway (Holgate and Polosa, 2006).

In another subgroup of asthma patients, pauci inflammatory asthma, no inflammation is reported (Wenzel, 2006; Holgate and Polosa, 2006). There is smooth muscle proliferation and hyperplasia. In general, smooth muscle cells become secretory in nature and their response to corticosteroids, bronchoconstrictors, and bronchodilators changes.

1.4.2 COPD

Chronic obstructive pulmonary disease is a chronic progressive inflammatory disease of the alveolar air sac and the small bronchiole. Inflammation of the small airway leads to narrowing of the passage through which air moves in and out of alveolar sac. Damage to the alveolar sac by proteolytic enzymes results in loss of surface area for gas exchange and loss of elastic recoil. Exchanged air is not thrown out of the sac. Mucus gland hyperplasia leads to excessive mucus secretion, which creates further blockade to airflow. COPD patients, depending upon disease severity, feel breathless while doing routine chore. Because the airway is damaged permanently by protease, airflow obstruction in COPD becomes irreversible and progressive. It is more or less established that cigarette smoke is one of the predominant cause of COPD. However, not all smokers are affected with COPD. There is also evidence that people exposed to smoke in a closed environment, as in indoor wood fire cooking, may get affected by the disease (GOLD, 2010; Salvi and Barnes, 2009).

At the cellular level, COPD is an interplay of macrophages, CD8$^+$ T-lymphocytes, and neutrophils. Noxious stimuli from cigarette smoke or the environment activate macrophages to release chemoattractants, cytokines, and chemokines such as TNFα, IL8, and leukotriene B$_4$ (LTB$_4$) as well as matrix metalloproteinase such as matrix metalloproteinase 12 (MMP12). Chemoattractants bring neutrophils carrying neutrophil elastase as well as MMP9. T-lymphocytes also release matrix metalloproteinases. Metalloproteinases cleave extracellular matrix proteins, prompting tissue repair and remodeling. Damage to alveolar structure by proteases results in the loss of elastic recoil of alveoli, which manifests as emphysema. Extracellular matrix fragments also act as chemotactic factors, bringing more neutrophils. Matrix metalloproteinases activate other metalloproteinases, cytokines, and chemokines. In fact, COPD is also considered to be a disease where there is imbalance between matrix metalloproteinases and their naturally occurring tissue inhibitor. Thus, the inflammation of COPD perpetuates. It is worth noting that once COPD inflammation sets in, it does not subside even when the individual stops smoking. Increased mucus secretion and loss of function of cilia lead to coughing and sputum production (Figure 1.3).

1.5 THERAPEUTICS

For a detailed reading on asthma and COPD therapeutics, some excellent reviews are available (Barnes, 2004; de Boer, 2005; Fitzgerald and Fox, 2007a,b). Asthma and COPD therapy can be grouped into controller and reliever medications. Controllers address inflammation and relievers offer symptomatic relief by providing bronchodilator action. In fact, there are very few drugs that act as a controller in COPD.

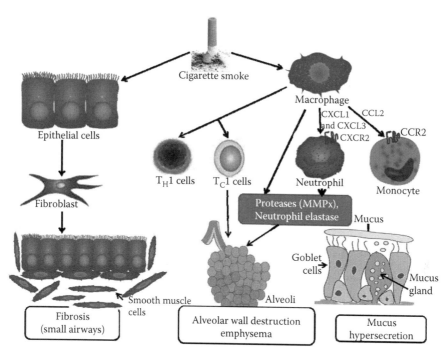

FIGURE 1.3 Cellular basis of COPD. Cigarette smoke activates alveolar macrophages to release cytokines and chemokines to bring neutrophils and lymphocytes. Macrophages release metalloelastase, and neutrophils and T-cells release metalloprotease as well as neutrophil elastase. Proteases activate other cytokines, chemokines, and proteases that perpetuate the cycle of inflammation. Proteolytic enzymes also promote tissue remodeling by alveolar destruction and block airway passage by promoting mucus hypersecretion. Cigarette smoke also promotes epithelial cells to release profibrotic mediator like TGFβ that promotes fibroblast proliferation and fibrosis of the small airway. (Reviewed in Barnes, P. J. 2008. *Nat Rev Immunol.* 8:183–92.)

There is also the option of dividing asthma and COPD therapies by the route of drug delivery. Inhaled therapy has the advantage of delivering drug to the site of action. The effect of inhaled delivery is rapid. As a result, for immediate effect, this route is preferred. Delivering drug to the site of action minimizes adverse effects by reducing the actual amount of drug and by restricting its distribution. However, inhaled delivery necessitates the use of a device that may increase the cost of medication. Patient compliance also becomes a problem with inhalers because certain inhalers require coordination between actuation and breathing for delivery of correct dose. Several types of inhalers are available, namely, dry powder inhaler, metered dose inhaler, breath-actuated pressurized metered dose inhaler, and soft mist inhaler. The most preferred delivery option remains the oral route. Not only this delivery option is less expensive, but patient compliance also is usually very high because of the ease of administration. However, only a fraction of the absorbed drug reaches the site of action, needing a large dose to be administered. This increases the propensity of adverse effects.

For the purpose of this chapter, we shall group asthma/COPD therapies by the route of delivery—inhaled and systemic.

1.5.1 INHALED THERAPY

1.5.1.1 Inhaled Corticosteroids

Inhaled corticosteroids are used commonly to treat mild persistent or controlled asthma. These agents improve the quality of life, reduce exacerbation severity, reduce asthma symptoms, improve lung function, reduce airway hyperresponsiveness, and reduce inflammation. Inhaled corticosteroids are not very effective to address the inflammation of COPD and no protective effect on the loss of lung function can be seen by the administration of a high dose of inhaled corticosteroid (Barnes, 2004; GOLD, 2010).

Many inhaled corticosteroids are available in the market; only a few new ones are shown in Table 1.6. For asthma treatment, up to a dose of 400 µg daily, these molecules are well tolerated in adult and children without any major systemic adverse effect (GINA, 2010). If it is difficult to control the asthma, the dose of inhaled corticosteroid has to be increased beyond 400 µg. Depending upon individual responsiveness, higher dose can range from 800 to 1200 µg daily.

The propensity of side effect increases with the dose of the steroid used. Higher the dose of inhaled corticosteroid, greater is the amount that goes into systemic circulation. Several approaches have been employed to minimize systemic exposure of corticosteroid. In one approach, steroids are designed to minimize their systemic exposure. For example, budesonide forms a conjugate with a protein in the airway from which it is slowly released, thus prolonging its duration of action and reducing plasma exposure. Ciclesonide on the other hand is cleaved into an active drug by the enzyme present in the airway. Thus, ciclesonide does not show an adverse effect even

TABLE 1.6
Corticosteroids for Asthma/COPD

Drug	Dose	Indication
Inhaled Corticosteroids		
Fluticasone	Up to 400 µg	Safe and used for controlled asthma
Budesonide		
Ciclesonide	400 µg and above	Used for severe asthma
Mometasone		Side effect happens depending upon
Triamcinolone		pharmacokinetic property
Beclomethasone		
Oral Corticosteroids		
Prednisone	5 mg for 3 months	Severe uncontrolled asthma
Prednisolone	50 mg for 10 days	Severe acute exacerbation

Source: Global Initiative for Asthma (GINA) 2010. Global Strategy for Asthma Management and Prevention. Available at www.ginaasthma.org, last updated 2010.

if it goes into the systemic circulation. A bulk of the inhaled dose is deposited in the gastrointestinal tract. As a result, as shown in Table 1.7, inhaled steroids have been designed to exhibit very poor gastrointestinal absorption. Many steroids exhibit rapid clearance once they reach systemic circulation.

1.5.1.1.1 Inhaled Corticosteroids in Children

According to GINA (2010), inhaled corticosteroids are efficacious in children 5 years and older as well as those younger than 5 years. Generally, utility, efficacy, and dosage remain very similar to that of adults. Side effects depend on the dose, the duration of treatment, the delivery device used, and the pharmacokinetic property of the steroid. High-dose inhaled glucocorticoids can cause growth retardation in children 5 years and older. This effect is dose dependent and steroid dependent. Steroids delay the onset of puberty and decelerate bone growth. However, at a later stage in life, children attain normal adult height. Unlike oral or systemic glucocorticoids, inhaled corticosteroid therapy does not increase the risk of fracture and reduce bone mineral density in a statistically significant manner (GINA, 2010).

1.5.1.2 Bronchodilators

1.5.1.2.1 Adrenoceptor Agonist

β-Adrenoceptor agonists act on β-adrenergic receptors in tracheal smooth muscle and cause muscle relaxation by elevating the cAMP level. Several different β-adrenoceptor agonists are available commercially. Depending upon the onset and duration of action of these agents, β-adrenoceptor agonists can be grouped into short-acting, long-acting, and ultralong-acting agents (Table 1.8).

Short-acting β-adrenoceptor agonists are used for acute relief of asthma/COPD symptoms by the inhaled route and by the intravenous route. As shown in Table 1.8, these agents have fast onset of action and the bronchodilator effect usually lasts for 4–6 h. Long-acting β-agonists (LABAs) are used in moderate to severe bronchial asthma and COPD. These agents improve lung function and quality-of-life parameters in patients suffering from asthma. LABAs offer prolonged protection against

---

TABLE 1.7

Pharmacokinetic Properties of Select Inhaled Corticosteroids

	Potency Relative to Dexamethasone	Bioavailability (%F)		$T_{1/2}$ (h)	Clearance (L/h)
		Oral	Inhaled		
Fluticasone	18	<1	17–29	7–8	66–90
Budesonide	9	11	18	2.8	183–300
Mometasone	23	<1	11	5.8	54
Ciclesonide	0.12	<1		0.36	207
Des Ciclesonide	12	<1	5.2	3.4	807

Source: Reprinted from Winkler, J., G. Hocchaus, and H. Derendorf 2004. *Proc Am Thrac Soc* 1: 356–63. With permission of the American Thracic Society, copyright American Thoracic Society.

TABLE 1.8
Bronchodilators for Asthma/COPD

	Onset (min)	Duration (h)
Short-Acting β-Adrenoceptor Agonist		
Albuterol	15	5–8
Metaprotarenol	1–5	2–6
Pirbuterol	5	5
Long-Acting β-Adrenoceptor Agonist		
Formoterol	15	12
Salmeterol	20	12
Arformoterol		
Ultralong-Acting β-Adrenoceptor Agonist		
Carmoterol	20	>36
Indacaterol	5	>24
Muscarinic Antagonists		
Ipratropium	15–30	3–6
Tiotropium	180	35

Source: Adapted from Holt, T. B. 2007. *Respir Care* 52:820–32.

exercise-induced bronchoconstriction and nighttime asthma attack, and protection from broncho-provocation challenge. The two most extensively used agents are formoterol and salmeterol. As shown in Table 1.8, formoterol has faster onset but comparable duration of action to salmeterol, which is in the range of 11 h. In general, LABA has been found to be safe upon repeated use (Moore et al., 1998). Side effects, although lower by the inhaled route, do occur in the form of increase in heart rate, muscle tremor, and agonist-mediated tachyphylaxis. LABA can produce adverse effect by acting on β-adrenoceptors upon reaching systemic circulation. LABA can cause receptor desensitization and down-regulation. As a result, β-agonists lose their efficacy, which may have consequence in terms of asthma aggravation. Because ₁eceptors have become nonfunctional, rescue short-acting β-agonists may not work in these patients. Studies have indicated that the use of LABA can increase the propensity of serious bronchial hyperreactivity upon prolonged use. In SMART Trial (Nelson et al., 2006) involving 26,355 subjects and in another study involving 16,787 patients (Castle et al., 1993), a small but statistically significant increase in asthma and respiratory-related death was reported in patients treated with salmeterol. In another meta-analysis study involving 33,826 participants (Salpeter et al., 2006a,b), it was observed that the use of salmeterol increased the risk of serious airway hyperreactivity and asthma-related death by two- to fourfold compared to placebo. The airway became resistant to rescue medication in salmeterol-treated patients. Glaxo Smith Kline, the originator company of salmeterol, included a black box warning stating that "salmeterol may increase the risk of asthma related death." Salmeterol should be used only when patients are not responding to rescue medication. The risk of bronchial hyperreactivity and asthma-related death were two- to fourfold higher in salmeterol-treated patients compared to placebo (Salpeter et al., 2006a). LABA can cause death due to desensitization or down-regulation of β-adrenoceptors upon repeated use.

Efforts are on to develop ultralong-acting β-adrenoceptor agonist—carmoterol and indacaterol (Table 1.8). Indacaterol has been approved for COPD treatment. This compound is shown to improve lung function and reduce breathlessness in COPD patients. The asthma clinical trial is underway.

1.5.1.2.2 Muscarinic Antagonist

Tiotropium and ipratropium are two commercially available muscarinic receptor antagonists used for COPD and asthma. According to GINA guidelines (GINA, 2008), muscarinic antagonists are not recommended for the long-term management of asthma. Short-acting muscarinic antagonists like ipratropium bromide are used as a reliever medicine in patients who have a problem with β-agonist therapy. However, muscarinic antagonists offer only modest efficacy and their benefit over β-agonist for the treatment of asthma has not been established.

Muscarinic antagonists improve lung function, breathlessness, exacerbation rate, exercise tolerance, and health-related quality of life in COPD patients (Scullion, 2007; Fitzgerald and Fox, 2007a). In most parameters, tiotropium offers better and more sustained protection compared to ipratropium (Casaburi et al., 2002). When compared with β- adrenoceptor agonists, muscarinic antagonists have shown to improve lung function better. Muscarinic antagonists also tend to cause less death compared to β-agonists (Salpeter et al., 2006b). In a recently published UPLIFT trial (Tashkin et al., 2008) involving 5993 patients, 2987 patients treated with tiotropium showed improved lung function, reduced exacerbation, and improved quality of life but did not show reduction of decline in lung function.

1.5.1.3 Inhaled Fixed Dose Combination

Fixed dose combination of β-agonist and corticosteroid is used in difficult-to-treat asthma that does not respond to medium-dose inhaled steroid alone. It has been shown that LABA addition reduces the dose of the steroid as well as improves nocturnal asthma attacks, lowers the use of rescue medication, reduces exacerbation, and improves lung function (GINA, 2010).

Studies have shown that inhaled β-agonists alone are inferior to inhaled corticosteroids as maintenance therapy in persistent asthma. On the other hand, a substantial number of patients with more severe disease are insufficiently controlled by inhaled corticosteroid alone. Increasing the dose of the steroid increases the propensity of a side effect. Steroids exhibit a flat dose–response relationship. It has been shown that when combined with a LABA, low-dose beclomethasone dipropionate produced faster and better improvement of symptom control compared to doubling the dose of inhaled corticosteroid (Kips and Pauwels, 2001). Based on this observation, fixed dose combinations of salmeterol and fluticasone (Advair®) and formoterol and budesonide (Symbicort®) were developed. For treatment of asthma, studies have shown that the efficacy observed with these formulations are superior to monotherapy and comparable to both agents given via separate inhalers (Kips and Pauwels, 2001) in adults as well as in children. This combination reduces the number of inhalations, minimizes the use of LABA alone, and improves patient compliance. Evidence also exists to suggest that fixed dose combination of inhaled LABA and inhaled corticosteroid (salmeterol and fluticasone) reduced exacerbation in COPD patients, improved lung function and quality of life, but did not reduce all cause mortality (Calverly et al., 2007). However, these are not substitutes for oral or inhaled glucocorticoids. The mechanism of synergy between an inhaled corticosteroid and a LABA has been investigated. It has been shown that glucocorticoids promote the expression of β-adrenoceptor gene and protein. As a result, β-agonist-mediated receptor desensitization, and down-regulation are miniminized in the presence of a steroid (Johnson, 2005).

Several additional fixed dose combinations are under development for the treatment of asthma and COPD (Fitzgerald and Fox, 2007b). In studies of salmeterol combined with ipratropium in COPD patients, improvement in lung function was more pronounced in patients on combination therapy compared to LABA alone (Johnson, 2005). Notably, a fixed dose combination of long-acting β-adrenoceptor agonist and long-acting muscarinic antagonist is undergoing clinical trial. LABA and inhaled steroid combination is being developed. In the future, the use of a triple combination of two bronchodilators and one anti-inflammatory agent and vice versa is also being considered.

1.5.2 SYSTEMIC THERAPY

1.5.2.1 Systemic Corticosteroids

Systemic steroids are used to address uncontrolled severe asthma that is not responsive to high-dose inhaled steroids and also to treat asthma exacerbation that needs hospitalization. For uncontrolled asthma, prednisone or prednisolone 5 mg is administered for 3 months. Long-term administration of steroids leads to systemic adverse effects. For treatment of asthma exacerbation, 50 mg prednisone or prednisolone is administered for 5–10 days.

Osteoporosis, arterial hypertension, diabetes obesity, adrenal suppression, cataract, etc. are side effects associated with long-term oral or systemic glucocorticoid use. For systemic therapy, oral route is preferred than parenteral route because of better dose titration, less effect on mineralocorticoid, short half-life, and less effect on skeletal muscle.

1.5.2.2 Leukotriene Modifiers

Two classes of drugs come under the category of leukotriene modifier—leukotriene receptor antagonist and 5-lipoxygenase inhibitor (Table 1.9). These agents reduce asthma symptoms, produce bronchodilation, improve lung function, reduce asthma exacerbation, exert some anti-inflammatory effect, and exhibit steroid sparing effect (Drazen, 1998; Garcia-Marcos et al., 2003). Leukotriene modifiers are more effective in addressing daytime asthma symptoms than nocturnal asthma. These agents also reduce urinary levels of leukotriene E4, a biomarker of arachidonic acid breakdown. Leukotriene receptor antagonists, montelukast and zafirlukast, have demonstrated efficacy in different types of asthma trials—aspirin-sensitive asthma,

TABLE 1.9

Nonsteroidal Anti-Inflammatory Agents

Category	Drug	Indication
Leukotriene receptor antagonist	Montelukast	Mild persistent asthma
	Zafirlukast	Add-on to inhaled steroid in moderate and severe asthma
Leukotriene biosynthesis inhibitor	Zileuton	Aspirin sensitive asthma
Anti IgE monoclonal antibody	Omalizumab	Add-on to inhaled steroid in severe allergic asthma
Miscellaneous	Theophylline	Add-on to inhaled steroid

exercise-induced asthma, as well as in the long-term management of asthma. Several placebo-controlled trials have been conducted for different durations of time, from 3 weeks to 12 weeks involving mild to moderate asthma patients in more than 1000 patients (Drazen, 1998; Garcia-Marcos et al., 2003). Montelukast is also effective in children with asthma. In the best of situation, leukotriene receptor antagonists exhibit modest efficacy and are less efficacious than inhaled corticosteroids (Barnes, 2008). Leukotriene receptor antagonists are used in mild persistent bronchial asthma and in asprin-sensitive asthma. They are used as add-ons to inhaled corticosteroid to treat difficult-to-treat asthma. Leukotriene receptor antagonists are considered to be very safe, in spite of the occasional occurrence of a systemic vasculitis, Churg Strauss syndrome (Garcia-Marcos et al., 2003). Efficacy by the oral route has made very high patient compliance possible. High safety record allows these agents to be prescribed to children 5 years and younger as well to those between 5 and 12 years.

Only one drug has been approved from the 5-lipoxygenase inhibitor category. Zileuton has shown efficacy in several clinical trials involving a similar category of asthma patients as leukotriene receptor antagonists (Drazen, 1998; Garcia-Marcos et al., 2003). Many zileuton clinical trials were also of identical duration as that of leukotriene receptor antagonists. In a 13-week trial conducted by Israel et al. (1996) involving patients where baseline FEV1 was <60% predicted, 400 and 600 mg four times daily, zileuton improved lung function and reduced the number of corticosteroid rescue medication during the course of treatment. Zileuton has low potency for the 5-lipoxygenase enzyme and poor pharmacokinetic property. As a result, this compound has to be administered four times a day up to a maximum of 2.4 g. Zileuton increases liver transaminase enzymes; that is why this compound has restricted utility and cannot be given to children. Zileuton exhibits similar efficacy as the leukotriene receptor antagonist, montelukast (Garcia-Marcos et al., 2003).

1.5.2.3 Phosphodiesterase-4 Inhibitor

This is a new class of anti-inflammatory agents (Fitzgerald and Fox, 2007a). Evidence suggests that the inhibitor of phosphodiesterase-4 may exhibit efficacy in steroid-resistant inflammation of COPD. Roflumilast has been approved by the European Medicines Agency (EMEA) and Health Canada for use in European Union countries and in Canada, respectively. Roflumilast at a dose of 500 µg has shown efficacy in severe COPD patients in two clinical trials published recently (Fabbri et al., 2009). In a 24-week clinical trial involving 1676 patients, roflumilast (837 patients) significantly improved lung function compared to placebo (839 patients). Phosphodiesterase-4 inhibitor class suffers from side effects that are an extension of the mechanism of action and include nausea, vomiting, and diarrhea, which becomes dose limiting. Adverse effects were also noted in the roflumilast trial. Data submitted to USFDA (http://www.fda.gov) indicated that 500 µg roflumilast causes weight loss, nausea, severe diarrhea, slightly higher incidence of cancer, psychiatric problems, and suicidal tendencies (three completed suicides and two suicide attempts). Long-term efficacy and adverse effects and patient acceptability need to be assessed.

1.5.2.4 Biologicals: Anti-IgE Monoclonal Antibody

Omalizumab is a humanized monoclonal antibody against IgE. Omalizumab binds IgE in the blood and prevents it from binding to high-affinity receptors in mast cells. Omalizumab (75–375 mg administered subcutaneously every 2–4 week) is recommended as an add-on therapy to inhaled corticosteroids and LABA in patients with severe persistent allergic asthma with a baseline IgE level of 30–750 IU/mL. These patients are poorly controlled by inhaled corticosteroid and long-acting bronchodilator. Patients exhibit daytime symptom, nighttime awakening, and multiple severe exacerbations. The clinical benefit of omalizumab has been established in several clinical trials (Busse et al., 2001; Soler et al., 2001; Holgate et al., 2004; Humbert et al., 2005; Bosquet et al., 2005). In INNOVATE trial (Humbert et al., 2005) involving 419 severe persistent asthma patients on a high dose of inhaled corticosteroid and LABA, mean exacerbation rate was reduced by omalizumab compared to placebo. Severe exacerbations and hospital visits were halved in the omalizumab-treated group. Greater number of patients reported improvement in the quality-of-life questionnaire. In total, nearly 2511 patients received omalizumab compared to 1797 receiving placebo in different clinical trials. Omalizumab statistically significantly reduced the number of clinically significant exacerbation characterized by worsening of asthma symptom that needed systemic corticosteroid or doubling the dose of inhaled corticosteroids. Omalizumab-treated groups exhibited higher percentage reduction of fluticasone dose over a 32-week period. Omalizumab reduced annual exacerbation rate, emergency room visit, and unscheduled physician visit.

Being a biological agent, treatment with omalzumab remains expensive (~US$10,000 per year). Omalizumab needs to be administered by a specialist in a facility where monitoring and resuscitation is possible in the case of any untoward event. Patients need to assess IgE level before initiating treatment with omalizumab. In general, in children 12 years and older, Omalizumab has been found to be well tolerated.

1.5.2.5 Miscellaneous: Theophylline and Chromones

Theophylline has some bronchodilator activity. Occasionally, it is used in asthma exacerbation. Theophylline is used as an add-on inhaled corticosteroid in asthma management (GINA, 2010). However, theophylline's utility is limited in short-term and long-term therapy because of adverse effects like gastrointestinal disturbance, cardiac arrhythmias, seizure, and death.

Chromones, sodium chromoglycate, nedocromil, etc. act as mast cell stabilizer and have modest efficacy in mild asthma in children. Studies have suggested that both nedocromil sodium (Sridhar and McKean, 2006) and chromolyn sodium (Tache et al., 2000) have very limited efficacy compared to placebo in mild asthma. Both agents are very similar and are still included in the National Asthma Education and Prevention Program Expert Panel Report 3 (http://www.nhlbi.nih.gov) guidelines for the management of mild asthma in children. Chromolyn and nedocromil are delivered by the inhalation route. These agents are not to be used for quick efficacy and need to be delivered for a long time to be effective.

1.6 SUMMARY AND CONCLUSIONS

In summary, the incidence of asthma and COPD is going to increase all over the world. Existing therapy is good to address mild to moderate asthma. Severe disease on the other hand is refractory to inhaled corticosteroid therapy. COPD has no disease-modifying therapeutic option available.

Inhaled therapy is effective for asthma, but there is the issue of cost of inhaler and patient compliance. Orally effective therapeutic options are preferred but have limited utility in asthma and COPD. Leukotriene antagonists are safe but their efficacy is limited to only mild bronchial asthma and as an add-on to inhaled corticosteroid. Phosphodiesterase inhibitors like roflumilast have recently been approved for COPD. Long-term adverse effect and efficacy data for this class are still awaited. Oral steroids are limited to patients suffering from severe asthma. Prolonged use is associated with side effects.

A lot of effort is being put to develop fixed dose combination. Inhaled steroid and LABA combination is very successful but this does not prevent mortality in COPD patients. Many combinations are under development where long-acting bronchodilators of different class are being combined. Long-term safety of these combinations needs to be evaluated. Bronchodilators under best case will only offer symptomatic relief.

In conclusion, more work needs to be done to bring about an effective disease-modifying therapy to address inflammation of severe asthma and COPD, prevent disease progression, and reverse the course of the disease.

REFERENCES

Aggarwal, A. N., K. Chaudhry, S. K. Chhabra, G. A. D'Souza, D. Gupta, S. K. Jindal, S. K., Katiyar, R. Kumar, B. Shah, and V. K. Vijayan 2006. Prevalence and risk factors for bronchial asthma in Indian adults: A multicentre study. *Indian J Chest Dis Allied Sci.* 48:13–22.

Barnes, P. J. 2004. New drugs for asthma. *Nat Rev Drug Discov.* 3:831–44.

Barnes, P. J. 2008. Immunology of asthma and chronic obstructive pulmonary disease. *Nat Rev Immunol.* 8:183–92.

Barnes, P. J., F. Chung, and C. Page 1998. Inflammatory mediators of asthma: An update. *Pharmacol Rev.* 50:515–96.

Bateman, E. D., S. S. Hurd, P. J. Barnes, J. Bousquet, J. M. Drazen, M. FitzGerald, P. Gibson et al. 2008. Global strategy for asthma management and prevention. GINA executive summary. *Eur Respir J.* 31:143–78.

Bosquet, J., P. Cabrera, N. Berkman, R. Buhl, S. Holgate, S. Wenzel, H. Fox, S. Hedgecock, M. Blogg, and G. D. Cioppa 2005. The effect of treatment with omalizumab, an anti-IgE antibody, on asthma exacerbations and emergency medical visits in patients with severe persistent asthma. *Allergy* 60:302–8.

Busse, W., J. Corren, B. Q. Lanier, M. McAlary, A. Fowler-Taylor, G. D. Cioppa, A. van As, and N. Gupta 2001. Omalizumab, anti-IgE recombinant humanized monoclonal antibody, for the treatment of severe allergic asthma. *J Allergy Clin Immunol.* 108:184–90.

Busse, W. W. and R. F. Lemanske 2001. Asthma. *New Eng J Med.* 344:350–62.

Calverly, P. M. A., J. A. Anderson, B. Celli, G. T. Ferguson, C. Jenkins, P. W. Jones, J. C. Yates, and J. Vestbo 2007. Salmeterol and fluticasone propionate and survival in chronic obstructive pulmonary disease. *New Eng J Med.* 356:775–89.

Casaburi, R., D. A. Mahler, P. W. Jones, A. Wanner, P. G. San, R. L. ZuWallack, S. S. Menjoge, C.W. Serby, and T. Witek Jr. 2002. A long term evaluation of once daily inhaled tiotropium in chronic obstructive pulmonary disease. *Eur Respir J.* 19:217–24.

Castle, W., R. Fuller, J. Hall, and J. Palmer 1993. Serevent nationwide surveillance study: Comparison of salmeterol with salbutamol in asthmatic patients who require regular bronchodilator treatment. *Br Med J.* 306:1034–37.

DATA Monitor. 2006. Pipeline insight asthma/COPD: Targeted therapies on the horizon Available at http://www.datamonitor.com.

de Boer, W. I. 2005. Perspectives for cytokine antagonist therapy for COPD. *Drug Discov Today* 10:93–106.

Drazen, J. 1998. Clinical pharmacology of leukotriene receptor antagonists and 5 lipoxygenase inhibitors. *Am J Respir Crit Care Med.* 157:S233–37.

Fabbri, L. M., P. M. Calverly, J. M. Izquierdo-Alonso, D. S. Bundschuh, M. Brose, F. Z. Martinez, and K. F. Rabe 2009. Roflumilast in moderate to severe chronic obstructive pulmonary disease treated with long acting bronchodilators: Two randomised clinical trials. *Lancet* 374:695–03.

Fitzgerald, M. F. and J. C. Fox 2007a. Emerging trends in the therapy COPD: Novel anti-inflammatory agents in clinical development. *Drug Discov Today.* 12:479–86.

Fitzgerald, M. F. and J. C. Fox 2007b. Emerging trends in the therapy COPD: Bronchodilators as mono- and combination therapies. *Drug Discov Today* 12:472–78.

Garcia-Marcos, L., A. Schuster, and E. Perez-Yarza 2003. Benefit-risk assessment of antileukotrienes in the management of asthma. *Drug Safety* 26:483–16.

Global Initiative for Asthma (GINA). 2010. Global strategy for asthma management and prevention. Available at www.ginaasthma.org, last updated 2010.

Global Initiative for Chronic Obstructive Lung Disease (GOLD). 2010. Global strategy for the diagnosis, management and prevention of chronic obstructive pulmonary disease. Available at www.goldcopd.org, last updated 2010.

Holgate, S. T., A. G. Chuchalin, J. Hebert, J. Lötvall, G. B. Persson, K. F. Chung, J. Bousquet et al. 2004. Efficacy and safety of a recombinant anti-immunoglobulin E antibody (omalizumab) in severe allergic asthma. *Clin Exp Allergy* 34:632–38.

Holgate, S. T. and R. Polosa 2006. The mechanisms, diagnosis and management of severe asthma in adults. *Lancet* 368:780–92.

Holt, T. B. 2007. Inhaled beta agonist. *Respir Care* 52:820–32.

Humbert, M., R. Beasley, J. Ayres, R. Slavin, J. Hébert, J. Bousquet, K. M. Beeh et al. 2005. Benefits of omalizumab as add-on therapy in patients with severe persistent asthma who are inadequately controlled despite best available therapy (GINA 2002 step 4 treatment): INNOVATE. *Allergy* 60:309–16.

Israel, E., J. Cohn, L. Dube, and J. M. Drazen 1996. Effect of treatment with zileuton, 5 lipoxygenase inhibitor in patients with asthma: A randomized controlled trial. *JAMA* 275:931–36.

Jindal, S. K., D. Gupta, and A. N. Aggarwal 2004. Guidelines for management of chronic obstructive pulmonary disease (COPD) in India: A guide for physicians. *Indian J Chest Dis Allied Sci.* 46:137–53.

Johnson, M. 2005. Corticosteroids: Potential β_2-agonist and anticholinergic interactions in chronic obstructive pulmonary disease. *Proc Am Thorac Soc.* 2:320–25.

Kips, J. C. and R. Pauwels 2001. Long acting inhaled β_2-agonist therapy in asthma. *Am J Respir Crit Care Med.* 164:923–32.

Leckie, M. J., A. Brinke, J. Khan, Z. Diamant, B. J. O'Connor, C. M. Walls, A. K. Mathur et al. 2000. Effects of an interleukin-5 blocking monoclonal antibody on Eosinophils, airway hyper-responsiveness, and the late Asthmatic response. *Lancet* 356:2144–48.

Masoli, M., D. Fabian, S. Holt, and Beasley. 2004. The global burden of asthma: Executive summary of GINA dissemination committee report. *Allergy* 59:469–78.

Moore, R. H., A. Khan, and B. F. Dickey 1998. Long acting inhaled β_2-agonists in asthma therapy. *Chest* 113:1095–08.

Murthy, K. J. R. and J. G. Sastry 2005. Economic burden of asthma. http://www.whoindia.org/LinkFiles/Commision_on_Macroeconomic_and_Health_Bg_P2_Economic_burden_of_asthma.pdf.

National asthma education and prevention program expert panel report 3: Guidelines for the diagnosis and management of asthma. Available at http://www.nhlbi.nih.gov/guidelines/asthma/asthsumm.pdf.

Nelson, H. S., S. T. Weiss, E. R. Bleeker, S. W. Yancey, and P. M. Dorinsky 2006. The salmeterol multicentre research trial: A comparison of usual pharmacotherapy for asthma or usual pharmacotherapy plus salmeterol. *Chest*:129:15–26.

Rabe, K. F., S. Hurd, A. Anzueto, P. J. Barnes, S. A. Buist, P. Calverley, Y. Fukuchi et al. 2007. Global strategy for the diagnosis, management and prevention of chronic obstructive pulmonary disease. Gold executive summary. *Am J Respir Crit Care Med.* 176:532–55.

Roflumilast new oral once daily antiinflammatory therapy for COPD, Pulmonary–allergy drug advisory committee Meeting. Available at http://www.fda.gov/downloads/AdvisoryCommittees/CommitteesMeetingMaterials/Drugs/Pulmonary-Allergy DrugsAdvisoryCommittee/ UCM208711.pdf.

Salvi, S. S. and P. J. Barnes 2009. Chronic obstructive pulmonary disease in nonsmokers. *Lancet* 374:733–43.

Salpeter, S. R., N. S. Buckley, T. M. Ormiston, and E. E. Salpeter 2006a. Meta analysis: Effect of long acting β-agonist on severe asthma exacerbations and asthma related death. *Ann Intern Med.* 144:904–12.

Salpeter, S. R., N. S. Buckley, and E. E. Salpeter 2006b. Meta analysis: Anticholinergics and not beta agonists reduce severe exacerbations and respiratory mortality in COPD. *J Gen Intern Med.* 21:1–9.

Scullion, J. E. 2007. The development of anticholinergics in the management of COPD. *Int J COPD* 2:33–40.

Soler, M., J. Matz, R. Townley, R. Buhl, J. O'Brien, H. Fox, J. Thirlwell, N. Gupta, and G. Della Cioppa 2001. The anti-IgE antibody omalizumab reduces exacerbations and steroid requirement in allergic asthmatics. *Eur Respir J.* 18:254–61.

Sridhar, A. V. and M. C. McKean 2006. Nedocromil sodium for chronic asthma in children. *Cochrane Database of Systematic Reviews*, Issue 3. Art. No.: CD004108. Available at http://www.thecochranlibrary.com.

Tache, M. J., J. H. Uijen, R. M. Bernsen, J. C. de Jongste, and J. C. van der Wouden 2000. Inhaled disodium chromoglycate as maintenance therapy in children with asthma: A systematic review. *Thorax* 55:913–20.

Tashkin, D. P., B. Celli, S. Senn, D. Burkhart, S. Kesten, S. Menjoge, and M. Decramer 2008. A 4-year trial of tiotropium in chronic obstructive pulmonary disease. *New Eng J Med.* 359:1543–54.

Wenzel, S. E. 2006. Asthma: Defining of the persistent adult phenotypes. *Lancet* 368:804–13.

Wills-Karp, M. 1999. Immunologic basis of antigen induced airway hyperresponsiveness. *Annu Rev Immunol.* 17:255–81.

Winkler, J., G. Hochhaus, and H. Derendorf 2004. How the lung handles drugs: Pharmacokinetics and pharmacodynamics of inhaled corticosteroids. *Proc Am Thorac Soc.* 1:356–63.

2 Lipids
Reservoir of Drug Targets

Abhijit Ray and Punit Kumar Srivastava

CONTENTS

2.1 INTRODUCTION

It is becoming increasingly clear that lipids act as a storehouse of molecules that exhibit biological activity. Lipids undergo metabolic transformation in the body to generate molecules that occupy and activate distinct cell surface receptors to exhibit pharmacological activity. Lipid mediators have an important role in biological systems under physiological and pathophysiological conditions. Quite a few clinically efficacious drugs have reached the market that work by modulating the pharmacology of bioactive lipids.

In this chapter, we shall overview important events in lipid metabolism that may be relevant to airway inflammation. Lipid metabolites introduced here will be discussed in detail in subsequent chapters with respect to their biology, signal transduction mechanism, role in airway inflammation, and drug discovery efforts around them. Although the theme of this book is lipid mediators, key enzymes that generate bioactive lipids will also be covered. We shall introduce nine different lipid mediators and four different enzymes covering four distinct pathways. The list of topics covered in this chapter will by no means be exhaustive. There is every possibility of key mediators of importance being omitted. However, keeping in view the scope of this book, we shall restrict our effort to the molecules described in this chapter because, in our assessment, biology around these molecules is better characterized than others.

2.2   LIPIDS

There are a wide variety of lipids in biological systems. These lipids can be divided into five broad categories: wax, triacylglycerol, fatty acids, phospholipids, and sphingolipids. Different lipids have distinct biological functions and it is beyond the scope of this chapter to cover all lipids. For the purpose of this chapter, we shall concentrate on phospholipids and sphingolipids. These lipids are integral proteins of biological membrane and there is emerging evidence in support of their role in airway inflammation.

Glycerolipids are esters of long-chain fatty acids and glycerol, where two hydroxyl groups of the glycerol are occupied by long-chain fatty acids, shown as R1 and R2 in Figure 2.1. One of the fatty acids in glycerolipid can be saturated and the other unsaturated, for example, arachidonic acid at R2 position. The third hydroxyl group of glycerol is coupled with phosphatidyl choline, phosphatidyl serine, phosphatidyl inositol, and phosphatidyl ethanolamine. R3 in Figure 2.1 can be choline, ethanolamine, serine, or inositol.

Sphingolipids on the other hand do not contain glycerol. These lipids contain an amino alcohol—sphingosine, linked with a long-chain fatty acid to form ceramide. In sphingomyelin, a phosphocholine group is attached to ceramide (Figure 2.2). In cerebroside, a galactoside group replaces a phosphocholine group, and other much more complex oligosaccharide moieties become part of ganglioside. Sphingolipids are a source of important biologically active lipids, namely, ceramide and sphingosine-1-phosphate (S1P).

2.3   LIPASES: KEY PLAYERS IN LIPID METABOLISM

Lipid metabolism starts with the breakdown of ester bond between fatty acids and alcohol. For the ease of understanding, we shall divide products of lipid metabolism

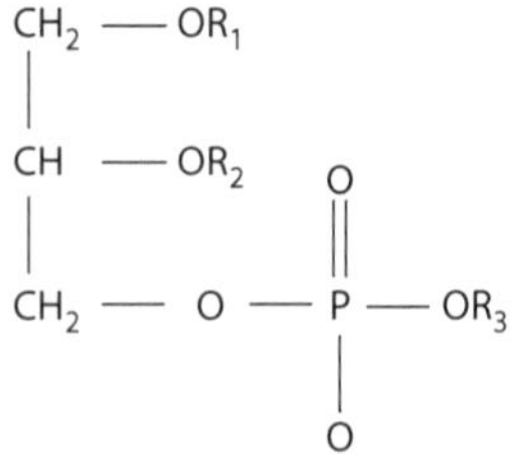

Phospholipid

FIGURE 2.1   Typical structure of a phospholipid where R1 and R2 can be any long-chain saturated and unsaturated fatty acid, respectively. R3 can be any one of the following—serine, choline, inositol, or ethanolamine.

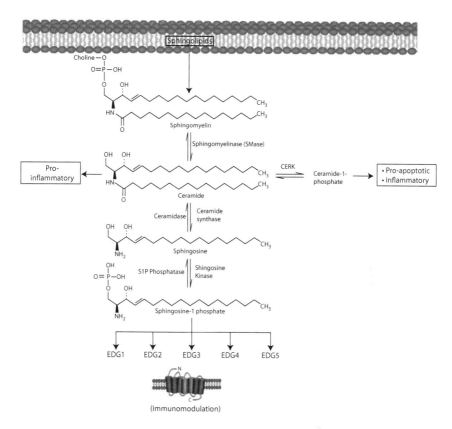

FIGURE 2.2 Sphingolipid metabolism. The breakdown of sphingomyelin starts with the formation of ceramide under the influence of sphingomyelinase. Ceramide is converted to ceramide-1-phosphate, sphingosine, and sphingosine-1-phosphate. Receptors have been identified for sphingosine-1-phosphate.

into two. Phospholipid breakdown is catalyzed by the phospholipase group of enzymes. Sphingomyelinase cleaves sphingolipids to generate ceramide.

Phospholipase enzymes can be grouped into three broad families—phospholipase A_2 (PLA$_2$) (Lambeau and Gelb, 2008), phospholipase C (PLC) (Rhee, 2001), and phospholipase D (PLD) (Jenkins and Frohman, 2005). Each class has numerous subclasses. PLC and PLD are part of G-protein-coupled seven transmembrane domain receptor systems. These enzymes are activated as a result of receptor occupancy. PLA$_2$ class of enzymes are activated as a result of change in intracellular environment and elevation of intracellular calcium levels brought out by hormones, cytokines, and growth factors (Lambeau and Gelb, 2008). Evidence exists to suggest that PLA$_2$ can also be functionally coupled to receptors. Different phospholipase enzymes cleave phospholipid at different sites. A range of different biologically important molecules, such as arachidonic acid, lysophospholipid, diacyl glycerol, phosphatidic acid etc., are generated as a result of phospholipid breakdown (Figure 2.3).

2.3.1   PHOSPHOLIPASE A₂ PATHWAY

PLA₂ enzymes can be grouped into three broad families—cytosolic, secretory, and calcium-independent PLA₂ (Murakami, 2004; Lambeau and Gelb, 2008; Hurley and McPherson, 2008). PLA₂ enzymes cleave the R2 group from the phospholipid molecule (Figure 2.3) to release a fatty acid, which is normally arachidonic acid (Lambeau and Gelb, 2008). Lysophospholipid thus formed can be converted to platelet-activating factor (PAF) or lysophosphatidic acid (LPA) (Figure 2.3). Using gene knockout

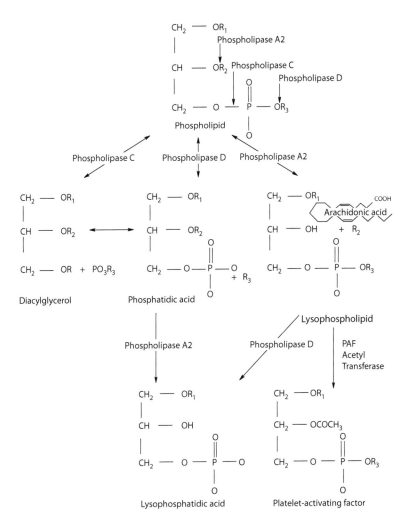

FIGURE 2.3   Effect of phospholipase enzymes on phospholipid. Phospholipase A₂ releases lysophospholipid and arachidonic acid (R2). Phospholipase D acts on phosphatidyl choline to release choline (R3) and phosphatidic acid. Phospholipase C acts on phosphatidyl inositol to release diacylglycerol and inositol 3,4,5-trisphosphate (R3). Phosphatidic acid and lysophospholipid (R3 is choline) are converted to lysophosphatidic acid. Lysophospholipid is acetylated to form platelet-activating factor.

animal models and selective PLA_2 inhibitors, involvement of both cytosolic PLA_2 (cPLA$_2$) and secretory PLA_2 (sPLA$_2$) family members have been demonstrated in experimental models of airway inflammation and lung surfactant metabolism (Hurley and McPherson, 2008). Potent inhibitors of cytosolic (McKew et al., 2008) and secretory (Draheim et al., 1996) PLA_2 enzymes have been designed. Several molecules have entered human trial (Draheim et al., 1996; McKew et al., 2008). Given the central role of PLA_2 enzymes in lipid metabolism and lung surfactant metabolism, PLA_2 family of enzymes become an important target for discussion.

2.3.1.1 Arachidonic Acid Metabolism

Three distinct and important products are generated as a result of phospholipid metabolism by PLA_2 enzyme/s. These include platelet-activating factor, lysophospholipid, and arachidonic acid (Figure 2.3). Of different products generated by PLA_2 action, arachidonic acid is the most well investigated. Arachidonic acid breakdown products have been targeted for drug discovery effort. Several clinically efficacious drugs work by inhibiting the formation or antagonizing pharmacology of arachidonic acid breakdown products. Key arachidonic acid metabolites introduced in this chapter include leukotrienes, oxoeicosatetraenoic acid (oxoETE), epieicosatrienoic acid (EET), prostaglandin D, and lipoxins (Figure 2.4).

2.3.1.1.1 Leukotrienes

Leukotrienes are lipid mediators generated by the action of 5-lipoxygenase (5-LO) enzyme on arachidonic acid (Figure 2.4). Two distinct families of leukotrienes exist—cysteinyl leukotrienes (cysLTs) and leukotriene B_4 (LTB$_4$). CysLTs have three members—leukotriene C_4 (LTC$_4$), leukotriene D_4 (LTD$_4$), and leukotriene E_4 (LTE$_4$). All cysLTs act on two distinct cell surface G-protein-coupled receptors—cysLT$_1$ and cysLT$_2$. CysLTs are synthesized by eosinophils, basophils, and mast cells and play a role in the recruitment of eosinophils, lymphocytes, and dendritic cells. Almost all inflammatory cells—neutrophils, macrophages, eosinophils, mast cells, lymphocytes, dendritic cells, and others—express cysLT. CysLTs are among the most powerful spasmogens known and also play a role in experimental airway inflammation and in human bronchial asthma (Busse, 1998; Drazen, 1998; Brink et al., 2006; Peters-Golden and Henderson, 2007). Antagonists of cysLT receptor have been designed. Molecules, most notably montelukast and zafirlukast, have been tested in double-blind placebo-controlled trials and have reached the bedside (Garcia-Marcos et al., 2003).

Leukotriene B_4 acts on Leukotriene B_4 receptor type 1 (BLT$_1$) and Leukotriene B_4 receptor type 2 (BLT$_2$). Neutrophils synthesize LTB$_4$ along with macrophages and monocytes. By acting on cell surface receptors present on most inflammatory cells, LTB$_4$ helps in the recruitment of different inflammatory cells—mast cells, lymphocytes, neutrophils, and macrophages (Brink et al., 2006; Peters-Golden and Henderson, 2007). Antagonists of BLT$_1$ receptor were designed and evaluated in experimental animal models. One molecule, LY 293111, has shown to lower neutrophil count and LTB$_4$ level in bronchoalveolar lavage fluid of asthma patients (Brooks and Summers, 1996; Kilfeather, 2002).

An alternative approach to alter leukotriene-mediated biological activity is to block their synthesis. Typical targets for blocking leukotriene biosynthesis are 5-LO

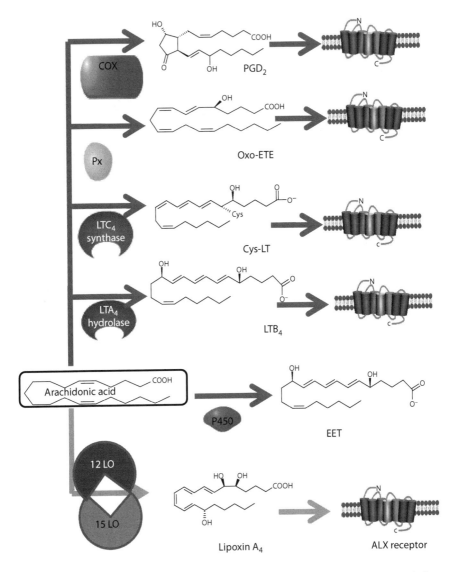

FIGURE 2.4 Arachidonic acid metabolic pathway. Arachidonic acid generates proinflammatory (filled arrow) and anti-inflammatory (unfilled arrow) mediators. Cyclooxygenase (COX) enzyme converts arachidonic acid to prostaglandins. Shown here is prostaglandin D2, known to play a role in airway inflammation. Peroxidase enzyme (Px) converts arachidonic acid to oxoETE. OxoETE exhibits a proinflammatory effect by acting on OxoETE receptor. Cysteinyl leukotrienes (CysLT) and LTB_4 are generated when arachidonic acid is acted upon by 5-lipoxygenase enzyme. CysLT and LTB_4 act on their distinct cell surface G-protein-coupled receptors. EETs are generated when arachidonic acid is acted upon by cytochrome P450 (P450). EETs have proinflammatory effects. The exact receptor of EET is not known. Under the influence of multiple lipoxygenases, arachidonic acid is converted to lipoxins. Lipoxin A_4 shown in the figure is a lipid that exhibits an anti-inflammatory effect by acting on its cell surface G-protein-coupled receptor, ALX receptor.

and 5-lipoxygenase-activating protein (FLAP). Theoretically, an inhibitor of 5-LO enzyme can knock out both cysLT and LTB$_4$ and exhibit better anti-inflammatory activity than a single agent targeting any one receptor alone. Despite intense effort to design 5-LO inhibitor, only one drug, zileuton, has been approved by USFDA so far (Brooks and Summers, 1996; Werz and Steinhilber, 2005; Pergola and Werz, 2010). Zileuton has been tested successfully in several placebo-controlled clinical trials in many different types of asthma patients (Brooks and Summers, 1996; Garcia-Marcos et al., 2003). In terms of clinical efficacy, zileuton was as efficacious as a leukotriene receptor antagonist despite being at least 1000 times less potent at inhibiting target at the molecular level.

As 5-LO is an iron-containing enzyme, the only chemical class that has been successful in inhibiting this enzyme in experimental setup as well as in clinical situations has been *N*-hydroxy urea chemotype. This chemotype, of which zileuton is also a member, has the attribute of poor pharmacokinetic property and serious adverse effects like liver damage, methemoglobinemia, etc. Zileuton had to be dosed four times a day at 400–600 mg dose to achieve clinical efficacy. Because of its effect on liver enzymes, zileuton is not prescribed for children (Brooks and Summers, 1996; Garcia-Marcos et al., 2003; Werz and Steinhilber, 2005).

5-LO enzyme needs FLAP protein for the metabolism of arachidonic acid. If FLAP can be blocked from binding arachidonic acid, the ability of 5-LO enzyme is seriously compromised. Several attempts have been made to design and synthesize FLAP inhibitors. Several highly potent FLAP inhibitors, which also inhibit leukotriene biosynthesis *in vitro* and *in vivo*, have been tested in humans. At least three different compounds, MK 886, MK 0591, and Bay x 1005, have shown efficacy in human asthma patients in short-term and medium-term studies (Brooks and Summers, 1996). However, FLAP inhibitors have never entered long-term efficacy trial for asthma. No serious adverse effect has been reported with this class of agents.

2.3.1.1.2　Oxoeicosatetraenoic Acid

The action of 5-LO on arachidonic acid also generates mediators like 5-hydroxy epieicosatrienoic acid (5HETE) (hydroxyeicosatetraenoic acid) and 5-hydroperoxy epieicosatrienoic acid (5HPETE) (hydroperoxyeicosatetraenoic acid), which have distinct pharmacology. 5HETE is metabolized to oxoETE under the influence of 5-hydroxyeicosanoid dehydrogenase (Figure 2.4). OxoETE acts on a distinct receptor and exerts chemoattractant and degranulating effect on inflammatory cells (Powell and Rokach, 2005). OxoETE may play an important role in allergic and inflammatory diseases (Grant et al., 2009). A 5-LO inhibitor can potentially block oxoETE-mediated responses. However, keeping in mind the tractability of 5-LO enzyme as a drug target (Brooks and Summers, 1996; Werz and Steinhilber, 2005), it may be worth looking at oxoETE as a potential intervention point for inflammatory diseases.

2.3.1.1.3　Epieicosatrienoic acid

The metabolism of arachidonic acid by cytochrome P450 monooxygenase enzyme family results in the formation of epieicosatrienoic acids (EET). Cytochrome P450 2B, 2C, and 2J families have been implicated in the generation of EETs. EETs are believed to activate calcium-activated potassium channels and regulate the tone of blood vessels and airway smooth muscle. In addition, EETs have been shown to inhibit the expression of cell adhesion molecules in blood vessel (Jacobs and Zeldin, 2001). Not much is known about the role of EETs in airway inflammation. It has been shown that blocking the formation of EETs using an inhibitor of soluble epoxide hydrolase enzyme exhibits anti-inflammatory effect in cigarette smoke-induced airway inflammation in a mouse model (Smith et al., 2004).

2.3.1.1.4 Prostaglandin D

Arachidonic acid, under the influence of cyclooxygenase enzyme (COX), generates prostaglandins and thromboxanes (Figure 2.4). Prostaglandins and thromboxanes act on distinct G-protein-coupled receptors. Different prostaglandins exhibit mutually antagonistic effect on bronchial smooth muscles and airway inflammation, so much so that a single prostaglandin may exhibit both bronchoconstriction and bronchodilation. Thus, it may be practically impossible to design a COX inhibitor that will show beneficial effect in bronchoconstriction and airway inflammation of asthma. The answer in all likelihood lies in designing selective receptor antagonists that will address a certain effect of a prostaglandin while sparing others. In recent years, a lot of attention has gone toward developing antagonist of prostaglandin D_2 (PGD_2) receptor. PGD_2 is produced by mast cells as a result of allergen challenge. Other cell types like dendritic cells, T-lymphocytes, etc. also produce prostaglandin D. PGD_2 acts on three distinct cell surface receptors, namely, TP, DP_1, and DP_2, also known as (CRT_H2—chemoattractant receptor homologous molecule expressed on T helper cell type 2). TP and DP_1 receptors are present in bronchial and vascular smooth muscle, platelets, and dendritic cells. DP_2 receptor is present in T helper cell type 2 (T_H2) cells, eosinophils, basophils, etc. TP and DP_1 receptors exhibit bronchodilation and bronchoconstriction in response to PGD_2, respectively. DP_2 receptors contribute toward chemotaxis and activation of inflammatory cells.

Inhalation of PGD_2 leads to bronchoconstriction in asthma patients, which can be blocked by TP receptor blockers. However, no protective effect on bronchial asthma can be demonstrated. Evidence is emerging in animal models in support of a role of DP_2 receptor in allergic asthma and cigarette smoke-induced airway neutrophilia (Spik et al., 2005; Uller et al., 2007; Pettipher et al., 2007; Stebbins et al., 2010). Selective inhibitors have shown inhibitory effect on neutrophil influx, lymphocyte migration, mucus cell metaplasia, and airway wall thickening (Stebbins et al., 2010).

There is evidence that the acetylated COX enzyme plays a role in the generation of epilipoxins and resolvin E series (Serhan, 2007). Also, PGD_2 may act as a molecular switch that initiates inflammation resolution by upregulating expression 15-LO enzyme and lipoxin synthesis (Serhan, 2007).

2.3.1.2 Platelet-Activating Factor

PAF is a lipid mediator generated by the acetylation of lysophosphatidyl choline (1-O-alkyl-2-acetyl-sn-glycero-3-phosphocholine) after PLA_2 has released arachidonic acid from the phosphatidyl choline (Prescott et al., 1990) (Figure 2.3). PAF acts on cell surface G-protein-coupled receptor. PAF is released from inflammatory cells. In the airway of experimental animals, PAF can promote bronchoconstriction, airway hyperresponsiveness, and airway inflammation, responses that were blocked by selective PAF antagonists (Sanjar et al., 1990; Gundel et al., 1992; Morooka et al., 1992; Arima et al., 1995; Nagase et al., 2002; Ishii et al., 2004). Studies involved guinea pig (Sanjar et al., 1990; Morooka et al., 1992), PAF receptor overexpressing (Ishii et al., 2004), and knockout mouse models (Nagase et al., 2002). Antagonists of PAF receptor have shown protective effect in animal models of asthma—PAF-induced airway hyperresponsiveness model in guinea pig (Sanjar et al., 1990), allergic guinea pig model (Arima et al., 1995), sheep model (Abraham et al., 1989), and monkey model (Gundel et al., 1992). At least four molecules have been tested in clinical settings of human asthma—SR 274174 (Gomez et al., 1998), UK 74505 (Kuitert et al., 1995), Y24180 (Hozawa et al., 1995), and Web 2086 (Freitag et al., 1993). Lack of clinical efficacy has diminished interest in exploring PAF receptor as a target for drug development.

2.3.1.3 Lysophosphatidic Acid

LPA is generated when phosphatidyl choline is metabolized by PLD followed by removal of arachidonic acid moiety under the influence of PLA_2 (Figure 2.3). Alternatively, lysophosphatidyl choline generated as a result of PLA_2 action on phosphatidyl choline can be converted to LPA by PLD (Figure 2.3). Evidence suggests the involvement of $sPLA_2$ and secreted lysophospholipase D toward the generation of LPA (Wang and Denis, 1999; Aoki et al., 2002; Moolenar et al., 2004). LPA has its distinct biology that is mediated through the activation of receptors (Moolenar et al., 2004). Receptors for LPA have been identified (Moolenar et al., 2004). By acting on receptors, LPA exhibits a wide variety of biological effects such as mobilization of cell calcium, activation of protein kinase C, and activation of transcription factors—AP1, NFKB, and C/EBPβ, expression of IL8 gene, eotaxin gene, and COX-2 gene expression (Saatian et al., 2006; He et al., 2008). Using a knockout mouse model, Zhao et al. (2009) have shown the involvement of LPA and LPA receptor in allergen-induced airway inflammation in a mouse model. Several attempts have been made to design agonists, and the role of these agonists in airway inflammation will be discussed.

2.3.1.4 Anti-Inflammatory Lipid Mediators: Lipoxins, Protectins, and Resolvins

Inflammation is a defense mechanism. Tissue injury signals inflammatory cells to reach the site of inflammation and attack the cause of injury. At some point in this battle, a signal emanates and instead of proinflammatory mediators, inflammatory cells release anti-inflammatory and proresolving chemicals. These proresolution molecules attempt to remove invading cells, normalize enhanced vascular permeability, and switch off the release of inflammatory cytokine, chemokine genes. In the absence of these resolution mechanisms, inflammation will continue unchecked and chronic inflammation may result in tissue remodeling and fibrosis. Several proresolution molecules have been identified, namely, lipoxins, resolvins, and protectins (Chiang et al., 2006; Serhan, 2007). These agents are derived from arachidonic acid, eicosapentaenoic acid, and docosahexaenoic acid (Figure 2.5). Proresolution agents and their

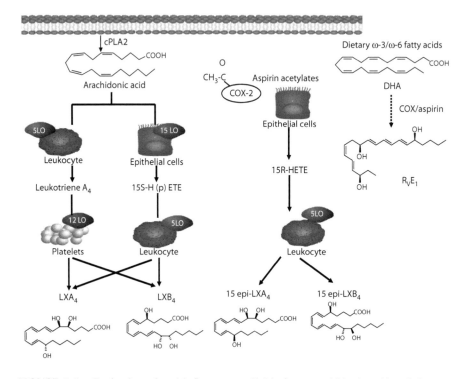

FIGURE 2.5 Generation of anti-inflammatory lipids from arachidonic acid and docosa-hexaenoic acid. The arachidonic acid released from membrane phospholipids under the influence of PLA$_2$ enzyme is converted to leukotriene A$_4$ by leukocyte 5-lipoxygenase enzyme (5 LO). Leukotriene A$_4$ is acted upon by platelet 12 LO enzyme to generate lipoxin A$_4$ (LXA$_4$) or its isomer lipoxin B$_4$ (LXB$_4$). Similarly, 15 LO enzyme in the epithelial cell acts on arachidonic acid to generate 15S-H (p) ETE, which is converted to LXB$_4$ or its isomer LXA$_4$ under the influence of 5 LO enzyme in leukocytes. By a separate mechanism, aspirin-induced acetylated cycloxygenase-2 (COX-2) enzyme in epithelial or endothelial cells converts arachidonic acid to 15R HETE, which in turn is converted to 15 epi LXA$_4$ or 15 epi LXB$_4$ in leukocytes under the influence of 5 LO enzyme. Resolvins are formed when docosahexaenoic acid is converted by acetylated COX-2 enzyme in endothelial cells.

signal transduction mechanism become important from the point of drug discovery for chronic inflammatory diseases like asthma and chronic obstructive pulmonary disease where uncontrolled inflammation often leads to tissue remodeling.

Lipoxins can be divided into two groups: (1) lipoxin A$_4$ and lipoxin B$_4$ and (2) aspirin-triggered lipoxins (ATLs)—epilipoxin A$_4$ and epilipoxin B$_4$. Lipoxins act on cell surface G-protein-coupled receptors—ALX. Lipoxin receptors are present in most inflammatory cells as well as structural cells like smooth muscle and fibroblasts (Chiang et al., 2006). Lipoxins block proinflammatory activity in many inflammatory cells—chemotaxis, adherence, and transmigration, release of proinflammatory cytokines, chemokines and gene expression. Using lipoxin receptor agonists, as well as lipoxin receptor overexpressing animals, Levy et al. (2002) demonstrated a protective effect on allergen-induced airway inflammation and reactivity in a mouse model. Efforts are underway to synthesize stable lipoxin analogs that are not removed from the body rapidly (Burli et al., 2006).

Resolvins and protectins are proresolution mediators generated from lipids other than arachidonic acid. Resolvin E is generated from eicosapentaenoic acid. Resolvin D series (D1–D4) and protectins are generated from docosahexaenoic acid. Acetylated cyclooxygenase and 5-LO enzymes play a role in the synthesis of resolvin E and D series, respectively. The receptor for resolvin has been reported. Protectin receptors are yet to be detected. Studies have shown that the administration of resolvin and protectins to experimental animals offer protection from allergen-induced airway inflammation (Gilroy et al., 2004; Haworth et al., 2008).

2.3.2 SPHINGOMYELINASE PATHWAY

Sphingolipid metabolism is initiated by sphingomyelinase enzyme. The two major bioactive lipids generated from sphingolipids are ceramide and sphingosine (Figure 2.2). An evidence of the involvement of these lipid molecules in airway inflammation comes from the protective effect of sphingomyelinase inhibitor in animal models (von Bismarck et al., 2008). Ceramide is the first mediator generated as a result of sphingomyelin metabolism. The role of ceramide in airway inflammation is less clear. In animal models of asthma, ceramide analogs promote airway hyperresponsiveness (Meyer et al., 2006) and enhance allergen-induced eosinophilia, cytokine and chemokine production, airway function, and histopathology (Bilenki et al., 2004; Masini et al., 2008). Based on the evidence generated using natural killer T (NKT) cell knockout mouse (Morishima et al., 2005), ceramide receptor knockout mouse model (Bilenki et al., 2004), and ceramide analogs (Lee et al., 2008), it is believed that ceramide acts on NKT cells, a key cellular mediator of asthma and COPD. Evidence also suggests that ceramide tips the T_H2 bias of allergic airway inflammation in favor of T helper cell type 1 (T_H1). Thus, in the absence of interferon gamma response, a principal T_H1 cytokine, ceramide response on airway inflammation is lost (Matsuda et al., 2005).

Sphingosine is formed from ceramide under the influence of ceramidase. Sphingosine kinase enzyme phosphorylates sphingosine to generate S1P. S1P is present in the plasma as well as secreted from inflammatory cells upon activation (Ryan and Spiegel, 2008). S1P acts on five distinct cell surface receptor types—$S1P_{1-5}$ (Uhlig and Gulbins, 2008; Kim et al., 2009; Rosen et al., 2009). $S1P_{1-3}$ receptor messages have been detected in lung tissue (Rosen et al., 2009). Using receptor knockout mouse and receptor-specific agonist and antagonist, it has been shown that $S1PR_1$ receptor plays a role in protecting lung permeability in response to LPS challenge in mouse (Sammani et al., 2009). S1P receptor plays a role in mast cell chemotaxis, mast cell degranulation, eosinophil influx, airway hyperresponsiveness, and bronchoconstriction (Uhlig and Gulbins, 2008; Ryan and Spiegel, 2008; Rosen et al., 2009).

Attempts have been made to design selective agonists and antagonist for S1P receptors (Kim et al., 2009; Rosen et al., 2009). S1P receptor agonist FTY720 and AUY 954, administered by the inhalation route, inhibited allergen-induced airway inflammation and airway reactivity in an experimental animal model (Idzko et al., 2006;

Ble et al., 2009). However, this effect has been attributed to inhibitory effect on cPLA$_2$ (Ryan and Spiegel, 2008). In clinical trials of FTY720, an adverse effect that has been reported is reduced lung function (Rosen et al., 2009). Thus, it is not clear in airway disease whether to develop agonist or antagonist of S1P receptors. Relative importance of each receptor subtype also needs exploration.

2.4 KINASES IN LIPID PATHWAY

Kinases play an important role in generating important biologically active lipid mediators. Other kinases are activated as a result of lipid metabolism. Two such kinases are the protein kinase C (PKC) family and the phosphatidyl inositol 3-kinase (PI3K) family.

2.4.1 PROTEIN KINASE C

As described earlier, PKC is activated by diacylglycerol formed as a result of lipid metabolism by phospholipases. Many subtypes of PKC have been identified. Using an inhibitor of PKC α, the role of this enzyme has been reported in cigarette smoke-induced expression of COX-2 gene expression in tracheal smooth muscle cell (Yang et al., 2009). The role of PKC δ (Langlois et al., 2009), PKC ζ (Martin et al., 2005), and PKC θ (Chaudhary and Kasain, 2006) in T$_H$2 cell function has been reported. PKC ζ null mice exhibit attenuation of T$_H$2 cytokine release and allergen-induced airway inflammation (Martin et al., 2005).

2.4.2 PHOSPHATIDYL INOSITOL 3-KINASE FAMILY

PI3K enzymes are activated as a result of chemokine and/or growth factor receptor occupancy. These enzymes phosphorylate phosphatidyl inositol 4,5-bisphosphate to inositol 3,4,5-trisphosphate. Inositol trisphosphate activates downstream substrates such as Akt, Btk, Tec kinases, etc. and regulates a variety of functions like cell growth, repair, inflammatory mediator release, inflammatory cell chemotaxis, etc. (Ito et al., 2007). Two members of the PI3K family—PI3K γ and δ—have been recognized to play a role in experimental models of airway inflammation, cytokine release, airway hyperreactivity, etc. PI3K δ has been suggested to play a role in steroid-resistant asthma (Ito et al., 2007). Inhibitors of PI3K γ and δ have been designed and tested in experimental systems (Duan et al., 2005; Lee et al., 2006).

2.5 LIPID MEDIATORS PROMOTING GENE TRANSCRIPTION

Evidence suggests that lipid molecules like HETE, LTB$_4$, and lysophosphatidic acid can switch on gene transcription. Notably, the activation of peroxisome proliferator-activated receptor α (PPAR α) and PPAR γ genes are switched on by PGJ2, HETE, and LTB$_4$ (Trifilieff et al., 2003; Ibabe et al., 2005). Activators of PPAR α and γ have been shown to inhibit airway inflammation in an experimental model (Trifilieff et al., 2003). Studies have also shown that LPA can regulate the transcription of several important inflammatory mediators like IL8 (Saatian et al.,

2006), COX-2 and prostaglandin E2 (He et al., 2008). Implication of lipid-mediated gene expression in the context of inflammatory airway disease will be discussed.

2.6 SUMMARY AND CONCLUSION

In summary, we have described the metabolic pathway for phospholipid and sphingolipids and identified key biologically active mediators for further discussion. Some of these mediators have been exploited to discover and develop drugs for asthma. A few others are in a more exploratory stage from the point of view of drug discovery effort. Lipid-mediated gene expression has also been evaluated. It is evident that lipids can alter the expression status of genes by activating second messengers. However, certain lipids have direct effect on gene expression as ligand for transcription factor activation. We have also discussed key enzymes that play an important role in the generation of lipid mediators as well as those activated by lipid metabolites. We have attempted to present information in the light of airway inflammation.

REFERENCES

Abraham, W. M., J. S. Stevenson, and R. Garrido. 1989. A possible role for PAF in allergen induced late responses: Modification by a selective antagonist. *J App Physiol.* 66:2351–57.

Arima, M., T. Yukawa, and S. Makino. 1995. Effect of YM264 on the airway hyperresponsiveness and the late asthmatic response in a guinea pig model of asthma. *Chest* 108:529–34.

Aoki, J., A. Taira, Y. Takanezawa Y. Kishi , K. Hama, T. Kishimoto, K. Mizuno, K. Saku, R. Taguchi, and H. Arai. 2002. Serum lysophosphatidic acid is produced through diverse phospholipase pathways. *J Biol Chem.* 277:48737–44.

Ble, F. X., C. Cannet, S. Zurbruegg, C. Gerard, N. Frossard, N. Beckmann, and A. Trifilieff. 2009. Activation of the lung S1P(1) receptor reduces allergen-induced plasma leakage in mice. *Br J Pharmacol.* 158:1295–301.

Bilenki, L., J. Yang, Y. Fan, S. Wang, and X. Yang. 2004. Natural killer T cells contribute to airway eosinophilic inflammation induced by ragweed through enhanced IL-4 and eotaxin production. *Eur J Immunol.* 34:345–54.

Brink, C., S.-E. Dahlen, J. Drazen, J. F. Evans, D. W. P. Hay, S. Nicosia, C. N. Serhan, T. Shimizu, and T. Yokomizo. 2006. International Union of Pharmacology XXXVII. Nomenclature for leukotriene and lipoxin receptors. *Pharmacol Rev.* 55:195–27.

Brooks, C. D. W. and J. B. Summers. 1996. Modulators of leukotriene biosynthesis and receptor activation. *J Med Chem.* 39:2629–54.

Burli, R. W., H. Xu, X. Zou, K. Muller, J. Golden, M. Frohn, M. Adlam et al. 2006. Potent hFPRL1 (ALXR) agonists as potential anti-inflammatory agents. *Bioorg Med Chem Lett.* 16:3713–18.

Busse, W. W. 1998. Leukotrienes and inflammation. *Am J Respir Crit Care Med.* 157:S210–13.

Chaudhary, D. and M. Kasaian. 2006. PKC theta: A potential therapeutic target for T-cell-mediated diseases. *Curr Opin Investig Drugs* 7:432–37.

Chiang, N, C. N. Serhan, S.-E. Dahlen, J. M. Drazen, D. P. W. Hay, G. Enrico G. Rovati, T. Shimizu, T. Yokomizo, and C. Brink. 2006. The lipoxin receptor ALX: Potent ligand-specific and stereoselective actions *in vivo. Pharmacol Rev.* 58:463–87.

Draheim S. E., N. J. Bach, R. D. Dillard, D. R. Berry, D. G. Carlson, N. Y. Chirgadze, D. K. Clawson et al. 1996. Indole inhibitors of human nonpancreatic secretory phospholipase A2. 3. Indole-3-glyoxamides. *J Med Chem.* 39:5159–75.

Drazen, J. 1998. Clinical pharmacology of leukotriene receptor antagonists and 5-lipoxygenase inhibitors. *Am J Respir Crit Care Med.* 157:S233–37.

Duan, W, A. M. Guinaldo Datiles, B. P. Leung, C. J. Vlahos, and W. S. Wong. 2005. An anti-inflammatory role for a phosphoinositide 3-kinase inhibitor LY294002 in a mouse asthma model. *Int Immunopharmacol* 5:495–502.

Freitag, A., R. M. Watson, G. C. Matsos, C. Eastwood, and P. M. O'Byrne. 1993. Effect of a platelet-activating factor antagonist, WEB 2086, on allergen induced asthmatic responses. *Thorax* 48:594–98.

Garcia-Marcos, L., A. Schuster, and E. Perez-Yarza. 2003. Benefit risk assessment of antileukotrienes in the management of asthma. *Drug Safety* 26:483–18.

Gilroy, D. W., T. Lawrence, M. Perreti, and A. Rossi. 2004. Inflammatory resolution: New opportunities for drug discovery. *Nat Rev Drug Discov.* 3:401–15.

Gomez, F. P., J. Roca, J. A. Barbera, K. F. Chung, V. I. Peinado, and R. Rodriguez-Roisin. 1998. Effect of a platelet-activating factor (PAF) antagonist, SR 27417A, on PAF-induced gas exchange abnormalities in mild asthma. *Eur Respir J.* 11:835–39.

Grant, G. E., J. Rokach, and W. S. Powell. 2009. 5-Oxo-ETE and the OXE receptor. *Prostaglandins Other Lipid Mediat.* 89:98–104.

Gundel, R. H., C. D. Wegner, H. O. Heuer, and L. G. Letts. 1992. A PAF receptor antagonist inhibits acute airway inflammation and late-phase responses but not chronic airway inflammation and hyperresponsiveness in a primate model of asthma. *Mediat Inflamm.* 1:379–84.

Haworth, O., M. Cernadas, R. Yang, C. N. Serhan, and B. D. Levy. 2008. Resolvin E1 regulates interleukin 23, interferon g and lipoxin A4 to promote resolution of allergic airway inflammation. *Nat Immunol.* 1–7, http://www.nature.com/natureimmunology.

He, D., V. Natarajan, R. Stern, I. A. Gorshkova, J. Solway, E. W. Spannhake, and Y. Zhao. 2008. Lysophosphatidic acid-induced transactivation of epidermal growth factor receptor regulates cyclo-oxygenase-2 expression and prostaglandin E(2) release via C/EBPbeta in human bronchial epithelial cells. *Biochem J.* 412:153–62.

Hozawa, S., Y. Haruta, S. Ishioka, and M. Yamakido. 1995. Effects of a PAF antagonist, Y-24180, on bronchial hyperresponsiveness in patients with asthma. *Am J Respir Crit Care Med.* 152:1198–202.

Hurley, B. P. and B. A. McCormack. 2008. Multiple roles of phospholipase A2 during lung infection and inflammation. *Infect Immun.* 76:2259–72.

Ibabe, A., A. Herrero, and M. P. Cajaraville. 2005. Modulation of peroxisome proliferator-activated receptors (PPARs) by PPAR (alpha)- and PPAR (gamma)-specific ligands and by 17 beta-estradiol in isolated zebrafish hepatocytes. *Toxicol In Vitro* 19:725–35.

Idzko, M., H. Hammad, M. van Nimwegen, M. Kool, T. Müller, T. Soullié, M. A. Willart, D. Hijdra, H. C. Hoogsteden, and B. N. Lambrecht. 2006. Local application of FTY720 to the lung abrogates experimental asthma by altering dendritic cell function. *J Clin Invest.* 116:2935–44.

Ishii, S., T. Nagase, H. Shindou, H. Takizawa, Y. Ouchi, and T. Shimizu. 2004. Platelet-activating factor receptor develops airway hyperresponsiveness independently of airway inflammation in a murine asthma model. *J Immunol.* 172:7095–102.

Ito, K., G. Caramori, and I. M. Adcock. 2007. Therapeutic potential of phosphatidylinositol 3-kinase inhibitors in inflammatory respiratory disease. *J Pharmacol Exp Ther.* 321:1–8.

Jacobs, E. R. and D. C. Zeldin. 2001. The lung HETEs (and EETs) up. *Am. J Physiol Heart Circ Physiol.* 280:H1–10.

Jenkins, G. M. and M. A. Frohman. 2005. Phospholipase D: A lipid centric review. *Cell Mol Life Sci.* 62:2305–16.

Kilfeather, S. 2002. 5-Lipoxygenase inhibitors for the treatment of COPD. *Chest* 121: 197S–200S.

Kim, R. H., K. Takabe, S. Milstein, and S. Spiegel. 2009. Export and function of sphingosine-1-phosphate. *Biochim Biophys Acta* 1791:692–96.

Kuitert, L. M., R. M. Angus, N. C. Barnes, P. J. Barnes, M. F. Bone, K. F. Chung, A. J. Fairfax, T. W. Higenbotham, B. J. O'Connor, and B. Piotrowska. 1995. Effect of a novel potent platelet-activating factor antagonist, modipafant, in clinical asthma. *Am J Respir Crit Care Med.* 151:1331–35.

Lambeau, G. and M. H. Gelb. 2008. Biochemistry and physiology of mammalian secreted phospholipases A2. *Annu Rev Biochem.* 77:495–20.

Langlois, A., F. Chouinard, N. Flamand, C. Ferland, M. Rola-Pleszczynski, and M. Laviolette. 2009. Crucial implication of protein kinase C (PKC)-delta, PKC-zeta, ERK-1/2, and p38 MAPK in migration of human asthmatic eosinophils. *J Leukoc Biol.* 85:656–63.

Lee, K. S., H. K. Lee, J. S. Hayflick, Y. C. Lee, and K. D. Puri. 2006. Inhibition of phosphoinositide 3-kinase delta attenuates allergic airway inflammation and hyperresponsiveness in murine asthma model. *FASEB J* 20:455–65.

Lee, K. A., M. H. Kang, Y. S. Lee, Y. J. Kim, D. H. Kim, H. J. Ko, and C. Y. Kang. 2008. A distinct subset of natural killer T cells produces IL-17, contributing to airway infiltration of neutrophils but not to airway hyperreactivity. *Cell Immunol.* 25:50–55.

Levy, B. D., G. T. Sanctis, P. R. Devchand, E. Kim, K. Ackerman, B. A. Schmiditt, W. Szczeklik, J. Drazen, and C. N. Serhan. 2002. Multipronged inhibition of airway hyperresponsiveness and inflammation by lipoxin A4. *Nat Med.* 8:1018–23.

Martin, P., R. Villares, S. Rodriguez-Mascarenhas, A. Zaballos, M. Leitges, J. Kovac, I. Sizing et al. 2005. Control of T helper 2 cell function and allergic airway inflammation by PKCzeta. *Proc Natl Acad Sci USA* 102:9866–71.

Masini, E., L. Giannini, S. Nistri, L. Cinci, R. Mastroianni, W. Xu, S. A. Comhair et al. 2008. Ceramide: A key signaling molecule in a guinea pig model of allergic asthmatic response and airway inflammation. *J Pharmacol Exp Ther.* 324:548–57.

Matsuda, H., T. Suda, J. Sato, T. Nagata, Y. Koide, K. Chida, and H. Nakamura. 2005. Alpha-galactosylceramide, a ligand of natural killer T cells, inhibits allergic airway inflammation. *Am J Respir Cell Mol Biol.* 33:22–31.

McKew, J. C., K. L. Lee, M. W. H. Shen, P. Thakker, M. A. Foley, M. L. Behnke, B. Hu et al. 2008. Indole cytosolic phospholipase A2 r inhibitors: Discovery and *in vitro* and *in vivo* characterization of 4-{3-[5-chloro-2-(2-{[(3,4-dichlorobenzyl)sulfonyl]amino}ethyl)-1-(diphenylmethyl)-1*H*-indol-3-yl]propyl}benzoic acid, efipladib. *J Med Chem.* 51:3388–413.

Meyer, E. H., S. Goya, O. Akbari, G. J. Berry, P. B. Savage, M. Kronenberg, T. Nakayama, R. H. DeKruyff, and D. T. Umetsu. 2006. Glycolipid activation of invariant T cell receptor+ NK T cells is sufficient to induce airway hyperreactivity independent of conventional CD4+ T cells. *Proc Natl Acad Sci USA* 103:2782–87.

Moolenar, W. H., L. A. van Meeteren, and B. N. G. Giepmans. 2004. The ins and outs of lyso-phosphatidic acid signaling. *Bioassays* 26:870–81.

Morooka, S., M. Uchida, and N. Imanishi. 1992. Platelet-activating factor (PAF) plays an important role in the immediate asthmatic response in guinea-pig by augmenting the response to histamine. *Br J Pharmacol.* 105:756–62.

Morishima, Y., Y. Ishii, T. Kimura, A. Shibuya, K. Shibuya, A. E. Hegab, T. Iizuka et al. 2005. Suppression of eosinophilic airway inflammation by treatment with alpha-galactosylceramide. *Eur J Immunol.* 35:2803–14.

Murakami, M. 2004. Hot topics in phospholipase A2 field. *Biol Pharm Bull.* 27:1179–82.

Nagase, T., S. Ishii, H. Shindou, Y. Ouchi, and T. Shimizu. 2002. Airway hyperresponsiveness in transgenic mice overexpressing platelet activating factor receptor is mediated by an atropine-sensitive pathway. *Am J Respir Care Med.* 165:200–5.

Pergola, C. and O. Werz. 2010. 5-Lipoxygenae inhibitors: A review of recent developments and patents. *Exp Opin Ther Pat.* 20:1–21.

Peters-Golden, M. and W. R. Henderson. 2007. Leukotrienes. *New Engl Med.* 357:1841–54.

Pettipher, R., T. Hansel, and R. Armer. 2007. Antagonism of the prostaglandin D2 receptors DP1 and CRTH2 as an approach to treat allergic diseases. *Nat Rev Drug Discov.* 6:313–25.

Powell, W. S. and J. Rokach. 2005. Biochemistry, biology and chemistry of the 5-lipoxygenase product 5-oxo-ETE. *Prog Lipid Res.* 44:154–8.

Prescott, S. M., G. A. Zimmerman, and T. M. McIntyre. 1990. Platelet activating factor. *J Biol Chem.* 265:17381–4.

Rhee, S. G. 2001. Regulation of phosphoinositide-specific phospholipase C. *Annu Rev Biochem.* 70:281–312.

Rosen, H., P. J. Gonzalez-Cabrera, M. G. Sanna, and S. Brown. 2009. Sphingosine 1-phosphate receptor signaling. *Annu Rev Biochem.* 78:743–68.

Ryan, J. J. and S. Spiegel. 2008. The role of sphingosine-1-phosphate and its receptor in asthma. *Drug News Perspect.* 21:89–96.

Saatian B, Y. Zhao, D. He, S. N. Georas, T. Watkins, E. W. Spannhake, and V. Natarajan. 2006. Transcriptional regulation of lysophosphatidic acid-induced interleukin-8 expression and secretion by p38 MAPK and JNK in human bronchial epithelial cells. *Biochem J* 393:657–68.

Sammani, S., L. Moreno-Vinasco, T. T. Mirzapoiazova, P. A. Singleton, E. T. Chiang, C. L. Evenoski, T. Wang et al. 2009. Differential effects of S1P receptors on airway and vascular barrier function in the murine lung. *Am J Respir Cell Mol Biol.* 43:394–402.

Sanjar, S., S. Aoki, K. Boubekeur, I. D. Chapman, D. Smith, M. A. Kings, and J. Morley. 1990. Eosinophil accumulation in pulmonary airways of guinea-pigs induced by exposure to an aerosol of platelet-activating factor: Effect of anti-asthma drugs. *Br J Pharmacol.* 99:267–72.

Serhan, C. N. 2007. Resolution phase of inflammation: Novel anti-inflammatory and pro-resolving lipid mediator pathway. *Annu Rev Immunol.* 25:101–37.

Smith, K. R., K. E. Pinkerton, T. Watanabe, T. L. Pederson, S. J. Ma, and B. D. Hammock. 2004. Attenuation of tobacco smoke induced lung inflammation by soluble epoxide hydrolase inhibitor. *Proc Natl Acad Sci USA* 102:2186–91.

Spik I., C. Brénuchon, V. Angéli, D. Staumont, S. Fleury, M. Capron, F. Trottein, and D. Dombrowicz. 2005. Activation of the prostaglandin D2 receptor DP2/CRTH2 increases allergic inflammation in mouse. *J Immunol.* 174:3703–8.

Stebbins K., A. R. Broadhead, C. S. Baccei, J. M. Scott, Y. P. Truong, A. M. Santini, P. Fagan et al. 2010. Pharmacological blockade of the DP2 receptor inhibits cigarette smoke-induced inflammation, mucus cell metaplasia and epithelial hyperplasia in the mouse lung. *J Pharmacol Exp Ther.* 332:764–75.

Trifilieff, A., A. Bench, M. Hanley, D. Bayley, E. Campbell, and P. Whittaker. 2003. PPAR-alpha and -gamma but not -delta agonists inhibit airway inflammation in a murine model of asthma: *In vitro* evidence for an NF-kappaB-independent effect. *Br J Pharmacol.* 139:163–71.

Uhlig, S. and E. Gulbins. 2008. Sphingolipids in the lungs. *Am J Respir Crit Care Med.* 178:1100–14.

Uller, L., J. M. Mathiesen, L. Alenmyr, M. Korsgren, T. Ulven, T. Högberg, G. Andersson, C. G. Persson, and E. Kostenis. 2007. Antagonism of the prostaglandin D2 receptor CRTH2 attenuates asthma pathology in mouse eosinophilic airway inflammation *Respir Res.* 8:16.

von Bismarck, P., C. F. Wistadt, K. Klemm, S. Winoto-Morbach, U. Uhlig, S. Schutze, D. Adam, B. Lachmann, S. Uhlig, and M. F. Krause. 2008. Improved pulmonary function by acid sphingomyelinase inhibition in a newborn piglet lavage model. *Am J Respir Crit Care Med.* 177:1233–41.

Wang, A. and E. A. Dennis. 1999. Mammalian lysophospholipases. *Biochim Biophys Acta* 1439:1–16.

Werz, O. and D. Steinhilber. 2005. Development of 5-lipoxygenase inhibitors—Lesson from cellular enzyme regulation. *Biochem Pharmacol.* 70:327–33.

Yang, C. M., I. T. Lee, C. C. Lin, Y. L. Yang, S. F. Luo, Y. R. Kou, and L. D. Hsiao. 2009. Cigarette smoke extract induces COX-2 expression via a PKCalpha/c-Src/EGFR, PDGFR/PI3K/Akt/NF-kappaB pathway and p300 in tracheal smooth muscle cells. *Am J Physiol Lung Cell Mol Physiol.* 297:L892–902.

Zhao, Y., J. Tong, D. He, S. Pendyala , B. Evgeny, J. Chun, A. I. Sperling, and V. Natarajan. 2009. Role of lysophosphatidic acid receptor LPA2 in the development of allergic airway inflammation in a murine model of asthma. *Respir Res.* 10:114.

3 Phospholipase A₂ as a Potential Drug Target for Airway Disorders

Yoshitaka Taketomi and Makoto Murakami

CONTENTS

Phospholipase A_2 (PLA_2) is a group of enzymes that liberate a fatty acid and a lysophospholipid from a glycerophospholipid, a major component in cell membranes. The PLA_2 reaction is considered as the first rate-limiting step for the production of lipid mediators derived from arachidonic acid (AA) or lysophospholipids, such as eicosanoids [prostaglandins (PGs) and leukotrienes (LTs)] and platelet-activating factor (PAF). Mammalian genomes encode a number of PLA_2 enzymes, which are subdivided into several subclasses based on their structures, localizations, and functions. Of these, cytosolic PLA_2 ($cPLA_2\alpha$) group IVA and secreted PLA_2s ($sPLA_2$s) group V and X have attracted much attention as regulators of the pathology in airway disorders. No doubt, $cPLA_2\alpha$ is a central player of AA release that globally provides airway eicosanoids, whereas $sPLA_2$s are likely to participate in spatiotemporal production of lipid mediators or in degradation of pulmonary surfactant, a noncellular phospholipid component essential for airway homeostasis and stability. Perturbation of the $cPLA_2\alpha$- and $sPLA_2$-dependent pathways causes detrimental airway disorders. In this chapter, we will highlight the basic biology of PLA_2s and their potential contributions to airway diseases on the basis of recent findings by pharmacological, cell biological, gene targeting, and clinical studies, as well as development of drugs that target particular PLA$_2$ subtypes for treatment of patients with airway diseases.

3.1 BASIC BIOLOGY OF PHOSPHOLIPASE A$_2$ ENZYMES

PLA$_2$ catalyzes the hydrolysis of the *sn*-2 ester bond of glycerophospholipids to liberate free fatty acids and lysophospholipids (Schaloske and Dennis, 2006) (Figure 3.1). In the view of signal transduction, the PLA$_2$ reaction is particularly important when the fatty acid liberated is AA, which can be metabolized by cyclooxygenases (COXs) and lipoxygenases (LOs) into the biologically active lipid mediators, PGs and LTs, which are referred to as eicosanoids (Austin and Funk, 1999; Funk, 2001). Lysophospholipids, the other products of PLA$_2$ hydrolysis, themselves are signal mediators in some cases [such as lysophosphatidic acid (LPA)] or serve as precursors of bioactive molecules (such as PAF) (Rivera and Chun, 2008). These lipid mediators exert numerous biological actions through their cognate G-protein-coupled receptors on target cells (Kobayashi and Narumiya, 2002; Ishii et al., 2004a,b). PLA$_2$ has also been implicated in maintaining cell membrane homeostasis by participating in the recycling of fatty acid moieties within membrane phospholipids or by helping regulate phospholipid mass (Kudo and Murakami, 2002). So far, more than 30 enzymes exhibiting PLA$_2$ or related activity have been identified in mammals and subdivided into several groups based on their structures, catalytic mechanisms, intracellular or extracellular localizations, and evolutionary relationships (Murakami and Kudo, 2002; Schaloske and Dennis, 2006). They cluster in five main categories: sPLA$_2$s, cPLA$_2$s, Ca^{2+}-independent PLA$_2$ (iPLA$_2$), PAF acetylhydrolases (PAF-AH), and lysosomal PLA$_2$s (LPLA$_2$).

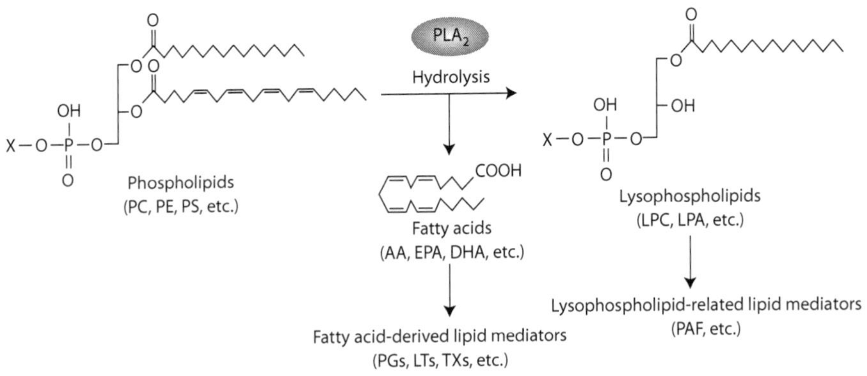

FIGURE 3.1 PLA$_2$ reaction. PLA$_2$ hydrolyzes the *sn*-2 ester of a glycerophospholipid to give rise to a free fatty acid and a lysophospholipid. Since the *sn*-2 position of glycerophospholipids is enriched in polyunsaturated fatty acid, such as arachidonic acid (AA), eicosapentaenoic acid (EPA), and docosahexaenoic acid (DHA), the PLA$_2$ reaction is regarded as the first rate-limiting step of fatty acid-derived lipid mediators (such as eicosanoids). Lysophospholipids act as another class of lipid mediators by themselves (such as LPC and LPA) or are metabolized to lysophospholipid-derived lipid mediators (such as PAF).

The cPLA$_2$ family (also designated group IV) comprises six intracellular enzymes commonly called cPLA$_2\alpha$, cPLA$_2\beta$, cPLA$_2\gamma$, cPLA$_2\delta$, cPLA$_2\epsilon$, and cPLA$_2\zeta$, which (except for cPLA$_2\gamma$) possess an N-terminal C2 domain for Ca^{2+}-dependent association with cellular membranes (Ghosh et al., 2006; Kita et al., 2006) (Figure 3.2). Among the cPLA$_2$ family, cPLA$_2\alpha$ has been most extensively studied (Sharp and White, 1993; Clark et al., 1995). This enzyme is ubiquitously expressed and has attracted much attention for its preference to AA as an *sn*-2 fatty acyl moiety of substrate phospholipids and for its activation by μM levels of Ca^{2+}, by phosphorylation directed by various kinases [e.g., mitogen-activated protein kinase (MAPK), protein

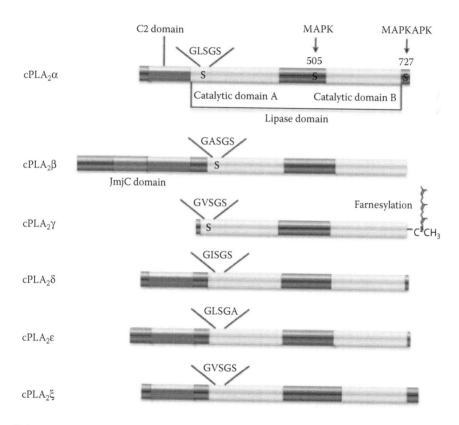

FIGURE 3.2 The cPLA$_2$ family. The enzymes belonging to this family typically possess an N-terminal C2 domain, which binds to two Ca^{2+} ions, followed by a catalytic domain with the lipase consensus motif GXSXG/A. cPLA$_2\alpha$ (group VIA) is a prototypic enzyme in this family, displays AA selectivity, and plays a central role in eicosanoid biosynthesis. Phosphorylation at two sites by mitogen-activated protein kinases (MAPK) and MAPK-activated protein kinases (MAPKAPK) is essential for the activation of cPLA$_2\alpha$ in cells. cPLA$_2\beta$ (group IVB) has a JimC domain prior to the C2 domain. Genes for cPLA$_2\beta$, δ, ε, and ζ are clustered in the same chromosomal locus, suggesting their closest evolutional relationship. cPLA$_2\gamma$ (group IVC) is unique in that it lacks the C2 domain and that human but not mouse enzyme undergoes farnesylation at the C-terminus. The functions of cPLA$_2\beta$–ζ *in vivo* are entirely unknown.

kinase C, and Ca^{2+}/calmodulin-dependent protein kinase], and by lipid phosphates, including phosphatidylinositol-4,5-bisphosphate and ceramide-1-phosphate (Leslie, 2004). An important step in cellular regulation of $cPLA_2\alpha$ involves its translocation from the cytosol to membranes (typically Golgi and perinuclear membranes) to access the substrate. The translocation is induced in response to an increase in intracellular Ca^{2+}, which binds to the C2 domain and increases the affinity of $cPLA_2\alpha$ for membranes (Clark et al., 1995; Leslie, 1997). Phosphorylation of Ser^{505} by MAPK is important for the membrane residence of $cPLA_2\alpha$, thereby increasing its enzymatic activity. Currently available information about the biochemical properties and tissue distribution of other $cPLA_2$ isoforms suggests that they may have distinct regulatory mechanisms and functional roles, as reviewed elsewhere (Ghosh et al., 2006; Kita et al., 2006).

The $sPLA_2$ family includes 11 isoforms identified so far (group IB, IIA, IIC, IID, IIE, IIF, III, V, X, XIIA, and XIIB) in mammals (Figure 3.3). These are structurally related, disulfide-rich, low-molecular-weight enzymes with a His–Asp catalytic dyad and a Ca^{2+}-binding loop (Murakami and Kudo, 2001; Lambeau and Gelb, 2008). $sPLA_2$s hydrolyze the *sn*-2 fatty acyl esters of glycerophospholipids in the presence of mM levels of Ca^{2+} without strict fatty acid selectivity but with some head group specificity. Individual $sPLA_2$s exhibit unique tissue and cellular localizations and specific enzymatic properties (Singer et al., 2002), suggesting their distinct pathophysiological roles (Lambeau and Gelb, 2008). Several $sPLA_2$s can contribute to the mobilization of AA that is converted to eicosanoids. The actions of group V and X $sPLA_2$s ($sPLA_2$-V and $sPLA_2$-X, respectively) on mammalian cell membranes are particularly potent among mammalian $sPLA_2$s, maybe due to their strong interfacial binding to phosphatidylcholine (PC) that is enriched in the outer leaflet of the plasma membrane (Murakami et al., 2001; Murakami and Kudo, 2003). In addition, $sPLA_2$s can act on various noncellular phospholipid components such as lipoproteins and pulmonary surfactants, as well as on foreign phospholipids such as microbe membranes and dietary phospholipids. For more detailed views of the $sPLA_2$ structure/function, mechanism, and signaling, see recent comprehensive reviews (Lambeau and Gelb, 2008; Murakami et al., 2010).

The $iPLA_2$ or patatin-like phospholipase domain-containing lipase (PNPLA) family (designated group VI) contains nine enzymes in mammals, some of which play fundamental roles in lipid remodeling and catabolism and are lipid hydrolases with specificities for diverse substrates such as triacylglycerol, phospholipids, and retinol ester (Williams and Galli, 2000; Balsinde and Balboa, 2005; Kienesberger et al., 2009). Analyses of mutant mouse models and clinical phenotypes of patients with the gene mutations have provided important insights into the physiological roles of several members of the $iPLA_2$/PNPLA family. For more details of this family, refer to other recent reviews (Hooks and Cummings, 2008; Cedars et al., 2009; Kienesberger et al., 2009).

Besides the big three families ($cPLA_2$, $sPLA_2$, and $iPLA_2$), several PLA_2-related enzymes belonging to other subclasses have been identified. The PAF-AH family, designated group VII and VIII PLA_2s, contains four enzymes that exhibit unique substrate specificity toward PAF and/or oxidized phospholipids, as reviewed in Arai (2002) and McIntyre et al. (2009). Additionally, two distinct lysosomal PLA_2s [acidic

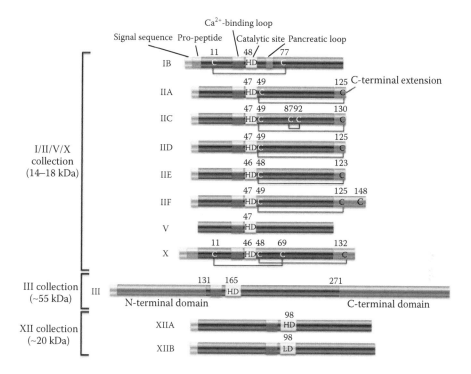

FIGURE 3.3   The sPLA$_2$ family. Mammalian sPLA$_2$s contain 11 isoforms, which are subdivided into three collections, namely group I/II/V/X (classical sPLA$_2$s), group III, and group XII. All enzymes have a conserved catalytic center with a His–Asp dyad and a Ca^{2+}-binding loop. sPLA$_2$-IB, a pancreatic sPLA$_2$, is characterized by an N-terminal propeptide whose proteolytic removal gives rise to a functional enzyme, the presence of a Cys11–Cys77 disulfide bond (group I-specific disulfide), and a unique pancreatic loop. The group II subfamily (IIA, IIC, IID, IIE, and IIF) is characterized by the absence of the propeptide and the presence of Cys49–Cys within the C-terminal extension (group II-specific disulfide). sPLA$_2$-IIC is absent in humans (pseudogene). sPLA$_2$-IIF has a longer C-terminal extension, which is Pro-rich. sPLA$_2$-V is evolutionarily close to the group II subfamily, but lacks the group II-specific disulfide and the C-terminal extension. sPLA$_2$-X has both group I and II properties since it has an N-terminal propeptide and both group I- and II-specific disulfides. sPLA$_2$-III is unique in that the central sPLA$_2$ domain, which is more similar to bee venom PLA$_2$ than to group I/II/V/X sPLA$_2$s, is flanked by unique and highly cationic N- and C-terminal domains. The N- and C-terminal domains are removed to produce a mature, sPLA$_2$ domain-only form. The group XII collection contains two isoforms, XIIA and XIIB, whose overall structures (except for the catalytic domain and Ca^{2+}-binding site) do not show any homology with other sPLA$_2$s. The catalytic center His is replaced by Leu in sPLA$_2$-XIIB, indicating that this enzyme is catalytically inactive.

lysosomal PLA$_2$ (aiPLA$_2$) and LPLA$_2$; the latter being designated group XV] (Abe et al., 2004; Fisher et al., 2005) and one adipose-specific phospholipase A (Ad-PLA; group XVI) (Duncan et al., 2008) have been identified as novel classes of PLA$_2$s. Human genome encodes at least five Ad-PLA-like genes, which are structurally similar to lecithin-rethinol acyltransferase.

Because of this molecular diversity, PLA_2 enzymes have been implicated in various biochemical processes. Mice with the $cPLA_2\alpha$ (*Pla2g4a*) gene disruption, which was first reported in 1997 (Bonventre et al., 1997; Uozumi et al., 1997), have provided unequivocal evidence for the central role of this enzyme in AA metabolism in many if not all biological events. Since the beginning of the twenty-first century, mice with transgenic overexpression and/or targeted disruption of several PLA_2 subtypes have been generated. The phenotypes displayed in individual PLA_2 gene-manipulated mice may not be simply the reflection of changes in lipid mediator signaling and more particularly eicosanoid signaling, but may be due to one or a combination of the hydrolysis of a variety of target membranes. In addition, several PLA_2s have been shown to link with human diseases. Given the central role of PLA_2 enzymes in phospholipid metabolism, each enzyme may represent an ideal target for anti-inflammatory drug discovery. The following section of this review will focus on the potential participations of the lipid networks mediated by several PLA_2 enzymes in airway disorders.

3.2 EVIDENCE FOR THE ROLES OF PLA_2 ENZYMES IN AIRWAY INFLAMMATION

3.2.1 LIPID MEDIATORS IN AIRWAY DISORDERS

Asthma is a chronic disease characterized by airway hyperresponsiveness and accompanied by airway structural changes, including goblet cell metaplasia, smooth muscle cell layer thickening, and subepithelial fibrosis (Holgate, 1999; Kay, 2001). The airway inflammation is associated with infiltration of inflammatory cells such as $CD4^+$ and $CD8^+$ T lymphocytes, eosinophils, macrophages, and mast cells (Barrett and Austen, 2009; Lloyd and Hawrylowicz, 2009). The biosynthesis of lipid mediators, which are the products/metabolites of PLA_2 hydrolysis, in the airway is a key component of asthma pathogenesis. Individual lipid mediators regulate airway inflammation in either positive or negative way. Figure 3.4 illustrates the proposed interplay of lipid mediators in the pathology of asthma, on the basis of studies using knockout mice for biosynthetic enzymes and receptors that act downstream of PLA_2 enzymes. Mice lacking PGD_2 receptors DP_1 or CRTH2 (DP_2) show resistance to allergic airway inflammation (Matsuoka et al., 2000; Shiraishi et al., 2008), whereas deficiency of the COX-1 and COX-2, the PGI_2/prostacyclin receptor IP, or the PGE_2 receptor EP3 worsens the disease states (Gavett et al., 1999; Zeldin et al., 2001; Kunikata et al., 2005; Nakata et al., 2005; Jaffar et al., 2007). Cysteinyl LTs (cys-LTs; LTC_4, LTD_4, and LTE_4) and LTB_4 promote the asthmatic responses since knockout mice for 5-LO, 5-LO-activating protein (FLAP), LTC_4 synthase (LTC_4S), the cys-LT receptor CysLT1, and the LTB_4 receptor BLT1 all show amelioration of the airway inflammation (Irvin et al., 1997; Tager et al., 2003; Kim et al., 2006). In contrast, 12/15-LO products, including AA-derived lipoxin A_4 (LXA_4) and ω-3 polyunsaturated fatty acid-derived resolvin E1 (RvE1) and protectin D1 (PD1), play a protective role against

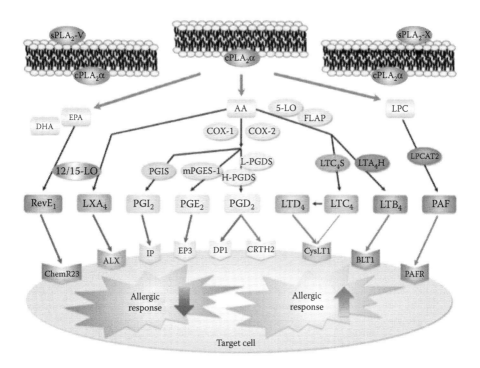

FIGURE 3.4 Lipid mediator network in asthma. Of the AA metabolites, the COX pathway product PGD₂ has both exacerbating and protective effects on asthma through two distinct receptors, whereas PGE₂ and PGI₂ display protective effects. The 5-LO pathway products LTB₄ and cysLTs (LTC₄ and LTD₄) as well as the LPC-derived mediator PAF promote the pathology of asthma. The 12/15-LO products of AA (lipoxin A₄; LXA₄) and of ω-3 lipids such as resolvin E1 (RevE1) are anti-inflammatory lipid mediators that play a role in the resolution of inflammation. cPLA₂α, which is expressed constitutively and ubiquitously, is a key player for the release of the precursors of these lipid mediators from membrane phospholipids. sPLA₂-V and sPLA₂-X, which are induced in distinct populations of bronchial epithelial cells and infiltrating leukocytes, may contribute to the supply of lipid mediator precursors in synergy with or independently of cPLA₂α. Alternatively, these sPLA₂s may contribute to the disease by excessively degrading pulmonary lipid surfactant, which has a protective role in alveolar tension (see text). PGIS (PGI synthase), H-PGDS (hematopoietic PGD synthase), L-PGDS (lipo-calin-type PGD synthase), mPGES-1 (microsomal PGE synthase), LTA₄H (LTA₄ hydrolase), LTC₄S (LTC₄ synthase), FLAP (5-LO activating protein), LPCAT2 (LPC acyltransferase 2), IP (PGI receptor), EP3 (PGE receptor 3), DP1 (PGD receptor 1), CRTH2 (chemoattractant receptor-homologous molecule expressed on Th2 cells, or DP2), BLT1 (LTB₄ receptor 1), CysLT1 (cysteinyl LT receptor 1), PAFR (PAF receptor), and ALX (lipoxin A₄ receptor).

inflammation in general (Levy et al., 2002, 2007; Haworth et al., 2008). PAF, a lysophospholipid-derived lipid mediator, also participates in the process of asthma, since mice lacking PAF receptor are resistant to this model (Ishii et al., 2004a,b). The roles of these lipid mediators in airway diseases are reviewed in other chapters in this book.

3.2.2   cPLA$_2$α IN AIRWAY DISORDERS

The first definitive evidence for the importance of PLA$_2$ enzymes in airway diseases *in vivo* was provided by studies using *Pla2g4a*$^{-/-}$ mice, which lack cPLA$_2$α. A marked reduction in alveolar lumen thickening and a complete attenuation of airway hyperresponsiveness to methacholine are observed in *Pla2g4a*-deficient mice, suggesting that cPLA$_2$α is important for the late phase of the asthmatic response (Uozumi et al., 1997). In the mouse models of acid-induced acute respiratory disorder acute respiratory distress syndrome (ARDS) and bleomycin-induced pulmonary fibrosis, *Pla2g4a* gene knockout significantly attenuates pulmonary inflammation and fibrosis (Nagase et al., 2000, 2002). In these lung injury models, the levels of proinflammatory lipid mediators, such as thromboxane A$_2$ (TXA$_2$), LTB$_4$, and cys-LTs, in bronchoalveolar lavage fluid (BALF) are markedly if not solely reduced in *Pla2g4a*-null mice (Nagase et al., 2000). Following intratracheal inoculation with live *Escherichia coli*, pulmonary PAF biosynthesis, inflammatory cell infiltration, and bacterial clearance are attenuated in *Pla2g4a*$^{-/-}$ mice compared with wild-type littermates (Rubin et al., 2005). In *ex vivo* experiments, cPLA$_2$α inhibition or gene disruption leads to complete ablation of AA release and PAF biosynthesis and thereby diminishes bacterial killing by neutrophils, although respiratory burst, granule secretion, and phagocytosis are unaffected. Macrophage production of PGE$_2$, cys-LTs, and PAF in response to A23187 or lipopolysaccharide (LPS) is markedly reduced in *Pla2g4a*-deficient mice (Uozumi et al., 1997). Immediate and late productions of lipid mediators, including cys-LTs, PGD$_2$, and PAF, in response to FcεRI (high-affinity IgE receptor) cross-linking or cytokines are ablated in bone marrow-derived mast cells (BMMC) of *Pla2g4a*-deficient mice (Fujishima et al., 1999; Nakatani et al., 2000). In addition to these inflammatory cells, primary lung fibroblasts obtained from *Pla2g4a*-deficient mice produce minimal levels of PGE$_2$ in response to serum, even though there still remains substantial PGE$_2$ generation that appears to depend on cPLA$_2$ζ (Ghosh et al., 2007). Nevertheless, the fact that cPLA$_2$α activation also results in the production of PGE$_2$ and PGI$_2$, which can exert anti-inflammatory or protective effects in the airway (Ghosh et al., 2004), suggests that cPLA$_2$α might also participate in the amelioration of airway inflammation. Thus, the final outcome of a given airway pathology could be determined by a spatiotemporal balance between pro- and anti-inflammatory lipid mediators generated by the cPLA$_2$α-dependent pathway.

3.2.3   sPLA$_2$S IN AIRWAY DISORDERS

As compared with cPLA$_2$α, the roles of sPLA$_2$s in airway pathology have been controversial until recently, even though a number of cell biological and pharmacological studies have pointed out the potential contribution of this extracellular group of enzymes to airway pathology. Although constitutive versus inducible secretion of sPLA$_2$s into extracellular spaces depends on cellular contexts, it is generally believed that, under pathologic conditions, they are mostly released into extracellular environments following appropriate cell activations, which explains their presence in the plasma and biologic fluids of patients with local or systemic inflammatory, autoimmune, or allergic diseases (e.g., bronchial asthma, rheumatoid arthritis, and allergic rhinitis) (Granata et al., 2003). Clinically, sPLA$_2$s are secreted at low levels in the airways of healthy individuals and tend to increase during inflammatory reactions as the result of both plasma extravasation and local production. Some sPLA$_2$ activity is detectable in the BALF and in the nasal lavage fluid of healthy individuals (Stadel et al., 1994; Samet et al., 1996). The levels of sPLA$_2$ dramatically increase in the airways of patients with acute inflammatory diseases such as pneumonia and ARDS as well as in chronic disorders such as sarcoidosis, interstitial lung fibrosis, and chronic obstructive pulmonary disease (COPD) (Granata et al., 2009). sPLA$_2$ activity is also greatly increased in the nasal fluid of patients with allergic rhinitis (Stadel et al., 1994) and in the BALF of patients with bronchial asthma following specific antigen challenge (Chilton et al., 1996; Bowton et al., 1997). Thus, sPLA$_2$ enzymes are released locally in both upper and lower airways during allergic reactions.

Immunohistochemistry of the lungs from human patients with airway diseases or from experimental animals subjected to airway disease models has provided some lines of evidence for the expression/localization of multiple sPLA$_2$ isoforms in airway epithelium and alveolar macrophages (Masuda et al., 2005; Henderson et al., 2007; Munoz et al., 2007). In ARDS models, sPLA$_2$-IIA is dramatically induced in rat and rabbit lungs (Furue et al., 1999), whereas the elevation of sPLA$_2$-V and sPLA$_2$-X is dominant in mouse lungs (Sato et al., 2010). In cultured human bronchial epithelial cells, sPLA$_2$-V is induced by IL-1β, while the expression of sPLA$_2$-X is constitutive and unaffected by proinflammatory stimuli (Masuda et al., 2005). In these cells, adenoviral overexpression of sPLA$_2$-V or sPLA$_2$-X increases IL-1β-stimulated PGE$_2$ generation. The roles of sPLA$_2$s in macrophages have been investigated using mouse macrophage-like cell lines (Balboa et al., 1996; Balsinde et al., 1998) and mouse peritoneal macrophages (Satake et al., 2004), where sPLA$_2$-V plays an augmentative role in eicosanoid production. However, considering that alveolar macrophages in the asthma pathology are polarized into the IL-4/IL-13-sensitized "M2" phenotype, which are distinct from peritoneal macrophages that are typically the proinflammatory "M1" phenotype (Gordon, 2003), the roles of sPLA$_2$-V or other sPLA$_2$ isoforms in alveolar macrophages remain entirely unknown. Interestingly, overexpression of sPLA$_2$-X in cultured macrophage-like cells results in more production of IL-10 and 15-deoxy-12d,14-prostaglandin J$_2$ (both are anti-inflammatory) and less production of tumor necrosis factor α and nitric oxide (both are proinflammatory) (Curfs et al., 2008). This result raises the intriguing possibility that sPLA$_2$-X has the ability to promote M2 polarization of macrophages leading to anti-inflammation. The *in vivo* relevance of these observations awaits a future study.

Because asthma has an allergic or atopic component characterized by allergic sensitivity to environmental allergens and increased serum levels of antigen-specific and total IgE antibodies (Wills-Karp, 1999; Busse and Lemanske, 2001; Umetsu et al., 2002), attempts to understand the pathophysiology of asthma often focus on the effector cells of allergic inflammation, namely mast cells. Mast cells are derivatives of hematopoietic progenitors that differentiate and mature locally in most tissues, especially those exposed to the external environments, such as the airway (Kitamura, 1989; Metcalfe et al., 1997; Galli et al., 2005). Mast cells are present in increased numbers in the airways of patients with asthma (Brightling et al., 2002, 2003; Marone et al., 2005), and histological signs of airway mast cell degranulation *in vivo* can be identified in biopsy specimens from asthmatic patients (Carroll et al., 2002; Boyce, 2003; Brightling et al., 2003). Furthermore, ovalbumin-induced airway hyperreactivity and inflammation are significantly reduced in mast cell-deficient $Kit^{W-sh/W-sh}$ and $Kit^{W/W-v}$ mice (Williams and Galli, 2000; Yu et al., 2006), implying that mast cells have a critical role in multiple features of allergic airway responses.

IgE-mediated activation of mast cells induces the release of $sPLA_2s$, suggesting that mast cells represent one of the main sources of extracellular $sPLA_2s$ during allergic reactions. Rat serosal mast cells contain $sPLA_2$-IIA substantially in their secretory granules (Murakami et al., 1992; Chock et al., 1994), and its transcript is also detectable in cytokine-stimulated BALB/c, but not C57BL/6, mouse mast cells (Bingham et al., 1996; Murakami et al., 1997). Note that the absence of $sPLA_2$-IIA in C57BL/6 mice is due to a frameshift mutation of the *Pla2g2a* gene (MacPhee et al., 1995). Addition of exogenous $sPLA_2$-IIA to rat serosal mast cells (Murakami et al., 1991, 1993; Fonteh et al., 1994) or forcible overexpression of $sPLA_2$-IIA, $sPLA_2$-V, or $sPLA_2$-X in the rat mast cell line RBL-2H3 (Bingham et al., 1996; Enomoto et al., 2000; Murakami et al., 2001) enhances degranulation and PGD_2/ LTC_4 production. An immunoelectron microscopic study shows distinct localization of $sPLA_2$-IIA and $sPLA_2$-V in secretory granules and perinuclear area, respectively, in mouse BMMC (Bingham et al., 1999). Human lung mast cells constitutively express several $sPLA_2s$ (i.e., IIA, IID, IIE, IIF, III, V, X, XIIA, and XIIB) (Triggiani et al., 2009; Granata et al., 2010), while $sPLA_2$-IID alone is detectable in human cord blood-derived mast cells (Murakami et al., 2002). It is intriguing that $sPLA_2$ inhibitors can attenuate degranulation or lipid mediator production of rat serosal mast cells (Murakami et al., 1992) and human lung mast cells (Triggiani et al., 2009). However, it still remains to be elucidated as to which $sPLA_2$ isoform(s) is indeed involved in mast cell activation and thereby in mast cell-associated asthma and other airway pathologies *in vivo*.

Granulocytes (neutrophils and eosinophils) also represent important effector cells in asthma. Deletion of IL-17A, a hallmark T_H17 cytokine that is important for recruitment of neutrophils, or deletion of its receptor shows impaired T_H2-type allergic airway inflammation, suggesting the importance of neutrophils in the asthma pathology (Nakae et al., 2002). $sPLA_2$-V and $sPLA_2$-X are stored in azurophilic granules of human neutrophils, and the former is rapidly released after Formyl-methionyl-leucyl-phenylalanine (FMLP) stimulation but fails to promote LTB_4 production (Degousee et al., 2002). On the other hand, $sPLA_2$-X is released by mouse neutrophils and participates in respiratory burst and LTB_4 generation (Fujioka et al., 2008). Eosinophils serve as an important proinflammatory effector cell population by releasing cys-LTs, T_H2 cytokines, and major basic proteins (Akuthota et al., 2008). Selective deletion of eosinophils in mice abrogates pulmonary mucus accumulation and airway hyperresponsiveness associated with asthma (Lee et al., 2004). Although information on $sPLA_2$ expression in eosinophils is limited, one report shows the expression of $sPLA_2$-IID in human eosinophils (Seedsz et al., 2003). Interestingly, exposure of human eosinophils to exogenous $sPLA_2$-V or sPLA$_2$-X triggers LT synthesis (Munoz et al., 2003; Lai et al., 2010), and coculture of human eosinophils with bronchial epithelial cells demonstrates that sPLA$_2$-V secreted from the epithelial cells can trigger the activation and adhesion of eosinophils (Munoz et al., 2006; Wijewickrama et al., 2006). These results suggest that eosinophils may be a better target of sPLA$_2$-V or sPLA$_2$-X released from the bronchial epithelium in an asthmatic situation.

A major breakthrough in the sPLA$_2$ biology has come over the past few years with the establishment of various sPLA$_2$ knockout and transgenic mice (Murakami et al., 2010). As described above, sPLA$_2$-V is expressed at low levels in the bronchial epithelium and is markedly induced in a T$_H$2-dependent mouse asthma model as well as in patients with asthma or severe pneumonia (Masuda et al., 2005; Munoz et al., 2007). In the asthma model induced by allergic sensitization with ovalbumin, aerosol ingestion of recombinant sPLA$_2$-V, but not sPLA$_2$-IIA or a sPLA$_2$-V mutant (W31A) with reduced activity, causes dose-related increase in airway resistance, persistent airway narrowing, and leukocyte migration. Bronchoconstriction elicited by sPLA$_2$-V in mice lacking cPLA$_2\alpha$ is comparable to that in wild-type mice, suggesting that the proasthmatic action of sPLA$_2$-V is cPLA$_2\alpha$-independent. Importantly, methacholine-induced airway hyperresponsiveness in allergic mice is reduced by anti-sPLA$_2$-V antibody or by disruption of the *Pla2g5* gene (Munoz et al., 2007). Furthermore, acute lung injury induced by intratracheal administration of LPS, a model of ARDS, is also markedly attenuated in *Pla2g5* knockout mice compared to wild-type mice (Munoz et al., 2009). *Ex vivo* TLR2-dependent PGD$_2$ production is partially reduced in BMMC from *Pla2g5*-null mice, even though this PGD$_2$ amplification depends on sPLA$_2$-V-dependent cPLA$_2\alpha$ activation rather than on the intrinsic capacity of sPLA$_2$-V to directly release AA (Diaz et al., 2006; Kikawada et al., 2007). Irrespective of this, PGD$_2$ production elicited by cross-linking of FcϵRI via IgE and its cognate antigen is unaffected in *Pla2g5*-null mast cells, suggesting that sPLA$_2$-V is not a major participant in lipid mediator production by mast cells in allergen-induced asthma.

The fact that sPLA$_2$-V is involved in both chronic (asthma) and acute (ARDS) airway disorders suggests that phospholipid hydrolysis by this enzyme in the airway environment is the primary cause of the sPLA$_2$-V action. However, since the above studies employing *Pla2g5* knockout mice did not assess the levels of lipid mediators in BALF in both models, it remains uncertain whether the deletion of sPLA$_2$-V reduces lipid mediators in the airway. Furthermore, given that the asthmatic phenotypes are markedly ameliorated in both *Pla2g5*-null and *Pla2g4a*-null mice, while the airway response to sPLA$_2$-V appears to be cPLA$_2\alpha$-independent (see above), it is possible that the action of sPLA$_2$-V occurs through an alternative mechanism independent of lipid mediator production in the airway. Interestingly, a recent study has shown that *Pla2g5*$^{-/-}$ mice have reduced serum IgE and splenic IL-4 levels (Giannattasio et al., 2010), suggesting that the absence of sPLA$_2$-V perturbs an early phase of acquired immunity prior to pathologic symptoms in the airway. Given that *Pla2g5*$^{-/-}$ macrophages show reduced phagocytosis of immune complexes (Balestrieri et al., 2006, 2009), the defective acquired immunity in *Pla2g5*$^{-/-}$ mice may be attributable to impaired antigen presentation by dendritic cells/macrophages to T cells.

Another sPLA$_2$ isoform that has been shown to participate in asthma is sPLA$_2$-X. In the ovalbumin-induced asthma model, the lungs of sPLA$_2$-X (*Pla2g10*) knockout mice, compared to those of wild-type littermates, have a significant reduction in infiltration by CD4$^+$ and CD8$^+$ T cells and eosinophils, goblet cell metaplasia, smooth muscle cell layer thickening, subepithelial fibrosis, and levels of T$_H$2 cell cytokines and eicosanoids (Henderson et al., 2007). sPLA$_2$-X is expressed by alveolar macrophages and airway epithelial cells in wild-type mice subjected to the asthmatic challenge. Likewise, sPLA$_2$-X is one of the major sPLA$_2$ isoforms detected in airway cells of human subjects with asthma, where the enzyme is localized to airway epithelial cells and bronchial macrophages (Seeds et al., 2000; Hallstrand et al., 2007). The level of sPLA$_2$-X protein present in the BALF increases in response to exercise challenge in the asthma group, whereas it is very low in nonasthmatic control subjects. These data direct attention to sPLA$_2$-X (as well as sPLA$_2$-V; see above) as a novel therapeutic target for asthma.

The fact that knockout mice for sPLA$_2$-X, sPLA$_2$-V, or cPLA$_2\alpha$ display a marked reduction of the asthma response raises the questions as to when, where, and how these distinct PLA$_2$ enzymes function and interplay in the pathogenesis of allergic airway inflammation. Spatial and temporal compartmentalization of distinct PLA$_2$ enzymes in distinct cell types in a given tissue could sequentially or synergistically affect the onset, duration, severity, and resolution of the pathology. Since cPLA$_2\alpha$ is ubiquitously expressed and plays a central role in AA metabolism, it is likely that its deficiency blunts or attenuates the synthesis of various lipid mediators during the entire period of the pathology. On the other hand, the fact that sPLA$_2$-X and sPLA$_2$-V are expressed temporally and locally in only a few populations of cells such as infiltrating leukocytes and epithelial cells in the airway, knockout of either of these two sPLA$_2$s may lead to the specific shutdown of some cell-specific pathways crucial for the propagation of the asthmatic response. Whether sPLA$_2$-V and sPLA$_2$-X act independently in distinct cells or in concert with cPLA$_2\alpha$ within the same cells remains to be addressed since both mechanisms have been proposed in *in vitro* studies (Munoz et al., 2003; Mounier et al., 2004; Saiga et al., 2005; Kikawada et al., 2007). Alternatively, sPLA$_2$s may exert their proasthmatic effects through the process involving the excessive hydrolysis of pulmonary surfactant, as described below.

Pulmonary surfactant is a lipid–protein complex, synthesized by alveolar type II epithelial cells, which lowers surface tension along the alveolar epithelium, thereby promoting alveolar stability. This complex is composed of 10% protein and 90% lipid, with a high proportion of dipalmitoyl-PC. Degradation of surfactant results in the loss of alveolar stability and severe impairment of gas exchange, leading to the alveolar collapse known as ARDS. BALF of patients with ARDS or with severe bronchial asthma contains high levels of sPLA$_2$ activity as well as lysophosphatidylcholine (LPC) as a degradation product of surfactant, the levels of which often positively correlate with the severity of disease (Chilton et al., 1996; Arbibe et al., 1998; Cupillard et al., 1999; Touqui and Alaoui-El-Azher, 2001). Moreover, accumulation of LPC damages type I alveolar cells, increases capillary permeability, reduces the surfactant tension activity, and recruits inflammatory cells into the lung. The symptoms of experimental ARDS are markedly attenuated by treatment of animals with sPLA$_2$ inhibitors that target several sPLA$_2$ members (Arbibe et al., 1998; Furue et al., 2001; Smart et al., 2006). Although sPLA$_2$-IIA is markedly induced in airway inflammation in rats, rabbits, and guinea pigs, intratracheal instillation of sPLA$_2$-IIA reduces the level of phosphatidylglycerol, a minor surfactant component, without altering the level of PC (Chabot et al., 2003), in agreement with the substrate specificity of this enzyme. Diagnostically, however, it is LPC that accumulates in the BALF of ARDS patients. Moreover, ARDS and asthmatic responses can occur normally in C57BL/6 mice, in which sPLA$_2$-IIA is absent. These facts argue for a role of sPLA$_2$ enzymes other than sPLA$_2$-IIA in surfactant hydrolysis in these pathogenic situations (Chabot et al., 2003; Hallstrand et al., 2007). *In vitro*, surfactant PC is susceptible to hydrolysis by sPLA$_2$-IB, sPLA$_2$-V, and sPLA$_2$-X, but not by sPLA$_2$-IIA and sPLA$_2$-IID, while the latter two enzymes can hydrolyze surfactant phosphatidylglycerol (Seeds et al., 2003; Hite et al., 2005).

Transgenic mice overexpressing sPLA$_2$-V under the β-actin promoter (*Pla2g5*-Tg mice) rapidly die during the neonatal period due to respiratory failure (Ohtsuki et al., 2006). The lungs of *Pla2g5*-Tg mice exhibit atelectasis with thickened alveolar walls and narrow air spaces, accompanied by infiltration of macrophages and only modest changes in eicosanoid levels. This severe pulmonary defect in *Pla2g5*-Tg mice is attributable to a marked reduction of the lung surfactant phospholipids, dipalmitoyl-PC and dipalmitoyl-phosphatidylglycerol. Given that the expression of sPLA$_2$-V is elevated in human lungs with severe inflammation, the results in *Pla2g5*-Tg mice argue that this sPLA$_2$ isoform may contribute to ongoing surfactant hydrolysis often observed in the lungs of patients with ARDS. In contrast, transgenic mice overexpressing human or mouse sPLA$_2$-IIA (Grass et al., 1996), sPLA$_2$-X (Ohtsuki et al., 2006), sPLA$_2$-III (Sato et al., 2008), sPLA$_2$-IID, sPLA$_2$-IIE, or sPLA$_2$-IIF (Murakami et al., unpublished observations) do not show neonatal death due to respiratory failure, suggesting that the surfactant represents a better target for sPLA$_2$-V than for other sPLA$_2$s *in vivo*. In this context, the amelioration of ARDS and asthma models in *Pla2g5* knockout mice (Munoz et al., 2007, 2009) could be accounted for by the protection of the airway from aberrant, disease-associated hydrolysis of surfactant by induced sPLA$_2$-V.

In contrast to the transgenic overexpression of sPLA$_2$-V that causes aberrant hydrolysis of lung surfactant leading to neonatal death (see above), that of human sPLA$_2$-X under the β-actin promoter (*PLA2G10*-Tg mice) does not induce any obvious phenotype in the lung (Ohtsuki et al., 2006). Unlike sPLA$_2$-V, sPLA$_2$-X has an N-terminal propeptide whose proteolytic cleavage by certain proteases leads to enzyme activation (Cupillard et al., 1997). One reason for the unaffected airway phenotype can be because a large fraction of the overexpressed sPLA$_2$-X remains as an enzymatically inactive zymogen that retains the N-terminal propeptide. Considering the fact that sPLA$_2$-X plays a key role in an inflammatory situation such as asthma (Henderson et al., 2007) and that sPLA$_2$-X gets activated by proteolytic processing after an inflammatory challenge in the *PLA2G10*-Tg mice (Ohtsuki et al., 2006), it is likely that the conversion of pro-sPLA$_2$-X to its mature form is accelerated under inflammatory conditions. In contrast to the above β-actin-promoter *PLA2G10*-Tg mice, transgenic mice specifically overexpressing human sPLA$_2$-X in macrophages by using a CD68 promoter die within a few weeks because of fatal pulmonary defects (Curfs et al., 2008). The reasons for the different phenotypes of the two transgenic mouse lines are unknown. Although no biochemical analyses were conducted on the macrophage-specific *PLA2G10*-Tg mice regarding sPLA$_2$-X expression and activation as well as its consequence on surfactant hydrolysis and lipid mediator levels, it is likely that differences in the design of the promoters could influence the mass expression and its spatiotemporal conversion into mature sPLA$_2$-X in pulmonary niches. A proposed model for the role of cPLA$_2$α and sPLA$_2$s (group V and X) in airway inflammation is shown in Figure 3.5.

COPD is a debilitating disorder characterized by aggravating dyspnea, impaired exercise tolerance, and weight loss. This unexplained weight loss is clinically relevant because it limits patients' physical performance, jeopardizes their quality of life, and is related to poor prognosis, independent of the severity of airflow obstruction (Schols et al., 1998; Landbo et al., 1999). Interestingly, a polymorphism in the human sPLA$_2$-IID (*PLA2G2D*) gene is associated with body weight loss in COPD patients (NCBI SNP reference: rs584367) (Takabatake et al., 2005). In a human pulmonary epithelial cell line, overexpression of the sPLA$_2$-IID G80S mutant, which is found in COPD patients, enhances the expression of IL-6 and IL-8 (Igarashi et al., 2009). This suggests important information on the mechanism of body weight loss in COPD patients and indicates that sPLA$_2$-IID may represent a novel therapeutic target

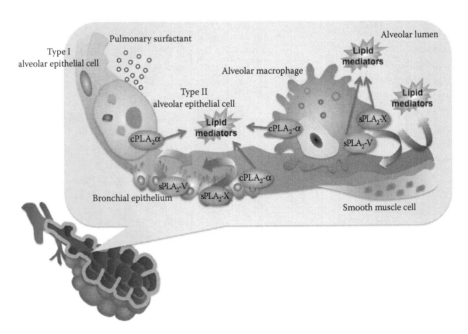

FIGURE 3.5   The roles of cPLA$_2$α and two sPLA$_2$s (V and X) in airway diseases. cPLA$_2$α is expressed in all cell types and is an essential component of eicosanoid generation. sPLA$_2$-V and sPLA$_2$-X are expressed in alveolar macrophages and bronchial epithelial cells and may participate in lipid mediator generation as well as in surfactant degradation. For details, see the text.

to this disease. A recent study has shown that sPLA$_2$-IID is localized in CD11b$^+$ and CD11c$^+$ cells (monocyte/macrophage and dendritic cell lineages) in the spleen and lymph nodes as well as in regulatory T (T$_{reg}$) cells and has the ability to suppress the proliferation of CD4$^+$ and CD8$^+$ T cells and to promote the differentiation of T$_{reg}$ cells (von Allmen et al., 2009).

3.2.4 LYSOSOMAL PLA$_2$S IN AIRWAY DISORDERS

Finally, besides sPLA$_2$s, two distinct lysosomal PLA$_2$s have been implicated in surfactant homeostasis. Studies of mice lacking or overexpressing aiPLA$_2$, which is identical to peroxiredoxin 6 (Prdx6), a bifunctional protein with PLA$_2$ and glutathione peroxidase activities, have shown that this unique PLA$_2$/antioxidant protein plays two major roles in the lung by controlling homeostatic surfactant phospholipid metabolism and oxidative stress leading to lung injury (Fisher et al., 2005; Wang et al., 2006). Mice with deficiency of LPLA$_2$, another lysosomal PLA$_2$ that is highly expressed in alveolar macrophages, delineate the role for this enzyme in phospholipid catabolism of pulmonary surfactant (Abe et al., 2004). Phospholipidosis is characterized by the appearance of concentric lamellar bodies within cells and the intralysosomal accumulation of phospholipids, which is suggestive of impaired surfactant catabolism (Trapnell et al., 2003). Gene ablation of LPLA$_2$ in mice leads to aberrant accumulation of PC and phosphatidylethanolamine (PE) in alveolar and peritoneal macrophages, which eventually causes pulmonary alveolar proteinosis (Abe et al., 2008; Hiraoka et al., 2006). This is consistent with the preference of the enzyme toward PC and PE (Abe and Shayman, 1998). Both alveolar and peritoneal macrophages from LPLA$_2$-deficient mice show foam cell formation characterized by lamellar inclusion bodies (Hiraoka et al., 2006). Overall, these observations suggest that aiPLA$_2$ and LPLA$_2$ play a role in homeostatic (or constitutive) surfactant catabolism in type II alveolar cells and macrophages, respectively, whereas sPLA$_2$-V and sPLA$_2$-X secreted from alveolar macrophages and airway epithelial cells promotes inducible surfactant degradation during the course of airway pathology.

3.3 DRUG DISCOVERY EFFORT AROUND THE TARGET FOR AIRWAY DISEASE

3.3.1 cPLA$_2$α INHIBITORS

As mention above, cPLA$_2$α is a central enzyme for the production of lipid mediators in asthmatic pathogenesis. It has been reported that the treatment with arachidonyl trifluoromethyl ketone (ATK), a potent inhibitor of cPLA$_2$α (Street et al., 1993), prevents allergen-induced airway hyperresponsiveness and airway inflammation as well as LPS-induced airway injury (Nagase et al., 2003; Malaviya et al., 2006). However, because ATK also exerts its inhibitory effect on COX (Reddy and Herschman, 1997) and iPLA$_2$ (Ackermann et al., 1995; Balsinde et al., 1999), the off-target inhibitions cannot be ruled out in the treatment with ATK.

Subsequently, several specific cPLA$_2$α inhibitors have been developed, such as pyrrolidine-based inhibitors by Shionogi (Seno et al., 2001; Ono et al., 2002), indole-based

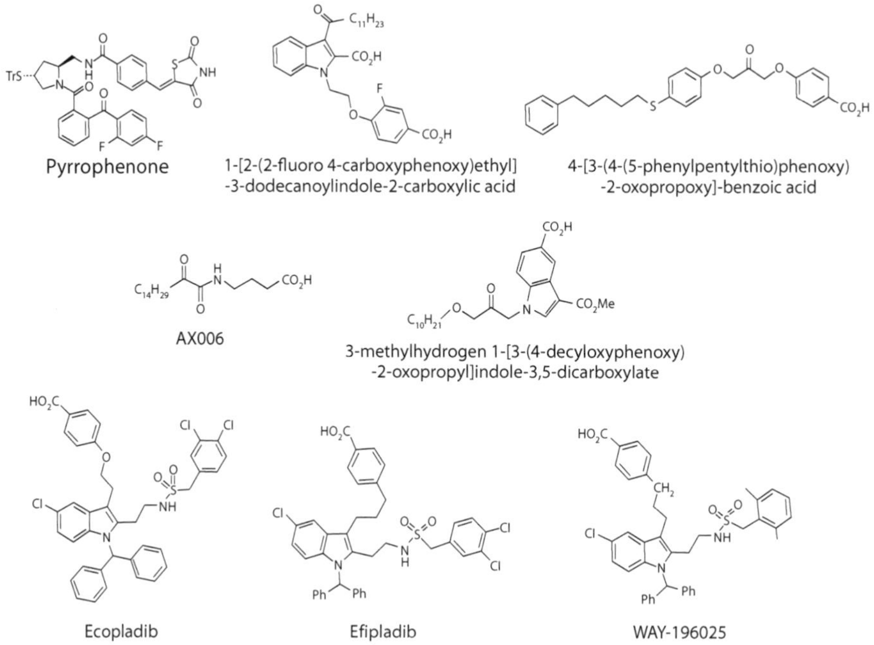

FIGURE 3.6 cPLA$_2\alpha$ inhibitors. Ecopladib and its derivatives Efipladib and WAY-196025 are potent cPLA$_2\alpha$-specific inhibitors that block eicosanoid synthesis in experimental animals following oral applications. For details, see the text.

inhibitors with a C2 carboxylic acid and indolylpropanone inhibitors by the Lehr group (Lehr et al., 2001; Griessbach et al., 2002; Ludwig et al., 2006), propanone inhibitors by Astra Zeneca (Connolly et al., 2002), and 2-oxoamide inhibitors by the Dennis group (Kokotos et al., 2002, 2004) (Figure 3.6). Ecopladib, an indole cPLA$_2\alpha$ inhibitor with a benzoyl sulfonamide substituent at C2 (from Wyeth), shows high specificity for cPLA$_2\alpha$ and displays oral efficacy in rat carrageenan-induced air pouch and paw edema models (Lee et al., 2007, 2008). Furthermore, optimization of the substituent at C3 and the substitution pattern of the phenylmethane sulfonamide region leads to the discovery of Efipladib and WAY-196025, which are shown to be potent and selective inhibitors of cPLA$_2\alpha$ in various *in vitro* assays and to be efficacious upon oral administration in animal models such as mouse collagen-induced arthritis and rat adjuvant-induced arthritis (models for human rheumatoid arthritis) (McKew et al., 2008). Importantly, when mice are treated with these cPLA$_2\alpha$ inhibitors before or even after antigen challenge, both inhibitors show a dramatic impact on the late asthmatic response. Trials to evaluate the efficacy of Efipladib in human diseases have been initiated, and WAY-196025 continues to be profiled in preclinical models.

3.3.2 sPLA$_2$ Inhibitors

Phenotypes of transgenic and knockout mice for sPLA$_2$-V or sPLA$_2$-X as described above suggest that these two sPLA$_2$s, and possibly some other sPLA$_2$ isoforms, can

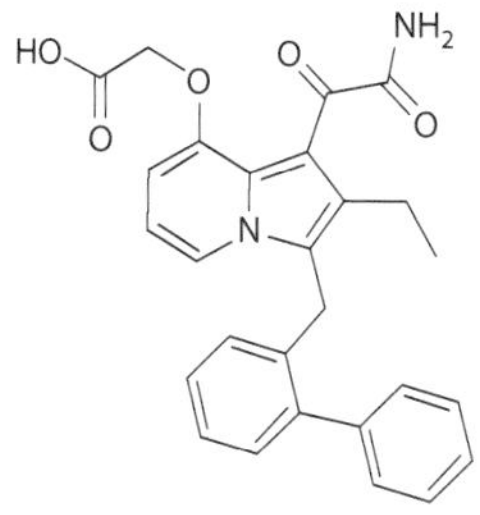

Indoxam

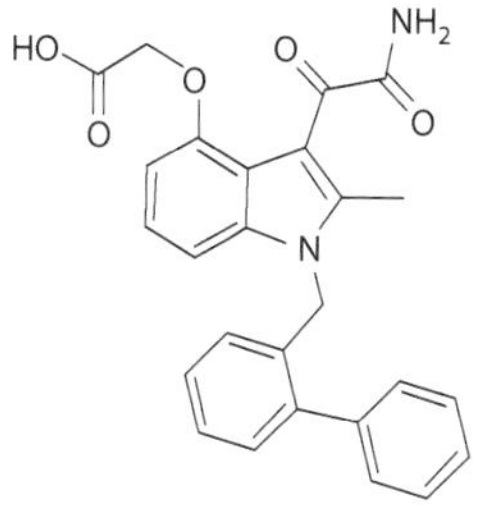

Me-Indoxam

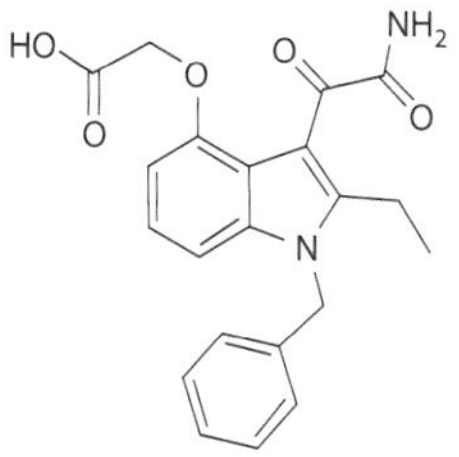

LY315920
(Varespladib; A-002)

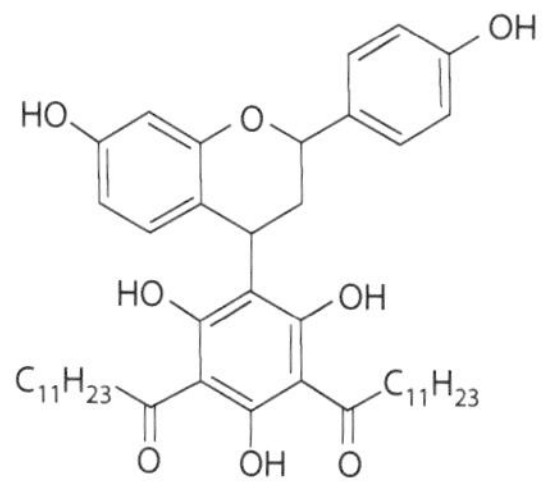

YM-26734

FIGURE 3.7   sPLA$_2$ inhibitors. Indoxam and related compounds are synthetic indole inhibitors that potently block group I/II/V/X sPLA$_2$s. A-002 (Varespladib), a lead compound of this group, is now under a clinical trial (Phase 3) for atherosclerosis. YM-26734 is a lead compound from a natural product that inhibits sPLA$_2$-IIA as well as sPLA$_2$-V and sPLA$_2$-X. For details, see the text.

be utilized as drug targets for airway diseases. Some of the potent sPLA$_2$ inhibitors, displaying low nM potency, include the functionalized indole scaffolds from Eli Lilly and Shionogi (see below) and the natural product analog YM-26734 from Yamanouchi Pharmaceuticals (Miyake et al., 1993) and its derivatives (Oslund et al., 2008) (Figure 3.7). Substituted indoles and indolizines, first reported by Lilly and Shionogi, are the most potent sPLA$_2$ inhibitors and represent those with drug potential in terms of excellent pharmacokinetic profiles. Compounds of this series include the indolizine Indoxam and the substituted indoles Me-Indoxam and LY315920 (Dillard et al., 1996; Draheim et al., 1996). The development of these compounds involves structure-guided improvement of binding capacity starting with a lead compound obtained through high-throughput screening and making use of the x-ray structure of human sPLA$_2$-IIA (Scott et al., 1991). Me-Indoxam inhibits sPLA$_2$-IIA, sPLA$_2$-IIC, sPLA$_2$-IIE, and sPLA$_2$-V with low nM potency, shows less potency against sPLA$_2$-IB and sPLA$_2$-X, and inhibits sPLA$_2$-IID only at μM concentrations (Singer et al., 2002). LY315920 potently inhibits sPLA$_2$-IIA, sPLA$_2$-IIE, and sPLA$_2$-X but is less potent on other sPLA$_2$s (Smart et al., 2006). Furthermore, a structure-guided approach using the x-ray structure of human sPLA$_2$-X (Pan et al., 2002) leads to the development of a highly specific inhibitor for this enzyme (Smart et al., 2006). Further screening of substituted benzo-fused indole inhibitors against sPLA$_2$s leads to identifications of a highly specific inhibitor that binds only to $sPLA_2$-IIA and $sPLA_2$-IIE and a broadly specific inhibitor that shows strong inhibition against all group I/II/V/X $sPLA_2$s (Oslund et al., 2008). However, none of these inhibitors block $sPLA_2$-III and $sPLA_2$-XIIA, whose active sites are predicted to be significantly different from those of group I/II/V/X $sPLA_2$s.

Unfortunately, LY315920 or its analog LY333013 fails to improve the symptoms of human sepsis, asthma, and rheumatoid arthritis in clinical trials (Bradley et al., 2005; Bowton et al., 2005; Zeiher et al., 2005). This might be because different $sPLA_2$s often exert opposite biological effects in a given disease; this is particularly evident for inflammatory arthritis, in which $sPLA_2$-IIA and $sPLA_2$-V display offensive and defensive roles, respectively (Boilard et al., 2010). Thus, blunting both enzymes together by a pan-$sPLA_2$ inhibitor would cancel both beneficial and detrimental effects and thereby give no desirable pharmacological impact. These observations provide the rationale for pursuing two distinct therapeutic approaches targeted at $sPLA_2$: the use of inhibitors highly selective for the proinflammatory $sPLA_2$ and the administration of the anti-inflammatory $sPLA_2$ protein (e.g., in the case of arthritis, $sPLA_2$-IIA-specific inhibitor or $sPLA_2$-V protein). Nevertheless, 1-H-indole-3-glyoxamide (A-002 or Varespladib), a pan-$sPLA_2$ inhibitor modified from the aforementioned indole inhibitors from Anthera (Figure 3.7), significantly attenuates atherosclerosis and aneurysm in experimental animal models, and more importantly, in a Phase 2 clinical trial for patients with cardiovascular diseases (Fraser et al., 2009; Karakas and Koenig, 2009; Leite et al., 2009; Rosenson et al., 2009; Shaposhnik et al., 2009). This is probably because the inhibitor blocks $sPLA_2$-IIA, $sPLA_2$-V, and $sPLA_2$-X, all of which have been implicated in the development of atherosclerosis by promoting the production of proatherogenic lipoprotein particles through the hydrolysis of lipoprotein-associated phospholipids (Ivandic et al., 1999; Hanasaki et al., 2002; Webb et al., 2003; Bostrom et al., 2007; Sato et al., 2008; Boyanovsky et al., 2009). In mice, $sPLA_2$-V expressed in macrophages appears to be a likely target of the drug, since *Pla2g5* deletion in macrophages leads to amelioration of atherosclerosis on the *LDLR*$^{-/-}$ background (Bostrom et al., 2007; Boyanovsky et al., 2009) and since $sPLA_2$-IIA is not intrinsically expressed in the mouse strain used (see above).

A-001 (Varespladib sodium) is now under a Phase 2 clinical study for the prevention of acute chest syndrome associated with sickle cell disease (personal communication, http://www.anthera.com/products_a001.asp). Sickle cell disease is a severe, life-shortening hemoglobin disorder. Its hallmark is the deformed, sickle-shaped red blood cell that is prone to cause organ damage by occluding small blood vessels (Styles et al., 1996). In many patients, the course of the disease includes episodes of fever, respiratory symptoms, and a new lung infiltrate on a chest x-ray; a combination termed acute chest syndrome. Acute chest syndrome is the second most common cause of hospitalization for patients with sickle cell disease and the most common cause of death. Patients suffer varying degrees of compromised lung function and may require mechanical ventilation. Serum $sPLA_2$ levels are higher in acute chest syndrome than in patients with pneumonia, reflecting the severity of the lung compromise (Styles et al., 1996; Bargoma et al., 2005). Correlation between $sPLA_2$ and acute chest syndrome is further supported by the finding that red blood cell transfusion lowers serum sPLA$_2$ level and prevents acute chest syndrome (Styles et al., 2007). Thus, inhibition of sPLA$_2$ to prevent acute chest syndrome may promise to be useful both in the acute setting and as maintenance therapy for patients with sickle cell disease. A phase trial is evaluating the safety and effectiveness of intravenous A-001 in preventing the development of acute chest syndrome in hospitalized sickle cell disease patients.

3.4   CONCLUDING REMARKS

It is now apparent that cPLA$_2\alpha$ and sPLA$_2$s (V and X), as well as lysosomal PLA$_2$s, are critical regulators of airway homeostasis and that the perturbation of the pathways regulated by each of these enzymes leads to airway disorders. Although a number of "old" PLA$_2$ inhibitors were applied to experimental animal models and some of them showed some therapeutic effects, these observations have not attracted attention because of the uncertain specificity of these molecules. In fact, many of the old inhibitors lacked a data showing direct inhibitor–PLA$_2$ interaction, and some inhibitors obviously exhibit broad specificity toward other hydrolases. Now, several "new" inhibitors that specifically bind and potently inhibit cPLA$_2\alpha$ (Efipladib or relatives) or sPLA$_2$s (Varespladib or relatives) display excellent therapeutic effects on various experimental animal models, including airway disease models, and are now coming into the phase trials in clinics. However, their efficacies on airway diseases should be carefully evaluated on the bases of the following concerns.

First, since cPLA$_2\alpha$ is expressed ubiquitously and participates in the production of not only pathogenic but also homeostatic eicosanoids, the inhibition of this enzyme could cause undesirable side effects such as gastric ulcer and renal damage, as has been known for nonsteroidal anti-inflammatory drugs (COX inhibitors) that shut off PG synthesis. Also, unlike cys-LT antagonists that blunt the proasthmatic LT pathway, cPLA$_2\alpha$ inhibitors could block both the LT and PG pathways, the latter of which (e.g., PGE$_2$ and PGI$_2$) have protective roles in the airway. Moreover, cPLA$_2\alpha$ reaction could also link to the anti-inflammatory lipid mediators such as lipoxins and resolvins. Thus, cPLA$_2\alpha$ inhibition could tip a balance in favor of untoward effect. Nevertheless, the fact that the oral application of the cPLA$_2\alpha$ inhibitor Efipladib to mice protects from an asthma model (McKew et al., 2008) suggests that the beneficial effects through blunting proinflammatory lipid mediators exceeds over the side effects predicted from the inhibition of anti-inflammatory lipid mediators.

Second, although the fact that the pan-sPLA$_2$ inhibitors prevent airway disease models in experimental animals (Arbibe et al., 1998; Furue et al., 2001; Smart et al., 2006; Sato et al., 2010) suggests their efficacies, inhibition of multiple sPLA$_2$s altogether could inhibit both offensive and defensive sPLA$_2$ isozymes and thereby cancel the therapeutic effect resulting from the inhibition of the proinflammatory one(s). It is now clear that at least two sPLA$_2$s (V and X) play offensive roles in the airway, yet we cannot rule out the possibility that some other sPLA$_2$ isoforms, whose *in vivo* functions remain ambiguous because of the current unavailability of knockout mice, may play a protective role in airway diseases. This might account for the reason why LY333013, a pan-sPLA$_2$ inhibitor, has no impact on allergen-induced bronchoconstriction following inhaled allergen challenge in human subjects with atopic asthma

(Bowton et al., 2005). In this view, an inhibitor that specifically blocks sPLA$_2$-V and sPLA$_2$-X but not other isozymes may be more desirable than the currently testing pan-sPLA$_2$ inhibitors, which block group I/II/V/X sPLA$_2$s altogether.

Third, it should be noted that sPLA$_2$s have beneficial actions as antimicrobial proteins. Indeed, several sPLA$_2$s (including IIA, V, and X) have the ability to effectively kill bacteria by directly degrading bacterial membranes (Weinrauch et al., 1998; Laine et al., 2000; Gronroos et al., 2001; Koduri et al., 2002; Mitsuishi et al., 2006), and sPLA$_2$-V exhibits an antifungal function by facilitating macrophage uptake of fungal particles (Balestrieri et al., 2009). Furthermore, sPLA$_2$-V and sPLA$_2$-X protect cultured cells from adenoviral infection, probably because plasma membrane lipid hydrolysis by these sPLA$_2$s could reduce the viral entry across the plasma membrane (Mitsuishi et al., 2006). In these standpoints, application of pan-sPLA$_2$ inhibitors to patients with infectious airway diseases might result in the expansion of microbial infection, thereby culminating in exacerbation of the diseases.

Nonetheless, we would like to draw attention to the fact that strong side effects should warrant strong beneficial effects also. Given that these PLA$_2$ inhibitors are being tested for clinical trial means that untoward effect may not be the limiting factor. We also believe that if these molecules exhibit strong enough efficacy in phase IIA clinical trials, a cost–benefit ratio could be assessed. We know that corticosteroids cause a lot of adverse effects, yet these are still used quite extensively. Last but not the least, all the above knowledge should help leading to the proper identification of certain PLA$_2$s as therapeutic targets or as novel biotherapeutic molecules in airway diseases as well as other diseases ranging from inflammatory diseases, metabolic diseases, to cancer.

ACKNOWLEDGMENTS

This chapter was supported in part by Grants-in-Aid for Scientific Research from the Ministry of Education, Science, Culture, Sports and Technology of Japan. The authors thank colleagues for their collaboration over the years in the original research reviewed here. In the interest of brevity, they have referenced other reviews whenever possible and apologize to the authors of the numerous primary papers that were not explicitly cited.

REFERENCES

Abe, A., M. Hiraoka, S. Wild, S.E. Wilcoxen, R. Paine, 3rd, and J.A. Shayman. 2004. Lysosomal phospholipase A$_2$ is selectively expressed in alveolar macrophages. *J Biol Chem.* 279:42605–11.

Abe, A., R. Kelly, J. Kollmeyer, M. Hiraoka, Y. Lu, and J.A. Shayman. 2008. The secretion and uptake of lysosomal phospholipase A$_2$ by alveolar macrophages. *J Immunol.* 181:7873–81.

Abe, A. and J. A. Shayman. 1998. Purification and characterization of 1-*O*-acylceramide synthase, a novel phospholipase A$_2$ with transacylase activity. *J Biol Chem.* 273:8467–74.

Ackermann, E. J., K. Conde-Frieboes, and E. A. Dennis. 1995. Inhibition of macrophage Ca^{2+}-independent phospholipase A$_2$ by bromoenol lactone and trifluoromethyl ketones. *J Biol Chem.* 270:445–50.

Akuthota, P., H. B. Wang, L. A. Spencer, and P.F. Weller. 2008. Immunoregulatory roles of eosinophils: A new look at a familiar cell. *Clin Exp Allergy.* 38:1254–63.

Arai, H. 2002. Platelet-activating factor acetylhydrolase. *Prostaglandins Other Lipid Mediat.* 68–69:83–94.

Arbibe, L., K. Koumanov, D. Vial, C. Rougeot, G. Faure, N. Havet, S. Longacre et al. 1998. Generation of lyso-phospholipids from surfactant in acute lung injury is mediated by type-II phospholipase A$_2$ and inhibited by a direct surfactant protein A-phospholipase A$_2$ protein interaction. *J Clin Invest.* 102:1152–60.

Austin, S. C. and C. D. Funk. 1999. Insight into prostaglandin, leukotriene, and other eicosanoid functions using mice with targeted gene disruptions. *Prostaglandins Other Lipid Mediat.* 58:231–52.

Balboa, M. A., J. Balsinde, M. V. Winstead, J.A. Tischfield, and E.A. Dennis. 1996. Novel group V phospholipase A$_2$ involved in arachidonic acid mobilization in murine P388D1 macrophages. *J Biol Chem.* 271:32381–4.

Balestrieri, B., V. W. Hsu, H. Gilbert, C. C. Leslie, W. K. Han, J. V. Bonventre, and J. P. Arm. 2006. Group V secretory phospholipase A$_2$ translocates to the phagosome after zymosan stimulation of mouse peritoneal macrophages and regulates phagocytosis. *J Biol Chem.* 281:6691–8.

Balestrieri, B., A. Maekawa, W. Xing, M. H. Gelb, H. R. Katz, and J. P. Arm. 2009. Group V secretory phospholipase A$_2$ modulates phagosome maturation and regulates the innate immune response against. *Candida albicans J Immunol.* 182:4891–8.

Balsinde, J. and M. A. Balboa. 2005. Cellular regulation and proposed biological functions of group VIA calcium-independent phospholipase A$_2$ in activated cells. *Cell Signal.* 17:1052–62.

Balsinde, J., M. A. Balboa, and E. A. Dennis. 1998. Functional coupling between secretory phospholipase A$_2$ and cyclooxygenase-2 and its regulation by cytosolic group IV phospholipase A$_2$. *Proc Natl Acad Sci USA.* 95:7951–6.

Balsinde, J., M. A. Balboa, P. A. Insel, and E. A. Dennis. 1999. Regulation and inhibition of phospholipase A$_2$. *Annu Rev Pharmacol Toxicol.* 39:175–89.

Bargoma, E. M., J. K. Mitsuyoshi, S. K. Larkin, L. A. Styles, F. A. Kuypers, and S. T. Test. 2005. Serum C-reactive protein parallels secretory phospholipase A$_2$ in sickle cell disease patients with vasoocclusive crisis or acute chest syndrome. *Blood.* 105:3384–5.

Barrett, N. A. and K. F. Austen. 2009. Innate cells and T helper 2 cell immunity in airway inflammation. *Immunity.* 31:425–37.

Bingham, C. O., 3rd, R. J. Fijneman, D. S. Friend, R. P. Goddeau, R. A. Rogers, K. F. Austen, and J. P. Arm. 1999. Low molecular weight group IIA and group V phospholipase A$_2$ enzymes have different intracellular locations in mouse bone marrow-derived mast cells. *J Biol Chem.* 274:31476–84.

Bingham, C. O., 3rd, M. Murakami, H. Fujishima, J. E. Hunt, K. F. Austen, and J. P. Arm. 1996. A heparin-sensitive phospholipase A$_2$ and prostaglandin endoperoxide synthase-2 are functionally linked in the delayed phase of prostaglandin D$_2$ generation in mouse bone marrow-derived mast cells. *J Biol Chem.* 271:25936–44.

Boilard, E., Y. Lai, K. Larabee, B. Balestrieri, F. Ghomashchi, D. Fujioka, R. Gobezie et al. 2010. A novel anti-inflammatory role for secretory phospholipase A$_2$ in immune complex-mediated arthritis. *EMBO Mol Med.* 2:172–87.

Bonventre, J. V., Z. Huang, M. R. Taheri, E. O'Leary, E. Li, M.A. Moskowitz, and A. Sapirstein. 1997. Reduced fertility and postischaemic brain injury in mice deficient in cytosolic phospholipase A$_2$. *Nature.* 390:622–5.

Bostrom, M. A., B. B. Boyanovsky, C. T. Jordan, M. P. Wadsworth, D. J. Taatjes, F. C. de Beer, and N. R. Webb. 2007. Group V secretory phospholipase A$_2$ promotes atherosclerosis: Evidence from genetically altered mice. *Arterioscler Thromb Vasc Biol.* 27:600–6.

Bowton, D. L., A. A. Dmitrienko, E. Israel, B. G. Zeiher, and G. D. Sides. 2005. Impact of a soluble phospholipase A_2 inhibitor on inhaled allergen challenge in subjects with asthma. *J Asthma.* 42:65–71.

Bowton, D. L., M. C. Seeds, M. B. Fasano, B. Goldsmith, and D. A. Bass. 1997. Phospholipase A_2 and arachidonate increase in bronchoalveolar lavage fluid after inhaled antigen challenge in asthmatics. *Am J Respir Crit Care Med.* 155:421–5.

Boyanovsky, B., M. Zack, K. Forrest, and N. R. Webb. 2009. The capacity of group V $sPLA_2$ to increase atherogenicity of $ApoE^{-/-}$ and $LDLR^{-/-}$ mouse LDL *in vitro* predicts its atherogenic role *in vivo*. *Arterioscler Thromb Vasc Biol.* 29:532–8.

Boyce, J. A. 2003. The role of mast cells in asthma. *Prostaglandins Leukot Essent Fatty Acids.* 69:195–205.

Bradley, J. D., A. A. Dmitrienko, A. J. Kivitz, O. S. Gluck, A. L. Weaver, C. Wiesenhutter, S. L. Myers, and G. D. Sides. 2005. A randomized, double-blinded, placebo-controlled clinical trial of LY333013, a selective inhibitor of group II secretory phospholipase A_2, in the treatment of rheumatoid arthritis. *J Rheumatol.* 32:417–23.

Brightling, C. E., P. Bradding, I. D. Pavord, and A. J. Wardlaw. 2003. New insights into the role of the mast cell in asthma. *Clin Exp Allergy.* 33:550–6.

Brightling, C. E., P. Bradding, F. A. Symon, S. T. Holgate, A. J. Wardlaw, and I. D. Pavord. 2002. Mast-cell infiltration of airway smooth muscle in asthma. *N Engl J Med.* 346:1699–705.

Busse, W. W. and R. F. Lemanske, Jr. 2001. Asthma. *N Engl J Med.* 344:350–62.

Carroll, N. G., S. Mutavdzic, and A. L. James. 2002. Increased mast cells and neutrophils in submucosal mucous glands and mucus plugging in patients with asthma. *Thorax.* 57:677–82.

Cedars, A., C. M. Jenkins, D. J. Mancuso, and R. W. Gross. 2009. Calcium-independent phospholipases in the heart: Mediators of cellular signaling, bioenergetics, and ischemia-induced electrophysiologic dysfunction. *J Cardiovasc Pharmacol.* 53:277–89.

Chabot, S., K. Koumanov, G. Lambeau, M. H. Gelb, V. Balloy, M. Chignard, J. A. Whitsett, and L. Touqui. 2003. Inhibitory effects of surfactant protein A on surfactant phospholipid hydrolysis by secreted phospholipases A_2. *J Immunol.* 171: 995–1000.

Chilton, F. H., F. J. Averill, W. C. Hubbard, N. Fonteh, M. Triggiani, M. C. Liu. 1996. Antigen-induced generation of lyso-phospholipids in human airways. *J Exp Med.* 183:2235–45.

Chock, S. P., E. A. Schmauder-Chock, E. Cordella-Miele, L. Miele, and A. B. Mukherjee. 1994. The localization of phospholipase A_2 in the secretory granule. *Biochem J.* 300 Pt 3:619–22.

Clark, J. D., A. R. Schievella, E. A. Nalefski, and L. L. Lin. 1995. Cytosolic phospholipase A_2. *J Lipid Mediat Cell Signal.* 12:83–117.

Connolly, S., C. Bennion, S. Botterell, P. J. Croshaw, C. Hallam, K. Hardy, P. Hartopp et al. 2002. Design and synthesis of a novel and potent series of inhibitors of cytosolic phospholipase A_2 based on a 1,3-disubstituted propan-2-one skeleton. *J Med Chem.* 45:1348–62.

Cupillard, L., K. Koumanov, M. G. Mattei, M. Lazdunski, and G. Lambeau. 1997. Cloning, chromosomal mapping, and expression of a novel human secretory phospholipase A_2. *J Biol Chem.* 272:15745–52.

Cupillard, L., R. Mulherkar, N. Gomez, S. Kadam, E. Valentin, M. Lazdunski, and G. Lambeau. 1999. Both group IB and group IIA secreted phospholipases A_2 are natural ligands of the mouse 180-kDa M-type receptor. *J Biol Chem.* 274:7043–51.

Curfs, D. M., S. A. Ghesquiere, M. N. Vergouwe, I. van der Made, M. J. Gijbels, D. R. Greaves, J. S. Verbeek, M. H. Hofker, and M. P. de Winther. 2008. Macrophage secretory phospholipase A_2 group X enhances anti-inflammatory responses, promotes lipid accumulation, and contributes to aberrant lung pathology. *J Biol Chem.* 283:21640–8.

Degousee, N., F. Ghomashchi, E. Stefanski, A. Singer, B. P. Smart, N. Borregaard, R. Reithmeier et al. 2002. Groups IV, V, and X phospholipases A$_2$s in human neutrophils: Role in eicosanoid production and Gram-negative bacterial phospholipid hydrolysis. *J Biol Chem.* 277:5061–73.

Diaz, B. L., Y. Satake, E. Kikawada, B. Balestrieri, and J. P. Arm. 2006. Group V secretory phospholipase A$_2$ amplifies the induction of cyclooxygenase 2 and delayed prostaglandin D$_2$ generation in mouse bone marrow culture-derived mast cells in a strain-dependent manner. *Biochim Biophys Acta.* 1761:1489–97.

Dillard, R. D., N. J. Bach, S. E. Draheim, D. R. Berry, D. G. Carlson, N. Y. Chirgadze, D. K. Clawson et al. 1996a. Indole inhibitors of human nonpancreatic secretory phospholipase A$_2$. 1. Indole-3-acetamides. *J Med Chem.* 39:5119–36.

Dillard, R. D., N. J. Bach, S. E. Draheim, D. R. Berry, D. G. Carlson, N. Y. Chirgadze, D. K. Clawson et al. 1996b. Indole inhibitors of human nonpancreatic secretory phospholipase A$_2$. 2. Indole-3-acetamides with additional functionality. *J Med Chem.* 39:5137–58.

Draheim, S. E., N. J. Bach, R. D. Dillard, D. R. Berry, D. G. Carlson, N. Y. Chirgadze, D. K. Clawson et al. 1996. Indole inhibitors of human nonpancreatic secretory phospholipase A$_2$. 3. Indole-3-glyoxamides. *J Med Chem.* 39:5159–75.

Duncan, R. E., E. Sarkadi-Nagy, K. Jaworski, M. Ahmadian, and H. S. Sul. 2008. Identification and functional characterization of adipose-specific phospholipase A$_2$ (AdPLA). *J Biol Chem.* 283:25428–36.

Enomoto, A., M. Murakami, E. Valentin, G. Lambeau, M. H. Gelb, and I. Kudo. 2000. Redundant and segregated functions of granule-associated heparin-binding group II subfamily of secretory phospholipases A$_2$ in the regulation of degranulation and prostaglandin D$_2$ synthesis in mast cells. *J Immunol.* 165:4007–14.

Fisher, A. B., C. Dodia, S. I. Feinstein, Y. S. Ho. 2005. Altered lung phospholipid metabolism in mice with targeted deletion of lysosomal-type phospholipase A$_2$. *J Lipid Res.* 46:1248–56.

Fonteh, A. N., D. A. Bass, L. A. Marshall, M. Seeds, J. M. Samet, and F. H. Chilton. 1994. Evidence that secretory phospholipase A$_2$ plays a role in arachidonic acid release and eicosanoid biosynthesis by mast cells. *J Immunol.* 152:5438–46.

Fraser, H., C. Hislop, R. M. Christie, H. L. Rick, C. A. Reidy, M. L. Chouinard, P. I. Eacho, K. E. Gould, and J. Trias. 2009. Varespladib (A-002), a secretory phospholipase A$_2$ inhibitor, reduces atherosclerosis and aneurysm formation in *ApoE$^{-/-}$* mice. *J Cardiovasc Pharmacol.* 53:60–5.

Fujioka, D., Y. Saito, T. Kobayashi, T. Yano, H. Tezuka, Y. Ishimoto, N. Suzuki et al. 2008. Reduction in myocardial ischemia/reperfusion injury in group X secretory phospholipase A$_2$-deficient mice. *Circulation.* 117:2977–85.

Fujishima, H., R. O. Sanchez Mejia, C. O. Bingham, 3rd, B. K. Lam, A. Sapirstein, J. V. Bonventre, K. F. Austen, and J. P. Arm. 1999. Cytosolic phospholipase A$_2$ is essential for both the immediate and the delayed phases of eicosanoid generation in mouse bone marrow-derived mast cells. *Proc Natl Acad Sci USA.* 96:4803–7.

Funk, C. D. 2001. Prostaglandins and leukotrienes: Advances in eicosanoid biology. *Science.* 294:1871–5.

Furue, S., K. Kuwabara, K. Mikawa, K. Nishina, M. Shiga, N. Maekawa, M. Ueno et al. 1999. Crucial role of group IIA phospholipase A$_2$ in oleic acid-induced acute lung injury in rabbits. *Am J Respir Crit Care Med.* 160:1292–302.

Furue, S., K. Mikawa, K. Nishina, M. Shiga, M. Ueno, Y. Tomita, K. Kuwabara et al. 2001. Therapeutic time-window of a group IIA phospholipase A$_2$ inhibitor in rabbit acute lung injury: Correlation with lung surfactant protection. *Crit Care Med.* 29:719–27.

Galli, S. J., J. Kalesnikoff, M. A. Grimbaldeston, A. M. Piliponsky, C. M. Williams, and M. Tsai 2005. Mast cells as "tunable" effector and immunoregulatory cells: Recent advances. *Annu Rev Immunol.* 23:749–86.

Gavett, S. H., S. L. Madison, P. C. Chulada, P. E. Scarborough, W. Qu, J. E. Boyle, H. F. Tiano et al. 1999. Allergic lung responses are increased in prostaglandin H synthase-deficient mice. *J Clin Invest.* 104:721–32.

Ghosh, M., R. Loper, F. Ghomashchi, D. E. Tucker, J. V. Bonventre, M. H. Gelb, and C. C. Leslie. 2007. Function, activity, and membrane targeting of cytosolic phospholipase $A_2\zeta$ in mouse lung fibroblasts. *J Biol Chem.* 282:11676–86.

Ghosh, M., A. Stewart, D. E. Tucker, J. V. Bonventre, R. C. Murphy, and C. C. Leslie. 2004. Role of cytosolic phospholipase A_2 in prostaglandin E_2 production by lung fibroblasts. *Am J Respir Cell Mol Biol.* 30:91–100.

Ghosh, M., D. E. Tucker, S. A. Burchett, and C. C. Leslie. 2006. Properties of the Group IV phospholipase A_2 family. *Prog Lipid Res.* 45:487–510.

Giannattasio, G., D. Fujioka, W. Xing, H. R. Katz, J. A. Boyce, and B. Balestrieri. 2010. Group V secretory phospholipase A2 reveals its role in house dust mite-induced allergic pulmonary inflammation by regulation of dendritic cell function. *J Immunol.* 185:4430–8.

Gordon, S. 2003. Alternative activation of macrophages. *Nat Rev Immunol.* 3:23–35.

Granata, F., B. Balestrieri, A. Petraroli, G. Giannattasio, G. Marone, and M. Triggiani. 2003. Secretory phospholipases A_2 as multivalent mediators of inflammatory and allergic disorders. *Int Arch Allergy Immunol.* 131:153–63.

Granata, F., V. Nardicchi, S. Loffredo, A. Frattini, R. Ilaria Staiano, C. Agostini, and M. Triggiani. 2009. Secreted phospholipases A_2: A proinflammatory connection between macrophages and mast cells in the human lung. *Immunobiology* 214:811–21.

Granata, F., R. I. Staiano, S. Loffredo, A. Petraroli, A. Genovese, G. Marone, and M. Triggiani. 2010. The role of mast cell-derived secreted phospholipases A_2 in respiratory allergy. *Biochimie.* 92:588–93.

Grass, D. S., R. H. Felkner, M. Y. Chiang, R. E. Wallace, T. J. Nevalainen, C. F. Bennett, and M. E. Swanson. 1996. Expression of human group II PLA_2 in transgenic mice results in epidermal hyperplasia in the absence of inflammatory infiltrate. *J Clin Invest.* 97:2233–41.

Griessbach, K., M. Klimt, A. Schulze Elfringhoff, and M. Lehr. 2002. Structure-activity relationship studies of 1-substituted 3-dodecanoylindole-2-carboxylic acids as inhibitors of cytosolic phospholipase A_2-mediated arachidonic acid release in intact platelets. *Arch Pharm (Weinheim)* 335:547–55.

Gronroos, J. O., V. J. Laine, M. J. Janssen, M. R. Egmond, and T. J. Nevalainen. 2001. Bactericidal properties of group IIA and group V phospholipases A_2. *J Immunol.* 166:4029–34.

Hallstrand, T. S., E. Y. Chi, A. G. Singer, M. H. Gelb, and W. R. Henderson, Jr. 2007. Secreted phospholipase A_2 group X overexpression in asthma and bronchial hyperresponsiveness. *Am J Respir Crit Care Med.* 176:1072–8.

Hanasaki, K., K. Yamada, S. Yamamoto, Y. Ishimoto, A. Saiga, T. Ono, M. Ikeda, M. Notoya, S. Kamitani, and H. Arita. 2002. Potent modification of low density lipoprotein by group X secretory phospholipase A_2 is linked to macrophage foam cell formation. *J Biol Chem.* 277:29116–24.

Haworth, O., M. Cernadas, R. Yang, C. N. Serhan, and B. D. Levy. 2008. Resolvin E1 regulates interleukin 23, interferon-gamma and lipoxin A4 to promote the resolution of allergic airway inflammation. *Nat Immunol.* 9:873–9.

Henderson, W. R., Jr., E. Y. Chi, J. G. Bollinger, Y. T. Tien, X. Ye, L. Castelli, Y.P. Rubtsov et al. 2007. Importance of group X-secreted phospholipase A_2 in allergen-induced airway inflammation and remodeling in a mouse asthma model. *J Exp Med.* 204:865–77.

Hiraoka, M., A. Abe, Y. Lu, K. Yang, X. Han, R. W. Gross, and J. A. Shayman. 2006. Lysosomal phospholipase A$_2$ and phospholipidosis. *Mol Cell Biol.* 26:6139–48.

Hite, R. D., M. C. Seeds, A. M. Safta, R. B. Jacinto, J. I. Gyves, D. A. Bass, and B. M. Waite. 2005. Lysophospholipid generation and phosphatidylglycerol depletion in phospholipase A$_2$-mediated surfactant dysfunction. *Am J Physiol Lung Cell Mol Physiol.* 288:L618–24.

Holgate, S. T. 1999. The epidemic of allergy and asthma. *Nature.* 402:B2–4.

Hooks, S. B. and B. S. Cummings. 2008. Role of Ca^{2+}-independent phospholipase A$_2$ in cell growth and signaling. *Biochem Pharmacol.* 76:1059–67.

Igarashi, A., Y. Shibata, K. Yamauchi, D. Osaka, N. Takabatake, S. Abe, S. Inoue et al. 2009. Gly80Ser polymorphism of phospholipase A$_2$-IID is associated with cytokine inducibility in A549 cells. *Respiration.* 78:312–21.

Irvin, C. G., Y. P. Tu, J. R. Sheller, J. R. Sheller, and C. D. Funk. 1997. 5-Lipoxygenase products are necessary for ovalbumin-induced airway responsiveness in mice. *Am J Physiol.* 272:L1053–8.

Ishii, I., N. Fukushima, X. Ye, and J. Chun. 2004a. Lysophospholipid receptors: Signaling and biology. *Annu Rev Biochem.* 73:321–54.

Ishii, S., T. Nagase, H. Shindou, H. Takizawa, Y. Ouchi, and T. Shimizu. 2004b. Platelet-activating factor receptor develops airway hyperresponsiveness independently of airway inflammation in a murine asthma model. *J Immunol.* 172:7095–102.

Ivandic, B., L. W. Castellani, X. P. Wang, J. H. Qiao, M. Mehrabian, M. Navab, A. M. Fogelman et al. 1999. Role of group II secretory phospholipase A$_2$ in atherosclerosis: 1. Increased atherogenesis and altered lipoproteins in transgenic mice expressing group IIa phospholipase A$_2$. *Arterioscler Thromb Vasc Biol.* 19:1284–90.

Jaffar, Z., M. E. Ferrini, M. C. Buford, G. A. Fitzgerald, and K. Roberts. 2007. Prostaglandin I$_2$-IP signaling blocks allergic pulmonary inflammation by preventing recruitment of CD4$^+$ Th2 cells into the airways in a mouse model of asthma. *J Immunol.* 179:6193–203.

Karakas, M. and W. Koenig. 2009. Varespladib methyl, an oral phospholipase A$_2$ inhibitor for the potential treatment of coronary artery disease. *IDrugs.* 12:585–92.

Kay, A. B. 2001. Allergy and allergic diseases. First of two parts. *N Engl J Med.* 344:30–7.

Kienesberger, P. C., M. Oberer, A. Lass, and R. Zechner. 2009. Mammalian patatin domain containing proteins: A family with diverse lipolytic activities involved in multiple biological functions. *J Lipid Res.* 50(Suppl):S63–8.

Kikawada, E., J. V. Bonventre, and J. P. Arm. 2007. Group V secretory PLA$_2$ regulates TLR2-dependent eicosanoid generation in mouse mast cells through amplification of ERK and cPLA$_2$α activation. *Blood* 110:561–7.

Kim, D. C., F. I. Hsu, N. A. Barrett, D. S. Friend, R. Grenningloh, I. C. Ho, A. Al-Garawi et al. 2006. Cysteinyl leukotrienes regulate Th2 cell-dependent pulmonary inflammation. *J Immunol.* 176:4440–8.

Kita, Y., T. Ohto, N. Uozumi, and T. Shimizu. 2006. Biochemical properties and pathophysiological roles of cytosolic phospholipase A$_2$s. *Biochim Biophys Acta.* 1761:1317–22.

Kitamura, Y. 1989. Heterogeneity of mast cells and phenotypic change between subpopulations. *Annu Rev Immunol.* 7:59–76.

Kobayashi, T. and S. Narumiya. 2002. Prostanoids in health and disease; lessons from receptor-knockout mice. *Adv Exp Med Biol.* 507:593–7.

Koduri, R. S., J. O. Gronroos, V. J. Laine, C. Le Calvez, G. Lambeau, T. J. Nevalainen, and M. H. Gelb. 2002. Bactericidal properties of human and murine groups I, II, V, X, and XII secreted phospholipases A$_2$. *J Biol Chem.* 277:5849–57.

Kokotos, G., S. Kotsovolou, D. A. Six, V. Constantinou-Kokotou, C. C. Beltzner, and E. A. Dennis. 2002. Novel 2-oxoamide inhibitors of human group IVA phospholipase A$_2$. *J Med Chem.* 45:2891–3.

Kokotos, G., D. A. Six, V. Loukas, T. Smith, V. Constantinou-Kokotou, D. Hadjipavlou-Litina, S. Kotsovolou, A. Chiou, C. C. Beltzner, and E. A. Dennis. 2004. Inhibition of group IVA cytosolic phospholipase A$_2$ by novel 2-oxoamides *in vitro*, in cells, and *in vivo*. *J Med Chem*. 47:3615–28.

Kudo, I. and M. Murakami. 2002. Phospholipase A$_2$ enzymes. *Prostaglandins Other Lipid Mediat*. 68–69:3–58.

Kunikata, T., H. Yamane, E. Segi, T. Matsuoka, Y. Sugimoto, S. Tanaka, H. Tanaka, H. Nagai, A. Ichikawa, and S. Narumiya. 2005. Suppression of allergic inflammation by the prostaglandin E receptor subtype EP3. *Nat Immunol*. 6:524–31.

Lai, Y., R. C. Oslund, J. G. Bollinger, W. R. Jr. Henderson, L. F. Santana, W. A. Altemeier, M. H. Gelb, and T. S. Hallstrand. 2010. Eosinophil cysteinyl leukotriene synthesis mediated by exogenous secreted phospholipase A2 group X. *J Biol Chem*. 285:41491–500.

Laine, V. J., D. S. Grass, and T. J. Nevalainen. 2000. Resistance of transgenic mice expressing human group II phospholipase A$_2$ to *Escherichia coli* infection. *Infect Immun*. 68:87–92.

Lambeau, G. and M. H. Gelb. 2008. Biochemistry and physiology of mammalian secreted phospholipases A$_2$. *Annu Rev Biochem*. 77:495–520.

Landbo, C., E. Prescott, P. Lange, J. Vestbo, and T. P. Almdal. 1999. Prognostic value of nutritional status in chronic obstructive pulmonary disease. *Am J Respir Crit Care Med*. 160:1856–61.

Lee, J. J., D. Dimina, M. P. Macias, S. I. Ochkur, M. P. McGarry, K. R. O'Neill, C. Protheroe et al. 2004. Defining a link with asthma in mice congenitally deficient in eosinophils. *Science*. 305:1773–6.

Lee, K. L., M. L. Behnke, M. A. Foley, L. Chen, W. Wang, R. Vargas, J. Nunez et al. 2008. Benzenesulfonamide indole inhibitors of cytosolic phospholipase A$_2$α: Optimization of *in vitro* potency and rat pharmacokinetics for oral efficacy. *Bioorg Med Chem*. 16:1345–58.

Lee, K. L., M. A. Foley, L. Chen, M. L. Behnke, F. E. Lovering, S. J. Kirincich, W. Wang et al. 2007. Discovery of Ecopladib, an indole inhibitor of cytosolic phospholipase A$_2$α. *J Med Chem*. 50:1380–400.

Lehr, M., M. Klimt, and A. S. Elfringhoff. 2001. Novel 3-dodecanoylindole-2-carboxylic acid inhibitors of cytosolic phospholipase A$_2$. *Bioorg Med Chem Lett*. 11:2569–72.

Leite, J. O., U. Vaishnav, M. Puglisi, H. Fraser, J. Trias, and M. L. Fernandez. 2009. A-002 (Varespladib), a phospholipase A$_2$ inhibitor, reduces atherosclerosis in guinea pigs. *BMC Cardiovasc Disord*. 9:7.

Leslie, C. C. 1997. Properties and regulation of cytosolic phospholipase A$_2$. *J Biol Chem*. 272:16709–12.

Leslie, C. C. 2004. Regulation of the specific release of arachidonic acid by cytosolic phospholipase A$_2$. *Prostaglandins Leukot Essent Fatty Acids*. 70:373–6.

Levy, B. D., G. T. De Sanctis, P. R. Devchand, E. Kim, K. Ackerman, B. A. Schmidt, W. Szczeklik, J. M. Drazen, and C. N. Serhan. 2002. Multi-pronged inhibition of airway hyper-responsiveness and inflammation by lipoxin A$_4$. *Nat Med*. 8:1018–23.

Levy, B. D., P. Kohli, K. Gotlinger, O. Haworth, S. Hong, S. Kazani, E. Israel, K. J. Haley, and C. N. Serhan. 2007. Protectin D1 is generated in asthma and dampens airway inflammation and hyperresponsiveness. *J Immunol*. 178:496–502.

Lloyd, C. M. and C. M. Hawrylowicz. 2009. Regulatory T cells in asthma. *Immunity*. 31:438–49.

Ludwig, J., S. Bovens, C. Brauch, A. S. Elfringhoff, and M. Lehr. 2006. Design and synthesis of 1-indol-1-yl-propan-2-ones as inhibitors of human cytosolic phospholipase A$_2$α. *J Med Chem*. 49:2611–20.

MacPhee, M., K. P. Chepenik, R. A. Liddell, K. K. Nelson, L. D. Siracusa, and A. M. Buchberg. 1995. The secretory phospholipase A$_2$ gene is a candidate for the Mom1 locus, a major modifier of ApcMin-induced intestinal neoplasia. *Cell*. 81:957–66.

Malaviya, R., J. Ansell, L. Hall, M. Fahmy, R. L. Argentieri, G. C. Olini, Jr., D. W. Pereira, R. Sur, and D. Cavender. 2006. Targeting cytosolic phospholipase A$_2$ by arachidonyl trifluoromethyl ketone prevents chronic inflammation in mice. *Eur J Pharmacol.* 539:195–204.

Marone, G., M. Triggiani, and A. de Paulis. 2005. Mast cells and basophils: Friends as well as foes in bronchial asthma? *Trends Immunol.* 26:25–31.

Masuda, S., M. Murakami, M. Mitsuishi, K. Komiyama, Y. Ishikawa, T. Ishii, and I. Kudo. 2005. Expression of secretory phospholipase A$_2$ enzymes in lungs of humans with pneumonia and their potential prostaglandin-synthetic function in human lung-derived cells. *Biochem J.* 387:27–38.

Matsuoka, T., M. Hirata, H. Tanaka, Y. Takahashi, T. Murata, K. Kabashima, Y. Sugimoto et al. 2000. Prostaglandin D$_2$ as a mediator of allergic asthma. *Science.* 287:2013–7.

McIntyre, T. M., S. M. Prescott, and D. M. Stafforini. 2009. The emerging roles of PAF acetylhydrolase. *J Lipid Res.* 50 Suppl:S255–9.

McKew, J. C., K. L. Lee, M. W. Shen, P. Thakker, M. A. Foley, M. L. Behnke, B. Hu et al. 2008. Indole cytosolic phospholipase A$_2$α inhibitors: Discovery and *in vitro* and *in vivo* characterization of 4-{3-[5-chloro-2-(2-{[(3,4-dichlorobenzyl)sulfonyl] amino}ethyl)-1-(diphenylmethyl)-1H-indol-3-yl]propyl}benzoic acid, efipladib. *J Med Chem.* 51:3388–413.

Metcalfe, D. D., D. Baram, and Y. A. Mekori. 1997. Mast cells. *Physiol Rev.* 77:1033–79.

Mitsuishi, M., S. Masuda, I. Kudo, E. Kubota, K. Hamaguchi, A. Kouda, K. Honda, and H. Kawashima. 2006. Group V and X secretory phospholipase A$_2$ prevents adenoviral infection in mammalian cells. *Biochem J.* 393:97–106.

Miyake, A., H. Yamamoto, E. Kubota, K. Hamaguchi, A. Kouda, K. Honda, and H. Kawashima. 1993. Suppression of inflammatory responses to 12-O-tetradecanoyl-phorbol-13-acetate and carrageenin by YM-26734, a selective inhibitor of extracellular group II phospholipase A$_2$. *Br J Pharmacol.* 110:447–53.

Mounier, C. M., F. Ghomashchi, M. R. Lindsay, S. James, A. G. Singer, R. G. Parton, and M. H. Gelb. 2004. Arachidonic acid release from mammalian cells transfected with human groups IIA and X secreted phospholipase A$_2$ occurs predominantly during the secretory process and with the involvement of cytosolic phospholipase A$_2$α. *J Biol Chem.* 279:25024–38.

Munoz, N. M., Y. J. Kim, A. Y. Meliton, K. P. Kim, S. K. Han, E. Boetticher, E. O'Leary et al. 2003. Human group V phospholipase A$_2$ induces group IVA phospholipase A$_2$-independent cysteinyl leukotriene synthesis in human eosinophils. *J Biol Chem.* 278:38813–20.

Munoz, N. M., A. Y. Meliton, J. P. Arm, J. V. Bonventre, W. Cho, and A. R. Leff. 2007. Deletion of secretory group V phospholipase A$_2$ attenuates cell migration and airway hyperresponsiveness in immunosensitized mice. *J Immunol.* 179:4800–7.

Munoz, N. M., A. Y. Meliton, A. Lambertino, E. Boetticher, J. Learoyd, F. Sultan, X. Zhu, W. Cho, and A. R. Leff. 2006. Transcellular secretion of group V phospholipase A$_2$ from epithelium induces β2-integrin-mediated adhesion and synthesis of leukotriene C$_4$ in eosinophils. *J Immunol.* 177:574–82.

Munoz, N. M., A. Y. Meliton, L. N. Meliton, S. M. Dudek, and A. R. Leff. 2009. Secretory group V phospholipase A$_2$ regulates acute lung injury and neutrophilic inflammation caused by LPS in mice. *Am J Physiol Lung Cell Mol Physiol.* 296:L879–87.

Murakami, M., N. Hara, I. Kudo, and K. Inoue. 1993. Triggering of degranulation in mast cells by exogenous type II phospholipase A$_2$. *J Immunol.* 151:5675–84.

Murakami, M., R. S. Koduri, A. Enomoto, S. Shimbara, M. Seki, K. Yoshihara, A. Singer et al. 2001. Distinct arachidonate-releasing functions of mammalian secreted phospholipase A$_2$s in human embryonic kidney 293 and rat mastocytoma RBL-2H3 cells through heparan sulfate shuttling and external plasma membrane mechanisms. *J Biol Chem.* 276:10083–96.

Murakami, M. and I. Kudo. 2001. Diversity and regulatory functions of mammalian secretory phospholipase A_2s. *Adv Immunol.* 77:163–94.

Murakami, M. and I. Kudo. 2002. Phospholipase A_2. *J Biochem.* 131:285–92.

Murakami, M. and I. Kudo. 2003. New phospholipase A_2 isozymes with a potential role in atherosclerosis. *Curr Opin Lipidol.* 14:431–6.

Murakami, M., I. Kudo, and K. Inoue. 1991. Eicosanoid generation from antigen-primed mast cells by extracellular mammalian 14-kDa group II phospholipase A_2. *FEBS Lett.* 294:247–51.

Murakami, M., I. Kudo, Y. Suwa, and K. Inoue. 1992. Release of 14-kDa group-II phospholipase A_2 from activated mast cells and its possible involvement in the regulation of the degranulation process. *Eur J Biochem.* 209:257–65.

Murakami, M., K. Tada, S. Shimbara, T. Kambe, H. Sawada, and I. Kudo. 1997. Detection of secretory phospholipase A_2s related but not identical to type IIA isozyme in cultured mast cells. *FEBS Lett.* 413:249–54.

Murakami, M., Y. Taketomi, C. Girard, K. Yamamoto, and G. Lambeau. 2010. Emerging roles of secreted phospholipase A_2 enzymes: Lessons from transgenic and knockout mice. *Biochimie.* 92:561–82.

Murakami, M., K. Yoshihara, S. Shimbara, M. Sawada, N. Inagaki, H. Nagai, M. Naito et al. 2002. Group IID heparin-binding secretory phospholipase A_2 is expressed in human colon carcinoma cells and human mast cells and up-regulated in mouse inflammatory tissues. *Eur J Biochem.* 269:2698–707.

Nagase, T., N. Uozumi, T. Aoki-Nagase, K. Terawaki, S. Ishii, T. Tomita, H. Yamamoto, K. Hashizume, Y. Ouchi, and T. Shimizu. 2003. A potent inhibitor of cytosolic phospholipase A_2, arachidonyl trifluoromethyl ketone, attenuates LPS-induced lung injury in mice. *Am J Physiol Lung Cell Mol Physiol.* 284:L720–6.

Nagase, T., N. Uozumi, S. Ishii, K. Kume, T. Izumi, Y. Ouchi, and T. Shimizu. 2000. Acute lung injury by sepsis and acid aspiration: A key role for cytosolic phospholipase A_2. *Nat Immunol.* 1:42–6.

Nagase, T., N. Uozumi, S. Ishii, Y. Kita, H. Yamamoto, E. Ohga, Y. Ouchi, and T. Shimizu. 2002. A pivotal role of cytosolic phospholipase A_2 in bleomycin-induced pulmonary fibrosis. *Nat Med.* 8:480–4.

Nakae, S., Y. Komiyama, A. Nambu, K. Sudo, M. Iwase, I. Homma, K. Sekikawa, M. Asano, and Y. Iwakura. 2002. Antigen-specific T cell sensitization is impaired in IL-17-deficient mice, causing suppression of allergic cellular and humoral responses. *Immunity* 17:375–87.

Nakata, J., M. Kondo, J. Tamaoki, T. Takemiya, M. Nohara, K. Yamagata, and A. Nagai. 2005. Augmentation of allergic inflammation in the airways of cyclooxygenase-2-deficient mice. *Respirology.* 10:149–56.

Nakatani, N., N. Uozumi, K. Kume, M. Murakami, I. Kudo, and T. Shimizu. 2000. Role of cytosolic phospholipase A_2 in the production of lipid mediators and histamine release in mouse bone-marrow-derived mast cells. *Biochem J.* 352 Pt 2:311–7.

Ohtsuki, M., Y. Taketomi, S. Arata, S. Masuda, Y. Ishikawa, T. Ishii, Y. Takanezawa et al. 2006. Transgenic expression of group V, but not group X, secreted phospholipase A_2 in mice leads to neonatal lethality because of lung dysfunction. *J Biol Chem.* 281:36420–33.

Ono, T., K. Yamada, Y. Chikazawa, M. Ueno, S. Nakamoto, T. Okuno, and K. Seno. 2002. Characterization of a novel inhibitor of cytosolic phospholipase $A_2\alpha$, pyrrophenone. *Biochem J.* 363:727–35.

Oslund, R. C., N. Cermak, C. L. Verlinde, and M. H. Gelb. 2008. Simplified YM-26734 inhibitors of secreted phospholipase A_2 group IIA. *Bioorg Med Chem Lett.* 18:5415–9.

Pan, Y. H., B. Z. Yu, A. G. Singer, F. Ghomashchi, G. Lambeau, M. H. Gelb, M. K. Jain, and B. J. Bahnson. 2002. Crystal structure of human group X secreted phospholipase A_2.

Electrostatically neutral interfacial surface targets zwitterionic membranes. *J Biol Chem.* 277:29086–93.

Reddy, S. T. and H. R. Herschman. 1997. Prostaglandin synthase-1 and prostaglandin synthase-2 are coupled to distinct phospholipases for the generation of prostaglandin D$_2$ in activated mast cells. *J Biol Chem.* 272:3231–7.

Rivera, R. and J. Chun. 2008. Biological effects of lysophospholipids. *Rev Physiol Biochem Pharmacol.* 160:25–46.

Rosenson, R. S., C. Hislop, D. McConnell, M. Elliott, Y. Stasiv, N. Wang, and D. D. Waters. 2009. Effects of 1-*H*-indole-3-glyoxamide (A-002) on concentration of secretory phospholipase A$_2$ (PLASMA study): A phase II double-blind, randomised, placebo-controlled trial. *Lancet.* 373:649–58.

Rubin, B. B., G. P. Downey, A. Koh, N. Degousee, F. Ghomashchi, L. Nallan, E. Stefanski et al. 2005. Cytosolic phospholipase A$_2\alpha$ is necessary for platelet-activating factor biosynthesis, efficient neutrophil-mediated bacterial killing, and the innate immune response to pulmonary infection: cPLA$_2\alpha$ does not regulate neutrophil NADPH oxidase activity. *J Biol Chem.* 280:7519–29.

Saiga, A., N. Uozumi, T. Ono, K. Seno, Y. Ishimoto, H. Arita, T. Shimizu, and K. Hanasaki. 2005. Group X secretory phospholipase A$_2$ can induce arachidonic acid release and eicosanoid production without activation of cytosolic phospholipase A$_2\alpha$. *Prostaglandins Other Lipid Mediat.* 75:79–89.

Samet, J. M., M. C. Madden, and A. N. Fonteh. 1996. Characterization of a secretory phospholipase A$_2$ in human bronchoalveolar lavage fluid. *Exp Lung Res.* 22:299–315.

Satake, Y., B. L. Diaz, B. Balestrieri, B. K. Lam, Y. Kanaoka, M. J. Grusby, and J. P. Arm. 2004. Role of group V phospholipase A$_2$ in zymosan-induced eicosanoid generation and vascular permeability revealed by targeted gene disruption. *J Biol Chem.* 279:16488–94.

Sato, H., R. Kato, Y. Isogai, G. Saka, M. Ohtsuki, Y. Taketomi, K. Yamamoto et al. 2008. Analyses of group III secreted phospholipase A$_2$ transgenic mice reveal potential participation of this enzyme in plasma lipoprotein modification, macrophage foam cell formation, and atherosclerosis. *J Biol Chem.* 283:33483–97.

Sato, R., S. Yamaga, K. Watanabe, S. Hishiyama, T. Kawabata, T. Kobayashi, D. Fujioka et al. 2010. Inhibition of secretory phospholipase A$_2$ activity attenuates lipopolysaccharide-induced acute lung injury in a mouse model. *Exp Lung Res.* 36:191–200.

Schaloske, R. H. and E. A. Dennis. 2006. The phospholipase A$_2$ superfamily and its group numbering system. *Biochim Biophys Acta.* 1761:1246–59.

Schols, A. M., J. Slangen, L. Volovics, and E. F. Wouters. 1998. Weight loss is a reversible factor in the prognosis of chronic obstructive pulmonary disease. *Am J Respir Crit Care Med.* 157:1791–7.

Scott, D. L., S. P. White, J. L. Browning, J. J. Rosa, M. H. Gelb, and P. B. Sigler. 1991. Structures of free and inhibited human secretory phospholipase A$_2$ from inflammatory exudate. *Science.* 254:1007–10.

Seeds, M. C., D. L. Bowton, R. D. Hite, J. I. Gyves, and D. A. Bass. 2003. Human eosinophil group IID secretory phospholipase A$_2$ causes surfactant dysfunction. *Chest.* 123:376S–7S.

Seeds, M. C., K. A. Jones, R. D. Hite, M. C. Willingham, H. M. Borgerink, R. D. Woodruff, D. L. Bowton, and D. A. Bass. 2000. Cell-specific expression of group X and group V secretory phospholipases A$_2$ in human lung airway epithelial cells. *Am J Respir Cell Mol Biol.* 23:37–44.

Seno, K., T. Okuno, K. Nishi, Y. Murakami, K. Yamada, S. Nakamoto, and T. Ono. 2001. Pyrrolidine inhibitors of human cytosolic phospholipase A$_2$. Part 2: Synthesis of potent and crystallized 4-triphenylmethylthio derivative "pyrrophenone". *Bioorg Med Chem Lett.* 11:587–90.

Shaposhnik, Z., X. Wang, J. Trias, H. Fraser, and A. J. Lusis. 2009. The synergistic inhibition of atherogenesis in *apoE*^{−/−} mice between pravastatin and the sPLA$_2$ inhibitor varespladib (A-002). *J Lipid Res.* 50:623–9.

Sharp, J. D. and D. L. White. 1993. Cytosolic PLA$_2$: mRNA levels and potential for transcriptional regulation. *J Lipid Mediat.* 8:183–9.

Shiraishi, Y., K. Asano, K. Niimi, K. Fukunaga, M. Wakaki, J. Kagyo, T. Takihara et al. 2008. Cyclooxygenase-2/prostaglandin D$_2$/CRTH2 pathway mediates double-stranded RNA-induced enhancement of allergic airway inflammation. *J Immunol.* 180:541–9.

Singer, A. G., F. Ghomashchi, C. Le Calvez, J. Bollinger, S. Bezzine, M. Rouault, M. Sadilek et al. 2002. Interfacial kinetic and binding properties of the complete set of human and mouse groups I, II, V, X, and XII secreted phospholipases A$_2$. *J Biol Chem.* 277:48535–49.

Smart, B. P., R. C. Oslund, L. A. Walsh, and M. H. Gelb. 2006. The first potent inhibitor of mammalian group X secreted phospholipase A$_2$: Elucidation of sites for enhanced binding. *J Med Chem.* 49:2858–60.

Stadel, J. M., K. Hoyle, R. M. Naclerio, A. Roshak, and F. H. Chilton. 1994. Characterization of phospholipase A$_2$ from human nasal lavage. *Am J Respir Cell Mol Biol.* 11:108–13.

Street, I. P., H. K. Lin, F. Laliberte, F. Ghomashchi, Z. Wang, H. Perrier, N. M. Tremblay, Z. Huang, P. K. Weech, and M. H. Gelb. 1993. Slow- and tight-binding inhibitors of the 85-kDa human phospholipase A$_2$. *Biochemistry* 32:5935–40.

Styles, L. A., M. Abboud, S. Larkin, M. Lo, and F. A. Kuypers. 2007. Transfusion prevents acute chest syndrome predicted by elevated secretory phospholipase A$_2$. *Br J Haematol.* 136:343–4.

Styles, L. A., C. G. Schalkwijk, E. P. Vichinsky, B. H. Lubin, and F. A. Kuypers. 1996. Phospholipase A$_2$ levels in acute chest syndrome of sickle cell disease. *Blood.* 87:2573–8.

Tager, A. M., S. K. Bromley, B. D. Medoff, S. A. Islam, S. D. Bercury, E. B. Friedrich, A. D. Carafone, R. E. Gerszten, and A. D. Luster. 2003. Leukotriene B$_4$ receptor BLT1 mediates early effector T cell recruitment. *Nat Immunol.* 4:982–90.

Takabatake, N., M. Sata, S. Inoue, Y. Shibata, S. Abe, T. Wada, J. Machiya et al. 2005. A novel polymorphism in secretory phospholipase A$_2$-IID is associated with body weight loss in chronic obstructive pulmonary disease. *Am J Respir Crit Care Med.* 172:1097–104.

Touqui, L. and M. Alaoui-El-Azher. 2001. Mammalian secreted phospholipases A$_2$ and their pathophysiological significance in inflammatory diseases. *Curr Mol Med.* 1:739–54.

Trapnell, B. C., J. A. Whitsett, and K. Nakata. 2003. Pulmonary alveolar proteinosis. *N Engl J Med.* 349:2527–39.

Triggiani, M., G. Giannattasio, C. Calabrese, S. Loffredo, F. Granata, A. Fiorello, M. Santini, M. H. Gelb, and G. Marone. 2009. Lung mast cells are a source of secreted phospholipases A$_2$. *J Allergy Clin Immunol.* 124:558–65, 565 e1–3.

Umetsu, D. T., J. J. McIntire, O. Akbari, C. Macaubas, and R. H. DeKruyff. 2002. Asthma: An epidemic of dysregulated immunity. *Nat Immunol.* 3:715–20.

Uozumi, N., K. Kume, T. Nagase, N. Nakatani, S. Ishii, F. Tashiro, Y. Komagata et al. 1997. Role of cytosolic phospholipase A$_2$ in allergic response and parturition. *Nature.* 390:618–22.

von Allmen, C. E., N. Schmitz, M. Bauer, H. J. Hinton, M. O. Kurrer, R. B. Buser, M. Gwerder et al. 2009. Secretory phospholipase A$_2$-IID is an effector molecule of CD4$^+$CD25$^+$ regulatory T cells. *Proc Natl Acad Sci USA.* 106:11673–8.

Wang, Y., S. I. Feinstein, Y. Manevich, Y. S. Ho, and A. B. Fisher. 2006. Peroxiredoxin 6 gene-targeted mice show increased lung injury with paraquat-induced oxidative stress. *Antioxid Redox Signal.* 8:229–37.

Webb, N. R., M. A. Bostrom, S. J. Szilvassy, D. R. van der Westhuyzen, A. Daugherty, and F. C. de Beer. 2003. Macrophage-expressed group IIA secretory phospholipase A$_2$

increases atherosclerotic lesion formation in LDL receptor-deficient mice. *Arterioscler Thromb Vasc Biol.* 23:263–8.

Weinrauch, Y., C. Abad, N. S. Liang, S. F. Lowry, and J. Weiss. 1998. Mobilization of potent plasma bactericidal activity during systemic bacterial challenge. Role of group IIA phospholipase A$_2$. *J Clin Invest.* 102:633–8.

Wijewickrama, G. T., J. H. Kim, Y. J. Kim, A. Abraham, Y. Oh, B. Ananthanarayanan, M. Kwatia, S. J. Ackerman, and W. Cho. 2006. Systematic evaluation of transcellular activities of secretory phospholipases A$_2$. High activity of group V phospholipases A$_2$ to induce eicosanoid biosynthesis in neighboring inflammatory cells. *J Biol Chem.* 281:10935–44.

Williams, C. M. and S. J. Galli. 2000. Mast cells can amplify airway reactivity and features of chronic inflammation in an asthma model in mice. *J Exp Med.* 192:455–62.

Wills-Karp, M. 1999. Immunologic basis of antigen-induced airway hyperresponsiveness. *Annu Rev Immunol.* 17:255–81.

Yu, M., M. Tsai, S. Y. Tam, C. Jones, J. Zehnder, and S. J. Galli. 2006. Mast cells can promote the development of multiple features of chronic asthma in mice. *J Clin Invest.* 116:1633–41.

Zeiher, B. G., J. Steingrub, P. F. Laterre, A. Dmitrienko, Y. Fukiishi, and E. Abraham. 2005. LY315920NA/S-5920, a selective inhibitor of group IIA secretory phospholipase A$_2$, fails to improve clinical outcome for patients with severe sepsis. *Crit Care Med.* 33:1741–8.

Zeldin, D. C., C. Wohlford-Lenane, P. Chulada, J. A. Bradbury, P. E. Scarborough, V. Roggli, R. Langenbach, and D. A. Schwartz. 2001. Airway inflammation and responsiveness in prostaglandin H synthase-deficient mice exposed to bacterial lipopolysaccharide. *Am J Respir Cell Mol Biol.* 25:457–65.

4 5-Lipoxygenase Pathway
A Validated Route to Drug Discovery

Carlo Pergola and Oliver Werz

CONTENTS

4.1   5-LIPOXYGENASE

5-Lipoxygenase (5-LO, EC 1.13.11.34) is a member of a family of dioxygenase enzymes containing a *nonheme* iron known as lipoxygenases, which catalyze the stereospecific insertion of molecular oxygen into polyunsaturated fatty acids containing a 1,4-pentadiene structure. 5-LO catalyzes the dioxygenation of arachidonic acid (AA) at the C-5 position, which represents the first step in the leukotriene (LT) biosynthetic pathway.

The first proof for the existence of a 5-LO enzyme was provided by Borgeat et al. (1976), who described the conversion of AA to 5-hydroxyeicosatetraenoic acid (5-HETE) by rabbit polymorphonuclear leukocytes (PMNL). This novel pathway of AA metabolism attracted even more attention as the formation of 5-HETE and LTB_4 was also observed in human leukocytes (Borgeat and Samuelsson, 1979). Soon it became evident that the unknown substances with contractile properties released after immunological challenge of sensitized lungs (the so-called slow-reacting substance of anaphylaxis) (Kellaway and Trethewie, 1940) consist of a mixture of 5-LO products (i.e., LTC_4, D_4, and E_4), which was the first evidence for a role of 5-LO products as key mediators of immediate hypersensitivity reactions and inflammation (Samuelsson, 1983).

4.2   5-LIPOXYGENASE STRUCTURE

5-LO is a monomeric soluble protein with an estimated molecular mass of 72–80 kDa (Radmark, 2000). The cDNAs for 5-LO has been cloned from human, mouse, rat, and hamster, and encode mature proteins of 672 or 673 amino acids (Funk, 1996). The crystal structure of 5-LO has been only recently clarified by replacement of a 5-LO-specific destabilizing sequence (Gilbert et al., 2011). Structural information derived thus far from *in silico* models based on the structure of rabbit reticulocyte 15-LO and from mutagenesis studies. Based on these data, 5-LO can be modeled as a monomeric enzyme with two domains.

4.2.1   A REGULATORY N-TERMINAL DOMAIN (RESIDUES 1–114)

This is a "PLAT" domain (polycystin-1, lipoxygenase, and α-toxin) (Allard and Brock, 2005), also known as "C2-like domain" because of similarities with the Ca^{2+}-dependent C2-like domain of phospholipases and protein kinase C (PKC). This domain is a ß-sandwich with typical ligand-binding loops (Hammarberg et al., 2000). Residues in these loops of 5-LO have been shown to bind Ca^{2+}, cellular membranes, and coactosin-like protein (CLP). Mutagenesis of residues 43–46 in this domain reduced Ca^{2+} binding and Trp13, 75, and 102 mediate effects of phosphatidylcholine (PC), glycerides (e.g., 1-oleoyl-2-acetyl-*sn*-glycerol, OAG), and CLP. As shown below, this C2-like domain may also serve as a functional site for pharmacological 5-LO inhibitors (e.g., for hyperforin) (Feisst et al., 2009).

4.2.2   A CATALYTIC C-TERMINAL DOMAIN (RESIDUES 121–673)

This domain is mainly α-helical in structure, and contains a prosthetic catalytic iron. The iron is positioned by five iron ligands, that is, His367, His372, His550, Asn554, and Ile673, whereas the sixth ligand is assumed to be a water molecule. The structural specificity of different LOs may depend on the space present in the pocket where the fatty acid substrate binds, since mutagenesis reducing the pocket space converted 5-LO to a 15-LO (Schwarz et al., 2001). This "active site" of 5-LO is also the locale where most of the pharmacological 5-LO inhibitors bind and thus confer their inhibitory activity (see below). The catalytic domain also contains three kinase motifs for mitogen-activated protein kinase (MAPK)-activating protein kinase (MK)-2 (*Leu–Glu–Arg–Gln–Leu–Ser271*), protein kinase A (PKA) (*Arg–Lys–Ser–Ser523*), and extracellular signal-regulated kinase (ERK)-2 (*Tyr–Leu–Ser663–Pro*) (Radmark et al., 2007).

Though the two domains appear independently folded, the structure of truncated 12*S*-LO suggested that the N-terminal domain may strongly influence the 3-D structure of the catalytic domain (Newcomer and Gilbert, 2010).

4.3 5-LIPOXYGENASE CATALYSIS AND LEUKOTRIENE BIOSYNTHESIS

5-LO possesses two distinct enzymatic activities: it catalyzes the incorporation of molecular oxygen into AA (oxygenase activity) and subsequently catalyzes the formation of the unstable epoxide LTA_4 (LTA_4 synthase activity) (Rouzer et al., 1986; Shimizu et al., 1984) (Figure 4.1). 5-LO first catalyzes the abstraction of the pro-S hydrogen at C-7 of the fatty acid, followed by the insertion of molecular oxygen at position C-5, leading to the hydroperoxide 5(*S*)-hydroperoxy-6-*trans*-8,11,14-*cis*-eicosatetraenoic acid (5-HpETE). The subsequent abstraction of the pro-R hydrogen from C-10 and the allylic shifts of the radical to C-6 results in the formation of the 5,6 epoxide LTA_4 (Shimizu et al., 1986). Besides conversion to LTA_4, 5-HpETE can be reduced to the corresponding alcohol 5-HETE by cellular peroxidases or by a pseudoperoxidase activity of 5-LO.

For 5-LO catalysis, the nonheme iron in the 5-LO structure acts as an electron acceptor or donor, and cycles between the ferrous (Fe^{2+}) and the ferric (Fe^{3+}) state (Radmark, 1995). 5-LO is inactive when the iron is in the ferrous state, and oxidation by lipid hydroperoxides (LOOH) is needed to yield the active, ferric form. Thus, the 5-LO reaction is characterized by an initiation phase (conversion of Fe^{2+} to Fe^{3+}), linear propagation (with high conversion rate), and irreversible inactivation (also called suicide-inactivation). Although LOOH are needed for iron oxidation in the initiation phase, such oxidants formed during catalysis may also be responsible for rapid irreversible inactivation.

The LTA_4 formed by 5-LO is an unstable intermediate and can be enzymatically converted to LTB_4 by LTA_4 hydrolase, or conjugated with glutathione to LTC_4 by LTC_4 synthase or by other members of the "membrane-associated proteins in eicosanoid and glutathione metabolism" (MAPEG) (Samuelsson et al., 1987) (Figure 4.1). The conversion of the LTA_4 intermediate to bioactive LTs can be performed by the same cells where LTA_4 is formed (depending on the enzymes present), or LTA_4 can be transferred from donor to acceptor cells for further conversion (transcellular biosynthesis of LTs) (Folco and Murphy, 2006; Zarini et al., 2009). The release of LTC_4 into the extracellular environment and the successive cleavage of the γ-glutamyl and glycine residues yield LTD_4 and LTE_4, respectively. LTC_4, D_4, and E_4 are collectively denominated cysteinyl-LTs (cys-LTs) due to the presence of a cysteinyl residue in their structure. The action of 5-LO combined with that of other LOs (namely, 12-LO or 15-LO) can also lead to different bioactive lipid mediators containing a trihydroxytetraene, known as lipoxins, which apparently can act as stop-signals

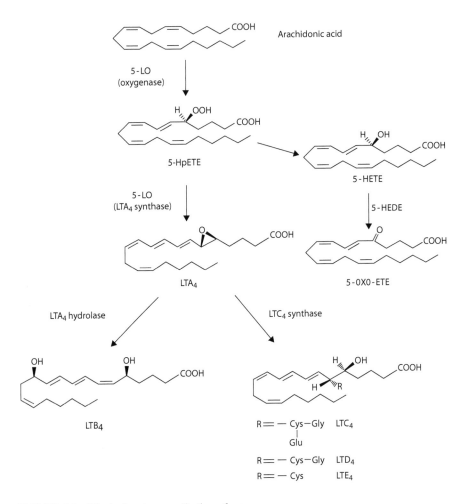

FIGURE 4.1   The leukotriene synthetic pathway.

during inflammation and promote repair and wound healing (Serhan et al., 2008). Similarly, the metabolism of eicosapentaenoic acid and docosahexaenoic acid (involving 5-LO) may lead to the formation of resolvins that actively promote cataba- sis via potent proresolving and anti-inflammatory actions (Bannenberg, 2009).

4.4   5-LIPOXYGENASE EXPRESSION

5-LO is expressed in a tissue- and cell differentiation-specific manner. 5-LO is pri- marily expressed in differentiated myeloid cells. Thus, it is constitutively present in blood leukocytes, such as granulocytes, monocytes/macrophages, mast cells, and dendritic cells (DCs), and in foam cells of human atherosclerotic tissue (Radmark et al., 2007). Also, 5-LO is expressed in mantle zone B cells but is low or absent in germinal B cells and plasma cells (Mahshid et al., 2009). In T lymphocytes, 5-LO expression on both transcriptional and translational levels is instead controversial.

Platelets, erythrocytes, and endothelial cells do not express 5-LO. In the skin, Langerhans cells (Spanbroek et al., 1998) and fibroblasts (Berg et al., 2006) express 5-LO while epidermal keratinocytes are 5-LO negative.

The regulation of 5-LO expression at the transcriptional level is complex and involves methylation of the 5-LO promoter. Various differentiation inducers such as dimethyl sulfoxide (DMSO), retinoic acid, $1\alpha,25$-dihydroxyvitamin D_3, and transforming growth factor-β upregulated 5-LO in immature myeloid cells, and granulocyte macrophage colony-stimulating factor (GM-CSF) can increase 5-LO expression in human mature granulocytes and in differentiating DCs. On the other hand, interleukin (IL)-4 decreases 5-LO expression in monocytes and maturating DCs. Interestingly, examination of genomic DNA from asthmatic as well as nonasthmatic subjects revealed the occurrence of natural mutations within the functional promoter regions (Silverman and Drazen, 2000). Also, overexpression of 5-LO has been observed after cytomegalovirus infection (Qiu et al., 2008) or in certain cancer cells/tissues (e.g., malignant prostate cancer) (Gupta et al., 2001).

4.5 CELLULAR REGULATION OF 5-LIPOXYGENASE AND OF LEUKOTRIENE BIOSYNTHESIS

Cellular biosynthesis of LTs requires an orchestrated interplay of crucial proteins/enzymes. Also, the activities of the enzymes involved in LT formation are compartmentalized and (spatially and temporally) dynamic. Besides 5-LO, enzymes and proteins involved in the early events of cellular 5-LO product biosynthesis are discussed below.

4.5.1 CYTOSOLIC PHOSPHOLIPASE A_2

The cytosolic phospholipase A_2 (cPLA$_2$) is a group IVA PLA$_2$ that catalyzes the hydrolysis of the *sn*-2 position of glycerophospholipids to release free AA, which is the substrate of 5-LO (Shimizu and Wolf, 1990). The liberation of free AA acid is tightly regulated and often the initial, rate-limiting step in the biosynthesis of LTs. In fact, cPLA$_2$ knockout mice fail to generate LTs (Uozumi et al., 1997). Purified cPLA$_2$ demonstrated considerable activity at micromolar or submicromolar concentrations of Ca^{2+} (Hirabayashi et al., 2004). Studies by direct mutagenesis and by active-site-directed inhibitor indicated that the catalytic center contains the catalytically active Ser228 (Sharp et al., 1994). Thiol-modifying reagents completely inactivate cPLA$_2$, and Cys331 has been involved in the loss of enzyme activity, suggesting a role for this amino acid for the catalytic activity of the enzyme (Li et al., 1994). The amino-terminal portion of the protein (~50 amino acids) is a C2-like domain and mediates the Ca^{2+}-dependent translocation to membranes (Clark et al., 1991). Upon membrane binding, conformational changes in the enzyme might take place to allow the fatty acyl chain of phospholipids to enter the active site. In most cases, Ca^{2+}-dependent translocation of cPLA$_2$ from the cytosol to membranes is necessary for its access to phospholipid substrates (Channon and Leslie, 1990). Ca^{2+}-independent localization of cPLA$_2$ on the perinuclear membranes may be regulated by phosphorylation, anionic lipids, and hydrophobic binding of the catalytic domain (Hirabayashi et al.,

2004). So far, three distinct phosphorylations on serine residues of this enzyme have been reported [by MAPKs on Ser505 (Kramer et al., 1996; Lin et al., 1993; Nemenoff et al., 1993; Casas et al., 2009); by Ca^{2+}/calmodulin-dependent protein kinase II (CaMKII) on Ser515 (Muthalif et al., 2001); by MAPK-interacting kinase Mnk1 on Ser727 (Hefner et al., 2000)]. Phosphorylation of $cPLA_2$ on Ser505, Ser515, or Ser727 fails to induce AA release in the absence of a concomitant increase in intracellular Ca^{2+}, but increases $cPLA_2$ intrinsic enzymatic activity two- to threefold at submicromolar intracellular Ca^{2+} concentrations (Hefner et al., 2000). However, the contribution of $cPLA_2$ phosphorylation to AA release is much less at high concentrations of intracellular Ca^{2+} (Hefner et al., 2000).

4.5.2 5-Lipoxygenase-Activating Protein

5-Lipoxygenase-activating protein (FLAP) is an 18-kDa membrane-bound protein. It is a member of the MAPEG superfamily. Unlike the other MAPEG members, however, it is apparently not catalytically active and does not have a binding site for glutathione. FLAP contains four transmembrane helices that are connected by two elongated cytosolic loops and one short lumenal loop, and crystallizes as a homotrimer (Ferguson et al., 2007). Expression of FLAP is consistent with the occurrence of 5-LO in myeloid cells (Steinhilber, 1994; Vickers, 1995) and the upregulation of FLAP often correlates with that of 5-LO. Though FLAP was initially assumed to function as a membrane anchor for 5-LO, it has then been suggested that FLAP mediates the transfer of released AA to 5-LO and functions as a 5-LO scaffold protein (Ferguson et al., 2007).

4.5.3 Coactosin-Like Protein

CLP is an F-actin-binding protein that can also bind 5-LO (Provost et al., 2001). CLP seemingly functions as a chaperoning scaffold factor for 5-LO upregulating Ca^{2+}-induced 5-LO activity but also stabilizes 5-LO, and these effects require protein interaction by Trp residues in ligand-binding loops of the 5-LO C2 domain (Esser et al., 2009).

In isolated cells, 5-LO occurs as a soluble enzyme, either in the cytosol or in the nucleus, depending on the cell type (Werz and Steinhilber, 2006). Thus, cytosolic 5-LO has been described in neutrophils, eosinophils, and peritoneal macrophages, whereas 5-LO is in a nuclear soluble compartment associated with the chromatin in alveolar macrophages, Langerhans cells, or rat basophilic leukemia cells. Cell stimulation by Ca^{2+}-mobilizing agents (ionophores or thapsigargin), soluble agonists [chemotactic PAF, LTB_4, formyl–methionyl–leucyl–phenylalanine (fMLP) and C5a], cytokines (IL-8), and phagocytic particles (zymosan, urate or phosphate crystals) activates 5-LO that translocates to the nuclear envelope, where FLAP resides. Similarly, $cPLA_2$ also accumulates in the perinuclear region and releases AA from phospholipids. AA is then transferred via FLAP to 5-LO for metabolism (Figure 4.2). CLP has been found to be associated with 5-LO and to comigrate after activation, functioning as a scaffold for Ca^{2+}-induced 5-LO activity (Rakonjac et al., 2006).

Besides natural cellular agonists, certain agents can upregulate LT generation after subsequent stimulation, although they are not able to induce LT synthesis by

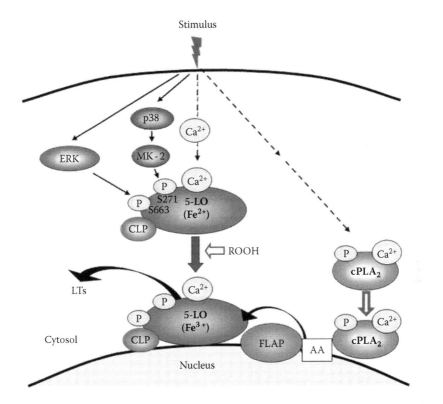

FIGURE 4.2 Scheme for the cellular activation of 5-lipoxygenase. In response to an adequate stimulation of the cells (e.g., neutrophils or monocytes), elevation of intracellular Ca^{2+} concentrations and activation of members of the MAPK family [i.e., ERKs and MAPKAPK-2 (MK-2)] are observed. Binding of Ca^{2+} and phosphorylations by these kinases activate 5-lipoxygenase (5-LO) and cytosolic phospholipase A_2 (cPLA$_2$), which translocate to the perinuclear region where the 5-LO activating protein (FLAP) is located. The coactosin-like protein (CLP) is bound to 5-LO and may act as 5-LO scaffold. When 5-LO is activated, lipid hydroperoxides (LOOH) oxidize the iron ($Fe^{+2} \rightarrow Fe^{3+}$). At the nuclear membrane, cPLA$_2$ releases AA from phospholipids, that is transferred via FLAP to 5-LO for metabolism.

themselves (Werz, 2002). These agents may prime the cells at multiple sites (enhanced AA availability, increased expression of FLAP, elevated levels of intracellular Ca^{2+}, increased accumulation of 5-LO at the nuclear membrane, and enhanced phosphorylation of 5-LO) and include growth factors, cytokines, phorbol esters, Epstein–Barr virus, lipopolysaccharide (LPS), and other Toll-like receptor ligands (Lefebvre et al., 2010).

5-LO activity is regulated by many factors. Since oxidation of the ferrous iron in the active site to the ferric state is needed for 5-LO catalysis, balanced cellular levels of oxidizing LOOH are required (Rouzer and Samuelsson, 1986). In fact, reduction of LOOH by selenium-dependent glutathione peroxidases (GPx) suppresses 5-LO product formation (Weitzel and Wendel, 1993). 5-LO activity is strongly stimulated by Ca^{2+}, which binds to the N-terminal C2-like domain of 5-LO (Radmark et al., 2007). Ca^{2+} increases the affinity of 5-LO toward AA and apparently also toward activating LOOH. It also causes 5-LO binding to PC vesicles and stimulates 5-LO translocation to the nuclear membrane (which is rich in PC), mediated by a putative PC-selective binding site within the C2 domain. Finally, 5-LO is stimulated by ATP (0.1–2 mM) and by various glycerides (e.g., OAG), and the intracellular formation of diacylglycerides (DAGs) via the phospholipase D/phosphatidic acid phosphatase pathway is required for full activation of 5-LO in leukocytes (Albert et al., 2008).

Subcellular localization of 5-LO as well as its activation for product synthesis are also modulated by phosphorylation events. The p38 MAPK-regulated MK-2/3 (Werz et al., 2000) and ERK1/2 (Werz et al., 2002a) phosphorylate 5-LO at Ser-271 and Ser-663 *in vitro*, respectively, which is strongly promoted by unsaturated fatty acids (i.e., AA) (Werz et al., 2002b). Interestingly, priming agents and agonists that induce LT synthesis also activate p38 MAPK and downstream MKs and/or ERKs in leukocytes. Moreover, cell stress (oxidative or genotoxic stress, osmotic shock) or the cytokines IL-1/tumor necrosis factor-α activates p38 MAPK and downstream MKs and thus strongly enhanced 5-LO activity in leukocytes (Werz et al., 2001a). Of interest, phosphorylation of 5-LO was connected to increased accumulation of 5-LO in the perinuclear region (Werz et al., 2001b). Recently, PKA was reported to phosphorylate Ser-523, which impairs the catalytic activity of 5-LO (Luo et al., 2004). In conclusion, Ca^{2+}, the cellular redox tone, cellular levels of DAGs, phosphorylation events, as well as 5-LO subcellular localization and interactions with FLAP are the key factors determining 5-LO activity in intact cells.

For downstream conversion of formed LTA_4 to effector LTs, additional enzymes are required, which are discussed below.

4.5.4 LTA_4 Hydrolase

LTA_4 hydrolase is a soluble enzyme and converts LTA_4 into LTB_4. It is a zinc-containing amino peptidase belonging to the M1 amino peptidase superfamily (Tholander et al., 2008), with two additional domains as compared to thermolysin: an N-terminal all-β domain and a C-terminal α-helical domain. LTA_4 hydrolase is widely expressed in different organs (intestine, spleen, lung, and kidney) and cell types. In the blood, it is present in neutrophils, monocytes, lymphocytes, and erythrocytes, whereas eosinophils have low levels and basophils and platelets are devoid of the enzyme. In the lung, epithelial cells and alveolar macrophages express high levels whereas mast cells express only low levels of LTA_4 hydrolase (Haeggstrom, 2000). The broader distribution of LTA_4 hydrolase in comparison to 5-LO might be functionally important for transcellular biosynthesis of LTB_4.

4.5.5 LTC_4 Synthase

As FLAP, LTC_4 synthase is a member of the MAPEG superfamily, contains four transmembrane helices, and crystallizes as a trimer, with active sites located at monomer interfaces. LTC_4 synthase contains a glutathione-binding site and catalyzes the conjugation of glutathione to LTA_4 to form LTC_4. LTC_4 synthase is a membrane-embedded enzyme localized only to the outer nuclear membrane. The location of LTA_4-binding sites suggests that the protein may receive its substrate from the bilayer (which could stabilize the LTA_4 intermediate) by lateral diffusion. The enzyme is expressed only in certain cell types such as eosinophils, mast cells, basophils, and monocyte-macrophages and also in platelets (which are devoid of 5-LO) (Lam and Austen, 2000).

Interestingly, *gender-related differences* in LT biosynthesis have been observed in human whole blood and neutrophils (the main source of LTs formed in whole blood) due to the downregulation of 5-LO product synthesis by androgens. This is reflected by a lower LT biosynthesis in stimulated blood derived from males versus females (Pergola et al., 2008). Such gender bias may explain well-recognized sex differences in the incidence of LT-related diseases such as asthma or allergic rhinitis that clearly dominate in females (Osman, 2003). Interestingly, knocking out of the BLT_1 receptor strongly protected the female but not the male mice in a model of PAF-induced lethality (Haribabu et al., 2000), and gender-specific attenuation of atheroma formation was observed in dual 5-LO and 12/15-LO-KO mice in an experimental model of atherosclerosis (Poeckel et al., 2009).

4.6 PHYSIOLOGICAL AND PATHOPHYSIOLOGICAL ROLES OF 5-LO PRODUCTS

The biological actions of all LTs as well as of lipoxin A_4 are mediated by specific G-protein-coupled receptors (GPCRs), which are located on the outer plasma membrane of structural or inflammatory cells. The binding of 5-LO products to the respective receptors results in increase in the intracellular Ca^{2+} concentrations and reduction of the intracellular levels of cyclic AMP (cAMP), which in turn regulate downstream pathways.

LTB_4 is a potent chemotactic and chemokinetic agent for leukocytes and stimulates migration and activation of leukocytes, leading to adherence of granulocytes to vessel walls, degranulation, and release of the cathelicidin LL-37 and superoxide. It also enhances the phagocytic and antimicrobial activity of neutrophils and macrophages and stimulates secretion of immunoglobulins by lymphocytes (Claesson and Dahlen, 1999; Peters-Golden, 2008). These properties imply a significant role for LTB_4 in the regulation of the immune response and in the pathogenesis of inflammatory diseases, such as arthritis and asthma and also in atherosclerosis (Back et al., 2007). Moreover, LTB_4 induces cell proliferation and promotes cell survival.

The actions of LTB_4 are mediated by the BLT_1 and BLT_2 receptors, which have high homology (36–45%) but distinct tissue distribution and affinity for the ligand (Brink et al., 2003). The BLT_1 receptor is mainly expressed on peripheral leukocytes, macrophages, and DCs and has a high affinity for LTB_4 ($K_d \sim 0.15$ nM). This receptor mediates most of the chemoattractant and proinflammatory actions of LTB_4. Data from different animal models suggest that LTB_4 signaling through this receptor is related to atherogenesis, bronchial asthma, glomerulonephritis, arthritis, and chronic inflammatory bowel diseases (Okuno et al., 2005). The physiological and pathophysiological roles of the BLT_2 receptor are instead not clear, although it has been hypothesized that BLT_2 receptor is responsible for LTB_4 signaling when LTB_4

concentrations are high and BLT_1 might be desensitized. The BLT_2 receptor is ubiquitously expressed, is a lower-affinity receptor for LTB_4 ($K_d \sim 23$ nM), and also binds other eicosanoids (as the cyclooxygenase metabolite 12(S)-hydroxy-5-*cis*-8,10-*trans*-heptadecatrienoic acid, 12-HHT) (Okuno et al., 2008). In addition, LTB_4 binds and activates the peroxisome proliferator-activated receptor-α (PPARα), a transcription factor that regulates genes related to inflammation resolution and lipid degradation, suggesting a pathway for inflammation control (Devchand et al., 1996).

The cys-LTs, foremost LTD_4, induce smooth muscle contraction, mucus secretion, plasma extravasation, vasoconstriction, recruitment of eosinophils, and fibrocyte proliferation (Claesson and Dahlen, 1999). They are thus assumed to play prominent roles in asthma and allergic rhinitis (see below) and also in chronic inflammation and in the regulation of the adaptive immune response (Austen, 2007).

The cys-LTs are recognized by two receptors ($CysLT_1$ and $CysLT_2$), which have been recently cloned, but there are indications for the existence of subclasses of $CysLT_1$ and $CysLT_2$ or of additional CysLT receptor types (as, e.g., the dual-uracil nucleotide-cys-LT receptor GPR17), but conclusive molecular/biological information are still not available (Brink et al., 2003; Capra et al., 2007). Also, a number of experimental evidences indicate the possibility that CysLT receptors might exist as homo- and/or heterodimers, but it is unknown how this can influence their pharmacology and function (Capra et al., 2007). Both $CysLT_1$ and $CysLT_2$ receptors are coupled with a G_q protein. The $CysLT_1$ is expressed in peripheral blood leukocytes, spleen, lung tissue, smooth muscle cells, and macrophages (Lynch et al., 1999), whereas the $CysLT_2$ is expressed more ubiquitously (eosinophils, peripheral blood monocytes, lung macrophages, endothelial cells, etc.). The agonist potency at the $CysLT_1$ receptor is $LTD_4 >> LTC_4 > LTE_4$, whereas LTC_4 and LTD_4 exhibit similar potency at the $CysLT_2$ receptor and LTE_4 is a weak agonist. Vascular leakage, edema in the airways, bronchoconstriction, mucus secretion, DC maturation, and migration are mediated by the $CysLT_1$ receptor, whereas both receptors contribute to macrophage activation, smooth muscle proliferation, and fibrosis. Interestingly, endothelial cell activation by cys-LTs seems to be predominantly related to $CysLT_2$ (Lotzer et al., 2003).

Also, for the monohydro(pero)xylated eicosanoids 5-H(p)ETE and 5-oxo-ETE, the existence of specific GPCRs (termed OXE receptors) has been described (Grant et al., 2009). Although the pathophysiological role of 5-oxo-ETE is not well understood, it may play important roles in asthma and allergic diseases, cancer, and cardiovascular disease.

Lipoxins bind to the FPR2/ALX receptor. They reduce neutrophil chemotaxis, edema, and pain signals and also stimulate nonphlogistic monocyte attraction and macrophage phagocytosis (Brink et al., 2003).

4.7   5-LO AND AIRWAY INFLAMMATION

Because of the multiple biological actions of 5-LO-derived lipid mediators, possible associations have been suggested with several pathological states characterized by inflammatory components (including local and systemic inflammatory diseases). Although large clinical trials are still needed to establish a clear role of LTs in certain diseases [e.g., chronic obstructive pulmonary disease (COPD), atherosclerosis, cardiovascular disease, rheumatoid arthritis, atopic dermatitis, and cancer], a close "mediator-disease" association has been validated for asthma and allergic rhinitis. In fact, the results of large-scale, double-blind, placebo-controlled clinical trials have supported the approval of anti-LT intervention in asthma.

Asthma is a syndrome characterized by a variable degree of airflow obstruction, bronchial hyperresponsiveness, and airway inflammation. Also, airway remodeling occurs, possibly as a result of the chronic airway inflammation, and structural changes entails thickening of airway walls (Busse and Lemanske, 2001). Asthma is however characterized by several clinical phenotypes in both adults and children, and genetic factors or environmental factors (as allergens, exercise, viruses, occupational exposure) may be involved in the onset and development of asthma.

Cys-LTs are important mediators of asthmatic responses. In fact, they are powerful bronchoconstrictors (Dahlen et al., 1980), particularly in small bronchi (inner diameter: 0.5–2 mm), and LTE_4 was more potent than histamine in eliciting a decrease in airflow when administered by inhalation to human subjects (Davidson et al., 1987). Release of cys-LTs is triggered by exposure to inhaled allergens (Manning et al., 1990) or exercise (Kikawa et al., 1992), or in patients with aspirin-sensitive asthma (Cowburn et al., 1998), and elevated levels of LTC_4 were found in bronchoalveolar lavage fluid (BAL) from atopic asthmatics after endobronchial allergen challenge (Wenzel et al., 1990). Also, the urinary excretion of LTE_4 was enhanced in subjects presenting to the emergency room for treatment of asthma exacerbations, indicating a bronchoconstrictor role for cys-LTs in spontaneous acute asthma (Drazen et al., 1992). Interestingly, the induction of allergen responses in passively sensitized airways is not only related to a higher cys-LT release, but also to an enhanced responsiveness of smooth muscle in the airways (Schmidt and Rabe, 2000). Moreover, cys-LTs increase microvascular permeability, stimulate mucus secretion, reduce mucus transport, and promote migration of inflammatory cells, in particular of eosinophils, into the airways. Interestingly, it has been suggested that airway eosinophilia may be a more prominent feature of childhood (atopic) asthma, while asthma in adults is often associated with neutrophilic infiltration (O'Byrne, 2009). Several evidences also indicate a role of cys-LTs in airway structural changes. In fact, they increase fibroblast proliferation and collagen synthesis, and exhaled breath condensate cys-LTs correlate with airway remodeling (Lex et al., 2006). Clinical trials have shown that antileukotriene therapy (with the 5-LO inhibitor zileuton or $CysLT_1$ receptor antagonists; see below) has bronchodilator and anti-inflammatory activities and is effective in reducing asthma symptoms and exacerbations, as monotherapy (in particular in mild asthma or in exercise-induced and aspirin-sensitive asthma) as well as add-on therapy to inhaled corticosteroids (Peters-Golden and Henderson, 2007).

Though studies on the association between LTs and asthma have essentially focused on cys-LTs, several evidences also suggest a role for LTB_4. An increase in the LTB_4 levels has in fact been observed in the sputum, plasma, and BAL of asthmatic patients (O'Driscoll et al., 1984; Shindo et al., 1990; Wardlaw et al., 1989), and ex vivo LTB_4 production by stimulated neutrophils is higher in asthmatic patients than in healthy controls (Pacheco et al., 1992). The involvement of LTB_4 in asthma however remains controversial since a BLT_1 antagonist did not produce clinical benefits in allergen-induced asthma, despite reduction of neutrophil influx in the airways

(Evans et al., 1996). Nonetheless, the LTB_4/BLT_1 pathway may be important in the development of cellular airway inflammation, in relation to the recruitment of early effector T cell and upregulation of effector cell function (Islam et al., 2006). Moreover, the interaction of LTB_4 with its receptors on the surface of airway smooth muscle cells induces migration and proliferation and may therefore contribute to the pathogenesis of airway remodeling (Watanabe et al., 2009).

The overlapping epidemiology and the similar cellular responses involved in the pathophysiology of asthma and allergic rhinitis as well as the finding that LTs are the most important mediators of allergen-induced contractile responses have also suggested a role of these lipid mediators in allergic diseases. High levels of cys-LTs have been found in the nasal secretions of patients with seasonal allergic rhinitis and the cys-LTs contribute to its symptoms. In fact, cys-LTs play a critical role in nasal congestion, which is likely related to an increased vascular permeability and accumulation of blood in the venous sinuses (Kim et al., 2008). Interestingly, a concomitant increase of LTB_4 and higher neutrophil and eosinophil numbers were observed in nasal lavage fluid from patients with allergic rhinitis after allergen challenge (Shaw et al., 1985). Though less effective than nasal corticosteroids, $CysLT_1$ receptor antagonists had some benefits in patients with allergic rhinitis in improving symptoms and quality of life (Wilson et al., 2004) and might be particularly indicated in patients presenting allergic rhinitis in concomitance with asthma, since they may confer beneficial effects in both upper and lower airways (Philip et al., 2004; Wilson et al., 2001).

Recent evidences also suggest LTs as mediators of COPD. COPD is characterized by airway remodeling and infiltration of neutrophils, $CD8^+$ T lymphocytes, and activated macrophages (Saetta et al., 2001). Airway neutrophilia (in particular in the upper airways) significantly contributes to the development of chronic inflammation and airway remodeling. Levels of LTB_4 were higher in the sputum of stable COPD patients and a further increase was observed during exacerbations (Biernacki et al., 2003; Kostikas et al., 2005; Montuschi et al., 2003). Also, LTB_4 correlated with neutrophilic inflammation (Corhay et al., 2009) suggesting a contribution of neutrophil chemotaxis and activation in the upper airways. However, results from clinical studies with a LT synthesis inhibitor (BAY-X1005, a FLAP antagonist) (Gompertz and Stockley, 2002) or BLT_1 receptor antagonist (Gronke et al., 2008) indicated only modest reduction of neutrophilic bronchial inflammation in patients with COPD. Promising results were instead obtained from preliminary studies with antagonists of cys-LT receptors, and an improvement of COPD control has been observed after long-term montelukast therapy in elderly patients with moderate to severe COPD (Rubinstein et al., 2004).

4.8   ANTILEUKOTRIENE DRUGS

Because LTs and other 5-LO-derived products have several pathophysiological functions and hence implicate potential benefit of an anti-LT therapy in a variety of diseases, strong efforts have been made within the past 30 years in order to develop selective and potent pharmacological agents that intervene with LTs. In principle, two different strategies can be distinguished to reach this aim: (1) inhibition of the biosynthesis of LTs by applying inhibitors of enzymatic reactions and (2) inhibition of the action of LTs by application of LT receptor antagonists.

To achieve reduction of LT formation, possible targets include PLA_2 enzymes, 5-LO, FLAP, LTA_4 hydrolase, and LTC_4 synthase/MAPEGs, with 5-LO being the preferred target. The first inhibitors of LT biosynthesis that were described were synthetic derivatives of LTs or prostaglandins (Orning and Hammarstrom, 1980; Bokoch and Reed, 1981). Although experiments in $cPLA_2$-deficient mice support the concept that *inhibition of PLA_2* enzymes prevents the formation of basically all eicosanoids (Uozumi et al., 1997), clinical studies using glucocorticoids that suppress $cPLA_2$ enzymes proved to be ineffective in reducing the levels of LTs (Claesson and Dahlen, 1999). Also, the recently reported beneficial effects of selective $cPLA_2$ inhibitors in experimental rodent models of arthritis (Tai et al., 2010), acute lung injury (Nagase et al., 2003), or ischemia–reperfusion injury (Bellido-Reyes et al., 2006) were always associated with the suppression of several mediators, cytokines, and proteases and thus, could not be clearly related to reduced levels of LTs.

The FLAP inhibitors may act as antagonists of FLAP by interfering with AA substrate transfer, and thus cause reduced availability of AA for 5-LO at the nuclear membrane as suggested by the recent elucidation of the structure of FLAP in complex with two different inhibitors (Ferguson et al., 2007). Such compounds are highly efficient in intact cells, in particular when 5-LO products are formed from endogenously derived AA (Werz, 2002). In contrast, AA supply from exogenous sources impairs the efficacy, and in cell-free systems, FLAP inhibitors fail to inhibit 5-LO product formation. Although the FLAP inhibitors MK886, MK0591, or BAY-X1005 (Figure 4.3) are efficient in isolated leukocytes, the drugs are 50- to 200-fold less potent in whole-blood assays, possibly due to plasma protein binding of the drugs and/or competition with AA and other *cis*-unsaturated fatty acids present in plasma. Several FLAP inhibitors (e.g., MK-886, MK-591, BAY-X1005) showed efficacy in

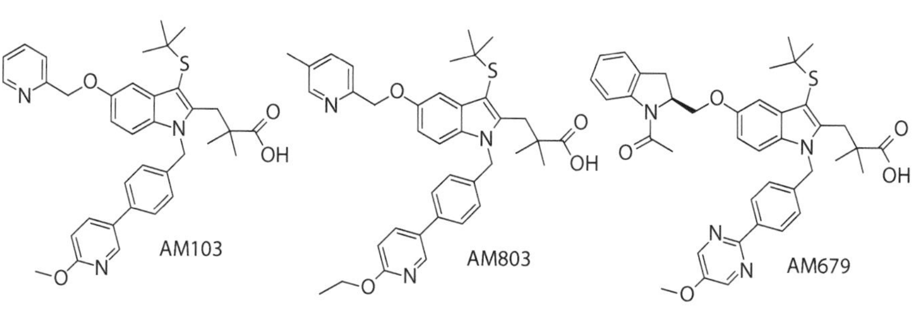

FIGURE 4.3 Inhibitors of FLAP.

early clinical trials in asthma (Evans et al., 2008) but were not developed commercially for unpublished reasons. However, recent developments led to AM103, AM803, and AM679 (Figure 4.3) that are novel, potent, and selective FLAP inhibitors with excellent pharmacodynamic properties *in vivo* and effectiveness in animal models of acute and chronic inflammation (Lorrain et al., 2009; Hutchinson et al., 2009; Lorrain et al., 2010; Bain et al., 2010). Ongoing clinical studies with these compounds will reveal their therapeutic potential. The interested reader is referred to an excellent review on FLAP inhibitors by Evans et al. (2008).

4.8.1 DIRECT 5-LO INHIBITORS

Most of the compounds that interfere with LT synthesis are direct 5-LO inhibitors, which advantageously block the formation of both LTB_4 and cys-LTs as well as the synthesis of 5-H(P)ETE. These direct 5-LO inhibitors are classified according to their molecular mode of action as:

1. Redox-active 5-LO inhibitors
2. Iron-ligand inhibitors
3. Nonredox-type 5-LO inhibitors
4. Compounds inhibiting 5-LO with unrecognized mechanism

4.8.1.1 Redox-Active 5-LO Inhibitors

Redox-active 5-LO inhibitors comprise lipophilic reducing agents, and in fact the first synthetic 5-LO inhibitors belong to the class of redox-type 5-LO inhibitors, represented by AA-861, phenidone, BW755C, or L-656,224 (Ford-Hutchinson et al., 1994; Yoshimoto et al., 1982) (Figure 4.4). These compounds act by keeping the active-site iron in the ferrous state, thereby uncoupling the catalytic cycle of the enzyme and are highly efficient inhibitors of 5-LO product formation *in vitro* and partially also *in vivo*. However, most of them lack suitable oral bioavailability, possess only poor selectivity for 5-LO, and thus, exert severe side effects (e.g., methemoglobin formation) due to interference with other biological redox systems or by the production of reactive radical species (McMillan and Walker, 1992). These detrimental features hampered the substances from entering the market and the development of redox-type 5-LO inhibitors have been essentially stopped. Nevertheless, among this class of 5-LO inhibitors, there are many plant-derived compounds with well-recognized anti-inflammatory properties such as flavonoids, coumarins, quinones, lignans, and other polyphenols administered through daily food.

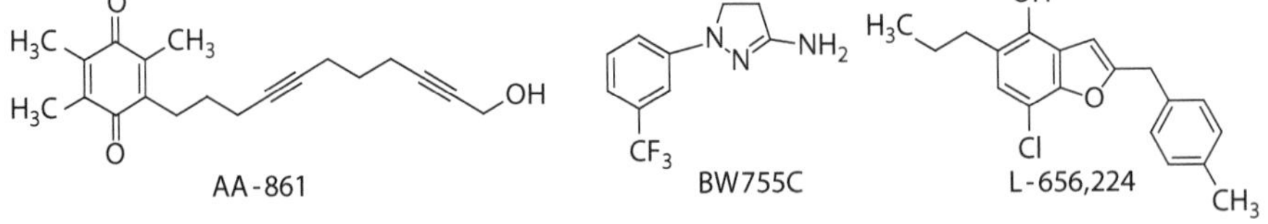

FIGURE 4.4 Redox-type 5-LO inhibitors.

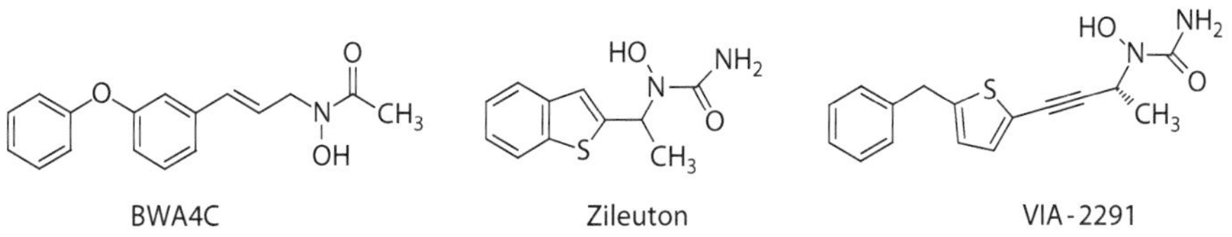

FIGURE 4.5 Iron ligand-type 5-LO inhibitors.

4.8.1.2 Iron-Ligand Inhibitors

Iron-ligand inhibitors chelate the active-site iron via a hydroxamic acid or an *N*-hydroxyurea moiety and also exert weak reducing properties. BWA4C, a hydroxamic acid, and zileuton, a hydrolytic stable *N*-hydroxyurea derivative (Figure 4.5), belong to this class of potent and orally active 5-LO inhibitors (McMilan and Walker, 1992). An advanced candidate is VIA-2291 (Figure 4.5), which is up to fivefold more potent than zileuton in animal models of bronchospasm with an oral half-life of 16 h and showed efficacy in exercise-induced bronchoconstriction in asthmatic patients (Brooks et al., 1995; Lehnigk et al., 1998). VIA-2291 has successfully completed phase II clinical trials for atherosclerosis and cardiovascular disease and is one of the leading 5-LO inhibitors in clinical development (Back, 2009).

4.8.1.3 Nonredox-Type 5-LO Inhibitors

The nonredox-type 5-LO inhibitors compete with AA or LOOH for binding to 5-LO without redox properties and encompass structurally diverse molecules. It is still unclear if the binding site of these compounds is in fact the AA substrate-binding cleft in the active site. Thus, experimental data from molecular and biochemical studies suggest an allosteric mode of action (Werz et al., 1998). Nonetheless, representatives of this class such as the orally active compounds ZD 2138 or its ethyl analog ZM 230487, L-739,010, or CJ-13,610 (Figure 4.6) are highly potent and selective for 5-LO in cellular assays (Werz and Steinhilber, 2005). We found that elevated peroxide levels and/or phosphorylation of 5-LO by MAPKAPK-2 and/or

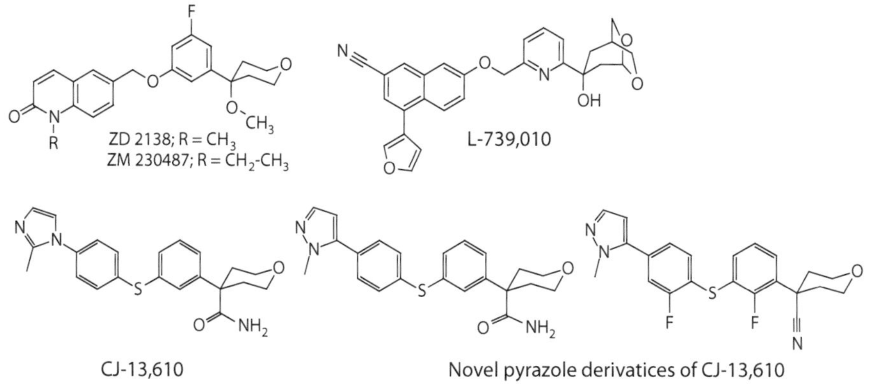

FIGURE 4.6 Nonredox-type 5-LO inhibitors with tetrahydropyrane structure.

ERKs strongly impaired the potency of nonredox-type 5-LO inhibitors in activated PMNL (Werz et al., 1998; Fischer et al., 2004; Fischer et al., 2003). Unfortunately, ZD2138, ZM230487, and L-739,010 showed inappropriate features in preclinical or clinical studies (toxicity, inefficacy), which stopped further development. CJ-13,610 (Figure 4.6), a more recent compound developed by Pfizer, was shown to be orally active and efficacious in preclinical models of pain, including the acute carrageenan model, the chronic inflammatory model using complete Freund's adjuvant, and against osteoarthritis like pain (Cortes-Burgos et al., 2009). Based on the structure of CJ-13,610, a series of novel pyrazoles (Figure 4.6) were disclosed that efficiently blocked LTB$_4$ formation in the submicromolar range in human whole blood and that suppressed carrageenan-induced LTB$_4$ formation in the rat air pouch model. Similarly, pyrazole derivatives where the carboxamide was exchanged by carbonitrile and carrying fluorinated benzenes (Figure 4.6) showed even improved potency (Masferrer et al., 2010). Clinical trials with these substances as antiasthmatics are ongoing.

Substituted, less toxic coumarins based on the structure of L-739,010 were obtained by replacement of the dioxabicyclooctanyl moiety by a hexafluorcarbinol substituent and exchange of the furyl-cyano-naphthyl moiety against a fluorophenyl-substituted coumarin (Figure 4.7). These derivatives exhibit comparable *in vitro* potency and high anti-inflammatory efficacy *in vivo* but devoid of the toxic side effects (i.e., covalent protein binding) as the former representatives (Grimm et al., 2006). The coumarin moiety could be exchanged against 2-cyanoquinoline (LTD MFc WO 2007a/038865), indoles, or benzothiophenes carrying an 1,2,3-triazolyl group (LTD MFc WO 2007b/016784) (Figure 4.7). Potent inhibition of 5-LO in cell-free assays as well as in human whole blood was demonstrated for related fluorophenyl-substituted coumarins where the thioaryl moiety carrying the hexafluorcarbinol was replaced by an amino-oxadiazol moiety (Co M WO 2009/042098).

The urea derivative RBx 7796 (Figure 4.7) with a dodecyl chain was characterized as competitive and selective, orally active 5-LO inhibitor with IC$_{50}$ values of

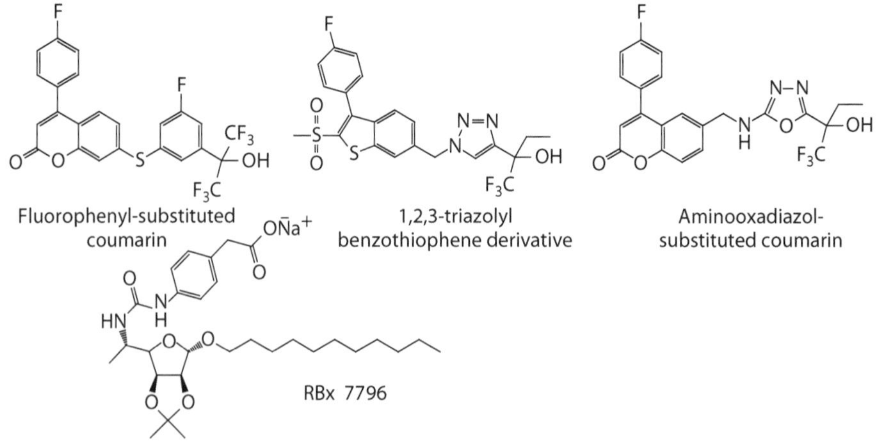

Fluorophenyl-substituted coumarin

1,2,3-triazolyl benzothiophene derivative

Aminooxadiazol-substituted coumarin

RBx 7796

FIGURE 4.7 Miscellaneous nonredox-type 5-LO inhibitors.

3.8 and 5 μM in cell-free and cell-based models, respectively (Shirumalla et al., 2006). The efficiency was independent of the redox status. The compound suppressed LTB_4 synthesis *ex vivo* in rat blood and was active in animal models of inflammation and of bronchoconstriction (Shirumalla et al., 2006). RBx 7796 was stable in liver microsomes and devoid of major cytochrome P450 inhibition potential, and showed sufficient oral bioavailability without unwanted metabolic features (Shirumalla et al., 2008).

4.8.1.4 Compounds Inhibiting 5-LO with Unrecognized Mechanism

Numerous compounds inhibiting 5-LO with unrecognized mechanism were identified in recent years. Many of these compounds act not solely on 5-LO, but also on other relevant targets, including the 12- and 15-LOs, COX enzymes, FLAP, or the microsomal prostaglandin E_2 synthase (mPGES)-1 (Koeberle and Werz, 2009; Pergola and Werz, 2010). Such multitarget properties are actually not surprising because all these enzymes share related substrates, that is, AA or metabolites thereof that may bind to a common, more or less conserved, fatty acid-binding pocket at the active site. Some of these compounds might have promising molecular mechanisms and favorable pharmacodynamics. For example, compounds that inhibit both COX and 5-LO pathways may possess higher anti-inflammatory efficacy accompanied by lower gastric toxicity in comparison with traditional COX inhibitors (Celotti and Laufer, 2001). Among such dual 5-LO/COX pathway inhibitors, the NSAID licofelone (Figure 4.8) has reached clinical phase III for the treatment of osteoarthritis (Kulkarni and Singh, 2008). Recent investigations indicated that licofelone primarily targets FLAP (Fischer et al., 2007), COX-1, and mPGES-1 (Koeberle et al., 2008a) but not 5-LO or COX-2. Along these lines, sulindac sulfide, a clinically used NSAID assumed to act by inhibition of COX enzymes, was found to inhibit 5-LO in various test systems at clinically relevant concentrations (Steinbrink et al., 2010). Finally, the COX-2 selective celecoxib (Figure 4.8) was also identified as an inhibitor of 5-LO (Maier et al., 2008) and thus, can be classified as dual COX/5-LO inhibitor. Celecoxib suppressed 5-LO product formation in human neutrophils and in human whole blood and inhibited 5-LO in cell-free assays in the low micromolar range, and reduced the blood LTB_4 levels in mice after oral administration (Maier et al., 2008). Derivatization of rofecoxib (inactive on 5-LO) by incorporation of a para-oxime moiety on the C3 phenyl ring (Figure 4.8) led to dual inhibitors of 5-LO and COX-2 with $IC_{50} = 1.4$ and 0.28 μM, respectively,

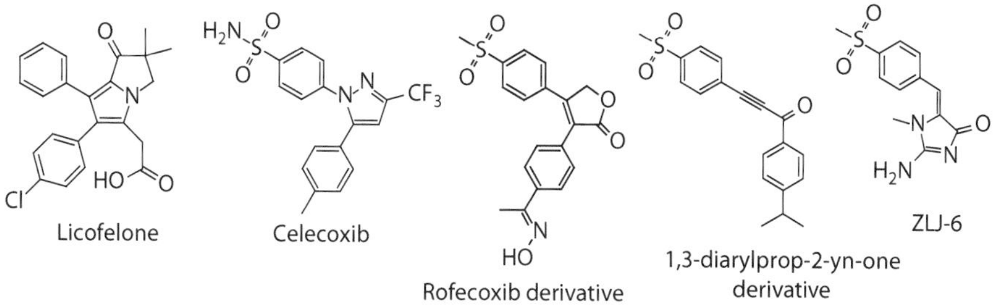

Licofelone   Celecoxib   Rofecoxib derivative   1,3-diarylprop-2-yn-one derivative   ZLJ-6

FIGURE 4.8 Dual 5-LO/COX pathway inhibitors.

connected to excellent anti-inflammatory activity in rats (Chen et al., 2006). A series of 1,3-diarylprop-2-yn-ones with a C3 p-SO$_2$Me COX-2 pharmacophore (Figure 4.8) was reported to inhibit of COX-1/2 and 5/15-LO with IC$_{50}$ values = 0.32–9 μM with good oral anti-inflammatory activity (ED$_{50}$ = 35 mg/kg, compared to celecoxib: ED$_{50}$ = 10.8 mg/kg) (Rao et al., 2006). Similarly, linking the p-SO$_2$Me moiety to an imidazol-4-one backbone (Figure 4.8), led to dual COX/5-LO inhibitors (IC$_{50}$ values 0.31–2.59 μM in cell-free and cell-based assays) with anti-inflammatory activity in rodents (Li et al., 2009).

Targeting 5-LO and mPGES-1 is assumed to be even superior to dual 5-LO/COX-2 inhibition because the cardiovascular risk of COX-2-selective drugs that negatively affect the prostacyclin/thromboxane balance might be reduced (Koeberle and Werz, 2009). Dual 5-LO/mPGES-1 inhibitors (Figure 4.9) include novel series of α-(n)-alkyl-substituted pirinixic acids (Koeberle et al., 2008b) and of benzo[g]indole-3-carboxylates (Karg et al., 2009) that exhibited anti-inflammatory efficacy in rats. But plant-derived natural compounds (e.g., curcumin from tumeric, myrtucommulone from myrtle, garcinol from Guttiferae species, and epigallocatechin-3-gallate from green tea) were also revealed to primarily act on 5-LO and mPGES-1 with minor effects on COX enzymes (Koeberle and Werz, 2009). Note that many of these natural compounds have been described before to suppress PGE$_2$ formation in cell-based assays or *in vivo*, but the identification of mPGES-1 as target was still missed. Moreover, the above-mentioned natural products or extracts of the respective plants possess well-known anti-inflammatory activity as demonstrated in various preclinical or clinical studies.

Finally, hyperforin (Figure 4.10), a polyprenylated acylphloroglucinol from *Hypericum perforatum* (St. John's wort), was identified as a dual inhibitor of COX-1 and 5-LO in the low micromolar range (Lehnigk et al., 1998). The compound may act by interference with the 5-LO C2-like domain, since 5-LO inhibition was abolished in the presence of PC, strongly reduced by mutation (W13A-W75A-W102A) of the C2-like domain, and the interaction of 5-LO with CLP and membrane translocation in activated neutrophils was impaired by hyperforin (Back, 2009). Such interference with the C2 domain represents a novel molecular approach for the design of a new class of selective and effective 5-LO inhibitors. In fact, in carrageenan-treated rats, hyperforin significantly suppressed LTB$_4$ levels in pleural exudates associated with potent anti-inflammatory effectiveness (Feisst et al., 2009).

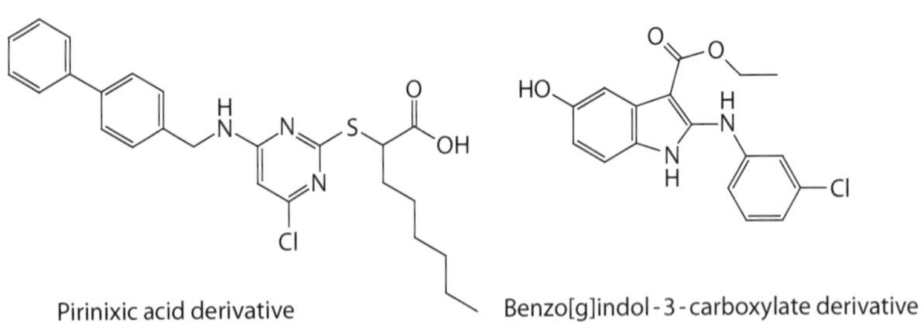

Pirinixic acid derivative   Benzo[g]indol-3-carboxylate derivative

FIGURE 4.9   Dual 5-LO/mPGES-1 inhibitors.

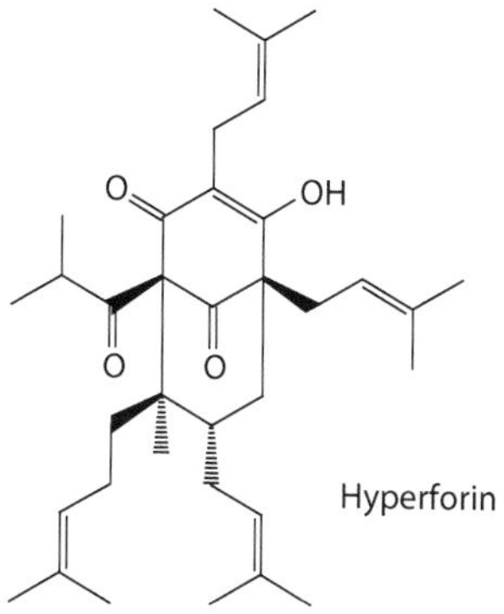

Hyperforin

FIGURE 4.10 Hyperforin, an inhibitor, acting at the 5-LO C2-like domain.

4.9 EXPERT OPINION

5-LO products exhibit proinflammatory and proallergic actions, and function in the host immune response, regulation of cell growth and apoptosis, and vascular homeostasis. The efficacy of CysLT receptor antagonists (e.g., montelukast) in asthma and allergic rhinitis validates the important role of cys-LTs in these diseases, whereas larger clinical trials are still needed to confirm a role of LTs in COPD, artherosclerosis, cardiovascular disease, rheumatoid arthritis, atopic dermatitis, and cancer. These conclusions are partly supported by data obtained from 5-LO- and FLAP-deficient mice. Thus, the development of anti-LT drugs remains a major challenge and inhibition of 5-LO may have a higher efficiency and broader spectrum of action than cys-LT receptor antagonism or inhibition of LTA_4 hydrolase and LTC_4 synthase. The recent discovery of a gender bias in LT formation, which correlates to the differences in the incidence of several LT-related diseases, provides novel aspects suggesting predominant therapeutic use of anti-LTs for treatment of female patients. Whether or not the gender difference in the regulation of LT biosynthesis can also affect the response of males and females to anti-LT drugs is unclear and remains a challenging issue. In the past 25 years, several types of compounds that intervene with LT biosynthesis (i.e., redox-, iron ligand-, and nonredox-type 5-LO inhibitors as well as FLAP inhibitors) have been developed and several preclinical and clinical studies were performed. Despite this intensive research, until today only one 5-LO inhibitor (zileuton) could enter the market and none of the FLAP inhibitors that had been the subject of intensive studies in advanced clinical phases could receive marketing authorization.

One of the leading 5-LO inhibitors in clinical development is currently represented by VIA-2291, a compound derived from the structural optimization of zileuton. Also, the demonstration of the *in vivo* activity of the tetrahydropyrane carboxamide CJ-13,610 as well as its refinement leading to related tetrahydropyranes with pyrazole and carbonitrile moieties with improved solubility, oral absorption, pharmacokinetics, and toxicological characteristics is a successful step forward. Promising scientific studies during the past 5 years have revealed new concepts for the pharmacological intervention with 5-LO (e.g., dual inhibition strategies) and valuable compounds were developed that may motivate for future research, such as novel celecoxib or rofecoxib derivatives that dually inhibit COX-2 and 5-LO. However, COX-2 inhibitors are limited by cardiovascular toxicity and the dual targeting of 5-LO and mPGES-1 has been purposed as alternative approach. Licofelone represents such a candidate that primarily inhibits FLAP, mPGES-1, and COX-1 but not COX-2, implying cardiovascular safety based on a favorable thromboxane/prostacyclin balance. Besides several natural anti-inflammatory compounds that suppress mPGES-1 and 5-LO, there are novel developments (e.g., pirinixic acid derivatives and benzo[g]indole-3-carboxylates) that act as dual 5-LO/mPGES-1 inhibitors with clear efficiency *in vivo*. Through the use of hyperforin, targeting of the regulatory C2 domain of 5-LO was suggested as a novel molecular strategy for inhibiting 5-LO activity. In this regard, the recent elucidation of the 5-LO structure (Gilbert et al., 2011) will allow the rational development of novel molecular strategies for intervention with 5-LO, in particular because the increasing knowledge of the manifold roles of LTs in various pathophysiologies stimulate the demand for effective anti-LT drugs.

ACKNOWLEDGMENTS

The authors acknowledge the financial support from the Deutsche Forschungs–gemeinschaft. Carlo Pergola received a Carl Zeiss stipend.

REFERENCES

Albert, D., C. Pergola, A. Koeberle, G. Dodt, D. Steinhilber, and O. Werz. 2008. The role of diacylglyceride generation by phospholipase D and phosphatidic acid phosphatase in the activation of 5-lipoxygenase in polymorphonuclear leukocytes. *J Leukoc Biol.* 83:1019–27.

Allard, J. B. and T. G. Brock. 2005. Structural organization of the regulatory domain of human 5-lipoxygenase. *Curr Protein Pept Sci.* 6:125–31.

Austen, K. F. 2007. Additional functions for the cysteinyl leukotrienes recognized through studies of inflammatory processes in null strains. *Prostaglandins Other Lipid Mediat.* 83:182–7.

Back, M. 2009. Inhibitors of the 5-lipoxygenase pathway in atherosclerosis. *Curr Pharm Des.* 15:3116–32.

Back, M., K. Sakata, H. Qiu, Haeggstrom, J. Z., and Dahlen, S. E. 2007. Endothelium-dependent vascular responses induced by leukotriene B4. *Prostaglandins Other Lipid Mediat.* 83:209–12.

Bain, G., C. D. King, M. Rewolinski, K. Schaab, A. M. Santini, D. Shapiro, M. Moran et al. 2010. Pharmacodynamics and pharmacokinetics of AM103, a novel inhibitor of 5-lipoxygenase-activating protein (FLAP). *Clin Pharmacol Ther.* 87:437–44.

Bannenberg, G. L. 2009. Resolvins: Current understanding and future potential in the control of inflammation. *Curr Opin Drug Discov Dev.* 12:644–58.

Bellido-Reyes, Y. A., H. Akamatsu, K. Kojima, H. Arai, H. Tanaka, and M. Sunamori. 2006. Cytosolic phospholipase A2 inhibition attenuates ischemia-reperfusion injury in an isolated rat lung model. *Transplantation* 81:1700–7.

Berg, C., S. Hammarstrom, H. Herbertsson, E. Lindstrom, A. C. Svensson, M. Soderstrom, P. Tengvall, and T. Bengtsson. 2006. Platelet-induced growth of human fibroblasts is associated with an increased expression of 5-lipoxygenase. *Thromb Haemost.* 96:652–9.

Biernacki, W. A., S. A. Kharitonov, and P. J. Barnes. 2003. Increased leukotriene B4 and 8-isoprostane in exhaled breath condensate of patients with exacerbations of COPD. *Thorax* 58:294–8.

Bokoch, G. M. and P. W. Reed. 1981. Evidence for inhibition of leukotriene A4 synthesis by 5,8,11,14-eicosatetraenoic acid in guinea pig polymorphonuclear leukocytes. *J Biol Chem.* 256:4156–9.

Borgeat, P., Hamberg, M. and B. Samuelsson. 1976. Transformation of arachidonic acid and homo-gamma-linolenic acid by rabbit polymorphonuclear leukocytes. Monohydroxy acids from novel lipoxygenases. *J Biol Chem.* 251:7816–20.

Borgeat, P. and B. Samuelsson. 1979. Metabolism of arachidonic acid in polymorphonuclear leukocytes. Structural analysis of novel hydroxylated compounds. *J Biol Chem.* 254:7865–9.

Brink, C., S. E. Dahlen, J. Drazen, J. F. Evans, D. W. Hay, S. Nicosia, C. N. Serhan, T. Shimizu, and T. Yokomizo. 2003. International Union of Pharmacology XXXVII. Nomenclature for leukotriene and lipoxin receptors. *Pharmacol Rev.* 55:195–227.

Brooks, C. D. W., A. O. Stewart, A. Basha, P. Bhatia, J. D. Ratajczyk, J. G. Martin, R. A. Craig, T. Kolasa, J.B. Bouska, and C. Lanni. 1995. *R*-(+)-*N*-[3-[5-(4-Fluorophenyl)methyl]-2-thienyl]-1-methyl-2-propynyl]-*N*-hydroxyurea (ABT-761), a second generation 5-lipoxygenase inhibitor. *J Med Chem.* 38:4768–75.

Busse, W. W. and R. F. Lemanske, Jr. 2001. Asthma. *N Engl J Med.* 344:350–62.

Capra, V., M. D. Thompson, and A. Sala. 2007. Cysteinyl-leukotrienes and their receptors in asthma and other inflammatory diseases: Critical update and emerging trends. *Med Res Rev.* 27:469–27.

Casas, J., C. Meana, E. Esquinas, M. Valdearcos, J. Pindado, J. Balsinde, and M. A. Balboa. 2009. Requirement of JNK-mediated phosphorylation for translocation of group IVA phospholipase A2 to phagosomes in human macrophages. *J Immunol.* 183:2767–74.

Celotti, F. and S. Laufer. 2001. Anti-inflammatory drugs: New multitarget compounds to face an old problem. The dual inhibition concept. *Pharmacol Res.* 43:429–36.

Channon, J. Y. and C. C. Leslie. 1990. A calcium-dependent mechanism for associating a soluble arachidonoyl-hydrolyzing phospholipase A2 with membrane in the macrophage cell line RAW 264.7. *J Biol Chem.* 265:5409–13.

Chen, Q. H., P. N. Rao, and E. E. Knaus. 2006. Synthesis and biological evaluation of a novel class of rofecoxib analogues as dual inhibitors of cyclooxygenases (COXs) and lipoxygenases (LOXs). *Bioorg Med Chem.* 14:7898–909.

Claesson, H. E. and S. E. Dahlen. 1999. Asthma and leukotrienes: Antileukotrienes as novel anti-asthmatic drugs. *J Intern Med.* 245:205–27.

Clark, J. D., L. L. Lin, R. W. Kriz, C. S. Ramesha, L. A. Sultzman, A. Y. Lin, N. Milona, and J. L. Knopf. 1991. A novel arachidonic acid-selective cytosolic PLA2 contains a Ca(2+)-dependent translocation domain with homology to PKC and GAP. *Cell* 65:1043–51.

Co, M. 2009. Chromenone derivatives and their use for leukotriene biosynthesis inhibition. WO 2009/042098.

Corhay, J. L., M. Henket, D. Nguyen, B. Duysinx, J. Sele, and R. Louis. 2009. Leukotriene B4 contributes to exhaled breath condensate and sputum neutrophil chemotaxis in COPD. *Chest* 136:1047–54.

Cortes-Burgos, L. A., B. S. Zweifel, S. L. Settle, R. A. Pufahl, G. D. Anderson, M. M. Hardy, D. E. Weir et al. 2009. CJ-13610, an orally active inhibitor of 5-lipoxygenase is efficacious in preclinical models of pain. *Eur J Pharmacol.* 617:59–67.

Cowburn, A. S., K. Sladek, J. Soja, L. Adamek, E. Nizankowska, A. Szczeklik, B. K. Lam et al. 1998. Overexpression of leukotriene C4 synthase in bronchial biopsies from patients with aspirin-intolerant asthma. *J Clin Invest.* 101:834–46.

Dahlen, S. E., P. Hedqvist, S. Hammarstrom, and B. Samuelsson 1980. Leukotrienes are potent constrictors of human bronchi. *Nature* 288:484–6.

Davidson, A. B., T. H. Lee, P. D. Scanlon, J. Solway, E. R. McFadden, Jr., R. H. Ingram, Jr., E. J. Corey, K. F. Austen, and J. M. Drazen. 1987. Bronchoconstrictor effects of leukotriene E4 in normal and asthmatic subjects. *Am Rev Respir Dis.* 135:333–7.

Devchand, P. R., H. Keller, J. M. Peters, M. Vazquez, F. J. Gonzalez, and W. Wahli 1996. The PPARalpha-leukotriene B4 pathway to inflammation control. *Nature* 384:39–43.

Drazen, J. M., J. O'Brien, D. Sparrow, S. T. Weiss, M. A. Martins, E. Israel, and C. H. Fanta. 1992. Recovery of leukotriene E4 from the urine of patients with airway obstruction. *Am Rev Respir Dis.* 146:104–8.

Esser, J., M. Rakonjac, B. Hofmann, L. Fischer, P. Provost, G. Schneider, D. Steinhilber, B. Samuelsson, and O. Radmark. 2009. Coactosin-like protein functions as a stabilizing chaperone for 5-lipoxygenase: Role of tryptophan 102. *Biochem J.* 425:265–74.

Evans, D. J., P. J. Barnes, S. M. Spaethe, E. L. van Alstyne, M. I. Mitchell, and B. J. O'Connor. 1996. Effect of a leukotriene B4 receptor antagonist, LY293111, on allergen induced responses in asthma. *Thorax* 51:1178–84.

Evans, J. F., A. D. Ferguson, R. T. Mosley, and J. H. Hutchinson. 2008. What's all the FLAP about?: 5-Lipoxygenase-activating protein inhibitors for inflammatory diseases. *Trends Pharmacol Sci.* 29:72–8.

Feisst, C., C. Pergola, M. Rakonjac, A. Rossi, A. Koeberle, G. Dodt, M. Hoffmann et al. 2009. Hyperforin is a novel type of 5-lipoxygenase inhibitor with high efficacy in vivo. *Cell Mol Life Sci.* 66:2759–71.

Ferguson, A. D., B. M. McKeever, S. Xu, D. Wisniewski, D. K. Miller, T. T. Yamin, R. H. Spencer et al. 2007. Crystal structure of inhibitor-bound human 5-lipoxygenase-activating protein. *Science* 317:510–2.

Fischer, L., M. Hornig, C. Pergola, N. Meindl, L. Franke, Y. Tanrikulu, G. Dodt, G. Schneider, D. Steinhilber, and O. Werz. 2007. The molecular mechanism of the inhibition by licofelone of the biosynthesis of 5-lipoxygenase products. *Br J Pharmacol.* 152:471–80.

Fischer, L., D. Steinhilber, and O. Werz. 2004. Molecular pharmacological profile of the nonredox-type 5-lipoxygenase inhibitor CJ-13,610. *Br J Pharmacol.* 142:861–8.

Fischer, L., D. Szellas, O. Radmark, D. Steinhilber, and O. Werz 2003. Phosphorylation- and stimulus-dependent inhibition of cellular 5-lipoxygenase activity by nonredox-type inhibitors. *FASEB J.* 17:949–51.

Folco, G. and R. C. Murphy. 2006 Eicosanoid transcellular biosynthesis: From cell-cell interactions to *in vivo* tissue responses. *Pharmacol Rev.* 58:375–88.

Ford-Hutchinson, A. W., M. Gresser, and R. N. Young. 1994. 5-Lipoxygenase. *Annu Rev Biochem.* 63:383–17.

Funk, C. D. 1996. The molecular biology of mammalian lipoxygenases and the quest for eicosanoid functions using lipoxygenase-deficient mice. *Biochim Biophys Acta.* 1304:65–84.

Gilbert, N. C., S. G. Bartlett, M. T. Waight, D. B. Neau, W. E. Boeglin, A. R. Brash, and M. E. Newcomer. 2011. The structure of human 5-lipoxygenase. *Science* 331:217–19.

Gompertz, S. and R. A. Stockley. 2002. A randomized, placebo-controlled trial of a leukotriene synthesis inhibitor in patients with COPD. *Chest* 122:289–94.

Grant, G. E., J. Rokach, and W. S. Powell. 2009. 5-Oxo-ETE and the OXE receptor. *Prostaglandins Other Lipid Mediat.* 89:98–104.

Grimm, E. L., C. Brideau, N. Chauret, C. C. Chan, D. Delorme, Y. Ducharme, D. Ethier et al. 2006. Substituted coumarins as potent 5-lipoxygenase inhibitors. *Bioorg Med Chem Lett.* 16:2528–31.

Gronke, L., K. M. Beeh, R. Cameron, O. Kornmann, J. Beier, M. Shaw, O. Holz, R. Buhl, H. Magnussen, and R. A. Jorres 2008. Effect of the oral leukotriene B4 receptor antagonist LTB019 on inflammatory sputum markers in patients with chronic obstructive pulmonary disease. *Pulm Pharmacol Ther.* 21:409–17.

Gupta, S., M. Srivastava, N. Ahmad, K. Sakamoto, D. G. Bostwick, and H. Mukhtar 2001. Lipoxygenase-5 is overexpressed in prostate adenocarcinoma. *Cancer* 91:737–43.

Haeggstrom, J. Z. 2000. Structure, function, and regulation of leukotriene A4 hydrolase. *Am J Respir Crit Care Med.* 161:S25–31.

Hammarberg, T., P. Provost, B. Persson, and O. Radmark. 2000. The N-terminal domain of 5-lipoxygenase binds calcium and mediates calcium stimulation of enzyme activity. *J Biol Chem.* 275:38787–93.

Haribabu, B., M. W. Verghese, D. A. Steeber, D. D. Sellars, C. B. Bock, and R. Snyderman. 2000. Targeted disruption of the leukotriene B(4) receptor in mice reveals its role in inflammation and platelet-activating factor-induced anaphylaxis. *J Exp Med.* 192:433–8.

Hefner, Y., A. G. Borsch-Haubold, M. Murakami, J. I. Wilde, S. Pasquet, D. Schieltz, F. Ghomashchi et al. 2000. Serine 727 phosphorylation and activation of cytosolic phospholipase A2 by MNK1-related protein kinases. *J Biol Chem.* 275:37542–51.

Hirabayashi, T., T. Murayama, and T. Shimizu. 2004. Regulatory mechanism and physiological role of cytosolic phospholipase A2. *Biol Pharm Bull.* 27:1168–73.

Hutchinson, J. H., Y. Li, J. M. Arruda, C. Baccei, G. Bain, C. Chapman, L. Correa et al. 2009. 5-Lipoxygenase-activating protein inhibitors: Development of 3-[3-*tert*-butylsulfanyl-1-[4-(6-methoxy-pyridin-3-yl)-benzyl]-5-(pyridin-2-ylmethoxy)-1H-indol-2-yl]-2,2-dimethyl-propionic acid (AM103). *J Med Chem.* 52:5803–15.

Islam, S. A., S. Y. Thomas, C. Hess, B. D. Medoff, T. K. Means, C. Brander, C. M. Lilly, A. M. Tager, and A. D. Luster. 2006. The leukotriene B4 lipid chemoattractant receptor BLT1 defines antigen-primed T cells in humans. *Blood* 107:444–53.

Karg, E. M., S. Luderer, C. Pergola, U. Buhring, A. Rossi, H. Northoff, L. Sautebin, R. Troschutz, and O. Werz. 2009. Structural optimization and biological evaluation of 2-substituted 5-hydroxyindole-3-carboxylates as potent inhibitors of human 5-lipoxygenase. *J Med Chem.* 52:3474–83.

Kellaway, C. and E. R. Trethewie. 1940. The liberation of a slow-reacting smooth muscle-stimulating substance in anaphylaxis. *Q J Exp Physiol.* 30:121–45.

Kikawa, Y., T. Miyanomae, Y. Inoue, M. Saito, A. Nakai, Y. Shigematsu, S. Hosoi, and M. Sudo. 1992. Urinary leukotriene E4 after exercise challenge in children with asthma. *J Allergy Clin Immunol.* 89:1111–9.

Kim, H., J. Bouchard, and P. M. Renzi. 2008. The link between allergic rhinitis and asthma: A role for antileukotrienes? *Can Respir J.* 15:91–8.

Koeberle, A., U. Siemoneit, U. Buhring, H. Northoff, S. Laufer, W. Albrecht, and O. Werz. 2008a. Licofelone suppresses prostaglandin E2 formation by interference with the inducible microsomal prostaglandin E2 synthase-1. *J Pharmacol Exp Ther.* 326:975–82.

Koeberle, A. and O. Werz. 2009. Inhibitors of the microsomal prostaglandin E(2) synthase-1 as alternative to non steroidal anti-inflammatory drugs (NSAIDs)—A critical review. *Curr Med Chem.* 16:4274–96.

Koeberle, A., H. Zettl, C. Greiner, M. Wurglics, M. Schubert-Zsilavecz, and O. Werz. 2008b. Pirinixic acid derivatives as novel dual inhibitors of microsomal prostaglandin E2 synthase-1 and 5-lipoxygenase. *J Med Chem.* 51:8068–76.

Kostikas, K., M. Gaga, G. Papatheodorou, T. Karamanis, D. Orphanidou, and S. Loukides. 2005. Leukotriene B4 in exhaled breath condensate and sputum supernatant in patients with COPD and asthma. *Chest* 127:1553–9.

Kramer, R. M., E. F. Roberts, S. L. Um, A. G. Borsch-Haubold, S. P. Watson, M. J. Fisher, and J. A. Jakubowski. 1996. p38 Mitogen-activated protein kinase phosphorylates cytosolic phospholipase A2 (cPLA2) in thrombin-stimulated platelets. Evidence that proline-directed phosphorylation is not required for mobilization of arachidonic acid by cPLA2. *J Biol Chem.* 271:27723–9.

Kulkarni, S. K. and V. P. Singh. 2008. Licofelone: The answer to unmet needs in osteoarthritis therapy? *Curr Rheumatol Rep.* 10:43–8.

Lam, B. K. and K. F. Austen. 2000. Leukotriene C4 synthase. A pivotal enzyme in the biosynthesis of the cysteinyl leukotrienes. *Am J Respir Crit Care Med.* 161:S16–9.

Lefebvre, J. S., S. Marleau, V. Milot, T. Levesque, S. Picard, N. Flamand, and P. Borgeat. 2010. Toll-like receptor ligands induce polymorphonuclear leukocyte migration: Key roles for leukotriene B4 and platelet-activating factor. *FASEB J.* 24:637–47.

Lehnigk, B., K. F. Rabe, G. Dent, R. S. Herst, P. J. Carpentier, and H. Magnussen. 1998. Effects of a 5-lipoxygenase inhibitor, ABT-761, on exercise-induced bronchoconstriction and urinary LTE4 in asthmatic patients. *Eur Respir J.* 11:617–23.

Lex, C., A. Zacharasiewicz, D. N. Payne, N. M. Wilson, A. G. Nicholson, S. A. Kharitonov, P. J. Barnes, and A. Bush. 2006. Exhaled breath condensate cysteinyl leukotrienes and airway remodeling in childhood asthma: A pilot study. *Respir Res.* 7:63.

LTD MFc. 2007a. Substituted quinolines as inhibitors of leukotriene biosynthesis. WO 2007/038865.

LTD MFc. 2007b. Novel substituted 1,2,3-triazolylmethyl-benzothiophene or -indole and their use as leukotriene biosynthesis inhibitors. WO 2007/016784.

Li, B., L. Copp, A. L. Castelhano, R. Feng, M. Stahl, Z. Yuan, and A. Krantz. 1994. Inactivation of a cytosolic phospholipase A2 by thiol-modifying reagents: Cysteine residues as potential targets of phospholipase A2. *Biochemistry* 33:8594–603.

Li, L., H. Ji, L. Sheng, Y. Zhang, Y. Lai, and X. Chen. 2009. The anti-inflammatory effects of ZLJ-6, a novel dual cyclooxygenase/5-lipoxygenase inhibitor. *Eur J Pharmacol.* 607:244–50.

Lin, L. L., M. Wartmann, A. Y. Lin, J. L. Knopf, A. Seth, and R. J. Davis. 1993. cPLA2 is phosphorylated and activated by MAP kinase. *Cell* 72:269–78.

Lorrain, D. S., G. Bain, L. D. Correa, C. Chapman, A. R. Broadhead, A. M. Santini, P. P. Prodanovich et al. 2009. Pharmacological characterization of AM103, a novel selective five-lipoxygenase—Activating protein (FLAP) inhibitor reduces acute and chronic inflammation. *J Pharmacol Exp Ther.* 331, 1042–50.

Lorrain, D. S., G. Bain, L. D. Correa, C. Chapman, A. R. Broadhead, A. M. Santini, P. P. Prodanovich et al. 2010. Pharmacology of AM803, a novel selective five-lipoxygenase-activating protein (FLAP) inhibitor in rodent models of acute inflammation. *Eur J Pharmacol.* 640:211–8.

Lotzer, K., R. Spanbroek, M. Hildner, A. Urbach, R. Heller, E. Bretschneider, H. Galczenski, J. F. Evans, and A. J. Habenicht. 2003. Differential leukotriene receptor expression and calcium responses in endothelial cells and macrophages indicate 5-lipoxygenase-dependent circuits of inflammation and atherogenesis. *Arterioscler Thromb Vasc Biol.* 23:e32–6.

Luo, M., S. M. Jones, S. M. Phare, M. J. Coffey, M. Peters-Golden, and T. G. Brock. 2004. Protein kinase A inhibits leukotriene synthesis by phosphorylation of 5-lipoxygenase on serine 523. *J Biol Chem.* 279:41512–20.

Lynch, K. R., G. P. O'Neill, Q. Liu, D. S. Im, N. Sawyer, K. M. Metters, N. Coulombe et al. 1999. Characterization of the human cysteinyl leukotriene CysLT1 receptor. *Nature* 399:789–93.

Mahshid, Y., M. R. Lisy, X. Wang, R. Spanbroek, J. Flygare, B. Christensson, M. Bjorkholm, B. Sander, A. J. Habenicht, and H. E. Claesson. 2009. High expression of 5-lipoxygenase in normal and malignant mantle zone B lymphocytes. *BMC Immunol.* 10:2.

Maier, T. J., L. Tausch, M. Hoernig, O. Coste, R. Schmidt, C. Angioni, J. Metzner et al. 2008. Celecoxib inhibits 5-lipoxygenase. *Biochem Pharmacol.* 76:862–72.

Manning, P. J., J. Rokach, J. L. Malo, D. Ethier, A. Cartier, Y. Girard, S. Charleson, and P. M. O'Byrne. 1990. Urinary leukotriene E4 levels during early and late asthmatic responses. *J Allergy Clin Immunol.* 86:211–20.

Masferrer, J. L., B. S. Zweifel, M. Hardy, G. D. Anderson, D. Dufield, L. Cortes-Burgos, R. A. Pufahl, and M. Graneto. 2010. Pharmacology of PF-4191834, a novel, selective non-redox 5-lipoxygenase inhibitor effective in inflammation and pain. *J Pharmacol Exp Ther.* 334:294–301.

McMillan, R. M. and E. R. H. Walker. 1992. Designing therapeutically effective 5-lipoxygenase inhibitors. *Trends Pharmacol Sci.* 13:323–30.

Montuschi, P., S. A. Kharitonov, G. Ciabattoni, and P. J. Barnes. 2003. Exhaled leukotrienes and prostaglandins in COPD. *Thorax* 58:585–8.

Muthalif, M. M., Y. Hefner, S. Canaan, J. Harper, H. Zhou, J. H. Parmentier, R. Aebersold, M. H. Gelb, and K. U. Malik. 2001. Functional interaction of calcium-/calmodulin-dependent protein kinase II and cytosolic phospholipase A(2). *J Biol Chem.* 276:39653–60.

Nagase, T., N. Uozumi, T. Aoki-Nagase, K. Terawaki, S. Ishii, T. Tomita, H. Yamamoto, K. Hashizume, Y. Ouchi, and T. Shimizu. 2003. A potent inhibitor of cytosolic phospholipase A2, arachidonyl trifluoromethyl ketone, attenuates LPS-induced lung injury in mice. *Am J Physiol Lung Cell Mol Physiol.* 284:L720–6.

Nemenoff, R. A., S. Winitz, N. X. Qian, V. Van Putten, G. L. Johnson, and L. E. Heasley. 1993. Phosphorylation and activation of a high molecular weight form of phospholipase A2 by p42 microtubule-associated protein 2 kinase and protein kinase C. *J Biol Chem.* 268:1960–4.

Newcomer, M. E. and Gilbert, N. C. 2010. Location, location, location: Compartmentalization of early events in leukotriene biosynthesis. *J Biol Chem.* 285:25109–14.

Okuno, T., Y. Iizuka, H. Okazaki, T. Yokomizo, R. Taguchi, and T. Shimizu. 2008. 12(S)-Hydroxyheptadeca-5Z, 8E, 10E-trienoic acid is a natural ligand for leukotriene B4 receptor 2. *J Exp Med.* 205:759–66.

Okuno, T., T. Yokomizo, T. Hori, M. Miyano, and T. Shimizu. 2005. Leukotriene B4 receptor and the function of its helix 8. *J Biol Chem.* 280:32049–52.

Osman, M. 2003. Therapeutic implications of sex differences in asthma and atopy. *Arch Dis Child.* 88:587–90.

O'Byrne, P. M. 2009. Obstructive lung disease from conception to old age: Differences in the treatment of adults and children with asthma. *Proc Am Thorac Soc.* 6:720–3.

O'Driscoll, B. R., O. Cromwell, and A. B. Kay. 1984. Sputum leukotrienes in obstructive airways diseases. *Clin Exp Immunol.* 55:397–404.

Orning, L. and S. Hammarstrom. 1980. Inhibition of leukotriene C and leukotriene D biosynthesis. *J Biol Chem.* 255:8023–6.

Pacheco, Y., R. Hosni, B. Chabannes, F. Gormand, P. Moliere, M. Grosclaude, D. Piperno, M. Lagarde, and M. Perrin-Fayolle. 1992. Leukotriene B4 level in stimulated blood neutrophils and alveolar macrophages from healthy and asthmatic subjects. Effect of beta-2 agonist therapy. *Eur J Clin Invest.* 22:732–9.

Pergola, C., G. Dodt, A. Rossi, E. Neunhoeffer, B. Lawrenz, H. Northoff, B. Samuelsson, O. Radmark, L. Sautebin, and O. Werz. 2008. ERK-mediated regulation of leukotriene biosynthesis by androgens: A molecular basis for gender differences in inflammation and asthma. *Proc Natl Acad Sci USA* 105:19881–6.

Pergola, C. and O. Werz. 2010. 5-Lipoxygenase inhibitors: A review of recent developments and patents. *Expert Opin Ther Pat.* 20:355–75.

Peters-Golden, M. 2008. Expanding roles for leukotrienes in airway inflammation *Curr Allergy Asthma Rep.* 8:367–73.

Peters-Golden, M. and W. R. Henderson, Jr. 2007. Leukotrienes. *N Engl J Med.* 357:1841–54.

Philip, G., A. S. Nayak, W. E. Berger, F. Leynadier, F. Vrijens, S. B. Dass, and T. F. Reiss. 2004. The effect of montelukast on rhinitis symptoms in patients with asthma and seasonal allergic rhinitis. *Curr Med Res Opin.* 20:1549–58.

Poeckel, D., K. A. Z. Berry, R. C. Murphy, and C. D. Funk. 2009. Dual 12/15- and 5-lipoxygenase deficiency in macrophages alters arachidonic acid metabolism and attenuates peritonitis and atherosclerosis in ApoE knock-out mice. *J Biol Chem.* 284:21077–89.

Provost, P., J. Doucet, T. Hammarberg, G. Gerisch, B. Samuelsson, and O. Radmark. 2001. 5-Lipoxygenase interacts with coactosin-like protein. *J Biol Chem.* 276:16520–7.

Qiu, H., K. Straat, A. Rahbar, M. Wan, C. Soderberg-Naucler, and J. Z. Haeggstrom. 2008. Human CMV infection induces 5-lipoxygenase expression and leukotriene B4 production in vascular smooth muscle cells. *J Exp Med.* 205:19–24.

Radmark, O. 1995. Arachidonate 5-lipoxygenase. *J Lipid Mediat Cell Signal.* 12:171–84.

Radmark, O. P. 2000. The molecular biology and regulation of 5-lipoxygenase. *Am J Respir Crit Care Med.* 161:S11–5.

Radmark, O., O. Werz, D. Steinhilber, and B. Samuelsson. 2007. 5-Lipoxygenase: Regulation of expression and enzyme activity. *Trends Biochem Sci.* 32:332–41.

Rakonjac, M., L. Fischer, P. Provost, O. Werz, D. Steinhilber, B. Samuelsson, and O. Radmark. 2006. Coactosin-like protein supports 5-lipoxygenase enzyme activity and up-regulates leukotriene A4 production. *Proc Natl Acad Sci USA* 103:13150–5.

Rao, P. N., Q. H. Chen, and E. E. Knaus. 2006. Synthesis and structure-activity relationship studies of 1,3-diarylprop-2-yn-1-ones: Dual inhibitors of cyclooxygenases and lipoxygenases. *J Med Chem.* 49:1668–83.

Rouzer, C. A., T. Matsumoto, and B. Samuelsson. 1986. Single protein from human leukocytes possesses 5-lipoxygenase and leukotriene A4 synthase activities. *Proc Natl Acad Sci USA* 83:857–61.

Rouzer, C. A. and B. Samuelsson. 1986. The importance of hydroperoxide activation for the detection and assay of mammalian 5-lipoxygenase. *FEBS Lett.* 204:293–6.

Rubinstein, I., B. Kumar, and C. Schriever. 2004. Long-term montelukast therapy in moderate to severe COPD—A preliminary observation. *Respir Med.* 98:134–8.

Saetta, M., G. Turato, P. Maestrelli, C. E. Mapp, and L. M. Fabbri. 2001. Cellular and structural bases of chronic obstructive pulmonary disease. *Am J Respir Crit Care Med.* 163:1304–9.

Samuelsson, B. 1983. Leukotrienes: Mediators of immediate hypersensitivity reactions and inflammation. *Science* 220:568–75.

Samuelsson, B., S. E. Dahlen, J. A. Lindgren, C. A. Rouzer, and C. N. Serhan. 1987. Leukotrienes and lipoxins: Structures, biosynthesis, and biological effects. *Science* 237:1171–6.

Schmidt, D. and K. F. Rabe. 2000. The role of leukotrienes in the regulation of tone and responsiveness in isolated human airways. *Am J Respir Crit Care Med.* 161:S62–7.

Schwarz, K., M. Walther, M. Anton, C. Gerth, I. Feussner, and H. Kuhn. 2001. Structural basis for lipoxygenase specificity. Conversion of the human leukocyte 5-lipoxygenase to a 15-lipoxygenating enzyme species by site-directed mutagenesis. *J Biol Chem.* 276:773–9.

Serhan, C. N., N. Chiang, and T. E. Van Dyke. 2008. Resolving inflammation: Dual anti-inflammatory and pro-resolution lipid mediators. *Nat Rev Immunol.* 8:349–61.

Sharp, J. D., R. T. Pickard, X. G. Chiou, J. V. Manetta, S. Kovacevic, J. R. Miller, A. D. Varshavsky et al. 1994. Serine 228 is essential for catalytic activities of 85-kDa cytosolic phospholipase A2. *J Biol Chem.* 269:23250–4.

Shaw, R. J., P. Fitzharris, O. Cromwell, A. J. Wardlaw, and A. B. Kay. 1985. Allergen-induced release of sulphidopeptide leukotrienes (SRS-A) and LTB4 in allergic rhinitis. *Allergy* 40:1–6.

Shimizu, T., T. Izumi, Y. Seyama, K. Tadokoro, O. Radmark, and B. Samuelsson. 1986. Characterization of leukotriene A4 synthase from murine mast cells: Evidence for its identity to arachidonate 5-lipoxygenase. *Proc Natl Acad Sci USA* 83:4175–9.

Shimizu, T., O. Radmark, and B. Samuelsson. 1984. Enzyme with dual lipoxygenase activities catalyzes leukotriene A4 synthesis from arachidonic acid. *Proc Natl Acad Sci USA* 81:689–93.

Shimizu, T. and L. S. Wolfe. 1990. Arachidonic acid cascade and signal transduction. *J Neurochem.* 55:1–15.

Shindo, K., Y. Matsumoto, Y. Hirai, M. Sumitomo, T. Amano, K. Miyakawa, M. Matsumura, and T. Mizuno. 1990. Measurement of leukotriene B4 in arterial blood of asthmatic patients during wheezing attacks. *J Intern Med.* 228:91–6.

Shirumalla, R. K., K. S. Naruganahalli, S. G. Dastidar, V. Sattigeri, G. Kaur, C. Deb, J. B. Gupta, M. Salman, and A. Ray. 2006. RBx 7796: A novel inhibitor of 5-lipoxygenase. *Inflamm Res.* 55:517–27.

Shirumalla, R. K., P. Sharma, S. G. Dastidar, J. K. Paliwal, S. Kakar, B. Varshney, G. S. Saini, V. Sattigeri, M. Salman, and A. Ray. 2008. Pharmacodynamic and pharmacokinetic characterisation of RBx 7796: A novel 5-lipoxygenase inhibitor. *Inflamm Res.* 57:135–43.

Silverman, E. S. and J. M. Drazen. 2000. Genetic variations in the 5-lipoxygenase core promoter. Description and functional implications. *Am J Respir Crit Care Med.* 161:S77–80.

Spanbroek, R., H. J. Stark, U. Janssen-Timmen, S. Kraft, M. Hildner, T. Andl, F. X. Bosch et al. 1998. 5-Lipoxygenase expression in Langerhans cells of normal human epidermis. *Proc Natl Acad Sci USA* 95:663–8.

Steinbrink, S. D., C. Pergola, U. Buhring, S. George, J. Metzner, A. S. Fischer, A. K. Hafner et al. 2010. Sulindac sulfide suppresses 5-lipoxygenase at clinically relevant concentrations. *Cell Mol Life Sci.* 67:797–806.

Steinhilber, D. 1994. 5-Lipoxygenase: Enzyme expression and regulation of activity. *Pharm Acta Helv.* 69:3–14.

Tai, N., K. Kuwabara, M. Kobayashi, K. Yamada, T. Ono, K. Seno, Y. Gahara, J. Ishizaki, and Y. Hori. 2010. Cytosolic phospholipase A2 alpha inhibitor, pyrroxyphene, displays antiarthritic and anti-bone destructive action in a murine arthritis model. *Inflamm Res.* 59:53–62.

Tholander, F., A. Muroya, B. P. Roques, M. C. Fournie-Zaluski, M. M. Thunnissen, and J. Z. Haeggstrom. 2008. Structure-based dissection of the active site chemistry of leukotriene A4 hydrolase: Implications for M1 aminopeptidases and inhibitor design. *Chem Biol.* 15:920–9.

Uozumi, N., K. Kume, T. Nagase, N. Nakatani, S. Ishii, F. Tashiro, Y. Komagata et al. 1997. Role of cytosolic phospholipase A(2) in allergic response and parturition. *Nature* 390:618–22.

Vickers, P. J. 1995. 5-Lipoxygenase-activating protein (FLAP). *J Lipid Mediat Cell Signal.* 12:185–94.

Wardlaw, A. J., H. Hay, O. Cromwell, J. V. Collins, and A. B. Kay. 1989. Leukotrienes, LTC4 and LTB4, in bronchoalveolar lavage in bronchial asthma and other respiratory diseases. *J Allergy Clin Immunol.* 84:19–26.

Watanabe, S., A. Yamasaki, K. Hashimoto, Y. Shigeoka, H. Chikumi, Y. Hasegawa, T. Sumikawa et al. 2009. Expression of functional leukotriene B4 receptors on human airway smooth muscle cells. *J Allergy Clin Immunol.* 124:59–65 e51–53.

Weitzel, F. and A. Wendel. 1993. Selenoenzymes regulate the activity of leukocyte 5-lipoxygenase via the peroxide tone. *J Biol Chem.* 268:6288–92.

Wenzel, S. E., G. L. Larsen, K. Johnston, N. F. Voelkel, and J. Y. Westcott. 1990. Elevated levels of leukotriene C4 in bronchoalveolar lavage fluid from atopic asthmatics after endobronchial allergen challenge. *Am Rev Respir Dis.* 142:112–9.

Werz, O. 2002. 5-Lipoxygenase: Cellular biology and molecular pharmacology. *Curr Drug Targets Inflamm Allergy* 1:23–44.

Werz, O., E. Burkert, L. Fischer, D. Szellas, D. Dishart, B. Samuelsson, O. Radmark, and D. Steinhilber. 2002a. Extracellular signal-regulated kinases phosphorylate 5-lipoxygenase and stimulate 5-lipoxygenase product formation in leukocytes. *FASEB J.* 16:1441–3.

Werz, O., J. Klemm, O. Radmark, and B. Samuelsson. 2001a. p38 MAP kinase mediates stress-induced leukotriene synthesis in a human B-lymphocyte cell line. *J Leukoc Biol.* 70:830–8.

Werz, O., J. Klemm, B. Samuelsson, and O. Radmark. 2001b. Phorbol ester up-regulates capacities for nuclear translocation and phosphorylation of 5-lipoxygenase in Mono Mac 6 cells and human polymorphonuclear leukocytes. *Blood* 97:2487–95.

Werz, O., J. Klemm, B. Samuelsson, and O. Radmark. 2000. 5-lipoxygenase is phosphorylated by p38 kinase-dependent MAPKAP kinases. *Proc Natl Acad Sci USA* 97:5261–6.

Werz, O. and D. Steinhilber. 2005. Development of 5-lipoxygenase inhibitors—Lessons from cellular enzyme regulation. *Biochem Pharmacol.* 70:327–33.

Werz, O. and D. Steinhilber. 2006. Therapeutic options for 5-lipoxygenase inhibitors. *Pharmacol Ther.* 112:701–18.

Werz, O., D. Szellas, M. Henseler, and D. Steinhilber. 1998. Nonredox 5-lipoxygenase inhibitors require glutathione peroxidase for efficient inhibition of 5-lipoxygenase activity. *Mol Pharmacol.* 54:445–51.

Werz, O., D. Szellas, D. Steinhilber, and O. Radmark. 2002b. Arachidonic acid promotes phosphorylation of 5-lipoxygenase at Ser-271 by MAPK-activated protein kinase 2 (MK2). *J Biol Chem.* 277:14793–800.

Wilson, A. M., P. M. O'Byrne, and K. Parameswaran. 2004. Leukotriene receptor antagonists for allergic rhinitis: A systematic review and meta-analysis. *Am J Med.* 116:338–44.

Wilson, A. M., O. J. Dempsey, E. J. Sims, and B. J. Lipworth. 2001. A comparison of topical budesonide and oral montelukast in seasonal allergic rhinitis and asthma. *Clin Exp Allergy* 31:616–24.

Yoshimoto, T., C. Yokoyama, K. Ochi, S. Yamamoto, Y. Maki, Y. Ashida, S. Terao, and M. Shiraishi. 1982. 2,3,5-Trimethyl-6-(12-hydroxy-5,10-dodecadiynyl)-1,4-benzoquinone (AA861), a selective inhibitor of the 5-lipoxygenase reaction and the biosynthesis of slow-reacting substance of anaphylaxis. *Biochim Biophys Acta* 713:470–3.

Zarini, S., M. A. Gijon, A. E. Ransome, R. C. Murphy, and A. Sala. 2009. Transcellular biosynthesis of cysteinyl leukotrienes *in vivo* during mouse peritoneal inflammation. *Proc Natl Acad Sci USA* 106:8296–301.

5 Leukotriene Receptor Antagonists and FLAP Inhibitors

Suman Gupta and Sanjay Malhotra

CONTENTS

5.1 INTRODUCTION

Leukotrienes are lipid mediators generated when arachidonic acid (AA) is metabolized by 5-lipoxygenase enzyme (5-LO). At the present time, leukotrienes are divided into cysteinyl leukotrienes (CysLTs) and leukotriene B_4 (LTB_4). CysLT and LTB_4 act on different cell surface receptors, namely CysLT receptor and LTB_4 receptor. Leukotrienes have potent pharmacological action on the airway. CysLTs exert potent bronchoconstrictor action whereas LTB_4 exhibits chemotactic activity on neutrophils. Leukotriene action can be modulated by antagonizing leukotriene receptors as well as by inhibiting leukotriene synthesis. A lot of effort has gone into designing leukotriene receptor antagonists. CysLT receptor antagonist montelukast has become a drug. However, not much success has been achieved in designing LTB_4 receptor antagonist. Synthesis of leukotrienes can be blocked using inhibitors of 5-LO as well as 5-lipoxygenase-activating protein (FLAP). 5-LO as a drug discovery target has been discussed in detail in Chapter 4. In this chapter, we will look into leukotriene receptor and FLAP as targets for airway inflammatory diseases such as asthma and chronic obstructive pulmonary disease.

5.2   LEUKOTRIENE BIOSYNTHESIS

CysLT and LTB_4 are derived from the ubiquitous membrane constituent, AA, and are members of a large family of molecules known as eicosanoids (Capra et al., 2007). These biologically active lipids are rapidly generated at the site of inflammation by a series of reactions initiated by cytosolic PLA_2 ($cPLA_2$), which releases AA from the phospholipids present at the nuclear envelope (Figure 5.1). AA becomes available to the bifunctional enzyme 5-LO after binding to the FLAP. 5-LO enzyme contains a non-heme iron at its active center, which undergoes transition from a divalent to a trivalent state during catalysis. 5-LO enzyme acts on AA bound to FLAP along with molecular oxygen, in a two-step reaction that catalyzes the formation of 5-hydroperoxy-eicosatetraenoic acid (5-HPETE) with the subsequent formation of an unstable, short-lived intermediate leukotriene A4 (LTA_4) (Brash, 1999). LTA_4 is converted to LTB_4 by the enzyme LTA_4 hydrolase, a cytosolic enzyme also exerting aminopeptidase activity and containing zinc at its catalytic center. Inflammatory cells, such as eosinophils, baso-phils, mast cells, and alveolar macrophages possessing the integral membrane protein leukotriene C4 (LTC_4) synthase, synthesize CysLTs from LTA_4 in response to biological and nonbiological stimuli (Thien et al., 1995). LTC_4 synthase conjugates reduced glutathione at position C6 of LTA_4 to form LTC_4, which is transferred to the extracel-lular space, where a γ-glutamyl transpeptidase cleaves the glutamic acid moiety to form leukotriene D4 (LTD_4). This, in turn, is transformed into LTE_4 by the cleavage of the glycine moiety by a variety of dipeptidases (Orning and Hammarstrom, 1980).

5.2.1   5-LIPOXYGENASE-ACTIVATING PROTEIN

In the late 1980s, scientists at Merck-Frosst identified an indole class of compound, designated MK-886, that inhibited leukotriene biosynthesis both *in vitro* and *in vivo* without affecting 5-LO, phospholipases, or other nonselective mechanisms of inhibiting leukotriene production (Gillard et al., 1989). An 18-kDa membrane pro-tein was identified that specifically bound MK-886 (Evans et al., 1991). This protein was purified and its cDNA cloned, and rabbit antisera raised against peptide epitopes derived from this protein was used to immunoprecipitate a photoaffinity-labeled protein from human leukocyte membranes (Dixon et al., 1990). This protein was named FLAP because it is required for the cellular activity of 5-LO.

FLAP belongs to the superfamily of membrane-associated proteins of the eicosanoid and glutathione metabolism (MAPEG) superfamily (Bresell et al., 2005) that includes three microsomal glutathione transferases (mGST-1, mGST-2, and mGST-3), LTC_4 synthase, and microsomal prostaglandin E synthase (mPGES-1).

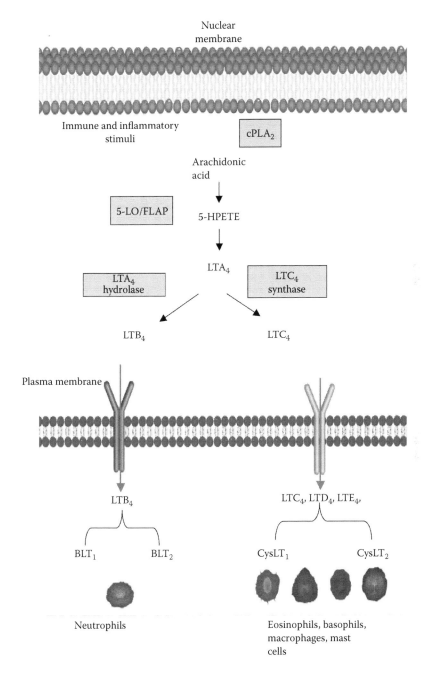

FIGURE 5.1 The arachidonic acid pathway and formation of leukotrienes. Arachidonic acid is catalyzed by the 5-LO enzyme. This reaction requires FLAP and results in the formation of LTA_4. The unstable epoxide LTA_4 is either acted upon by epoxide hydrolase to form LTB_4 or conjugated with glutathione by LTC_4 synthase and yields LTC_4. LTC_4 is metabolized by γ-glutamyl transpeptidase to LTD_4, which is, in turn, metabolized by dipeptidase to LTE_4.

Unlike other members of the family, FLAP does not have any defined enzymatic activity and its ability to bind AA and transfer it to 5-LO is not modulated by gluta-thione. Comparison of primary amino acid sequence of 5-LO proteins from different species with other plant and mammalian lipoxygenases revealed a high level of simi-larity, particularly in their carboxy terminal residues. This similarity suggested that common structural characteristics might underlie their common functions.

Significant expression of FLAP is confined to bone-marrow-derived cells, includ-ing neutrophils, eosinophils, monocytes/macrophages, mast cells, certain lymphocyte population, and dendritic cells (Table 5.1). Tissue-specific differences in expression levels of FLAP have also been described in mature macrophages. For example, human alveolar macrophages contain ~35-fold more FLAP protein (per unit of cellular pro-tein) than do their precursors, the peripheral blood monocyte (Coffey et al., 1994). Increased expression of 5-LO and/or FLAP has been observed in various leukocyte populations in asthmatic subjects (Seymour et al., 2001) or in allergic animal models of asthma (Chu et al., 2000). Although precedence exists for posttranscriptional regu-latory mechanisms (Pouliot et al., 1994), transcriptional regulation appears to be the dominant means by which expression of FLAP is controlled.

5.3 LEUKOTRIENE RECEPTORS

5.3.1 Receptors for CysLTs

CysLTs exert their effect through cell surface receptors. According to the International Union of Pharmacology (IUPHAR), CysLT receptor nomenclature was originally

TABLE 5.1
Expression of LT Receptor Subtype and FLAP in Different Cell Types

Cell Type	CysLTR$_1$ Receptor	CysLTR$_2$ Receptor	BLTR$_1$ Receptor	BLTR$_2$ Receptor	FLAP
Eosinophils	+	+	−	−	+
Basophils	+	+	−	−	+
Mast cells	+	+	+	+	+
Monocytes	+	−	−	−	+
Macrophages	+	−	−	−	+
Neutrophils	+	−	−	−	+
T-lymphocytes	+	−	+	−	−
Hematopoietic stem cells	+	−	−	−	−
Endothelial cells	+	+	−	−	−
Glandular cells	+	−	−	−	−
Smooth muscle cells	+	−	−	−	−
Fibroblasts	+	−	+	−	+
Platelets	−	−	−	−	−
Erythrocytes	−	−	−	−	−
Epithelial cells	−	−	−	−	−

Note: +, presence of receptors or enzymes; −, absence or lack of evidence of receptors or enzymes.

based on the sensitivity to the so-called classical antagonists, the most widely used of which include montelukast, zafirlukast, pranlukast, pobilukast, and others (Holgate et al., 1996).

CysLT receptors have been mainly divided into two classes: $CysLT_1$, receptors that are sensitive to blockade by the classical antagonists, and $CysLT_2$, receptors that are not inhibited by the classical antagonists. Both $CysLT_1$ and $CysLT_2$ receptors are reported to be coupled to their effectors through $G\alpha_q$-protein. Their activation results in an increase in intracellular calcium through the PLC pathway (Yoshihide and Joshua, 2004; Singh et al., 2010).

Most of the effects of CysLTs that are relevant to the pathophysiology of asthma are mediated by activation of the $CysLT_1$ receptors, which are expressed on monocytes and macrophages, eosinophils, basophils, mast cells, neutrophils, T cells, B-lymphocytes, pluripotent hematopoietic stem cells (CD34+), interstitial cells of the nasal mucosa, airway smooth muscle cells, bronchial fibroblasts, and vascular endothelial cells (Table 5.1). The $CysLT_2$ receptor is expressed in human peripheral basophils, endothelial cells, and cultured mast cells, and in nasal eosinophils and mast cells in patients with active seasonal allergic rhinitis. The $CysLT_2$ activation might enable the CysLTs to elicit IL-8 generation (Kanaoka and Boyce, 2004). At the present time, the role played by $CysLT_2$ receptors in allergic inflammation is poorly understood.

The existence of a third receptor type of CysLT has been suggested. This suggestion is based on two distinct lines of evidence. First, in human and porcine pulmonary arteries, leukotriene response is poorly antagonized by dual antagonist of CysLT receptors, BAYu9773. A second line of evidence came from using agonist response to CysLT receptors. LTE_4, an agonist that poorly binds to leukotriene receptors, exhibits comparable potency to other CysLT agonists for promoting contraction of asthmatic airway and swelling of human skin. Similarly, in $CysLT_1$ and $CysLT_2$ knockout mouse, LTC_4 and LTD_4 produce ear edema. This data suggested the existence of a third receptor type for CysLT. This receptor belongs to G-protein receptor class and is called $CysLT_ER$ (Walch et al., 2002).

5.3.2 RECEPTORS FOR LTB_4

Two GPCR subtypes of receptor for LTB_4 (BLT_1 and BLT_2) have been identified. Both BLT_1 and BLT_2 are G-protein-coupled seven transmembrane domain receptors. Coding genes for both these receptors are located on chromosome 14. cDNA for BLT_1 encodes a cell surface protein of 352 amino acids belonging to the superfamily of seven transmembrane GPCR. BLT_2 receptor shows a high homology with BLT_1 (36–45% amino acid identity). These receptors differ in their affinity and specificity for LTB_4 and their expression pattern. BLT_1, a specific high-affinity receptor for LTB_4, is expressed predominantly on leukocytes, including granulocytes, monocytes, macrophages, mast cells, dendritic cells, and effector T cells. BLT_2, a low-affinity receptor, also binds other eicosanoids. BLT_2 is expressed ubiquitously and its biological role in humans is unknown. In native systems, the major signal transduction pathway for BLT_1 receptor has been identified to be through activation of phospholipase C (PLC), coupled to phosphoinositide (PI) turnover, and mobilization of intracellular calcium. A pertussis toxin-sensitive G-protein has been implicated in signal transduction. It mediates LTB_4-induced chemotaxis, Ca^{2+} flow, and inhibition of adenylate cyclase. It is insensitive to several BLT_1 antagonists and thus it is a pharmacologically distinct receptor subtype (Yokomizo et al., 2001; Hicks et al., 2007). In inflammatory cells, ligation of BLT_1 and/or BLT_2 by LTB_4 triggers a variety of intracellular signals such as intracellular Ca^{2+} mobilization, activation of extracellular signal-regulated kinase, activation of phosphoinositide 3-kinase and Akt and stimulates cellular events such as chemotaxis, degranulation, and the production of inflammatory proteins.

5.4   LEUKOTRIENES IN ASTHMA

Asthma is a chronic inflammatory disease of the airway that is precipitated upon exposure to allergens. An asthma attack is characterized by acute bronchoconstriction due to the release of bronchoconstrictors. The acute response is followed by a late-phase bronchoconstriction, which is believed to be due to the inflammation of the airway. Subsequent to asthma attack, the airway remains hyperresponsive to bronchoconstrictors and noxious stimuli in the environment.

5.4.1   Cysteinyl Leukotrienes in Asthma

Pathogenesis of asthma involves several different cellular and molecular mediators. CysLTs play an important role in the pathogenesis of asthma as they are one of the most potent bronchoconstrictors found to date. However, it appears that CysLTs may contribute toward all components of airway response following allergen exposure. Data indicate that CysLT levels are increased in bronchoalveolar lavage fluid and sputum of patients with asthma. CysLT levels also increase with asthma severity, after allergen challenge or exercise challenge, and decrease with effective treatment (Drazen and Austen, 1987).

5.4.1.1   CysLTs and Bronchoconstriction

Contraction of the smooth muscle of the airway was the first biological property recognized for CysLTs. LTC_4 and LTD_4 are potent inducers of bronchoconstriction in guinea pig airways *in vitro* and *in vivo* and cause contraction of isolated human bronchi. *In vivo* experiments have clearly shown that CysLTs are able to induce bronchoconstriction when inhaled, both in healthy and in asthmatic individuals (Barnes et al., 1984; Drazen, 1988). In chronic asthma, it appears that human bronchi possess an increased contractile tone, due to constitutive release of CysLTs. Administration of a $CysLT_1$ receptor antagonist induces an apparent bronchodilation, despite the fact that these compounds are devoid of direct effects on bronchial tone.

The relative potencies of CysLTs—LTC_4, LTD_4, and LTE_4—have been a matter of debate for a long time. It has become clear that LTE_4 is less potent than the other two CysLTs, which are almost equipotent. Contraction induced by LTD_4 is mediated through a pathway that promotes phosphoinositide turnover and increases cytosolic calcium, as well as through a pathway that may involve the activation of protein kinase C isoforms.

In susceptible individuals, montelukast has been shown to inhibit bronchoconstriction provoked by aspirin, inhaled allergen, and exercise (both in children and in adults). Furthermore, maintenance therapy for 8 weeks provided greater and sustained inhibition of bronchoconstriction than salmeterol (50 μg twice daily), a long-acting β-adrenoceptor agonist, used alone or in combination with ICS treatment (Diamant et al., 1999; Dryden et al., 2010; de Benedictis et al., 2008; Hallstrand and Henderson, 2009).

5.4.1.2 CysLTs and Inflammation

Experimental evidence suggests that CysLTs contribute toward airway inflammation in asthma. Based on data generated by direct administration of CysLTs as well as using CysLT receptor antagonist, it has become apparent that CysLTs increase microvascular permeability leading to pulmonary edema, increase mucus secretion, and impair ciliary activity.

Airway inflammation in bronchial asthma is an interplay of different inflammatory cells, where the inflammatory response starts with allergen-induced degranulation of mast cells and results in the recruitment of inflammatory and immune cells such as eosinophils, basophils, lymphocytes, etc. CysLTs are formed by inflammatory cells such as eosinophils, basophils, mast cells, and alveolar macrophages (Figure 5.2). $CysLT_1$ receptor is expressed on a variety of airway mucosal inflammatory cells, eosinophils, neutrophils, mast cells, macrophages, B-lymphocytes, and plasma cells, but not T-lymphocytes. The number of cells expressing $CysLT_1$ receptor was investigated by Zhu et al. (2005) in subjects suffering from asthma. It was observed that $CysLT_1$ receptor mRNA and protein-positive inflammatory cells were found in significantly greater number in stable

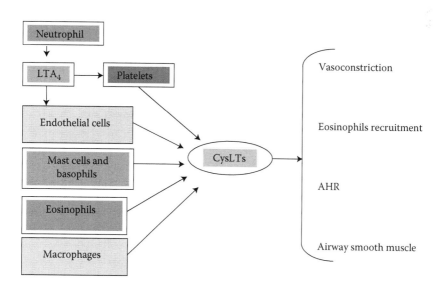

FIGURE 5.2 Cellular sources of cysteinyl leukotrienes and summary of their effects on airway and inflammatory cells.

asthmatics and in subjects hospitalized for asthma exacerbation compared with controls. A strong positive correlation was observed between CysLT receptor positive inflammatory cells and the increased numbers of CD45[+] progenitors. In human nasal mucosa, $CysLT_1$ receptor has been localized at the gene level as well as at the protein level in blood vessels and in the interstitial cells such as vascular endothelial cells, eosinophils, mast cells, macrophages, and neutrophils. It has been suggested (Sousa et al., 2002) that the increased expression of $CysLT_1$ receptors on inflammatory cells may be responsible for higher incidence of exaggerated airway responsiveness to inhaled CysLTs in subjects with aspirin-induced asthma compared to those with aspirin-tolerant asthmatics.

Eosinophils are considered to be an important cell type that contributes toward inflammation associated with asthma. CysLTs increase eosinophil survival in response to paracrine signals from mast cells and lymphocytes. CysLTs promote leukocyte maturation and migration from the bone marrow into the circulatory system. CysLTs act as a chemoattractant for eosinophils, and increase their cellular adhesion and transendothelial migration across the vessel wall into the airways. The number and activity of the eosinophils are increased, resulting in the increased concentration of CysLTs in the bronchoalveolar lavage fluid and urine.

5.4.1.3   CysLTs and Vascular Permeability

It has long been known that CysLTs can increase vascular permeability and promote extravasation of plasma proteins, resulting in localized edema. LTC_4 and LTD_4 cause exudation of plasma proteins in postcapillary venules in the hamster cheek pouch (Hedqvist et al., 1980) and their intradermal application produces a flare and wheal reaction in humans (Camp et al., 1983; Soter et al., 1983). Local injection of these two CysLTs also increased accumulation of intravenously injected Evans blue in guinea pig airway. Plasma exudation occurred in all airway segments, ranging from the most peripheral small bronchi to trachea, and there was evidence of Evans blue accumulation in superficial as well as deep layers of the airway mucosa. This is very likely to be due to the retraction of endothelial cells at the level of postcapillary venules, as shown for LTE_4 (Joris et al., 1987). It is interesting to note that increased veno-permeability and plasma extravasation *in vivo* might be a result of the cooperation between LTB_4 and CysLTs. In fact, LTB_4, being a potent chemoattractant, can participate in margination, rolling, and adhesion of leukocytes (Hedqvist et al., 1994); the proximity of leukocytes to the vascular wall facilitates the transfer of LTA_4 from the former to endothelial cells, thus triggering the biosynthesis of LTC_4 precisely where this and/or the other CysLTs can induce the opening of gaps between adjacent endothelial cells.

5.4.2 LTB_4 IN ASTHMA AND COPD

Leukocytes are found to be the primary targets for the biological activity of LTB_4. LTB_4 is a potent stimulus for various leukocyte functions, eliciting aggregation, chemotaxis and chemokinesis (Ford-Hutchinson, 1990), and lysosomal enzyme release (Hafstrom et al., 1981), and promoting rolling and adhesion to endothelium

(Dahlén et al., 1981), resulting in emigration into the extravascular space. LTB_4 is also a chemokinetic factor and a weak chemoattractant for eosinophils (Wardlaw et al., 1986). Accordingly, 5-LO inhibitors and BLT receptor antagonists inhibit allergen-induced eosinophil infiltration in airways. However, LTB_4 is only a weak activator of eosinophils.

During a short-lasting exposure to LTB_4, polymorphonuclear leukocytes are mainly recruited. It has been shown that LTB_4 can act as a chemoattractant for eosinophils and may stimulate production of IL-5 in T-lymphocytes. In addition to effects on leukocyte adhesion and migration, LTB_4 as other secretagogues stimulates secretion of superoxide anion and release of different granule constituents from leukocytes. With respect to effects of LTB_4 on the lung, it has been shown that LTB_4 causes contraction of guinea pig lung parenchyma (Hansson et al., 1981; Piper and Samhoun, 1982). This response is believed to be due to the release of thromboxane A_2 and histamine, possibly from pulmonary mast cells (Sirois et al., 1982; Dahlén et al., 1983). Although inhalation of LTB_4 by healthy human volunteers was followed by distinct cellular changes in the airways, and possibly also some plasma exudation, the lack of immediate bronchoconstrictive property was confirmed in healthy volunteers as well as in subjects with asthma.

LTB_4 is believed to be associated with the development of airway hyperresponsiveness during an asthma attack. Inhaled methacholine stimulates LTB_4 release in patients with asthma, but not in healthy subjects, without affecting the number of inflammatory cells in BAL fluid. The LTB_4 receptor antagonist, LY293111, prevented development of airway hyperresponsiveness and neutrophil accumulation but failed to inhibit accumulation of eosinophils in sensitized and challenged animals (Evans et al., 1996). However, recent studies using another BLT_1 antagonist, CP-105,696, and BLT_1-deficient mice have confirmed the role of the LTB_4/ BLT_1 pathway in the recruitment of not only neutrophils but also effector T cells, including effector memory $CD8^+$ T cells, into the lungs of allergen-induced allergic airway inflammatory responses in mice. Inhibition of the LTB_4/BLT_1 pathway resulted in decrease in airway hyperresponsiveness and allergic airway inflammation, including accumulation of eosinophils and lymphocytes in the airway. This suggests that the LTB_4/BLT_1 pathway appears to play an important role in the pathophysiology of asthma along with other mediators, including CysLTs, cytokines, and chemokines.

LTB_4 levels were increased in the plasma, sputum, and BAL fluid of asthma patients but not in the healthy subjects. Increased synthesis of LTB_4 was accompanied by increased transcriptional upregulation of 5-LO and LTA_4 hydrolase in peripheral blood leukocytes of asthmatics. Generation of LTB_4 by calcium ionophore-stimulated alveolar macrophages and peripheral blood neutrophils was increased in asthma patients. Taken together, these data suggest that an upregulation of the LTB_4 synthesis in the circulating leukocytes and lungs is associated with asthma (Holloway et al., 2008).

Chronic obstructive pulmonary disease (COPD) is also characterized by neutrophilic inflammation and elevated levels of LTB_4. Recently, it was reported that a novel dual BLT_1 and BLT_2 receptor antagonist, RO5101576, potently inhibited LTB_4-evoked calcium mobilization in HL-60 cells and chemotaxis of human neutrophils. It also significantly attenuated LTB_4-evoked pulmonary eosinophilia in guinea pigs.

In nonhuman primates, RO5101576 inhibited allergen and ozone-evoked pulmonary neutrophilia with comparable efficacy to budesonide (Hicks et al., 2010). However, the exact role of LTB_4 pathways in mediating pulmonary neutrophilia and the potential therapeutic application of LTB_4 receptor antagonists in these diseases remain controversial.

5.5 DRUG DISCOVERY EFFORTS: MODULATING LEUKOTRIENE LEVELS AND LEUKOTRIENE FUNCTION

5.5.1 CysLT RECEPTOR ANTAGONISTS

Leukotriene receptor antagonists (LTRAs) have emerged as a distinct class of drugs over the past decade. During the 1990s, the leukotriene approach moved from concept to proof. Three cysteinyl leukotriene antagonists ICI204219 (Zafirlukast, Accolate), MK0476 (Montelukast, Singulair), and ONO1078 (Pranlukast) (Figure 5.3) were launched as drugs for clinical use in asthma (Table 5.2). Zafirlukast is the first oral leukotriene receptor antagonist approved in the United States and the United Kingdom for the treatment of mild to moderate asthma (Adkins and Brodgen, 1998). It is a selective and competitive receptor antagonist of leukotriene D4 and E4. In humans, it inhibited bronchoconstriction caused by several kinds of inhalational challenges. Montelukast is an orally active and highly selective LTRA. It causes inhibition of airway cysteinyl leukotriene receptors as demonstrated by the ability to inhibit bronchoconstriction due to inhaled LTD_4 in asthmatics. Doses as low as 5 mg cause substantial blockage of LTD_4-induced bronchoconstriction. Pranlukast (ONO1078) was identified through random screening of series of carboxylic acids. During the optimization of these early leads, two intervening compounds were synthesized that demonstrated links to other chemical series that enhanced CysLT receptor antagonist drug discovery efforts. First, restricting rotational freedom via

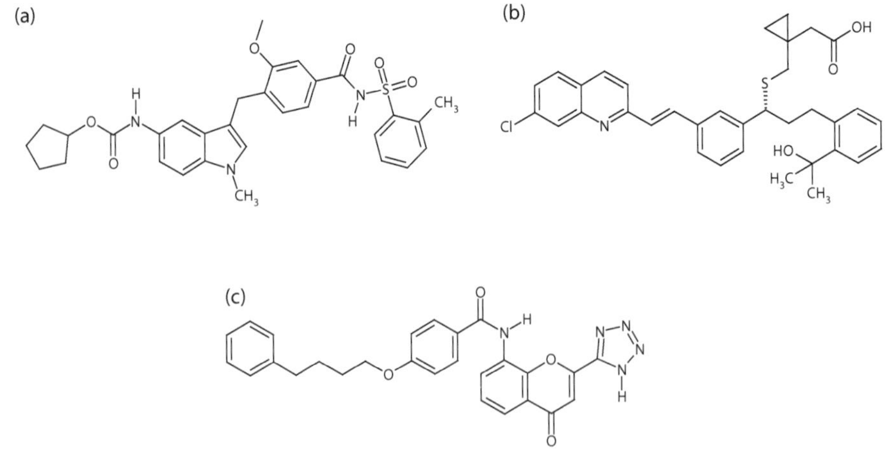

FIGURE 5.3   Structures of (a) zafirlukast, (b) montelukast, and (c) pranlukast.

TABLE 5.2
Comparison of Leukotriene Receptor Antagonists

	Montelukast (Singulair®)	Zafirlukast (Accolate®)	Pranlukast (Onon®)
Indication	Asthma, allergic rhinitis	Asthma	Asthma, allergic rhinitis
Mechanism of action	$CystLT_1$ receptor antagonism	$CystLT_1$ receptor antagonism	$CystLT_1$ receptor antagonism
Potency	0.5 nM (IC_{50}); 9.3 (pA_2)	0.3 nM (K_i); 9.5 (pA_2)	4–7 nM (IC_{50}); $pK_B = 7.5$
Duration of action	>50-fold shift in LTD_4 response 24 h post 40 mg oral dose	5-fold shift in LTD_4 response 24 h post 40 mg oral dose	7-fold shift in LTD_4 response 24 h post 450 mg dose
Pharmacokinetics	Bioavailability 64% Protein binding 99% Extensively metabolized by CYP 3A4 and 2C9	Rapidly absorbed T_{max}: 3 h Protein binding 99% Extensively metabolized by CYP 2C9	T_{max}: 6–9 h Substrate of CYP3A4
Benefits and Indications	Monotherapy; suitable for children; mild persistent asthma; exercise-induced asthma; aspirin-sensitive asthma; allergen-induced asthma; add-on therapy with ICS	Exercise-induced asthma; aspirin-sensitive asthma; allergen-induced asthma; add-on therapy with ICS	Exercise-induced asthma; aspirin-sensitive asthma; allergen-induced asthma; add-on therapy with ICS
Side effects	Headache, abdominal pain; Churg–Strauss syndrome	Headache, abdominal pain; elevation of liver enzymes; Churg–Strauss syndrome	Headache, abdominal pain; elevation of liver enzymes; Churg–Strauss syndrome
Dose	Adults: 10 mg od	Adults: 20 mg bid	Adults: 225 mg bid
Comments	Most widely prescribed $CysLT_1$ receptor antagonist	First $CysLT_1$ receptor antagonist to be approved	Marketed only in Asia

Note: ASA, aspirin-sensitive asthma; ICS, inhaled corticosteroid.

incorporation of a ring-containing substructure led to the discovery of the chromone carboxylic acid segment that resulted in a 90-fold increase in potency. Second, replacement of the carboxylic acid with tetrazole led to further increase in potency. Optimization of the lipid backbone (Figure 5.4) led to pranlukast, the first CysLT receptor antagonist approved for marketing (Obase et al., 2004).

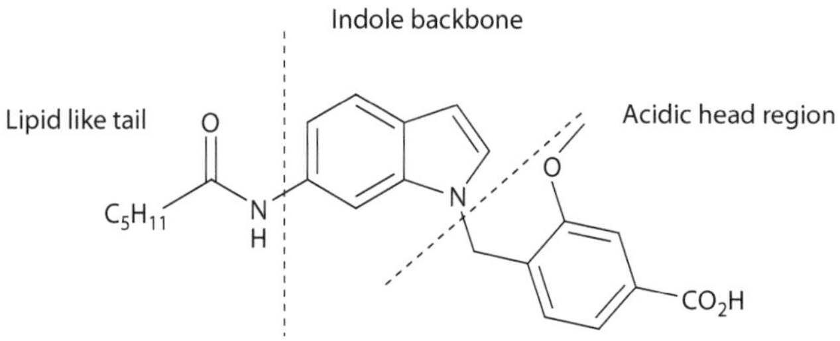

FIGURE 5.4 Initial SAR on the lipid analogs led to the fabrication of indole-containing lead compound.

5.5.1.1 Clinical Evidence

The discovery of leukotriene receptor antagonists introduced a new era in asthma research. $CysLT_1$ receptor antagonists improve pulmonary function and quality of life (QOL), reduce β-agonist use, reduce eosinophilia and asthma exacerbations, and reduced the dose of inhaled glucocorticoids in asthmatic patients (Price et al., 2003; Dempsey, 2000). $CysLT_1$ antagonists that are currently commercially available are zafirlukast, pranlukast, and montelukast. Although there are differences in potency and pharmacokinetics, LTRAs have similar effects. As they are administered orally, LTRAs are of particular value in improving patient compliance. LTRAs are effective bronchoprotective agents blocking the bronchoconstrictive response to a variety of specific or nonspecific bronchial challenges (Riccioni et al., 2007). Zafirlukast can blunt bronchoconstriction induced by inhaled LTD_4 in normal individuals and in patients with asthma. It has also been able to prevent early bronchial reactions to several allergens (Garćia-Marcos et al., 2003). In addition, zafirlukast has also demonstrated anti-inflammatory properties by showing a reduction in both peripheral blood and sputum eosinophils in long-term trials (Nathan et al., 1998). In the prevention of exercise-induced asthma, zafirlukast has been found to be efficacious in various clinical trials. In adult asthmatic patients receiving β2-adrenoceptor agonists, inhaled corticosteroids, or both, 14 days of treatment with zafirlukast, 20 or 80 mg twice daily, reduced both the area under the forced expiratory volume in 1 s (FEV1)-time curve and the maximum fall in FEV1 (Dessanges et al., 1999). In asthmatic children aged 6–14 years, zafirlukast at different doses has also shown its superiority over placebo regarding the maximum FEV1 fall, the area under the curve, and recovery time for return to basal FEV1 (Pearlman et al., 1999). It has also been quite extensively studied as a preventer of asthma symptoms. Most of the trials have included patients with moderate asthma and have studied several endpoints. Three large trials, including a total of 1362 patients who were treated with zafirlukast 20 mg twice daily or placebo for 13 weeks, have resulted in significant improvement in symptom scores, FEV1, and peak expiratory flow rate coupled with a need to use β2-agonist as a rescue medicine (Suissa et al., 1997).

Montelukast, the most commonly prescribed drug, is administered once daily at approved doses of 10 mg for adults, 5 mg for children (aged 6–14 years), and 4 mg for preschool children (aged 2–5 years). A sprinkle formulation is also approved in some countries for use in children from the age of 1 year. It has been the most extensively studied antileukotriene in exercise-induced asthma. Montelukast proved to be as potent as salmeterol and slightly more efficient than zafirlukast after a single dose of each drug, at 1, 2, 4, 8, and 12 h after drug administration (Coreno et al., 2000). There have been a number of trials that support the efficacy of montelukast in comparison to placebo. In a clinical trial of 681 patients with moderate asthma, 408 were treated with montelukast 10 mg twice daily and 273 with placebo. The follow-up period was 12 weeks. Superiority of montelukast compared with placebo regarding FEV1 and QOL score was established (Reiss et al., 1998). Montelukast has also proven its efficacy in children. Two large trials have been conducted in two pediatric age groups: 6–14 and 2–5 years. The beneficial effects of montelukast were found to be significant when compared with placebo, and of a magnitude comparable to those seen in the adult trials. Three clinical trials involving 139, 373, and 401 mild to moderate asthma patients showed a significant increase in FEV1 and PEFR (Garćia-Marcos et al., 2003). In asthmatics inadequately controlled by ICS, addition of montelukast led to a significant reduction in peripheral blood and induced sputum eosinophil counts, suggesting that leukotriene-driven inflammation is not fully suppressed by ICS treatment (Bjermer et al., 2003). One potential explanation for this effect is the observation that in addition to inhibiting the growth of eosinophils locally, montelukast has similar effect centrally on bone marrow progenitor cells. In children with asthma, montelukast has been shown to reduce the levels of exhaled nitric acid, another marker for airway inflammation, both as monotherapy and on top of inhaled corticosteroids. It has been reported that 12 weeks of treatment with montelukast (10 mg once daily) both as monotherapy and in combination with ICS (budesonide, 400 µg, twice daily) decreased airway hyperresponsiveness in patients with mild to moderate persistent asthma (Riccioni et al., 2003).

Pranlukast has mainly been studied in Japan. The twice daily doses of 450 mg were administered to 11 patients for 14 days, and the authors found a reduction of the maximum fall in FEV1 from 50% below baseline before treatment to 30% below baseline after treatment. In another clinical trial conducted in Asia, pranlukast 225 mg twice daily was evaluated versus placebo for 4 weeks in a group of 197 patients with mild to moderate asthma (Yoo et al, 2001). At the end of the treatment period, most endpoints were significantly better for the active treatment group, although the size of the effect over the placebo group was modest.

LTRAs are well tolerated with few side effects, suggesting that the endogenous CysLTs are not important in regulating normal physiological functions. Headaches and gastrointestinal side effects are the most common, but these are rarely severe enough to cause discontinuation of the therapy. The major concern is Churg–Strauss syndrome, with circulating eosinophilia, cardiac failure, and associated eosinophilic vasculitis. It has been reported in a few patients treated with LTRAs. There are several case reports of Churg–Strauss syndrome developing in patients who have not been treated with inhaled or oral corticosteroids, suggesting that it may be a direct effect of the LTRAs. As these drugs are metabolized by the liver, the possibility for significant drug interactions with other drugs metabolized by the cytochrome P450 enzyme system may exist. However, LTRAs have an excellent safety profile and have a good therapeutic index and limited toxicity.

5.5.2 BLT$_1$ AND BLT$_2$ RECEPTOR ANTAGONISTS

LTB$_4$ produces its biological effects via specific G-protein-coupled receptors known as BLT$_1$ and BLT$_2$ (Yokomizo et al., 1997, 2000; Qiu et al., 2006). Till date, most studies of LTB$_4$ receptors have focused on the high-affinity LTB$_4$ receptor, BLT$_1$, expressed exclusively in leukocytes, especially its role in inflammatory responses. Two chemical classes of LTB$_4$ receptor antagonists were developed: benzophenone dicarboxylic acids (Gapinski et al., 1990a, b) and hydroxyacetophenones (Herron et al., 1992). Using inhibition of the specific binding of radiolabeled LTB$_4$ to isolated human neutrophils as a marker of activity, LY223982 (IC$_{50}$ = 13.2 nM) and LY255283 (IC$_{50}$ = 87 nM) were the most potent compounds identified from their respective chemical classes (Figure 5.5) (Silbaugh et al., 1992). Because of the poor oral bioavailability, the uses of these compounds were confined to topical application or i.v. injection. Further modifications of LY223982 led to xanthone dicarboxylic acids (LY210073, William et al., 1993), which mimicked different conformation states of benzophenone dicarboxylic acids. However, it was also not orally active. Finally, efforts were directed toward studying derivatives of LY255283. LY255283 was divided conceptually into three regions: a lipophilic region containing 2-acetophenone, an acid region encompassing the tetrazole moiety, and a section linking the two regions. Structural modifications at either end of the molecule and SAR eventually led to LY293111 (Figure 5.5), which was found to be a more potent and selective LTB$_4$ receptor antagonist with greatly improved potency in cell function assays over the first-generation antagonist. It blocked LTB$_4$-induced activation of neutrophils in whole blood as assessed by the upregulation of the expression of the cell surface adhesion molecule CD11b/CD18 (IC$_{50}$:3.9 nM) (Marder et al., 1995). It had improved selectivity, being about 10,000-folds more potent in inhibiting LTB$_4$-induced CD11b upregulation relative to the corresponding fMLP-mediated response in human neutrophils. It was also highly effective in LTB$_4$-induced acute airway obstruction in guinea pigs when administered orally or intravenously with ED$_{50}$ values of 0.4 and 0.04 mg/kg, respectively. It blocked LTB$_4$-induced pulmonary granulocyte infiltration at doses as low as 0.3 mg/kg and calcium ionophore-induced lung

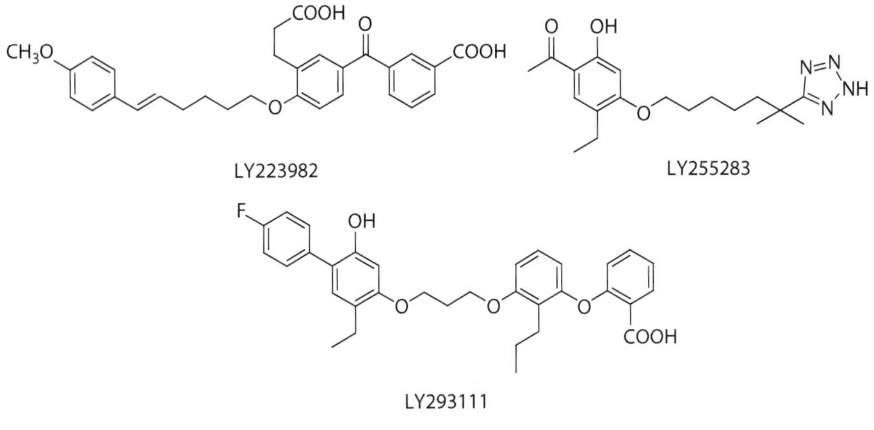

FIGURE 5.5 Structures of BLT$_1$ receptor antagonist.

TABLE 5.3

BLTR Antagonist LY293111 versus Placebo in Asthmatic Patients and the Effect on Content of BAL Fluid after Antigen Challenge[a]

Variable	LY293111	Placebo
Neutrophil cell count	0.04	0.09 million/mL
Myeloperoxidase activity	3.5	16.0 ng/mL
LTB$_4$	2.2	4.6 pg/mL

Source: Adapted from Evans, D. J. et al. 1996. *Thorax* 51:1178–84.

[a] Eosinophils, macrophages, and lymphocytes in BAL fluid did not differ between treatments.

inflammation in guinea pigs 1 h after dosing. The ability of LTB$_4$ to upregulate the expression of CD11b/CD18 has been used to assess the activity of LY293111 in clinical trials. When blood from volunteers receiving 200 mg bid of LY293111 was challenged with LTB$_4$, CD11b/CD18 expression was inhibited by >73% at 4 h.

The ability of LTB$_4$ to amplify neutrophil activity has supported the drive to develop compounds that act by blocking LTB$_4$ response. There are long-acting and potent LTB$_4$ receptor antagonists (CP-195543, biil 284, etc.) that have shown efficacy against inflammation and neutrophilia in primates (Kilfeather, 2002). LTB$_4$ antagonist LY293111 inhibits antigen-induced airway neutrophilia in asthma patients (Table 5.3).

Modifications of LTD$_4$ antagonists and subsequent optimization resulted in the discovery of Ro 25-4094 where the acetophenone moiety was cyclized with the phenol to yield a chromanone group. The compound displayed an ED$_{50}$ of <1 mg/kg even when administered orally 20 h prior to challenge. Development of Ro 25-4094 was discontinued due to animal liver toxicity. Recently, RO5101576 (Figure 5.6) has been identified by Roche.

RO5101576 potently inhibited LTB$_4$-evoked calcium mobilization in HL-60 cells and chemotaxis of human neutrophils. Studies using human BLT$_1$ and BLT$_2$

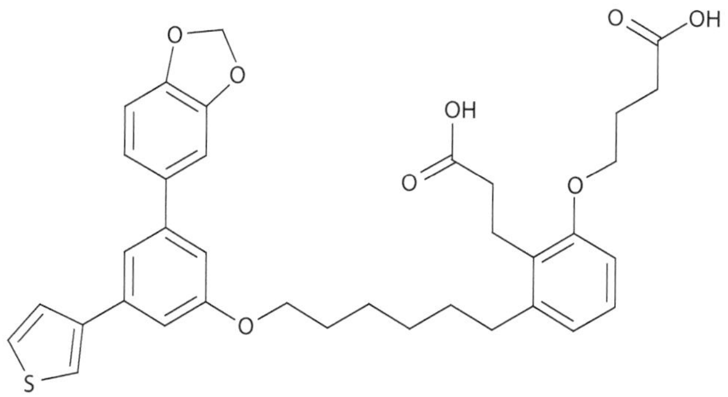

RO5101576

FIGURE 5.6 Structure of dual BLT$_1$ and BLT$_2$ receptor antagonist.

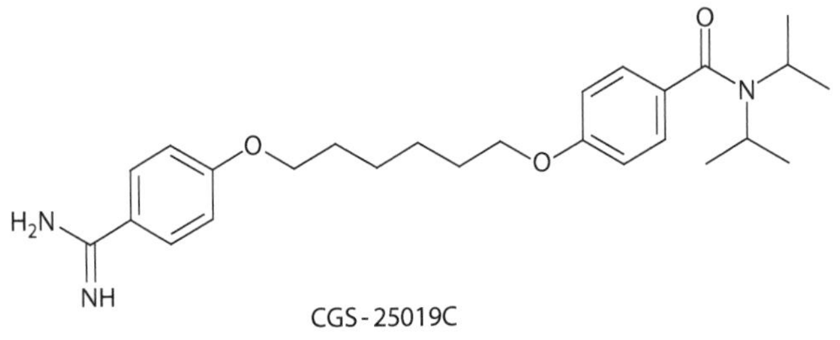

FIGURE 5.7 LTB$_4$ antagonist.

receptors showed that RO5101576 is a dual antagonist. The IC$_{50}$ for RO5101576 at human BLT$_1$ and BLT$_2$ receptors was higher than in HL-60 cells and neutrophil chemotaxis (187 and 379 nM vs. 0.42 and 8 nM, respectively). In nonhuman primates, RO5101576 inhibited allergen and ozone-evoked pulmonary neutrophilia, with comparable efficacy to budesonide allergic responses. When administered to rats and dogs for 2 weeks, RO5101576 was found to be safe and well tolerated with the no observed adverse effect level (NOAEL) of RO5101576 in rats and dogs of 40 and 1000 mg/kg, respectively. Thus, it may represent a potential new treatment for pulmonary neutrophilia in asthma (Hicks et al., 2010).

CGS-25019C is structurally different from the other LTB$_4$ antagonists (Figure 5.7). Instead of an acidic polar group, the molecule has a strong basic group. It inhibits [^{3}H] LTB$_4$ binding on human neutrophils with an IC$_{50}$ of 4 nM. The compound blocked a variety of LTB$_4$-induced cell functions (calcium mobilization, chemotaxis, aggregation, degranulation, and CD11b/CD18 upregulation) at low nanomolar concentration. CGS-25019C inhibited LTB$_4$-induced neutropenia in rats when dosed orally with an ED50 of 4 mg/kg (Ray Chaudhuri et al., 1995). AA-induced edema and neutrophil migration in mice were also inhibited, indicating that the compound blocked responses of endogeneously produced LTB$_4$. Clinical evaluation of the compound was done by measuring *ex vivo* the ability to inhibit upregulation of CD11b/CD18 on human neutrophils. In a rising single oral dose study, the compound inhibited the upregulation of the integrin (ID$_{50}$ = 50 mg). Maximum inhibition was observed 3 h post dosing and $t_{1/2}$ was 3–5 h. Measurement of blood levels in a 7-day multidose study indicated no accumulation of drug or induction of tolerance. The duration of action of the compound was long. A single 300 mg oral dose provided >50% inhibition for 24 h. Doses >500 mg provided even longer periods of inhibition but were accompanied with symptoms of gastrointestinal intolerance such as diarrhea (Morgan et al., 1995).

5.5.3 FLAP INHIBITOR

Although there are other antiasthmatic compounds that target the LT pathway (e.g., LT receptor antagonists), a FLAP inhibitor with improved efficacy, safety, and tolerability to existing LT receptor antagonists would probably capture a significant portion of this market.

MK-886 was the lead compound for the development of the indole series of FLAP inhibitors. It is a potent inhibitor of leukotriene biosynthesis with an IC_{50} of 36 nM but was less potent in whole-blood assay (IC_{50}:2.1 μM). MK-0591 represents a second generation of FLAP inhibitor with an oral half-life of 6 h that strongly inhibits leukotriene biosynthesis in human neutrophils (IC_{50}: 3 nM) and whole blood (IC_{50}: 500 nM). MK-866 was the first FLAP inhibitor to be taken to clinical trials. It was evaluated (500 mg po) in eight atopic asthmatics 1 h before allergen challenge and at 250 mg 2 h afterward. Pulmonary function as compared to the control response was improved by 58% in EAR and 44% in the LAR (Friedman et al., 1993). This combined 750 mg oral dose of MK-866 provided ~50% inhibition of *ex vivo* A23187-stimulated whole-blood LTB_4 biosynthesis and urinary LTE_4 excretion. The degree of LT inhibition measured in stimulated blood and in the urine was disappointing in view of the excellent *in vitro* activity of MK-866. Clinical investigation of MK-866 was subsequently terminated. Clinical studies were also conducted with MK-0591, another potent FLAP inhibitor. It was given to eight atopic asthmatics (given in three 250 mg po doses at 24, 12, and 1.5 h prior to inhaled allergen) where it showed reduction in bronchoconstriction in both early-phase (79%) and late-phase (39%) responses (Diamant et al., 1995).

In clinical studies with asthmatics, the degree of improvement of MK-0591 was not good, so the development was stopped. Activity of *N*-benzyl indole-type translocation inhibitors is structurally associated with (a) hydrophobic *N*-benzyl-*p*-substitution; (b) a 2,2'-dimethyl-substituted propanoic or butanoic acid at C-2; (c) a lipophilic sulfide at C-3; and (d) a nonpolar sterically hindered group at C-5. The minimally active indole is qualitatively represented in Figure 5.8.

More structurally diverse compounds have been found to bind to FLAP (Figure 5.9). Among them, quinoline L-674,573 is a nanomolar FLAP inhibitor (human PMN $IC_{50} = 6$ nM). Quindole L-689,037 possessing both indole and quinoline rings was more potent (human PMN $IC_{50} = 2$ nM) when compared to MK-886 and L-674,573. Thus, three classes of compounds share common binding sites for FLAP, a membrane protein target for structurally diverse compounds (Mancini

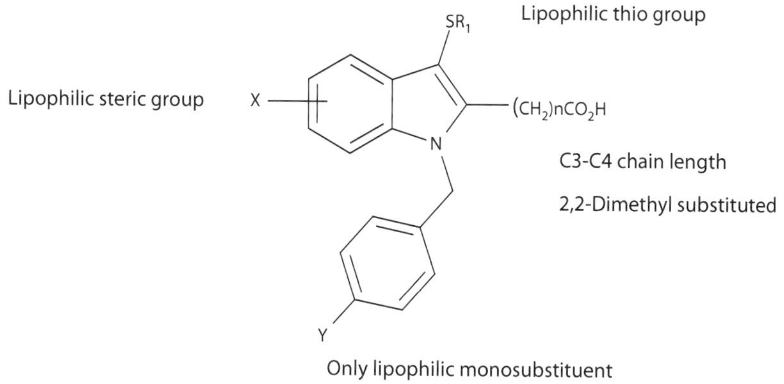

FIGURE 5.8 Structural diversity points in *N*-benzyl indole containing FLAP inhibitors.

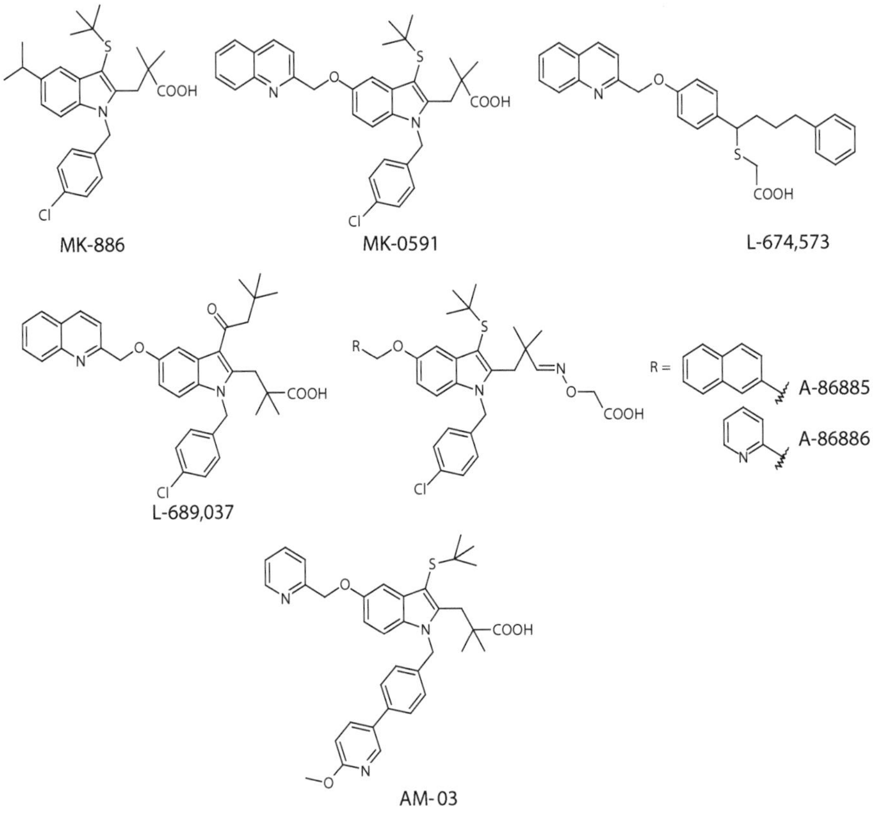

FIGURE 5.9 Some representative examples of potent FLAP inhibitors.

et al., 1992). Other derivatives of MK-0591 with good oral bioavailability were A-86885 and A-86886, which inhibited leukotriene biosynthesis of human neutrophils with an IC_{50} value of 21 and 9 nM, respectively (Kolasa et al., 1997).

Another compound, chiral cycloalkyl-substituted (quinolylmethoxy) phenylacetic acid FLAP inhibitor BAY X1005, (Hatzelmann et al., 1994a) inhibited leukotriene biosynthesis in rat leukocytes with an IC_{50} value of 26 nM. However, BAY X1005 was 10-fold less potent in human leucocytes (IC_{50} = 220 nM) and only weakly active in human whole blood (IC_{50} = 11.6 μM) (Hatzelmann et al., 1994b).

Compared with $CysLT_1$ receptor antagonists, there are few published clinical trials of FLAP inhibitors. Phase I studies at single doses from 50 to 750 mg showed no clinically significant adverse events. BAY X1005 was well absorbed, achieving maximum plasma levels at about 2–3 h and had an estimated oral half-life ranging from 4 to 8 h in humans. Greater than 50% reduction in urinary LTE_4 was found with a single 500 mg po dose. In allergen-induced challenge studies in asthmatics, BAY X1005 (750 mg po) 4 h prior to allergen inhalation attenuated both early and late responses. In a study of 10 severe, chronic, steroid-dependent asthmatics, BAY X1005 (250 mg bid po) in combination with daily steroid treatment (10–30 mg po) in a double-blind placebo-controlled crossover study provided mean FEV1 improvement of 8.5% over baseline for FLAP inhibitor treatment phase compared to 5.7% in the placebo phase. Compound BAY X1005 also provided acute bronchodilatory improvement in asthmatics (Meltzer et al., 1994).

While no clinical trials have directly compared FLAP inhibitors with CysLT$_1$ receptor antagonists, the degree of inhibition of allergen-induced bronchoconstriction by FLAP inhibitors is broadly comparable to that observed in similar trials with CysLT$_1$ receptor antagonists (Holgate et al., 1996) and, physiologically, is likely to reflect the blockade of CysLT synthesis from airway mast cells and infiltrating eosinophils. In a UK population, SNPs identified in the FLAP gene (*ALOX5AP*) and in the gene encoding the terminal enzyme for LTB$_4$ synthesis (*LTA4H*) increased the risk of developing asthma by twofold.

A problem with FLAP inhibitors could be that there is no absolute requirement of FLAP for cellular leukotriene biosynthesis when exogenous AA is available. Thus, FLAP inhibitors might work well under conditions of limited substrate availability, for example, under the usual *in vitro* conditions when cells are stimulated with ionophore only or under physiological and mild inflammatory conditions *in vivo*. In vitro addition of exogenous AA to ionophore stimulated B-lymphocytes or HL-60 cells impair the inhibition of leukotriene biosynthesis by MK-886 (Steinhilber et al., 1993; Jackobsson et al., 1995). The reduced potency of MK-886, MK-0591, and BAY X1005 in whole-blood assays is not only due to plasma protein binding of drugs but also due to the extensive release of AA under the whole-blood assay conditions. Recently, AM-103 has been shown to be a novel, potent, and selective FLAP inhibitor with IC$_{50}$ values of 350, 113, and 117 nM against human, rat, and mouse whole-blood ionophore-stimulated LTB$_4$ production, respectively, and is in development for the treatment of respiratory conditions such as asthma.

There is a worldwide increase in the prevalence of asthma and, despite the success of treatment with inhaled β-agonists and corticosteroids, there is still a significant need for alternative therapies. Although there are other antiasthmatic compounds that target the LT pathway (e.g., LT receptor antagonists), a FLAP inhibitor with improved efficacy, safety, and tolerability to existing LT receptor antagonists, in both asthma and rhinitis, would probably capture a significant portion of this market. A FLAP inhibitor has the potential to inhibit a broad spectrum of LTs and, therefore, such compounds are also being investigated for the management of other respiratory indications as well as having potential for the treatment of other inflammatory diseases.

5.6 SUMMARY AND PERSPECTIVE

AA metabolism through 5-LO pathway has emerged as one of the most extensively researched areas in airway inflammation. It has emerged that LTs generated through 5-LO pathway play an important role in the pathophysiology of inflammatory airway diseases. Improved understanding of LT biology has revealed that it is possible to interfere with LT biology by blocking its effect at the receptor level as well as by inhibiting its biosynthesis. At the present time, it is known that CysLTs as well as LTB$_4$ act on cell surface receptors. A lot of success has been achieved in designing CysLT receptor antagonist. Several molecules, such as montelukast, zafirlukast, and pranlukast, have completed clinical trials and have become drugs. It has become evident that CysLT antagonists such as montelukast are very safe, can be administered orally, needs to be taken once daily, and can be given to children. Several potent antagonists of BLT receptors have been designed, and several of them have been tested in preclinical development and evaluated in human trials. However, unlike CysLT antagonists, BLT antagonists have not been successful in the clinic. In another approach, biosynthesis of LTs was interfered with using FLAP inhibitor. Conceptually, a FLAP inhibitor is expected to inhibit biosynthesis of both CsyLT and LB4 and exhibit better efficacy compared to CysLT antagonists. However, efficacy observed for several potent FLAP inhibitors in the clinical trial was not satisfactory.

REFERENCES

Adkins, J. C. and R. N. Brodgen. 1998. Zafirlukast: A review of its pharmacology and therapeutic potential in the management of asthma. *Drugs* 55:121–44.

Barnes, N. C., P. J. Piper, and J. F. Costello. 1984. Comparative effects of inhaled leukotriene C4, leukotriene D4, and histamine in normal human subjects. *Thorax* 39:500–4.

Bjermer, L., H. Bisgaard, J. Bousquet, L. M. Fabbri, A. P. Greening, T. Haahtela, S. T. Holgate et al. 2003. Montelukast and fluticasone compared with salmeterol and fluticasone in protecting against asthma exacerbation in adults: One year, double blind, randomised, comparative trial. *Br Med J.* 327:891–97.

Brash, A. 1999. Lipoxygenases: Occurrence, functions, catalysis, and acquisition of substrate. *J Biol Chem.* 274:23679–82.

Bresell, A., R. Weinander, G. Lundqvist, H. Raza, M. Shimoji, T. H. Sun, L. Balk et al. 2005. Bioinformatic and enzymatic characterization of the MAPEG superfamily. *FEBS J.* 272:1688–03.

Camp, R. D., A. A. Coutts, M. W. Greaves, A. B. Kay, and M. J. Walport. 1983. Responses of human skin to intradermal injection of leukotrienes C4, D4 and B4. *Br J Pharmacol.* 80:497–502.

Capra, V., M. D. Thompson, A. Sala, D. E. Cole, G. Folco, and G. E. Rovati. 2007. Cysteinyl-leukotrienes and their receptors in asthma and other inflammatory diseases: Critical update and emerging trends. *Med Res Rev.* 27:469–527.

Coffey, M. J., S. E. Wilcoxen, and M. Peters-Golden. 1994. Increases in 5-lipoxygenase activating protein expression account for enhanced capacity for 5-lipoxygenase metabolism that accompanies differentiation of peripheral blood monocytes into alveolar macrophages. *Am J Respir Cell Mol Biol.* 11:153–58.

Coreno, A., M. Skowronski, C. Kotaru, and E. R. McFadden, Jr. 2000. Comparative effects of long-acting beta2-agonists, leukotriene receptor antagonists, and a 5-lipoxygenase inhibitor on exercise-induced asthma. *J Allergy Clin Immunol.* 106:500–6.

Chu, S. J., L. O. Tang, E. Watney, E. Y. Chi, and W. R. Henderson, Jr. 2000. *In situ* amplification of 5-lipoxygenase and 5-lipoxygenase-activating protein in allergic airway inflammation and inhibition by leukotriene blockade. *J Immunol.* 165:4640–48.

Dahlén, S. E., J. Björk, P. Hedqvist, K. E. Arfors, S. Hammarström, J. A. Lindgren, and B. Samuelsson. 1981. Leukotrienes promote plasma leakage and leukocyte adhesion in postcapillary venules: *In vivo* effects with relevance to the acute inflammatory response. *Proc Natl Acad Sci USA* 78:3887–91.

Dahlén, S.-E., P. Hedqvist, P. Westlund, E. Granström, S. Hammarström, J. Å. Lindgren, and O. Rådmark. 1983. Mechanisms for leukotriene-induced contractions of guinea pig airways: Leukotriene C4 has a potent direct action whereas leukotrieneB4 acts indirectly. *Acta Physiol Scand.* 118:393–403.

De Benedictis, F. M., S. Vaccher, and D. de Benedictis. 2008. Montelukast sodium for exercise-induced asthma. *Drugs Today (Barc).* 44:845–55.

Dempsey, O. J. 2000. Leukotriene receptor antagonist therapy. *Postgrad Med J.* 76:767–73.

Dessanges, J. F., C. Prefaut, A. Taytard, R. Matran, I. Naya, A. Compagnon, and A. T. Dinh-Xuan. 1999. The effect of zafirlukast on repetitive exercise-induced bronchoconstriction: The possible role of leukotrienes in exercise-induced refractoriness. *J Allergy Clin Immunol.* 104:1155–61.

Diamant, Z., M. C. Timmers, H. Van der Veen, B. S. Friedman, M. D. Smet, M. Depre, D. Hillard, E. H. Bel, and P. J. Sterk. 1995. The effect of MK-0591, a novel 5-lipoxygenase activating protein inhibitor, on leukotriene biosynthesis and allergeninduced airway responses in asthmatic subjects *in vivo. J Allergy Clin Immunol.* 95:42–51.

Dixon, R. A. F., R. E. Diehl, E. Opas, E. Rands, P. J. Vickers, J. F. Evans, J. W. Gillard, and D. K. Miller. 1990. Requirement of a 5-lipoxygenase-activating protein for leukotriene synthesis. *Nature* 343:282–84.

Drazen, J. M. 1988. Comparative contractile responses to sulfidopeptide leukotrienes in normal and asthanatic human subjects. *Ann NY Acad Sci.* 524:289–97.

Drazen, J. M. and F. K. Austen. 1987. Leukotrienes and airway responses. *Am Rev Respir Dis.* 136:985–98.

Diamant, Z., D. C. Grootendorst, M. Veselic-Charvat, M.C. Timmers, M. De Smet, J. A. Leff, B.C. Seidenberg, A. H. Zwinderman, I. Peszek, and P.J. Sterk. 1999. The effect of montelukast (MK-0476), a cysteinyl leukotriene receptor antagonist, on allergen-induced airway responses and sputum cell counts in asthma. *Clin Exp Allergy* 29:42–51.

Dryden, D. M., C. H. Spooner, M. K. Stickland, B. Vandermeer, L. Tjosvold, L. Bialy, K. Wong, and B. H. Rowe. 2010. Exercise-induced bronchoconstriction and asthma. *Evid Rep Technol Assess (Full Rep).* 189:1–154, v–vi.

Evans, D. J., P. J. Barnes, S. M. Spaethe, E. L. Van Alstyne, M. I. Mitchell, and B. J. O'Connor. 1996. Effects of a leukotriene B4 receptor antagonist, LY293111, on allergen-induced responses in asthma. *Thorax* 51:1178–84.

Evans, J. F., C. Léville, J. A. Mancini, P. Prasit, M. Thérien, R. Zamboni, J.Y. Gauthier, R. Fortin, P. Charleson, and D. E. MacIntyre. 1991. 5-Lipoxygenase-activating protein is the target of a quinoline class of leukotriene synthesis inhibitors. *Mol Pharmacol.* 40:22–7.

Ford-Hutchinson, A. W. 1990. Leukotriene B4 in inflammation. *Crit Rev Immunol.* 10:1–12.

Friedman, B. S., E. H. Bel, A. Buntinx, W. Tanaka, Y. H. Han, S. Shingo, R. Spector, and P. Sterk. 1993. MK-886, an effective oral leukotriene biosynthesis inhibitor on antigen-induced early and late asthmatic reactions in man. *Am Rev Respir Dis.* 147:839–44.

Gapinski, D. M., B. E. Mallet, L. L. Froelich, and W. T. Jackson. 1990a. Benxzophenone dicarboxylic acid antagonists of leukotriene B4. 1. Structure-activity relationships of the benzophenone nucleus. *J Med Chem.* 33:2798–807.

Gapinski, D. M., B. E. Mallet, L. L. Froelich, and W. T. Jackson. 1990b. Benxzophenone dicarboxylic acid antagonists of leukotriene B4. 2. Structure-activity relationships of the lipophilic side chain. *J Med Chem.* 33:2807–13.

Garćia-Marcos, L., A. Schuster, and G. Pérez-Yarza. 2003. Benefit-risk assessment of antileukotrienes in the management of asthma. *Drug Safety* 26:483–518.

Gillard, J., A. W. Ford-Hutchinson, C. Chan, S. Charleson, D. Denis, A. Foster, R. Fortin, S. Leger, C. S. McFarlane, and H. Morton. 1989. L-663,536 (MK-886) (3-[1-(4-chlorobenzyl)-3-tbutyl-thio-5-isopropylindol-2-yl]-2,2-dimethylpropanoic acid), a novel, orally active leukotriene biosynthesis inhibitor. *Can J Physiol Pharmacol.* 67:456–64.

Hafstrom, I., J. Palmblad, C. L. Malmsten, O. Rådmark, and B. Samuelsson. 1981. Leukotriene B4—A stereospecific stimulator for release of lysosomal enzymes from neutrophils; *FEBS Lett.* 130:146–48.

Hallstrand, T. S. and W. R. J. Henderson. 2009. Role of leukotrienes in exercise-induced bronchoconstriction. *Curr Allergy Asthma Rep.* 9:18–25.

Hansson, G., J. Å. Lindgren, S. E. Dahlén, P. Hedqvist, and B. Samuelsson. 1981. Identification and biological activity of novel v-oxidized metabolites of leukotriene B4 from human leukocytes. *FEBS Lett.* 130:107–12.

Hatzelmann, A., R. Fruchtmann, K. H. Mohrs, S. Raddatz, M. Matzke, U. Pleiss, J. Keldenich, and R. Muller-Peddinghaus. 1994a. Mode of action of the leukotriene synthesis (FLAP) inhibitor BAY X 1005: Implications for biological regulation of 5-lipoxygenase. *Agents Actions* 43:64–8.

Hatzelmann, A., R. Fruchtmann, K. H. Mohrs, S. Raddatz, M. Matzke, U. Pleiss, J. Keldenich, and R. Muller-Peddinghaus. 1994b. Mode of action of the leukotriene synthesis (FLAP) inhibitor BAY X1005. Implications for biological regulation of 5-lipoxygenase. *Adv Prostaglandin Thromboxane Leukotriene Res.* 22:23–31.

Hedqvist, P., S. E. Dahlén, L. Gustafsson, S. Hammarström, and B. Samuelsson. 1980. Biological profile of leukotrienes C4 and D4. *Acta Physiol Scand.* 110:331–33.

Hedqvist, P., L. Lindbom, U. Palmertz, and J. Raud. 1994. Microvascular mechanisms in inflammation. *Adv Prostaglandin Thromboxane Leukot Res.* 22:91–9.

Herron, D. K., T. Goodson, N. G. Bollinger, D. Swanson-Bean, I. G. Wright, G. S. Staten, A. R. Thompson, L. L. Froelich, and W. T. Jackson. 1992. Leukotriene B4 receptor antagonists: The LY255283 series of hydroxyacetophenones. *J Med Chem.* 15:1818–28.

Hicks, A., R. Goodnow Jr., G. Cavallo, S. A. Tannu, J. D. Ventre, D. Lavelle, J. M. Lora et al. 2010. Effects of LTB4 receptor antagonism on pulmonary inflammation in rodents and non-human primates. *Prostaglandins Other Lipid Mediat.* 92:33–43.

Hicks, A., S. P. Monkarsh, A. F. Hoffman, and R. Goodnow, Jr. 2007. Leukotriene B4 receptor antagonists as therapeutics for inflammatory disease: Preclinical and clinical developments. *Expert Opin Investig Drugs* 16:1909–20.

Holgate, S. T., P. Bradding, and A. P. Sampson. 1996. Leukotriene antagonists and synthesis inhibitors: New directions in asthma therapy. *J Allergy Clin Immunol.* 98:1–13.

Holloway, J. W., S. J. Barton, S. T. Holgate, M. J. Rose-Zerilli, and I. Sayers. 2008. The role of *LTA4H* and *ALOX5AP* polymorphism in asthma and allergy susceptibility. *Allergy* 63:1046–53.

Jackobsson, P. J., P. Shaskin, P. Larsson, S. Feltenmark, B. Odlander, M. Aguilar-santelises, M. Jondal, P. Biberfield, and H. E. Claesson. 1995. Studies on the regulation and localization of 5-lipoxygenase in human B-lymphocytes. *Eur J Biochim.* 232:37–46.

Joris, I., G. Majno, E. J. Corey, and R. A. Lewis. 1987. The mechanism of vascular leakage induced by leukotriene E4. Endothelial contraction. *Am J Pathol.* 126:19–24.

Kanaoka, Y. and J. A. Boyce. 2004. Cysteinyl leukotrienes and their receptors: Cellular distribution and function in immune and inflammatory responses. *J Immunol.* 173:1503–10.

Kilfeather, S. 2002. 5-Lipoxygenase inhibitors for the treatment of COPD. *Chest* 121:197S–200S.

Kolasa, T., P. Bhatia, C. D. Brooks, K. I. Hulkower, J. B. Bouska, R. R. Harris, and R. L. Bell. 1997. Synthesis of indolylalkoxyiminoalkylcarboxylates as leukotriene biosynthesis inhibitors. *Bioorg Med Chem Lett.* 5:507–14.

Mancini, J. A., P. Prasit, M. G. Coppolino, P. Charleson, S. Leger, J. F. Evans, J. W. Gillard, and P. J. Vickers. 1992. 5-Lipoxygenase-activating protein is the target of a novel hybrid of two classes of leukotriene biosynthesis inhibitors *Molecular Pharmacol.* 41:267–72.

Marder, P., J. S. Sawyer, L. L. Froelich, L. L. Mann, and S. E. Spacthe. 1995. Blockade of human neutrophil activation by 2-[2-propyl-3-[3-[2-ethyl-4-(4-fluorophenyl)-5-hydroxyphenoxy]propoxy]phephenoxy] benzoic acid (LY293111), a novel LTB4 receptor antagonist. *Biochem Phamacol.* 49:1683–90.

Meltzer, S. S., M. A. Johns, E. A. Rechsteiner, S. Jungerwirth, J. M. D'Amico, and E. R. Bleeker. 1994. Bronchodilatory effects of BAY x1005, a 5-lipoxygenase inhibitor, in mild to moderate asthma. *J Allergy Clin Immunol.* 93:294.

Morgan, J., R. Stevens, S. Uziel-Fusi, B. Seligmann, W. Haston, H. Lau, M. Hayes, W. L. Hirschhorn, S. Saris, and A. Piraino. 1995. Multiple dose pharmacokinetics of a mono-aryl-amidine compound (CGS-25019 C) and its inhibition of dihydroxyleukotri-enes (LTB$_4$) induced CD11beta expression. *Clin Pharmacol Ther.* 57:153.

Nathan, R. A., J. A. Bernstein, L. Bielory, C. M. Bonuccelli, W. J. Calhoun, S. P. Galant, L. A. Hanby et al. 1998. Zafirlukast improves asthma symptoms and quality of life in patients with moderate reversible airflow obstruction. *J Allergy Clin Immunol.* 102:935–42.

Obase, Y., T. Shimoda, H. Matsuse, Y. Kondo, I. Machida, T. Kawano, S. Saeki et al. 2004. The position of Pranlukast, a cysteinyl leukotriene receptor antagonist, in the long-term treat-ment of asthma. *Respiration* 71:225–32.

Orning, L. and S. Hammarstrom. 1980. Inhibition of leukotriene C4 and D4 biosynthesis. *J Biol Chem.* 255:8023–6.

Pearlman, D. S., N. K. Ostrom, E. A. Bronsky, C. M. Bonuccelli, and L. A. Hanby. 1999. The leukotriene D4-receptor antagonist zafirlukast attenuates exercise-induced bronchocon-striction in children. *J Pediatr.* 134:273–9.

Piper, P. J. and M. N. Samhoun. 1982. Stimulation of arachidonic acid metabolism and genera-tion of thromboxane A2 by leukotrienes B4, C4 and D4 in guinea-pig lung in vitro. *Br J Pharmacol.* 77:267–75.

Pouliot, M., P. P. McDonald, P. Borgeat, and S. R. G. McColl. 1994. Granulocyte/macrophage colonystimulating factor stimulates the expression of the 5-lipoxygenase-activating pro-tein (FLAP) in human neutrophils. *J Exp Med.* 79:1225–32.

Price, D. B., D. Hernandez, P. Magyar, J. Fiterman, K. M. Beeh, I. G. James, S. Konstantopoulos et al. 2003. Randomised controlled trial of montelukast plus inhaled budesonide versus double dose inhaled budesonide in adult patients with asthma. *Thorax* 58:211–16.

Qiu, H., A. S. Johansson, M. Sjostrom, M. Wan, O. Schroder, J. Palmblad, and J. Z. Haeggstrom. 2006. Differential induction of BLT receptor expression on human endothelial cells by lipopolysaccharide, cytokines, and leukotriene B4. *Proc Nat Acad Sci USA* 103:6913–18.

Ray Chaudhuri, A., B. Kotyik, T. C. Pellas, G. Pastor, L. R. Fryer, M. Morrisey, and A. J. Main. 1995. Effects of CGS25019C and other LTB4 antagonists in the mouse ear edema and rat neutropenia models. *Inflamm Res.* 44:S141–42.

Reiss, T. F., P. Chervinsky, R. J. Dockhorn, S. Shingo, B. Seidenberg, and T. B. Edwards. 1998. Montelukast, a once-daily leukotriene receptor antagonist, in the treatment of chronic asthma: A multicenter, randomized, double-blind trial Montelukast Clinical Research Study Group. *Arch Intern Med.* 158:1213–20.

Riccioni, G., T. Bucciarelli, B. Mancini, C. Di Ilio, and N. D'Orazio. 2007. Anteleukotriene drugs: Clinical application, effectiveness and safety. *Curr Med Chem.* 14:1966–77.

Riccioni, G., R. D. Vecchia, N. D'Orazio, S. Sensi, and M. T. Guagnano. 2003. Comparison of montelukast and budesonide on bronchial reactivity in subjects with mild-moderate per-sistent asthma. *Pulm Pharmacol Ther.* 16:111–14.

Seymour, M. L., S. Rak, D. Aberg, G. C. Riise, J. F. Penrose, Y. Kanaoka, K. F. Austen, S. T. Holgate, and A. P. Sampson. 2001. Leukotriene and prostanoid pathway enzymes in bronchial biopsies of seasonal allergic asthmatics. *Am J Respir Crit Care Med.* 164:2051–56.

Silbaugh, S. A., P. W. Stengel, S. L. Cockerham, C. R. Roman, D. L. Saussy, Jr., S. M. Spaethe, T. Goodson, Jr., D. K. Herron, and J. H. Fleisch. 1992. Pulmonary actions of LY255283, a leukotriene B4 receptor antagonist. *Eur J Pharmacol.* 223:57–64.

Singh, R. K., S. Gupta, S. Dastidar, and A. Ray. 2010. Cysteinyl leukotrienes and their recep-tors: Molecular and functional characteristics. *Pharmacology* 85:336–49.

Sirois, P., S. Roy, P. Borgeat, and S. Picard, and P. Vallerand. 1982. Evidence for a mediator role of thromboxane A2 in the myotropic action of leukotriene B4 (LTB4) on the guinea-pig lung. *Prostaglandins Leukot Med.* 8:157–70.

Soter, N. A., R. A. Lewis, E. J. Corey, and K. F. Austen. 1983. Local effects of synthetic leukot-rienes (LTC4, LTD4, LTE4, and LTB4) in human skin. *J Invest Dermatol.* 80:115–19.

Sousa, A. R., A. Parikh, G. Scadding, C. J. Corrigan, and T. H. Lee. 2002. Leukotriene-receptor expression on nasal mucosal inflammatory cells in aspirin-sensitive rhinosinusitis. *N Engl J Med.* 347:1493–9.

Steinhilber, D., S. Hoshiko, J. Grunewald, O. Radmark, and B. Samuelsson. 1993. Serum factors regulate 5-lipoxygenase activity in maturing HL60 cells. *Biochim Biophys Acta* 1178:1–8.

Suissa, S., R. Dennis, P. Ernst, O. Sheehy, and S. Wood-Dauphinee. 1997. Effectiveness of the leuko-triene receptor antagonist zafirlukast for mild-to-moderate asthma: A randomized, double-blind, placebo-controlled trial. *Ann Intern Med.* 126:177–83.

Thien, F. and E. Walters. 1995. Eicosanoids and asthma: An update. *Prostaglandins Leukot Essent Fatty Acids* 52:271–88.

Walch, L., X. Norel, M. Bäck, J. P. Gascard, S. E. Dahlén, and C. Brink. 2002. Pharmacological evidence for a novel cysteinyl-leukotriene receptor subtype in human pulmonary artery smooth muscle. *Br J Pharmacol.* 137:1339–45.

Wardlaw, A. J., R. Moqbel, O. Cromwell, and A. B. Kay. 1986. Platelet-activating factor. A potent chemotactic and chemokinetic factor for human eosinophils. *J Clin Invest.* 78:1701–06.

William, T. J., J. B. Robert, L. F. Larry, D. M. Gapinski, E. M. Barbara, and J. Scott Sawyer. 1993. Design, synthesis, and pharmacological evaluation of potent xanthone dicarbox-ylic acid leukotriene B4 receptor antagonists. *J Med Chem.* 36:1726–34.

Yokomizo, T., T. Izumi, K. Chang, and T. T. Shimizu. 1997. Hydroxyeicosanoids bind to and activate the low affinity leukotriene B4 receptor, BLT2. *Nature* 387:620–24.

Yokomizo, T., T. Izumi, and T. Shimizu. 2001. Leukotriene B4: Metabolism and signal trans-duction. *Arch Biochem Biophys.* 385:231–41.

Yokomizo, T., K. Kato., K. Terawaki, T. Izumi, and T. Shimizu. 2000. A second leukotriene B(4) receptor, BLT2. A new therapeutic target in inflammation and immunological dis-orders. *J Exp Med.* 192:421–32.

Yoo, S. H., S. H. Park, J. S. Song, K. H. Kang, C. S. Park, J. H. Yoo, B. W. Choi, and M. H. Hahn. 2001. Clinical effects of pranlukast, an oral leukotriene receptor antagonist, in mild-to-moderate asthma: A 4 week randomized multicentre controlled trial. *Respirology* 6:15–21.

Yoshihide, K. and A. B. Joshua. 2004. Cysteinyl leukotrienes and their receptors: Cellular distribution and function in immune and inflammatory responses1. *J Immunol.* 173:1503–10.

Zhu, J., Y. S. Qiu, D. J. Figueroa, V. Bandi, H. Galczenski, K. Hamada, K. K. Guntupalli, J. F. Evans, and P. K. Jeffery. 2005. Localization and upregulation of cysteinyl leukotriene-1 receptor in asthmatic bronchial mucosa. *Am J Respir Cell Mol Biol.* 33:531–40.

6 EETs and Oxo-ETE in Airway Diseases

Jun Yang, Hua Dong, and Bruce D. Hammock

CONTENTS

6.1 EPOXYEICOSATRIENOIC ACIDS

Epoxyeicosatrienoic acids (EETs) are monoepoxides of arachidonic acid and fall into the general class of chemical mediators termed eicosanoids along with prostaglandins (PGs) and thromboxanes (TXs). There are several excellent reviews describing the biosynthesis of EETs, their metabolic pathways, their role as endothelium-derived hyperpolarizing factors, and other biological activities (Fleming 2001; Spector 2009; Spector et al. 2004; Spector and Norris 2007; Zeldin 2001).

In this chapter, we briefly introduce the biosynthesis, metabolism, and function of the EETs, and then focus on biochemical responses to EETs in the airway. In addition, the therapeutic efforts related to the lung diseases will also be discussed, including efforts to alter the endogenous roles of EETs.

6.1.1   BASIC BIOLOGY OF EETs

6.1.1.1   Production of EETs

EETs are mainly formed by cytochrome P450 (CYP) epoxygenases, primarily the CYP 2C and 2J isoforms in humans. They may also be produced by rearrangement of oxidized eicosanoids (Jiang et al. 2004). CYP epoxygenases produce four EET regioisomers by adding an epoxide group across one of the four double bonds of arachidonic acid to yield 5,6-, 8,9-, 11,12-, 14,15-EET as shown in Figure 6.1. In the airway, many CYP 450 epoxygenases were shown and characterized in different species, including guinea pig, dog, rabbit, rat, and human (Knickle and Bend 1994; Stephenson et al. 1996; Zeldin et al. 1995a,b,c, 1996). Immunohistochemical experiments demonstrate that CYP 2J expression in rat and human lung is localized to both ciliated and nonciliated airway epithelial cells, bronchial and vascular smooth muscle cells, and endothelium and alveolar macrophages (Knickle and Bend 1994; Scarborough et al. 1999; Stephenson et al. 1996; Zeldin et al. 1996). Results from studies on rabbits indicate that CYP2B4 likely contributes to pulmonary P450 arachidonic acid epoxygenase activity (Zeldin et al. 1995a,b,c). The CYP2C is an inducible form very important in the vasculature, including pulmonary vasculature. Many other P450 isoforms will produce epoxides from unsaturated fatty acids but their role in chemical mediation is not understood.

Each CYP epoxygenase produces several regioisomers, with one form usually predominating. These enzymes also produce other oxidation products of arachidonic and other fatty acids in varying amounts. Each regioisomer contains two *R/S* enantiomeric forms in different proportions (Capdevila et al. 2000; Spector and Norris 2007; Zeldin et al. 1995a,b,c, 1996). P450 epoxygenases retain the geometry of the precursor olefin; thus, most EETs are *cis-* in geometry. EETs are generally considered as a single class of compounds because they have a number of similar metabolic and functional properties. However, we should keep in mind that there are quantitative and even qualitative differences in the actions of the various regioisomers (Spector et al. 2004). Also, particularly the diols of linoleic acid (18:2 fatty acid) and the epoxides of the ω-3 lipids eicosapentaenoic acid (EPA) and docosahexaenoic acid (DHA) are biologically active (Moghaddam et al. 1997; Morisseau et al. 2010).

Another source for the EETs is membrane lipids from where free EETs can be released directly. In rat plasma, more than 90% of the EETs are contained in lipoprotein phospholipids (Spector et al. 2004). *In vitro* experiments showed that more EETs are released into the culture medium when cells are incubated with a Ca^{2+} ionophore than under basic conditions (Bernstrom et al. 1992; Fang et al. 2003; Weintraub et al. 1997). The stimulated release is also believed to be mediated by cytosolic phospholipase A_2 (cPLA$_2$). Thus, the monitoring of EETs can be a complex question of looking at free or bound EETs: multiple tissues and body fluids to examine, fractionation of protein-bound EETs before analysis, and separation of the multiple classes of glycerides before releasing EETs.

6.1.1.2   Metabolism of EETs

The main pathway of EET metabolism in many cells is conversion to dihydroxyeicosatrienoic acid (DHET) by epoxide hydrolases, primarily sEH (Fretland and

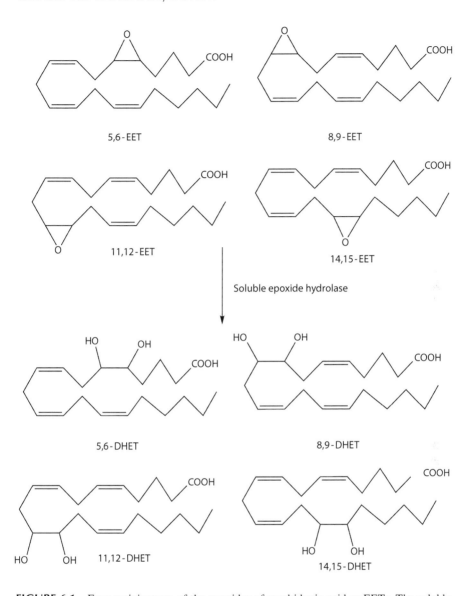

FIGURE 6.1 Four regioisomers of the epoxides of arachidonic acid or EETs. The soluble epoxide hydrolase converts EETs to the corresponding diols or DHETs. The epoxides and particularly the diols of 18:2 fatty acids as well as the epoxides of the ω-3 fatty acids EPA and DHA are biologically active.

Omiecinski 2000; Morisseau and Hammock 2005; Newman et al. 2005; Zeldin 2001; Zeldin et al. 1993). The mammalian sEHs are homodimers of ~62 kDa monomeric subunits (Beetham et al. 1993). The epoxide hydrolase activity resides in the ~35 kDa C-terminal domain, which contains an α/β-hydrolase fold structure distantly homologous to the bacterial haloalkane dehalogenase, the plant soluble EHs, and the microsomal EH (Beetham et al. 1995). The N-terminal domain is a Mg^{2+}-dependent lipid phosphatase (Cronin et al. 2003; Newman et al. 2003). The sEH is broadly distributed in mammalian tissues. The activity has been detected in the liver, kidney, lungs, heart, brain, spleen, adrenals, intestine, urinary bladder, vascular endothelium and smooth muscle, placenta, skin, mammary gland, testis, and leukocytes (Newman et al. 2005). The specific activity of sEH is highest in the liver, followed by the kidney. Outside of the liver, the distribution of the sEH suggests a role in chemical mediation. A study shows that sEH appears to colocalize with CYP 450 2C9 in many tissues (Enayetallah et al. 2004). In the lung, sEH appears localized to vascular tissues (Zheng et al. 2001) and pneumocytes (Enayetallah et al. 2004).

The catalytic site in C-terminal domain of the sEH is responsible for its epoxide hydrolase activity. The reported K_m for epoxy lipids with rodent sEHs range from ~3 to 40 µM with maximum velocities ranging from not detectable to 9 µmol product/min/mg of protein (Zeldin et al. 1993, 1995a,b,c). 14,15-EET is a better substrate for sEH than either 11,12-EET or 8,9-EET, and 5,6-EET is a relatively poor substrate (Zeldin et al. 1993). A simulated analysis performed using DynaFit with the production of 1 µM EET at a rate of 0.01 s^{-1} indicated complete conversion to DHET in 6 min (Widstrom et al. 2001). The result suggests that sEH activity is rapid enough to hydrolyze all the EETs generated under physiological conditions very quickly. Inhibition of sEH or its genetic removal has proven to be beneficial in many disease models because it increases and prolongs the functional effects of EETs. This observation will be discussed later.

Other metabolic pathways for EETs include incorporation to cell lipids by coenzyme A (CoA) (Fang et al. 1995; Shivachar et al. 1995; VanRollins et al. 1993), binding to fatty acids binding protein (FABP) (Glatz and Storch 2001; Veerkamp and Zimmerman 2001; Widstrom et al. 2001), β-oxidation to 16-carbon epoxy-fatty acids (Fang et al. 2000), chain elongation to 22-carbon products (Fang et al. 1995, 1996, 2001), metabolism by cyclooxygenase (COX) to epoxy PGs (Carroll et al. 1992, 1993; Oliw 1984a,b), metabolism by lipoxygenase (LOX) to hepoxilins (HXs) (Pace-Asciak et al. 1983a,b), metabolism by CYP 450 ω-oxidase (4A) to ω-hydroxy compounds (Cowart et al. 2002; Ma et al. 1994), and conjugation with glutathione by glutathione S-transferases although this pathway is quite slow (Murphy and Zarini 2002; Spearman et al. 1985). Figure 6.2 shows an overview of the EET metabolic pathways.

6.1.1.3 *In Vitro* Effects of EETs

EETs have a number of diverse *in vitro* effects in a variety of tissues and cells. Here we emphasize effects on the pulmonary system.

Vasodilation is the most extensively studied EET function. The most potent effects of EETs occur in small resistance vessels. EETs function as an endothelium-derived hyperpolarization factor (EDHF) in a number of vascular beds (Campbell et al. 1996; Campbell and Harder 1999; Fisslthaler et al. 1999). Based on this effect, many studies were performed to target EETs' role in disease models of hypertension (Certikova Chabova et al. 2007; Ellis et al. 2009; Hercule et al. 2009; Imig 2010; Imig et al. 2001, 2002, 2005; Lee et al. 2010; Quilley et al. 1997; Sacerdoti et al. 2003; Wang et al. 2003; Zhao and Imig 2003). Although EETs are broadly active in many models

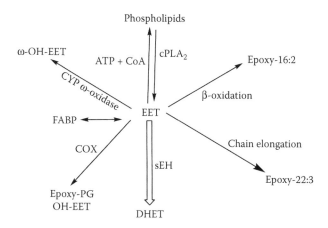

FIGURE 6.2 The main metabolic pathways of EET. The two main metabolic pathways of the EETs are incorporation to the membrane by CoA and hydrolyzation to diols by sEH. (Modified from Figure 2a of Spector, A. A. and A. W. Norris. 2007. *Am J Physiol Cell Physiol* 292(3):C996–1012.)

of hypertension, Dorrance et al. (2005) pointed out that there are strains of SHR rats where the hypertension is resistant to increasing EETs. EETs also appear to modulate the blood pressure back toward the normotensive state, and may not have the dramatic effects observed with some other drugs and chemical mediators.

EETs are also found to be able to activate several ion channels, including the BK_{Ca} channel, which is believed to contribute to the hyperpolarization of the vascular smooth muscle to produce vasorelaxation (Archer et al. 2003; Behm et al. 2009; Benoit et al. 2001; Campbell et al. 2006; Dimitropoulou et al. 2007; Fukao et al. 2001; Krotz et al. 2004; Larsen et al. 2006; Lauterbach et al. 2002; Lee et al. 2006; Lu et al. 2001; Sun et al. 2009; Weston et al. 2005; Wu et al. 2000; Yang et al. 2007; Yousif and Benter 2007; Zhang et al. 2001). In the airway smooth muscle (ASM), it produces bronchodilation through hyperpolarization (Spector and Norris 2007). The inhibition of smooth muscle Cl⁻ channels is also shown to be related to the relaxation of the ASM by EETs (Benoit et al. 2001; Dumoulin et al. 1998; Salvail et al. 1998, 2002). EETs are also reported to affect other ion channels, including K^{+}-ATP, Na^{+}, and L-type Ca^{2+} channels (Spector and Norris 2007).

EETs produce anti-inflammatory effects in a number of ways (Liu et al. 2005; Node et al. 1999; Revermann et al. 2009; Schmelzer et al. 2005; Smith et al. 2005). For example, they transcriptionally down-regulate inflammatory pathways in the arachidonate cascade and other predominantly inflammatory enzymes. They are directly anti-inflammatory by inhibiting NFKB nuclear translocation (Node et al. 1999; Spiecker and Liao 2005). EETs have also been proven to increase peroxisome proliferator-activated receptor-γ (PPAR-γ) transcription activity in endothelial cells and 3T3-L1 preadipocytes, which contributes to the anti-inflammatory effect of laminar flow (Liu et al. 2005). Since inflammation is commonly involved in many diseases, there are many researches targeting on stabilizing the EETs level *in vivo* based on their anti-inflammatory property. We will discuss about the ones related to the airway diseases in detail later in this chapter.

The angiogenic (Cheranov et al. 2008; Fleming 2007; Michaelis et al. 2003, 2008; Pozzi et al. 2005; Webler et al. 2008; Yan et al. 2008; Yang et al. 2009; Zhang et al. 2006) and mitogenic (Chen et al. 1998, 2000; Harris et al. 1990) effects of EETs are also reported although the involved signaling pathways appear very different depending on the species, cell types, and the EET regioisomers. EETs are clearly angiogenic in the presence of vascular endothelial growth factor (VEGF) and this may be important in wound repair and normal development.

6.1.1.3.1   Pulmonary Vascular Tone

Since EETs are important components of the EDHF, it is not surprising that they play important roles in pulmonary vascular tone. In the pulmonary literature, there are numerous reports of the involvement of EETs in the regulation of the pulmonary vascular tone.

In PGF2α-contracted, isolated pulmonary venous rings, 5,6-EET induced relaxation in a concentration-dependent manner. This action of 5,6-EET was prevented by indomethacin (10^{-5} M). These results suggest that 5,6-EET may serve as the COX-dependent endogenous pulmonary vasodilator (Stephenson et al. 1996). Tan et al. (1997) reported that 50 nM 5,6-EET decreased the perfusion pressure of isolated rabbit lungs preconstricted with the thromboxane mimetic U-46619. The dilation of 5,6-EET was evident only when the pulmonary vascular tone was increased. The mechanism of vasodilation to 5,6-EET in the rabbit pulmonary circulation is via both endothelium-derived relaxing factor (EDRF)-NO and PG pathways, and vasodilation is largely EDRF-NO dependent (Tan et al. 1997). 5,6-EET was also found to be a potent vasodilator in newborn piglet pulmonary resistance arteries (PRAs). This dilation is mediated by redundant pathways that include the release of nitric oxide (NO) and COX metabolites and the activation of K_{Ca}^{2+} channels. The endothelium dependence of this response suggests that 5,6-EET is not itself an EDHF but that it may induce the release of EDRF and/or EDHF (Fuloria et al. 2002). In a canine study of intrapulmonary vessels, 5,6-EET decreased active tension in veins contracted with U-46619. 5,6-EET decreased active tension in arteries but not veins contracted with 5-HT. These results support that 5,6-EET is a vasodilator in intact pulmonary circulation, while its dilator activity depends on the constrictor agent present, segmental resistance, and COX activity (Stephenson et al. 1998).

While the same group claimed in their later study that although 5,6-EET dilates large extralobar pulmonary arteries (PA) segments in a COX-dependent manner, in the intact rabbit lung, 5,6-EET produces constriction that requires synthesis of a COX-dependent agonist of the TP receptor other than TX (Stephenson et al. 2003). In another study using isolated pressurized rabbit PA, Zhu et al. demonstrated that all four EETs contract pressurized rabbit PA in a concentration-dependent manner (Zhu et al. 2000). Constriction of isolated rabbit PA to EETs is nonregioselective and depends on intact endothelium and COX, consistent with the formation of a presser prostanoid compound. In other two studies using PA, the results suggest that 5,6-EET-induced contraction is primarily dependent on COX-1 activity (Moreland et al. 2007) and by increasing Rho-kinase activity, phosphorylating myosin light chain, and increasing the Ca^{2+} sensitivity of the contractile apparatus (Losapio et al. 2005). It is common among eicosanoids to have one action in the periphery and the opposite action in the pulmonary system. However, these observations raise a caution regarding the involvement of EETs in pulmonary hypertension, particularly under hypoxic conditions.

Taken together, it is likely that the effect of EETs on PA and veins depends on many factors such as species, preexisting tension, the constrictor agents, hypoxia, or the other eicosanoid enzymes like COX. However, all these reports support that EETs contribute to the tone of pulmonary arteries and veins.

6.1.1.3.2   Effects on Tracheal Cl⁻ Channel

11,12-EET was shown to cause a concentration-dependent decrease in transepithelial voltage, short circuit current, and electrical resistance. These effects were highly enantioselective for $11(R)$, $12(S)$-EET and may be mediated through inhibition of a chloride conductive pathway (Pascual et al. 1998). Salvail (Salvail et al. 1998, 2002) also showed that EETs directly inhibit a Ca^{2+}-insensitive Cl⁻ channel from bovine ASM. These data suggest that EETs cause a net reduction in Cl⁻ secretion by airway epithelium (Jacobs and Zeldin 2001).

6.1.1.3.3   Effects on Airway Smooth Muscle and Bronchomotor Tone

5,6-EET, 11,12-EET, and 14,15-EET act through an epithelium-derived hyperpolarizing factor-like effect (Benoit et al. 2001; Dumoulin et al. 1998; Morin et al. 2007). They activate large-conductance, Ca^{2+}-activated K^+ currents in lipid bilayer reconstitution experiments. 5,6-EET methyl ester and 8,9-EET have been shown to relax histamine-precontracted guinea pig and human bronchi (Zeldin et al. 1995a,b,c).

6.1.2   CLINICAL INDICATIONS AND RESEARCHES OF *IN VIVO* AIRWAY DISEASE MODELS

Two *ex vivo* studies using human lungs showed that EETs are the dominant eicosanoids detected in human lungs upon treatment with the ionophore A23187 (Kiss et al. 2000) and a microbial challenge (Kiss et al. 2010). These results suggest that EETs play an important role in the human airway diseases. In addition, the human gene polymorphisms of EET-related enzymes—CYP 2C8, 2C9, 2J2, and sEH—are also associated with many diseases such as myocardial infarction (Kirchheiner et al. 2008; Liu et al. 2007; Rodenburg et al. 2010), hypertension (Kirchheiner et al. 2008), stroke (Koerner et al. 2007), coronary artery diseases (Spiecker et al. 2004), and type II diabetes (Wang et al. 2010). Although the polymorphism studies published to date are not with the airway diseases, there are many studies targeting the effects of EETs in *in vivo* airway disease models.

6.1.2.1   COPD

Cigarette smoking is frequently associated with several pulmonary diseases: bronchitis, airway obstruction, and emphysema, all of which are referred to as chronic obstructive pulmonary disease (COPD). COPD is a preventable but poorly treatable disease that is characterized by chronic airflow limitation and is associated with abnormal inflammatory response to noxious particles or gases (Qaseem et al. 2007). In the United Sates, COPD is the fourth leading cause of chronic morbidity and mortality and affects more than 5% of the adult population (Global strategy for the diagnosis, management and prevention of COPD, global initiative for chronic obstructive lung disease (GOLD) 2008). Changes in the lung from this disease are thought to be related to smoking-induced inflammation, which leads to cell injury and cellular hyperplasia (Jeffery 1998; Di Stefano et al. 1998). The common symptoms of COPD are chronic and progressive dyspnea, sputum production, and increased number of inflammatory cells, including neutrophils, macrophages, and lymphocytes (Saetta et al. 1997, 1998; Grashoff et al. 1997).

Smith and coworkers demonstrated that EETs have anti-inflammatory properties in tobacco smoke-induced lung inflammation (Smith et al. 2005). In their study, spontaneously hypertensive rats were exposed to tobacco smoke for 3 days (6 h/day). The soluble epoxide hydrolase (sEH) inhibitor 12-(3-adamantan-1-yl-ureido)-dodecanoic acid (AUDA)-nBE (Figure 6.3), which inhibits the enzymatic reaction from EETs to DHETs, was used alone or with exogenous EETs to treat rats exposed to tobacco smoke. Acute exposure to tobacco smoke significantly increased the number of inflammatory cells recovered by bronchoalveolar lavage (BAL), and arachidonate and linoleate diols concentrations in plasma. Administration of the sEH inhibitor alone or combined with exogenous EETs significantly lowered the total BAL cells and macrophages, decreased some arachidonate and linoleate diol concentrations, and increased endogenous 14,15-EET level in plasma. They concluded that the sEH inhibitor, in the presence or absence of EETs, can attenuate acute tobacco smoke-induced lung inflammation and that inhibition of sEH with or without EETs might be used to therapeutically modulate pulmonary epoxy and diol levels for COPD patients.

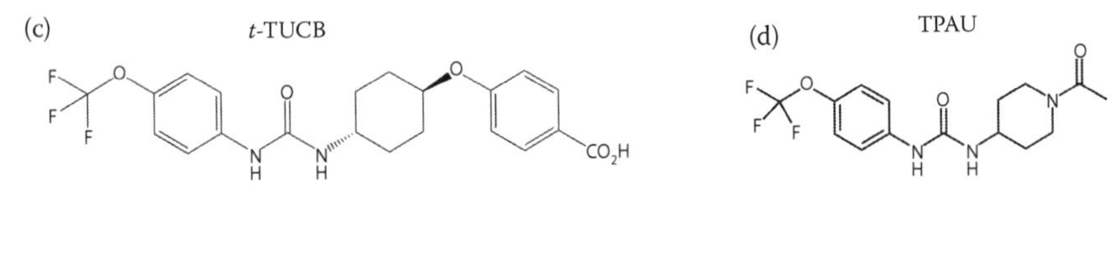

FIGURE 6.3 Structures of several inhibitors of the soluble epoxide hydrolase (sEHI). (a) AUDA, a highly potent but metabolically unstable sEHI. (b) APAU, the IND candidate of Arête Therapeutics. The compound is of moderate potency and short half-life. (c) *t*-TUCB, a cyclohexyl ether inhibitor of high potency and long half-life. (d) TPAU, a potent piperidyl amide inhibitor of high potency and long half-life.

6.1.2.2   Pulmonary Hypertension

As mentioned above, EETs regulate pulmonary vascular tone depending on the species and other vascular contexts. Several studies assessed the roles of EETs by chemical knockout of sEH or by using an EET action in the *in vivo* hypoxia-induced pulmonary hypertension model. Keseru et al. assessed the influence of the sEH and 11,12-EET on pulmonary artery pressure and hypoxic pulmonary vasoconstriction (HPV) in the isolated mouse lung (Keseru et al. 2008a,b). In lungs from wild-type mice, HPV was significantly increased by sEH inhibition but abolished by a selective CYP epoxygenase inhibitor and EET antagonist 14,15-EEZE. Moreover, acute hypoxia and 11,12-EET increased pulmonary pressure in lungs from TRPC6 (+/−) mice; the lungs from TRPC6 (−/−) mice did not respond to either stimuli. The result suggests that EETs are involved in HPV, and EET-induced pulmonary contraction under normoxic and hypoxic conditions involves a TRPC6-dependent pathway (Keseru et al. 2008b; Loot and Fleming 2011). Interestingly, in a subsequent study, after prolonged hypoxia, the acute HPV and the sensitivity to the EET antagonist were increased, but potentiation of vasoconstriction following sEH inhibition was not evident. Chronic hypoxia also stimulated the muscularization of PA and decreased sEH expression in WT mice. In addition, the sEH was expressed in PA of human lungs and was absent in the samples from patients with pulmonary hypertension. All these results suggest that a decrease in sEH expression is intimately linked to pathophysiology of hypoxia-induced pulmonary remodeling and hypertension. However, the N-terminal of sEH, which is a lipid phosphatase, may play a role in this model since sEH inhibitors in the absence of hypoxia did not promote the development of pulmonary hypertension (Keseru et al. 2010).

In another pulmonary hypertension study, Revermann reported that the inhibition of the soluble epoxide hydrolase attenuates monocrotaline-induced pulmonary hypertension in rats (Revermann et al. 2009). The sEH inhibition attenuated the monocrotaline-induced increase in pulmonary artery medial wall and the degree of vascular muscularization (Revermann et al. 2009).

Again, the *in vivo* results of the hypertension models also suggest that the role of EETs in pulmonary hypertension depends on many factors such as the inducer, the length of time span, and the extent of sEH inhibition. The poorly studied N-terminal phosphatase activity of sEH makes the interpretation of these data still more complex. The interpretation of the limited studies to date indicates that sEH inhibitors or EET mimics could be therapeutically useful for treating pulmonary hypertension. However, these studies caution that the sEH may provide protection from pulmonary hypertension, particularly in a hypoxic state.

6.1.2.3   Cancer

In contrast to the reports that EETs are protective in vascular tone and anti-inflammatory effects, EETs were found to be associated with tumor promotion, metastasis due to angiogenesis, atherogenesis (Pritchard et al. 1990), and mitogenesis characteristics. Since EETs are mildly angiogenic in the presence of VEGF, they can be anticipated to increase the vascularization of any solid tumor producing significant VEGF. The effects of EETs are likely to vary dramatically among tumor types.

CYP2J2, which converts arachidonic acid to four regioisomeric EETs, was reported to be elevated in mRNA and protein levels in a set of human-derived cancer cell lines and human cancer tissues but not in noncancer cell lines or adjacent normal tissues. Addition of exogenous EETs or recombinant adeno-associated viral vector (rAAV)-mediated delivery of CYP2J2 or CYP102 F87V (a selective 14,15-EETepoxygenase) markedly promoted the proliferation of cancer cells *in vitro* and *in vivo*. EETs are also found to inhibit carcinoma cell apoptosis. Taken together, the CYP2J2 epoxygenase promotes the neoplastic phenotype and the pathogenesis of a variety of human cancers in animal models of cell culture (Jiang et al. 2005).

Another study by the same group showed that CYP epoxygenase overexpression or EET treatment markedly enhanced the migration, invasion, and prometastatic gene expression profiles in a variety of cancer cell lines *in vitro*. CYP epoxygenase overexpression enhanced tumor metastasis of MDA-MB-231 human breast carcinoma cells to lungs of athymic BALB/C mice (Jiang et al. 2007). In a follow-up study, inhibitors of CYP2J2 exhibited strong activity against human cancers *in vitro* and *in vivo* (Chen et al. 2009).

Using the cultured lung endothelial cells as the *in vitro* system, CYP 2C44, an arachidonic acid epoxygenase, was shown to be a component of the signaling pathways associated with VEGF-stimulated angiogenesis and play a role in growth factor-induced changes in the activation states of the extracellular signal-regulated kinases 1/2 (ERK1/2) and Akt kinase pathways (Yang et al. 2009).

A multikinase inhibitor, sorafenib, which targets in part the VEGF receptor kinase, was found to be a potent sEH inhibitor ($K_i = 17$ nmol/L). The expected *in vivo* shifts in oxylipin profiles resulting from sEH inhibition and the reduction in the acute inflammatory response were found after sorafenib was administered to the lipopolysaccharide (LPS)-induced inflammatory model. In contrast, another VEGF receptor inhibitor, sunitinib, without sEH inhibitory activity did not show the same beneficial effect. This suggests that sEH inhibition, which increases EET levels *in vivo*, contributes the beneficial effects from the VEGF receptor inhibitor—sorafenib (Liu et al. 2009a,b). The similarity of sorafenib to other potent sEH inhibitors (such as *trans*-4-[4-(3-trifluoromethoxy-1-yl ureido) cyclohexyloxy]-benzoic acid (*t*-TUCB) (Figure 6.3) also suggests that one can combine varying degrees of potency of sEH inhibitors on VEGF receptor kinase with inhibition of the sEH itself to balance pro- and antiangiogenic effects of sEH inhibitors.

It demonstrated a pronounced protective effect of sEH inhibitors and thus, likely EETs and related epoxy lipids against the cytotoxic effects of cisplatin. These data suggest the benefit of an increase in EETs in reducing the side effects of chemotherapeutic agents. They also introduce the concept of not only optimizing the relation inhibition of various kinases such as Raf-1 and B-Raf kinase with sorafenib-like molecules, but also of optimizing sEH inhibition.

6.1.2.4 *In Vivo* Studies of Other Airway Diseases

In a series studies using a rat model of acute pneumonia, EETs are formed at significantly lower rates in microsomes prepared from lungs of rats with acute pneumonia caused by *Pseudomonas* (Yaghi et al. 2001). The pulmonary CYP2J4, a prominent enzyme involved in the biosynthesis of EETs and hydroxyeicosatetraenoic acids (HETEs) in rat lungs, is reduced in pneumonia (Yaghi et al. 2003), possibly accounting for the EET reduction. Using the iNOS inhibitor 1400W, the

CYP2J4 protein content and CYP activity were restored in acute pneumonia, indicating an important NO–CYP interaction in pulmonary response to infection (Yaghi et al., 2004). Taken together, all the results implicate the involvement of EETs in bacterial pneumonia.

In a lung ischemia–reperfusion injury study, inhibition of endogenous EET degradation or administration of exogenous 11,12- or 14,15-EET at reperfusion significantly limited the permeability response to ischemia–reperfusion (Townsley et al. 2010). The mechanism of the beneficial effect remains to be investigated. Neither blockade of K_{ATP} nor blockade of TRPV4 impaired the beneficial effect of 11,12-EET. In addition, the expression of the adhesion molecules (VCAM and ICAM) in lung after ischemia–reperfusion was similar to that in controls (Townsley et al. 2010).

6.1.3 DRUG DISCOVERY EFFORTS

Increasing EET levels have proven to be beneficial in numerous disease models. Since EETs broadly seem to return an organism to its homeostasis, it is attractive to stabilize endogenous EETs as a therapeutic approach. Since altering P450 activities has proven difficult therapeutically and because of the diversity and lack of specificity of biosynthetic enzymes and other factors, the regulation of the biosynthesis of EETs is currently not feasible, although recently some phosphodiesterase inhibitors were shown to increase EETs dramatically *in vivo* (Inceoglu et al. 2011). Thus, more drug discovery efforts have been put on the inhibition of the degradation metabolism of EETs by sEH inhibitors or the development of EET agonists. In certain applications, the metabolically unstable but chemically quite stable natural EETs could be used therapeutically. This is particularly attractive in the lung where they could be delivered directly. The use of natural EETs or biochemically stable EET agonists is attractive since it does not rely on endogenous EET production for activity. This approach has potential disadvantages since EET agonists can act where no EETs are produced and unlike sEH inhibitors, overdoses are in theory more likely. Still EET agonists are very attractive, but since there are only a few applications of the EETs antagonists to airway disease research (Keseru et al. 2008a,b, 2010), our focus will be on inhibition of the sEH.

As mentioned in Section 6.1.2, sEH is the major intracellular route of degradative metabolism of EETs. The development of stable, highly potent, and selective sEH inhibitors (Morisseau et al. 1999) allows the stabilization of EETs *in vivo* and to elucidate the biology associated with the sEH (Fretland and Omiecinski 2000; Morisseau and Hammock 2005).

The similarity of the mammalian sEH sequence to a bacterial haloalkane dehalogenase (Verschueren et al. 1993) suggested that sEH might have a similar catalytic mechanism to the bacterial enzyme, which is a member of α/β-hydrolase fold family (Arand et al. 1994). This proved to be the case and Figure 6.4 shows the mechanism of sEH hydration of an epoxide (Morisseau and Hammock 2005). In the first step, the epoxide quickly binds to the active site of the enzyme. Crystal structures show that the murine and human sEHs have a 25-Å-deep L-shaped hydrophobic tunnel, with the nucleophilic aspartate located near the bend of the "L." Either both ends of the tunnel open to the solvent or possibly the enzyme breathes to allow substrates to enter a cleft (Argiriadi et al. 1999). The substrate epoxide is polarized by two acidic tyrosine residues (382 and 465), which hydrogen bond with the epoxide oxygen. The nucleophilic carboxylic acid of Asp334, present on the opposite side of the catalytic cavity from the tyrosines, makes a backside attack on the epoxide, usually at the least sterically hindered and most reactive carbon. The nucleophilic acid is orientated and activated by His523, a second carboxylic amino acid (Asp495) and possibly other amino acids in the catalytic site for this attack. The opening of the epoxide results in a covalent ester bond between the enzyme carboxylic acid and one oxygen functionality of the future diol. This is termed the hydroxyl acyl–enzyme intermediate, and it can be isolated in near quantitative yield. Once the covalent hydroxyl acyl–enzyme is formed, the histidine moves far enough from the nucleophilic acid (now ester) to allow a water molecule to be activated by the acid–histidine pair. This very basic water attacks the carbonyl of the ester, releasing the diol product and regenerating the original enzyme (Morisseau and Hammock 2005).

The first inhibitors discovered for the sEH were kinetically irreversible epoxide-containing compounds with a relatively low turnover. The compounds suffered from rapid metabolism and rapid reactivation of the target enzyme. 1,3-Disubstituted ureas, carbamates, and amides were more recently described as potent and relatively stable inhibitors of sEH (Morisseau et al. 1999). These compounds are competitive tight-binding transition-state inhibitors with nanomolar or picomolar K_i values (Morisseau et al. 1999, 2002). Using classical quantitative structure–activity relationship (QSAR), and medicinal chemistry approaches, the potency of the inhibitors was improved (Kim et al. 2004; Morisseau et al. 2002). A greater challenge was maintaining potency while improving physical properties and the PK and ADME of the inhibitors. Most of the new generation of sEH inhibitors display on one side of the urea functionality a secondary and sometimes a tertiary pharmacophore at 7 and 11 atoms away from the urea carbonyl group, respectively (Kim et al. 2004). This secondary polar pharmacophore hydrogen bonds in the enzyme thus increasing water solubility while maintaining potency (Rose et al. 2010; Tsai et al. 2010; Liu et al. 2009a,b; Hwang et al. 2007; Kim et al. 2007; Wolf et al. 2006; Jones et al. 2005). In Figure 6.4, AUDA represents an early inhibitor lacking the secondary pharmacophore but with the tertiary pharmacophore. Although its physical properties are poor and it is rapidly metabolized, AUDA is a weak EET agonist as well as a very potent sEH inhibitor. Both 1-(1-acetypiperidin-4-yl)-3-adamantanylurea (APAU) and trifluoromethoxyphenyl-3-(1-acetylpiperidin-4-yl) urea (TPAU) have an amide as a secondary pharmacophore. t-TUCB has an ether as a secondary pharmacophore and carboxylic acid as a tertiary pharmacophore.

APAU was the investigational new drug (IND) candidate of a small pharmaceutical company (Anandan et al. 2011). The compound has some desirable physical properties and has shown high potency in some limited disease models. However, it is only a moderate inhibitor of the sEH and its half-life is very short. In contrast, TPAU and t-TUCB are potent inhibitors with good pharmacokinetic properties. Both are very potent compounds with rodent and human enzymes and t-TUCB is potent with the sEH from nonhuman primates, horses, dogs, cats, and other species. There are also several additional classes of sEH inhibitors with potent compounds reported by Boehringer Ingelheim, Merck, and other laboratories (Shen 2010).

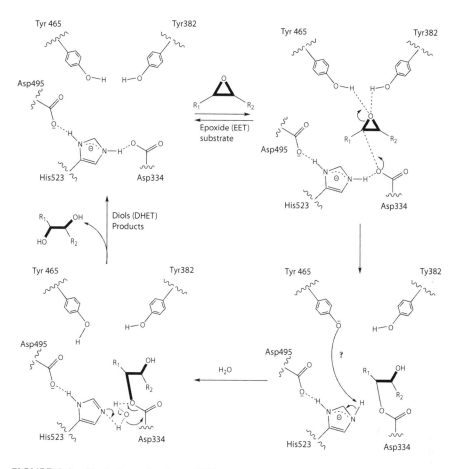

FIGURE 6.4 Catalytic mechanism of EH. The amino acid residue numbers correspond to human sEH. (Adapted from Figure 3 of Morisseau, C. and B. D. Hammock. 2005. *Annu Rev Pharmacol Toxicol* 45:311–33.)

In many disease models, sEH inhibitors (sEHIs) showed beneficial effects. Examples include hypertension (Chiamvimonvat et al. 2007; Huang et al. 2007; Imig 2005, 2010; Jung et al. 2005; Imig et al. 2002), kidney diseases (Imig et al. 2005), inflammation (Liu et al. 2009a,b, 2010; Schmelzer et al. 2005; Smith et al. 2005), heart failure (Qiu et al. 2010), pain (Wagner et al. 2010; Inceoglu et al. 2006, 2008; Schmelzer et al. 2006), stroke (Koerner et al. 2007; Simpkins et al. 2009; Zhang et al. 2007; Dorrance et al. 2005), and others. Presumably, these effects are due at least in part to stabilizing endogenous fatty acid epoxides, including EETs (Morisseau et al. 2010). Moreover, sEHIs were shown to not only have better anti-inflammatory effects but also to synergize with COX inhibitors (Celebrex, aspirin) (Liu et al. 2010; Schmelzer et al. 2006) and 5-lipoxygenase activation protein (FLAP) inhibitors (Liu et al. 2010) in the murine model of acute systemic inflammation. The synergetic effects allow lowering the COX inhibitor doses and thus side effects while improving the anti-inflammatory effects. In addition, the sEHIs appear to reduce the thrombotic side effects associated with the use of some COX inhibitors.

In the airway disease research field, the sEHIs have already been used in the pulmonary hypertension models (Keseru et al. 2008a,b, 2010; Revermann et al. 2009) and tobacco smoke-induced airway inflammation model (Smith et al. 2005). In a tobacco smoke-exposed spontaneous hypertensive rat inflammation model, sEHI was found to be able to attenuate the increases of the airway resistance (Yang et al. unpublished observation). Moreover, in an HDM-induced guinea pig asthmatic model, EETs were found decreased compared to control, which suggests that inhibition of sEH could be beneficial in this model (Yang et al. unpublished observation).

6.1.4 PERSPECTIVE

EETs have been shown to play very important roles in the vascular tone, anti-inflammatory processes, angiogenesis, and mitogenesis. Thus, the sEH enzyme is an attractive therapeutic target for many diseases, including airway diseases, although the multiple functions of EETs and other epoxy fatty acids in the airway are still poorly understood. The multiple roles of epoxy fatty acids in pulmonary biology certainly will be complicated and will likely depend on their context of production. Still the P450 branch of the arachidonate cascade appears promising to develop new diagnostics and treatments for pulmonary diseases.

6.2 OXO-ETE

5-Oxo-ETE (5-oxo-6,8,11,14-eicosatetraenoic acid) is an arachidonic acid metabolite formed by the oxidation of 5S-hydroxy-6,8,11,4-eicosatetraenoic acid (5-HETE) by 5-hydroxyeicosanoid dehydrogenase (5-HEDH), a microsomal enzyme found in leukocytes and platelets. Powell and coworkers (Grant et al. 2009; Powell and Rokach 2005) have excellent reviews on 5-oxo-ETE biochemistry, biology, and chemistry. We will briefly introduce the basic biochemistry and biology of 5-oxo-ETE and its potential in asthma research and treatment rather than duplicating these recent comprehensive reviews.

As mentioned above, 5-oxo-ETE is formed by the oxidation of 5(S)-HETE by 5-HEDH. The synthesis requires $NADP^+$ (Zimpfer et al. 1998) and can be stimulated by the activation of the respiratory burst and by oxidative stress. 5-Oxo-ETE is found to be a very potent chemoattractant for eosinophils and neutrophils and elicits a variety of responses in these cells, including actin polymerization (Powell et al. 1997, 1999), calcium mobilization (Mercier et al. 2004; Powell et al. 1999), integrin expression (Powell et al. 1997, 1999), and degranulation (Iikura et al. 2005; O'Flaherty et al. 1996a,b). The eosinophil cells seem to be the primary target since 5-oxo-ETE is the strongest chemoattractant (Powell et al. 2001; Guilbert et al. 1999; Stamatiou et al. 1998) for eosinophil and the G_i-protein-coupled receptor (OXE receptor) (Grant et al. 2009; Patel et al. 2008; Hosoi et al. 2005; Jones et al. 2003) is highly expressed by eosinophils.

Moreover, 5-oxo-ETE synergize with eotaxin, RANTES (Powell et al. 2001), and platelet-activating factor (PAF) (Powell et al. 1995) in inducing eosinophil migration. 5-Oxo-ETE strongly enhances eosinophil degranulation in responses to a variety of other mediators, including PAF, C5a, LTB4, and FMLP (O'Flaherty et al. 1996a,b). In addition, 5-oxo-ETE stimulates human monocytes to release granulocyte macrophage colony-stimulating factor (GM-CSF) (Stamatiou et al. 2004), which enhances the responsiveness of eosinophils to 5-oxo-ETE and stimulates the formation of 5-LOX products at several levels (Grant et al. 2009).

Given the wide-ranging effects of 5-oxo-ETE on eosinophils, blocking the actions of this lipid mediator may be a useful strategy for the treatment of asthma. A 5-HETE analog containing only one double bond has been synthesized and it can selectively inhibit 5-oxo-ETE formation in stimulated monocytes (Patel et al. 2009). More work is needed to apply to test if this OXE antagonist efficiency in the future *in vivo* model works.

ACKNOWLEDGMENT

The authors' work was supported in part by NIEHS SBRP Grant p42 ES004699, NIEHS Grant R37 ES02710, and NIH/NIEHS Grant R01 ES013933. Partial support was provided by the American Asthma Association #09-0269. Jun Yang was supported by the Elizabeth Nash Memorial fellowship from the Cystic Fibrosis Research, Inc. Bruce D. Hammock is a George and Judy Marcus Senior Fellow of the American Asthma Foundation.

REFERENCES

Anandan, S. K., H. K. Webb, D. Chen , Y. X. Wang, B. R. Aavula, S. Cases, Y. Cheng et al. 2011. 1-(1-Acetyl-piperidin-4-yl)-3-adamantan-1-yl-urea (AR9281) as a potent, selective, and orally available soluble epoxide hydrolase inhibitor with efficacy in rodent models of hypertension and dysglycemia. *Bioorg Med Chem Lett* 21(3):983–8.

Arand, M., D. F. Grant, J. K. Beetham, T. Friedberg, F. Oesch, and B. D. Hammock. 1994. Sequence similarity of mammalian epoxide hydrolases to the bacterial haloalkane dehalogenase and other related proteins. Implication for the potential catalytic mechanism of enzymatic epoxide hydrolysis. *FEBS Lett* 338(3):251–6.

Archer, S. L., F. S. Gragasin, X. Wu, S. Wang, S. McMurtry, D. H. Kim, M. Platonov et al. 2003. Endothelium-derived hyperpolarizing factor in human internal mammary artery is 11,12-epoxyeicosatrienoic acid and causes relaxation by activating smooth muscle BK(Ca) channels. *Circulation* 107(5):769–76.

Argiriadi, M. A., C. Morisseau, B. D. Hammock, and D. W. Christianson. 1999. Detoxification of environmental mutagens and carcinogens: Structure, mechanism, and evolution of liver epoxide hydrolase. *Proc Natl Acad Sci USA* 96(19):10637–42.

Beetham, J. K., D. Grant, M. Arand, J. Garbarino, T. Kiyosue, F. Pinot, F. Oesch, W. R. Belknap, K. Shinozaki, and B. D. Hammock. 1995. Gene evolution of epoxide hydrolases and recommended nomenclature. *DNA Cell Biol* 14(1):61–71.

Beetham, J. K., T. Tian, and B. D. Hammock. 1993. cDNA cloning and expression of a soluble epoxide hydrolase from human liver. *Arch Biochem Biophys* 305(1):197–201.

Behm, D. J., A. Ogbonna, C. Wu, C. L. Burns-Kurtis, and S. A. Douglas. 2009. Epoxyeicosatrienoic acids function as selective, endogenous antagonists of native thromboxane receptors: Identification of a novel mechanism of vasodilation. *J Pharmacol Exp Ther* 328(1):231–9.

Benoit, C., B. Renaudon, D. Salvail, and E. Rousseau. 2001. EETs relax airway smooth muscle via an EpDHF effect: BK(Ca) channel activation and hyperpolarization. *Am J Physiol Lung Cell Mol Physiol* 280(5):L965–73.

Bernstrom, K., K. Kayganich, R. C. Murphy, and F. A. Fitzpatrick. 1992. Incorporation and distribution of epoxyeicosatrienoic acids into cellular phospholipids. *J Biol Chem* 267(6):3686–90.

Campbell, W. B., D. Gebremedhin, P. F. Pratt, and D. R. Harder. 1996. Identification of epoxyeicosatrienoic acids as endothelium-derived hyperpolarizing factors. *Circ Res* 78(3):415–23.

Campbell, W. B. and D. R. Harder.. 1999. Endothelium-derived hyperpolarizing factors and vascular cytochrome P450 metabolites of arachidonic acid in the regulation of tone. *Circ Res* 84(4):484–8.

Campbell, W. B., B. B. Holmes, J. R. Falck, J. H. Capdevila, and K. M. Gauthier. 2006. Regulation of potassium channels in coronary smooth muscle by adenoviral expression of cytochrome P-450 epoxygenase. *Am J Physiol Heart Circ Physiol* 290(1):H64–71.

Capdevila, J. H., J. R. Falck, and R. C. Harris.. 2000. Cytochrome P450 and arachidonic acid bioactivation. Molecular and functional properties of the arachidonate monooxygenase. *J Lipid Res* 41(2):163–81.

Carroll, M. A., M. Balazy, P. Margiotta, J. R. Falck, and J. C. McGiff. 1993. Renal vasodilator activity of 5,6-epoxyeicosatrienoic acid depends upon conversion by cyclooxygenase and release of prostaglandins. *J Biol Chem* 268(17):12260–6.

Carroll, M. A., M. P. Garcia, J. R. Falck, and J. C. McGiff. 1992. Cyclooxygenase dependency of the renovascular actions of cytochrome P450-derived arachidonate metabolites. *J Pharmacol Exp Ther* 260(1):104–9.

Chabova, V. C., H. J. Kramer, I. Vaneckova, M. Thumova, P. Skaroupkova, V. Tesar, J. R. Falck, J. D. Imig, and L. Cervenka. 2007. The roles of intrarenal 20-hydroxyeicosatetraenoic and epoxyeicosatrienoic acids in the regulation of renal function in hypertensive Ren-2 transgenic rats. *Kidney Blood Press Res* 30(5): 335–46.

Chen, C., G. Li, W. Liao, J. Wu, L. Liu, D. Ma, J. Zhou et al. 2009. Selective inhibitors of CYP2J2 related to terfenadine exhibit strong activity against human cancers *in vitro* and *in vivo*. *J Pharmacol Exp Ther* 329(3):908–18.

Chen, J. K., J. Capdevila, and R. C. Harris. 2000. Overexpression of C-terminal Src kinase blocks 14, 15-epoxyeicosatrienoic acid-induced tyrosine phosphorylation and mitogenesis. *J Biol Chem* 275(18):13789–92.

Chen, J. K., J. R. Falck, K. M. Reddy, J. Capdevila, and R. C. Harris. 1998. Epoxyeicosatrienoic acids and their sulfonimide derivatives stimulate tyrosine phosphorylation and induce mitogenesis in renal epithelial cells. *J Biol Chem* 273(44):29254–61.

Cheranov, S. Y., M. Karpurapu, D. Wang, B. Zhang, R. C. Venema, and G. N. Rao. 2008. An essential role for SRC-activated STAT-3 in 14,15-EET-induced VEGF expression and angiogenesis. *Blood* 111(12):5581–91.

Chiamvimonvat, N., C. M. Ho, H. J. Tsai, and B. D. Hammock. 2007. The soluble epoxide hydrolase as a pharmaceutical target for hypertension. *J Cardiovasc Pharmacol* 50(3):225–37.

Cowart, L. A., S. Wei, M. H. Hsu, E. F. Johnson, M. U. Krishna, J. R. Falck, and J. H. Capdevila. 2002. The CYP4A isoforms hydroxylate epoxyeicosatrienoic acids to form high affinity peroxisome proliferator-activated receptor ligands. *J Biol Chem* 277(38):35105–12.

Cronin, A., S. Mowbray, H. Durk, S. Homburg, I. Fleming, B. Fisslthaler, F. Oesch, and M. Arand. 2003. The N-terminal domain of mammalian soluble epoxide hydrolase is a phosphatase. *Proc Natl Acad Sci USA* 100(4):1552–7.

Di Stefano, A., A. Capelli, M. Lusuardi, P. Balbo, C. Vecchio, P. Maestrelli, C. E. Mapp, L. M. Fabbri, C. F. Donner, and M. Saetta. 1998. Severity of airflow limitation is associated with severity of airway inflammation in smokers. *Am J Respir Crit Care Med* 158(4):1277–85.

Dimitropoulou, C., L. West, M. B. Field, R. E. White, L. M. Reddy, J. R. Falck, and J. D. Imig. 2007. Protein phosphatase 2A and Ca^{2+}-activated K^+ channels contribute to 11,12-epoxyeicosatrienoic acid analog mediated mesenteric arterial relaxation. *Prostaglandins Other Lipid Mediat* 83(1–2):50–61.

Dorrance, A. M., N. Rupp, D. M. Pollock, J. W. Newman, B. D. Hammock, and J. D. Imig. 2005. An epoxide hydrolase inhibitor, 12-(3-adamantan-1-yl-ureido)dodecanoic acid (AUDA), reduces ischemic cerebral infarct size in stroke-prone spontaneously hypertensive rats. *J Cardiovasc Pharmacol* 46(6):842–8.

Dumoulin, M., D. Salvail, S. B. Gaudreault, A. Cadieux, and E. Rousseau. 1998. Epoxyeicosatrienoic acids relax airway smooth muscles and directly activate reconstituted KCa channels. *Am J Physiol* 275(3 Pt 1):L423–31.

Ellis, A., K. Goto, D. J. Chaston, T. D. Brackenbury, K. R. Meaney, J. R. Falck, R. J. Wojcikiewicz, and C. E. Hill.. 2009. Enalapril treatment alters the contribution of epoxyeicosatrienoic acids but not gap junctions to endothelium-derived hyperpolarizing factor activity in mesenteric arteries of spontaneously hypertensive rats. *J Pharmacol Exp Ther* 330(2):413–22.

Enayetallah, A. E., R. A. French, M. S. Thibodeau, and D. F. Grant. 2004. Distribution of soluble epoxide hydrolase and of cytochrome P450 2C8, 2C9, and 2J2 in human tissues. *J Histochem Cytochem* 52(4):447–54.

Fang, X., T. L. Kaduce, M. VanRollins, N. L. Weintraub, and A. A. Spector. 2000. Conversion of epoxyeicosatrienoic acids (EETs) to chain-shortened epoxy fatty acids by human skin fibroblasts. *J Lipid Res* 41(1):66–74.

Fang, X., T. L. Kaduce, N. L. Weintraub, M. VanRollins, and A. A. Spector. 1996. Functional implications of a newly characterized pathway of 11,12-epoxyeicosatrienoic acid metabolism in arterial smooth muscle. *Circ Res* 79(4):784–93.

Fang, X., T. L. Kaduce, N. L. Weintraub, S. Harmon, L. M. Teesch, C. Morisseau, D. A. Thompson, B. D. Hammock, and A. A. Spector. 2001. Pathways of epoxyeicosatrienoic acid metabolism in endothelial cells. Implications for the vascular effects of soluble epoxide hydrolase inhibition. *J Biol Chem* 276(18):14867–74.

Fang, X., M. VanRollins, T. L. Kaduce, and A. A. Spector. 1995. Epoxyeicosatrienoic acid metabolism in arterial smooth muscle cells. *J Lipid Res* 36(6):1236–46.

Fang, X., N. L. Weintraub, and A. A. Spector.. 2003. Differences in positional esterification of 14,15-epoxyeicosatrienoic acid in phosphatidylcholine of porcine coronary artery endothelial and smooth muscle cells. *Prostaglandins Other Lipid Mediat* 71(1–2): 33–42.

Fisslthaler, B., R. Popp, L. Kiss, M. Potente, D. R. Harder, I. Fleming, and R. Busse. 1999. Cytochrome P450 2C is an EDHF synthase in coronary arteries. *Nature* 401(6752):493–7.

Fleming, I. 2001. Cytochrome p450 and vascular homeostasis. *Circ Res* 89(9):753–62.

Fleming, I. 2007. Epoxyeicosatrienoic acids, cell signaling and angiogenesis. *Prostaglandins Other Lipid Mediat* 82(1–4):60–7.

Fretland, A. J. and C. J. Omiecinski. 2000. Epoxide hydrolases: Biochemistry and molecular biology. *Chem Biol Interact* 129(1–2):41–59.

Fukao, M., H. S. Mason, J. L. Kenyon, B. Horowitz, and K. D. Keef. 2001. Regulation of BK(Ca) channels expressed in human embryonic kidney 293 cells by epoxyeicosatrienoic acid. *Mol Pharmacol* 59(1):16–23.

Fuloria, M., T. K. Smith, and J. L. Aschner. 2002. Role of 5,6-epoxyeicosatrienoic acid in the regulation of newborn piglet pulmonary vascular tone. *Am J Physiol Lung Cell Mol Physiol* 283(2):L383–9.

Glatz, J. F. and J. Storch. 2001. Unravelling the significance of cellular fatty acid-binding proteins. *Curr Opin Lipidol* 12(3):267–74.

Global strategy for the diagnosis, management and prevention of COPD, global initiative for chronic obstructive lung disease (GOLD). 2008. http://www.goldcopd.org.

Grant, G. E., J. Rokach, and W. S. Powell. 2009. 5-Oxo-ETE and the OXE receptor. *Prostaglandins Other Lipid Mediat* 89(3–4):98–104.

Grashoff, W. F., J. K. Sont, P. J. Sterk, P. S. Hiemstra, W. I. de Boer, J. Stolk, J. Han, and J. M. van Krieken. 1997. Chronic obstructive pulmonary disease: Role of bronchiolar mast cells and macrophages. *Am J Pathol* 151(6):1785–90.

Guilbert, M., C. Ferland, M. Bosse, N. Flamand, S. Lavigne, and M. Laviolette. 1999. 5-Oxo-6,8,11,14-eicosatetraenoic acid induces important eosinophil transmigration through basement membrane components: Comparison of normal and asthmatic eosinophils. *Am J Respir Cell Mol Biol* 21(1):97–104.

Harris, R. C., T. Homma, H. R. Jacobson, and J. Capdevila. 1990. Epoxyeicosatrienoic acids activate Na^+/H^+ exchange and are mitogenic in cultured rat glomerular mesangial cells. *J Cell Physiol* 144(3):429–37.

Hercule, H. C., W. H. Schunck, V. Gross, J. Seringer, F. P. Leung, S. M. Weldon, ACh da Costa Goncalves, Y. Huang, F. C. Luft, and M. Gollasch. 2009. Interaction between P450 eicosanoids and nitric oxide in the control of arterial tone in mice. *Arterioscler Thromb Vasc Biol* 29(1):54–60.

Hosoi, T., E. Sugikawa, A. Chikada, and T. Ohnuki. 2005. TG1019/OXE, a Galpha(i/o)-protein-coupled receptor, mediates 5-oxo-eicosatetraenoic acid-induced chemotaxis. *Biochem Biophys Res Commun* 334(4):987–95.

Huang, H., C. Morisseau, J. Wang, T. Yang, J. R. Falck, B. D. Hammock, and M. H. Wang. 2007. Increasing or stabilizing renal epoxyeicosatrienoic acid production attenuates abnormal renal function and hypertension in obese rats. *Am J Physiol Renal Physiol* 293(1):F342–9.

Hwang, S. H., H. J. Tsai, J. Y. Liu, C. Morisseau, and B. D. Hammock. 2007. Orally bioavailable potent soluble epoxide hydrolase inhibitors. *J Med Chem* 50(16):3825–40.

Iikura, M., M. Suzukawa, M. Yamaguchi, T. Sekiya, A. Komiya, C. Yoshimura-Uchiyama, H. Nagase, K. Matsushima, K. Yamamoto, and K. Hirai. 2005. 5-Lipoxygenase products regulate basophil functions: 5-Oxo-ETE elicits migration, and leukotriene B(4) induces degranulation. *J Allergy Clin Immunol* 116(3):578–85.

Imig, J. D. 2005. Epoxide hydrolase and epoxygenase metabolites as therapeutic targets for renal diseases. *Am J Physiol Renal Physiol* 289(3):F496–503.

Imig, J. D. 2010. Targeting epoxides for organ damage in hypertension. *J Cardiovasc Pharmacol* 56(4):329–35.

Imig, J. D., X. Zhao, J. H. Capdevila, C. Morisseau, and B. D. Hammock. 2002. Soluble epoxide hydrolase inhibition lowers arterial blood pressure in angiotensin II hypertension. *Hypertension* 39(2 Pt 2):690–4.

Imig, J. D., X. Zhao, J. R. Falck, S. Wei, and J. H. Capdevila. 2001. Enhanced renal microvascular reactivity to angiotensin II in hypertension is ameliorated by the sulfonimide analog of 11,12-epoxyeicosatrienoic acid. *J Hypertens* 19(5):983–92.

Imig, J. D., X. Zhao, C. Z. Zaharis, J. J. Olearczyk, D. M. Pollock, J. W. Newman, I. H. Kim, T. Watanabe, and B. D. Hammock. 2005. An orally active epoxide hydrolase inhibitor lowers blood pressure and provides renal protection in salt-sensitive hypertension. *Hypertension* 46(4):975–81.

Inceoglu, B., S. L. Jinks, K. R. Schmelzer, T. Waite, I. H. Kim, and B. D. Hammock. 2006. Inhibition of soluble epoxide hydrolase reduces LPS-induced thermal hyperalgesia and mechanical allodynia in a rat model of inflammatory pain. *Life Sci* 79(24):2311–19.

Inceoglu, B., S. L. Jinks, A. Ulu, C. M. Hegedus, K. Georgi, K. R. Schmelzer, K. Wagner, P. D. Jones, C. Morisseau, and B. D. Hammock. 2008. Soluble epoxide hydrolase and epoxyeicosatrienoic acids modulate two distinct analgesic pathways. *Proc Natl Acad Sci USA.* 105(48):18901–6.

Inceoglu, B., K. Wagner, N. H. Schebb, C. Morisseau, S. L. Jinks, A. Ulu, C. Hegedus, T. Rose, R. Brosnan, and B. D. Hammock. 2011. Analgesia mediated by soluble epoxide hydrolase inhibitors is dependent on cAMP. *Proc Natl Acad Sci USA* 108(12):5093–7.

Jacobs, E. R. and D. C. Zeldin. 2001. The lung HETEs (and EETs) up. *Am J Physiol Heart Circ Physiol* 280(1):H1–10.

Jeffery, P. K. 1998. Structural and inflammatory changes in COPD: A comparison with asthma. *Thorax* 53(2):129–36.

Jiang, H., J. C. McGiff, J. Quilley, D. Sacerdoti, L. M. Reddy, J. R. Falck, F. Zhang, K. M. Lerca, and P. Y. Wong. 2004. Identification of 5,6-trans-epoxyeicosatrienoic acid in the phospholipids of red blood cells. *J Biol Chem* 279(35):36412–8.

Jiang, J. G., C. L. Chen, J. W. Card, S. Yang, J. X. Chen, X. N. Fu, Y. G. Ning, X. Xiao, D. C. Zeldin, and D. W. Wang. 2005. Cytochrome P450 2J2 promotes the neoplastic phenotype of carcinoma cells and is up-regulated in human tumors. *Cancer Res* 65(11):4707–15.

Jiang, J. G., Y. G. Ning, C. Chen, D. Ma, Z. J. Liu, S. Yang, J. Zhou et al. 2007. Cytochrome p450 epoxygenase promotes human cancer metastasis. *Cancer Res* 67(14):6665–74.

Jones, C. E., S. Holden, L. Tenaillon, U. Bhatia, K. Seuwen, P. Tranter, J. Turner et al. 2003. Expression and characterization of a 5-oxo-6E,8Z,11Z,14Z-eicosatetraenoic acid receptor highly expressed on human eosinophils and neutrophils. *Mol Pharmacol* 63(3):471–7.

Jones, P. D., N. M. Wolf, C. Morisseau, P. Whetstone, B. Hock, and B. D. Hammock. 2005. Fluorescent substrates for soluble epoxide hydrolase and application to inhibition studies. *Anal Biochem* 343(1):66–75.

Jung, O., R. P. Brandes, I. H. Kim, Schweda, R. Schmidt, B. D. Hammock, R. Busse, and I. Fleming. 2005. Soluble epoxide hydrolase is a main effector of angiotensin II-induced hypertension. *Hypertension* 45(4):759–65.

Keseru, B., E. Barbosa-Sicard, R. T. Schermuly, R. T. Schermuly, H. Tanaka, B. D. Hammock, N. Weissmann, B. Fisslthaler, and I. Fleming. 2010. Hypoxia-induced pulmonary hypertension: Comparison of soluble epoxide hydrolase deletion vs. inhibition. *Cardiovascular Research* 85(1):232–40.

Keseru, B., B. Fissithaler, and I. Fleming. 2008a. Role of the soluble epoxide hydrolase in hypoxic pulmonary vasoconstriction and pulmonary vascular remodeling. *Circulation* (18):S754–S754.

Keseru, B., E. Barbosa-Sicard, R. Popp, B. Fisslthaler, A. Dietrich, T. Gudermann, B. D. Hammock et al. 2008b. Epoxyeicosatrienoic acids and the soluble epoxide hydrolase are determinants of pulmonary artery pressure and the acute hypoxic pulmonary vasoconstrictor response. *FASEB J* 22(12):4306–15.

Kim, I. H., C. Morisseau, T. Watanabe, and B. D. Hammock. 2004. Design, synthesis, and biological activity of 1,3-disubstituted ureas as potent inhibitors of the soluble epoxide hydrolase of increased water solubility. *J Med Chem* 47(8):2110–22.

Kim, I. H., K. Nishi, H. J. Tsai, T. Bradford, Y. Koda, T. Watanabe, C. Morisseau, J. Blanchfield, I. Toth, and B. D. Hammock. 2007. Design of bioavailable derivatives of 12-(3-adamantan-1-yl-ureido)dodecanoic acid, a potent inhibitor of the soluble epoxide hydrolase. *Bioorg Med Chem* 15(1):312–23.

Kirchheiner, J., I. Meineke, U. Fuhr, C. Rodriguez-Antona, E. Lebedeva, and J. Brockmoller. 2008. Impact of genetic polymorphisms in CYP2C8 and rosiglitazone intake on the urinary excretion of dihydroxyeicosatrienoic acids. *Pharmacogenomics* 9(3):277–88.

Kiss, L., H. Schutte, K. Mayer, H. Grimm, W. Padberg, W. Seeger, and F. Grimminger. 2000. Synthesis of arachidonic acid-derived lipoxygenase and cytochrome P450 products in the intact human lung vasculature. *Am J Respir Crit Care Med* 161(6):1917–23.

Kiss, L., H. Schutte, W. Padberg, N. Weissmann, K. Mayer, T. Gessler, R. Voswinckel, W. Seeger, and F. Grimminge. 2010. Epoxyeicosatrienoates are the dominant eicosanoids in human lungs upon microbial challenge. *Eur Respir J* 36(5):1088–98.

Knickle, L. C. and J. R. Bend. 1994. Bioactivation of arachidonic acid by the cytochrome P450 monooxygenases of guinea pig lung: The orthologue of cytochrome P450 2B4 is solely responsible for formation of epoxyeicosatrienoic acids. *Mol Pharmacol* 45(6):1273–80.

Koerner, I. P., R. Jacks, A. E. DeBarber, D. Koop, P. Mao, D. F. Grant, and N. J. Alkayed. 2007. Polymorphisms in the human soluble epoxide hydrolase gene EPHX2 linked to neuronal survival after ischemic injury. *J Neurosci* 27(17):4642–9.

Krotz, F., T. Riexinger, M. A. Buerkle, K. H. Nithipatikom, T. Gloe, H. Y. Sohn, W. B. Campbell, and U. Pohl. 2004. Membrane-potential-dependent inhibition of platelet adhesion to endothelial cells by epoxyeicosatrienoic acids. *Arterioscler Thromb Vasc Biol* 24(3):595–600.

Larsen, B. T., H. Miura, O. A. Hatoum, W. B. Campbell, B. D. Hammock, D. C. Zeldin, J. R. Falck, and D. D. Gutterman. 2006. Epoxyeicosatrienoic and dihydroxyeicosatrienoic acids dilate human coronary arterioles via BK(Ca) channels: Implications for soluble epoxide hydrolase inhibition. *Am J Physiol Heart Circ Physiol* 290(2):H491–9.

Lauterbach, B., E. Barbosa-Sicard, M. H. Wang, H. Honeck, E. Kargel, J. Theuer, M. L. Schwartzman et al. 2002. Cytochrome P450-dependent eicosapentaenoic acid metabolites are novel BK channel activators. *Hypertension* 39(2 Pt 2):609–13.

Lee, C. R., J. D. Imig, M. L. Edin, J. Foley, L. M. DeGraff, J. A. Bradbury, J. P. Graves et al. 2010. Endothelial expression of human cytochrome P450 epoxygenases lowers blood pressure and attenuates hypertension-induced renal injury in mice. *FASEB J* 24(10):3770–81.

Lee, K. M., S. W. Son, G. Babnigg, and M. L. Villereal. 2006. Tyrosine phosphatase and cytochrome P450 activity are critical in regulating store-operated calcium channels in human fibroblasts. *Exp Mol Med* 38(6):703–17.

Liu, J. Y., S. H. Park, C. Morisseau, S. H. Hwang, B. D. Hammock, and R. H. Weiss. 2009a. Sorafenib has soluble epoxide hydrolase inhibitory activity, which contributes to its effect profile in vivo. *Mol Cancer Ther* 8(8):2193–203.

Liu, J. Y., H. J. Tsai, S. H. Hwang, P. D. Jones, C. Morisseau, and B. D. Hammock. 2009b. Pharmacokinetic optimization of four soluble epoxide hydrolase inhibitors for use in a murine model of inflammation. *Br J Pharmacol* 156(2):284–96.

Liu, J. Y., J. Yang, B. Inceoglu, H. Qiu, A. Ulu, S. H. Hwang, N. Chiamvimonvat, and B. D. Hammock. 2010. Inhibition of soluble epoxide hydrolase enhances the anti-inflammatory effects of aspirin and 5-lipoxygenase activation protein inhibitor in a murine model. *Biochem Pharmacol* 79(6):880–7.

Liu, P. Y., Y. H. Li, T. H. Chao, H. L. Wu, L. J. Lin, L. M. Tsai, and J. H. Chen. 2007. Synergistic effect of cytochrome P450 epoxygenase CYP2J2*7 polymorphism with smoking on the onset of premature myocardial infarction. *Atherosclerosis* 195(1):199–206.

Liu, Y., Y. Zhang, K. Schmelzer, T. S. Lee, X. Fang, Y. Zhu, A. A. Spector et al. 2005. The antiinflammatory effect of laminar flow: The role of PPARgamma, epoxyeicosatrienoic acids, and soluble epoxide hydrolase. *Proc Natl Acad Sci USA* 102(46):16747–52.

Loot, A. E. and I. Fleming. 2011. Cytochrome P450-derived epoxyeicosatrienoic acids and pulmonary hypertension: Central role of transient receptor potential (TRP) C6 channels. *J Cardiovasc Pharmacol* 57(2):140–7.

Losapio, J. L., R. S. Sprague, A. J. Lonigro, and A. H. Stephenson. 2005. 5,6-EET-induced contraction of intralobar pulmonary arteries depends on the activation of Rho-kinase. *J Appl Physiol* 99(4):1391–6.

Lu, T., P. V. Katakam, M. VanRollins, N. L. Weintraub, A. A. Spector, and H. C. Lee. 2001. Dihydroxyeicosatrienoic acids are potent activators of Ca(2+)-activated K(+) channels in isolated rat coronary arterial myocytes. *J Physiol* 534(Pt 3):651–67.

Ma, Y. H., M. L. Schwartzman, and R. J. Roman. 1994. Altered renal P-450 metabolism of arachidonic acid in Dahl salt-sensitive rats. *Am J Physiol* 267(2 Pt 2):R579–89.

Mercier, F., C. Morin, M. Cloutier, S. Proteau, J. Rokach, W. S. Powell, and E. Rousseau. 2004. 5-Oxo-ETE regulates tone of guinea pig airway smooth muscle via activation of Ca^{2+} pools and Rho-kinase pathway. *Am J Physiol Lung Cell Mol Physiol* 287(4):L631–40.

Michaelis, U. R., B. Fisslthaler, M. Medhora, D. Harder, I. Fleming, and R. Busse. 2003. Cytochrome P450 2C9-derived epoxyeicosatrienoic acids induce angiogenesis via cross-talk with the epidermal growth factor receptor (EGFR). *FASEB J* 17(6):770–2.

Michaelis, U. R., N. Xia, E. Barbosa-Sicard, J. R. Falck, and I. Fleming. 2008. Role of cytochrome P450 2C epoxygenases in hypoxia-induced cell migration and angiogenesis in retinal endothelial cells. *Invest Ophthalmol Vis Sci* 49(3):1242–7.

Moghaddam, M. F., D. F. Grant, J. M. Cheek, J. F. Greene, K. C. Williamson, and B. D. Hammock. 1997. Bioactivation of leukotoxins to their toxic diols by epoxide hydrolase. *Nat Med* 3(5):562–6.

Moreland, K. T., J. D. Procknow, R. S. Sprague, J. L. Iverson, A. J. Lonigro, and A. H. Stephenson. 2007. Cyclooxygenase (COX)-1 and COX-2 participate in 5,6-epoxyeicosatrienoic acid-induced contraction of rabbit intralobar pulmonary arteries. *J Pharmacol Exp Ther* 321(2):446–54.

Morin, C., M. Sirois, V. Echave, M. M. Gomes, and E. Rousseau. 2007. Epoxyeicosatrienoic acid relaxing effects involve Ca^{2+}-activated K$^+$ channel activation and CPI-17 dephosphorylation in human bronchi. *Am J Respir Cell Mol Biol* 36(5):633–641.

Morisseau, C., M. H. Goodrow, D. Dowdy, J. Zheng, J. F. Greene, J. R. Sanborn, and B. D. Hammock. 1999. Potent urea and carbamate inhibitors of soluble epoxide hydrolases. *Proc Natl Acad Sci USA* 96(16):8849–54.

Morisseau, C., M. H. Goodrow, J. W. Newman, C. E. Wheelock, D. L. Dowdy, and B. D. Hammock. 2002. Structural refinement of inhibitors of urea-based soluble epoxide hydrolases. *Biochem Pharmacol* 63(9):1599–608.

Morisseau, C. and B. D. Hammock. 2005. Epoxide hydrolases: Mechanisms, inhibitor designs, and biological roles. *Annu Rev Pharmacol Toxicol* 45:311–33.

Morisseau, C., B. Inceoglu, K. Schmelzer, H. J. Tsai, S. L. Jinks, C. M. Hegedus, and B. D. Hammock. 2010. Naturally occurring monoepoxides of eicosapentaenoic acid and docosahexaenoic acid are bioactive antihyperalgesic lipids. *J Lipid Res* 51(12):3481–90.

Murphy, R. C. and S. Zarini. 2002. Glutathione adducts of oxyeicosanoids. *Prostaglandins Other Lipid Mediat* 68–69:471–82.

Newman, J. W., C. Morisseau, and B. D. Hammock. 2005. Epoxide hydrolases: Their roles and interactions with lipid metabolism. *Prog Lipid Res* 44(1):1–51.

Newman, J. W., C. Morisseau, T. R. Harris, and B. D. Hammock. 2003. The soluble epoxide hydrolase encoded by EPXH2 is a bifunctional enzyme with novel lipid phosphate phosphatase activity. *Proc Natl Acad Sci USA* 100(4):1558–63.

Node, K., Y. Huo, X. Ruan, B. Yang, M. Spiecker, K. Ley, D. C. Zeldin, and J. K. Liao. 1999. Anti-inflammatory properties of cytochrome P450 epoxygenase-derived eicosanoids. *Science* 285(5431):1276–9.

O'Flaherty, J. T., M. Kuroki, A. B. Nixon, J. Wijkander, E. Yee, S. L. Lee, P. K. Smitherman, R. L. Wykle, and L. W. Daniel. 1996a. 5-Oxo-eicosanoids and hematopoietic cytokines cooperate in stimulating neutrophil function and the mitogen-activated protein kinase pathway. *J Biol Chem* 271(30):17821–8.

O'Flaherty, J. T., M. Kuroki, A. B. Nixon, Wijkander, E. Yee, S. L. Lee, P. K. Smitherman, R. L. Wykle, and L. W. Daniel. 1996b. 5-Oxo-eicosatetraenoate is a broadly active, eosinophil-selective stimulus for human granulocytes. *J Immunol* 157(1):336–42.

Oliw, E. H. 1984a. Isolation and chemical conversion of two novel prostaglandin endoperoxides: 5(6)-epoxy-PGG1 and 5(6)-epoxy-PGH1. *FEBS Lett* 172(2):279–83.

Oliw, E. H. 1984b. Metabolism of 5(6)oxidoeicosatrienoic acid by ram seminal vesicles. Formation of two stereoisomers of 5-hydroxyprostaglandin I1. *J Biol Chem* 259(5):2716–21.

Pace-Asciak, C. R., E. Granstrom, and B. Samuelsson. 1983a. Arachidonic acid epoxides. Isolation and structure of two hydroxy epoxide intermediates in the formation of 8,11,12- and 10,11,12-trihydroxyeicosatrienoic acids. *J Biol Chem* 258(11):6835–40.

Pace-Asciak, C. R., K. Mizuno, S. Yamamoto, E. Granstrom, and B. Samuelsson. 1983b. Oxygenation of arachidonic acid into 8,11,12- and 10,11,12-trihydroxyeicosatrienoic acid by rat lung. *Adv Prostaglandin Thromboxane Leukot Res* 11:133–9.

Pascual, J. M. S., A. McKenzie, J. R. Yankaskas, J. R. Falck, and D. C. Zeldin. 1998. Epoxygenase metabolites of arachidonic acid affect electrophysiologic properties of rat tracheal epithelial cells. *J Pharmacol Exp Ther* 286(2):772–79.

Patel, P., C. Cossette, J. R. Anumolu, S. Gravel, A. Lesimple, O. A. Mamer, J. Rokach, and W. S. Powell. 2008. Structural requirements for activation of the 5-oxo-6*E*,8*Z*,11*Z*,14*Z*-eicosatetraenoic acid (5-oxo-ETE) receptor: Identification of a mead acid metabolite with potent agonist activity. *J Pharmacol Exp Ther* 325(2):698–707.

Patel, P., C. Cossette, J. R. Anumolu, K. R. Erlemann, G. E. Grant, J. Rokach, and W. S. Powell. 2009. Substrate selectivity of 5-hydroxyeicosanoid dehydrogenase and its inhibition by 5-hydroxy-Delta6-long-chain fatty acids. *J Pharmacol Exp Ther* 329(1):335–41.

Powell, W. S., S. Ahmed, S. Gravel, and J. Rokach. 2001. Eotaxin and RANTES enhance 5-oxo-6,8,11,14-eicosatetraenoic acid-induced eosinophil chemotaxis. *J Allergy Clin Immunol* 107(2):272–8.

Powell, W. S., D. Chung, and S. Gravel. 1995. 5-Oxo-6,8,11,14-eicosatetraenoic acid is a potent stimulator of human eosinophil migration. *J Immunol* 154(8):4123–32.

Powell, W. S., S. Gravel, F. Halwani, C. S. Hii, Z. H. Huang, A. M. Tan, and A. Ferrante. 1997. Effects of 5-oxo-6,8,11,14-eicosatetraenoic acid on expression of CD11b, actin polymerization, and adherence in human neutrophils. *J Immunol* 159(6):2952–9.

Powell, W. S., S. Gravel, and F. Halwani. 1999. 5-Oxo-6,8,11,14-eicosatetraenoic acid is a potent stimulator of L-selectin shedding, surface expression of CD11b, actin polymerization, and calcium mobilization in human eosinophils. *Am J Respir Cell Mol Biol* 20(1):163–70.

Powell, W. S. and J. Rokach. 2005. Biochemistry, biology and chemistry of the 5-lipoxygenase product 5-oxo-ETE. *Prog Lipid Res* 44(2–3):154–83.

Pozzi, A., I. Macias-Perez, T. Abair, S. Wei, Y. Su, R. Zent, J. R. Falck, and J. H. Capdevila. 2005. Characterization of 5,6- and 8,9-epoxyeicosatrienoic acids (5,6- and 8,9-EET) as potent *in vivo* angiogenic lipids. *J Biol Chem* 280(29):27138–46.

Pritchard, K. A., Jr., R. R. Tota, M. B. Stemerman, and P. Y. Wong. 1990. 14,15-Epoxyeicosatrienoic acid promotes endothelial cell dependent adhesion of human monocytic tumor U937 cells. *Biochem Biophys Res Commun* 167(1):137–42.

Qaseem, A., V. Snow, P. Shekelle, K. Sherif, T. J. Wilt, S. Weinberger, and D. K. Owens. 2007. Diagnosis and management of stable chronic obstructive pulmonary disease: A clinical practice guideline from the American College of Physicians. *Ann Intern Med* 147(9):633–8.

Qiu, H., N. Li, J. Y. Liu, T. R. Harris, B. D. Hammock, and N. Chiamvimonvat. 2010. Soluble epoxide hydrolase inhibitors and heart failure. *Cardiovasc Ther* 29(2):99–111.

Quilley, J., D. Fulton, and J. C. McGiff. 1997. Hyperpolarizing factors. *Biochem Pharmacol* 54(10):1059–70.

Revermann, M., E. Barbosa-Sicard, E. Dony, R. T. Schermuly, C. Morisseau, G. Geisslinger, I. Fleming, B. D. Hammock, and R. P. Brandes. 2009. Inhibition of the soluble epoxide hydrolase attenuates monocrotaline-induced pulmonary hypertension in rats. *J Hypertens* 27(2):322–31.

Rodenburg, E. M., L. E. Visser, A. H. Danser, A. Hofman, C. van Noord, J. C. Witteman, A. G. Uitterlinden, and B. H. Stricker. 2010. Genetic variance in CYP2C8 and increased risk of myocardial infarction. *Pharmacogenet Genomics* 20(7):426–34.

Rose, T. E., C. Morisseau, J. Y. Liu, B. Inceoglu, P. D. Jones, J. R. Sanborn, and B. D. Hammock. 2010. 1-Aryl-3-(1-acylpiperidin-4-yl)urea inhibitors of human and murine soluble epoxide hydrolase: Structure-activity relationships, pharmacokinetics, and reduction of inflammatory pain. *J Med Chem* 53(19):7067–75.

Sacerdoti, D., A. Gatta, and J. C. McGiff. 2003. Role of cytochrome P450-dependent arachidonic acid metabolites in liver physiology and pathophysiology. *Prostaglandins Other Lipid Mediat* 72(1–2):51–71.

Saetta, M., A. Di Stefano, G. Turato, F. M. Facchini, L. Corbino, C. E. Mapp, P. Maestrelli, A. Ciaccia, and L. M. Fabbri. 1998. CD8+ T-lymphocytes in peripheral airways of smokers with chronic obstructive pulmonary disease. *Am J Respir Crit Care Med* 157(3 Pt 1):822–6.

Saetta, M., G. Turato, F. M. Facchini, L. Corbino, R. E. Lucchini, G. Casoni, P. Maestrelli, C. E. Mapp, A. Ciaccia, and L. M. Fabbri. 1997. Inflammatory cells in the bronchial glands of smokers with chronic bronchitis. *Am J Respir Crit Care Med* 156(5):1633–9.

Salvail, D., M. Cloutier, and E. Rousseau. 2002. Functional reconstitution of an eicosanoid-modulated Cl⁻ channel from bovine tracheal smooth muscle. *Am J Physiol Cell Physiol* 282(3):C567–77.

Salvail, D., M. Dumoulin, and E. Rousseau. 1998. Direct modulation of tracheal Cl⁻ channel activity by 5,6- and 11,12-EET. *Am J Physiol Lung Cell Mol Physiol* 275(3):L432–41.

Scarborough, P. E., J. Ma, W. Qu, and D. C. Zeldin. 1999. P450 subfamily CYP2J and their role in the bioactivation of arachidonic acid in extrahepatic tissues. *Drug Metab Rev* 31(1):205–34.

Schmelzer, K. R., B. Inceoglu, L. Kubala, I. H. Kim, S. L. Jinks, J. P. Eiserich, and B. D. Hammock. 2006. Enhancement of antinociception by coadministration of non-steroidal anti-inflammatory drugs and soluble epoxide hydrolase inhibitors. *Proc Natl Acad Sci USA* 103(37):13646–51.

Schmelzer, K. R., L. Kubala, J. W. Newman, I. H. Kim, J. P. Eiserich, and B. D. Hammock. 2005. Soluble epoxide hydrolase is a therapeutic target for acute inflammation. *Proc Natl Acad Sci USA* 102(28):9772–7.

Shen, H. C. 2010. Soluble epoxide hydrolase inhibitors: A patent review. *Expert Opin Ther Pat* 20(7):941–56.

Shivachar, A. C., K. A. Willoughby, and E. F. Ellis. 1995. Effect of protein kinase C modulators on 14,15-epoxyeicosatrienoic acid incorporation into astroglial phospholipids. *J Neurochem* 65(1):338–46.

Simpkins, A. N., R. D. Rudic, D. A. Schreihofer, S. Roy, M. Manhiani, H. J. Tsai, B. D. Hammock, and J. D. Imig. 2009. Soluble epoxide inhibition is protective against cerebral ischemia via vascular and neural protection. *Am J Pathol* 174(6):2086–95.

Smith, K. R., K. E. Pinkerton, T. Watanabe, T. L. Pedersen, S. J. Ma, and B. D. Hammock. 2005. Attenuation of tobacco smoke-induced lung inflammation by treatment with a soluble epoxide hydrolase inhibitor. *Proc Natl Acad Sci USA* 102(6):2186–91.

Spearman, M. E., R. A. Prough, R. W. Estabrook, J. R. Falck, S. Manna, K. C. Leibman, R. C. Murphy, and J. Capdevila. 1985. Novel glutathione conjugates formed from epoxyeicosatrienoic acids (EETs). *Arch Biochem Biophys* 242(1):225–30.

Spector, A. A. 2009. Arachidonic acid cytochrome P450 epoxygenase pathway. *J Lipid Res* 50(Suppl):S52–6.

Spector, A. A., X. Fang, G. D. Snyder, and N. L. Weintraub. 2004. Epoxyeicosatrienoic acids (EETs): Metabolism and biochemical function. *Prog Lipid Res* 43(1):55–90.

Spector, A. A. and A. W. Norris. 2007. Action of epoxyeicosatrienoic acids on cellular function. *Am J Physiol Cell Physiol* 292(3):C996–1012.

Spiecker, M., H. Darius, T. Hankeln, M. Soufi, A. M. Sattler, J. R. Schaefer, K. Node et al. 2004. Risk of coronary artery disease associated with polymorphism of the cytochrome P450 epoxygenase CYP2J2. *Circulation* 110(15): 2132–6.

Spiecker, M. and J. K. Liao. 2005. Vascular protective effects of cytochrome p450 epoxygenase-derived eicosanoids. *Arch Biochem Biophys* 433(2):413–20.

Stamatiou, P. B., C. C. Chan, G. Monneret, D. Ethier, J. Rokach, and W. S. Powell. 2004. 5-Oxo-6,8,11,14-eicosatetraenoic acid stimulates the release of the eosinophil survival factor granulocyte/macrophage colony-stimulating factor from monocytes. *J Biol Chem* 279(27):28159–64.

Stamatiou, P., Q. Hamid, R. Taha, W. Yu, T. B. Issekutz, J. Rokach, S. P. Khanapure, and W. S. Powell. 1998. 5-Oxo-ETE induces pulmonary eosinophilia in an integrin-dependent manner in Brown Norway rats. *J Clin Invest* 102(12):2165–72.

Stephenson, A. H., R. S. Sprague, and A. J. Lonigro. 1998. 5,6-Epoxyeicosatrienoic acid reduces increases in pulmonary vascular resistance in the dog. *Am J Physiol* 275(1 Pt 2):H100–9.

Stephenson, A. H., R. S. Sprague, J. L. Losapio, and A. J. Lonigro. 2003. Differential effects of 5,6-EET on segmental pulmonary vasoactivity in the rabbit. *Am J Physiol Heart Circ Physiol* 284(6):H2153–61.

Stephenson, A. H., R. S. Sprague, N. L. Weintraub, L. McMurdo, and A. J. Lonigro. 1996. Inhibition of cytochrome P-450 attenuates hypoxemia of acute lung injury in dogs. *Am J Physiol* 270(4 Pt 2):H1355–62.

Sun, P., W. Liu, D. H. Lin, P. Yue, R. Kemp, L. M. Satlin, and W. H. Wang. 2009. Epoxyeicosatrienoic acid activates BK channels in the cortical collecting duct. *J Am Soc Nephrol* 20(3):513–23.

Tan, J. Z., G. Kaley, and G. H. Gurtner. 1997. Nitric oxide and prostaglandins mediate vasodilation to 5,6-EET in rabbit lung. *Adv Exp Med Biol* 407:561–6.

Townsley, M. I., C. Morisseau, B. Hammock, and J. A. King. 2010. Impact of epoxyeicosatrienoic acids in lung ischemia-reperfusion injury. *Microcirculation* 17(2):137–46.

Tsai, H. J., S. H. Hwang, C. Morisseau, J. Yang, P. D. Jones, T. Kasagami, I. H. Kim, and B. D. Hammock. 2010. Pharmacokinetic screening of soluble epoxide hydrolase inhibitors in dogs. *Eur J Pharm Sci* 40(3):222–38.

VanRollins, M., T. L. Kaduce, H. R. Knapp, and A. A. Spector. 1993.. 14,15-Epoxyeicosatrienoic acid metabolism in endothelial cells. *J Lipid Res* 34(11):1931–42.

Veerkamp, J. H. and A. W. Zimmerman. 2001. Fatty acid-binding proteins of nervous tissue. *J Mol Neurosci* 16(2–3):133–42; discussion 151–7.

Verschueren, K. H., F. Seljee, H. J. Rozeboom, K. H. Kalk, and B. W. Dijkstra. 1993. Crystallographic analysis of the catalytic mechanism of haloalkane dehalogenase. *Nature* 363(6431):693–8.

Wagner, K., B. Inceoglu, S. S. Gill, and B. D. Hammock. 2010. Epoxygenated fatty acids and soluble epoxide hydrolase inhibition: Novel mediators of pain reduction (dagger). *J Agric Food Chem* 59(7):2816–24.

Wang, C. P., W. C. Hung, T. H. Yu, C. A. Chiu, L. F. Lu, F. M. Chung, C. H. Hung, S. J. Shin, H. J. Chen, and Y. J. Lee. 2010. Genetic variation in the G-50T polymorphism of the cytochrome P450 epoxygenase CYP2J2 gene and the risk of younger onset type 2 diabetes among Chinese population: Potential interaction with body mass index and family history. *Exp Clin Endocrinol Diabetes* 118(6):346–52.

Wang, M. H., A. Smith, Y. Zhou, H. H. Chang, S. Lin, X. Zhao, J. D. Imig, and A. M. Dorrance. 2003. Downregulation of renal CYP-derived eicosanoid synthesis in rats with diet-induced hypertension. *Hypertension* 42(4):594–9.

Webler, A. C., R. Popp, T. Korff, U. R. Michaelis, C. Urbich, R. Busse, and I. Fleming. 2008. Cytochrome P450 2C9-induced angiogenesis is dependent on EphB4. *Arterioscler Thromb Vasc Biol* 28(6):1123–9.

Weintraub, N. L., X. Fang, T. L. Kaduce, M. VanRollins, P. Chatterjee, and A. A. Spector. 1997. Potentiation of endothelium-dependent relaxation by epoxyeicosatrienoic acids. *Circ Res* 81(2):258–67.

Weston, A. H., M. Feletou, P. M. Vanhoutte, J. R. Falck, W. B. Campbell, and G. Edwards. 2005. Bradykinin-induced, endothelium-dependent responses in porcine coronary arteries: Involvement of potassium channel activation and epoxyeicosatrienoic acids. *Br J Pharmacol* 145(6):775–84.

Widstrom, R. L., A. W. Norris, and A. A. Spector. 2001. Binding of cytochrome P450 mono-oxygenase and lipoxygenase pathway products by heart fatty acid-binding protein. *Biochemistry* 40(4):1070–6.

Wolf, N. M., C. Morisseau, P. D. Jones, B. Hock, and B. D. Hammock. 2006. Development of a high-throughput screen for soluble epoxide hydrolase inhibition. *Anal Biochem* 355(1):71–80.

Wu, S. N., H. F. Li, and H. T. Chiang. 2000. Actions of epoxyeicosatrienoic acid on large-conductance $Ca(2+)$-activated $K(+)$ channels in pituitary GH(3) cells. *Biochem Pharmacol* 60(2):251–62.

Yaghi, A., J. R. Bend, C. D. Webb, D. C. Zeldin, S. Weicker, S. Mehta, and D. G. McCormack. 2004. Excess nitric oxide decreases cytochrome P-450 2J4 content and P-450-dependent arachidonic acid metabolism in lungs of rats with acute pneumonia. *Am J Physiol Lung Cell Mol Physiol* 286(6):L1260–7.

Yaghi, A., J. A. Bradbury, D. C. Zeldin, S. Mehta, J. R. Bend, and D. G. McCormack. 2003. Pulmonary cytochrome P-450 2J4 is reduced in a rat model of acute *Pseudomonas* pneumonia. *Am J Physiol Lung Cell Mol Physiol* 285(5):L1099–105.

Yaghi, A., C. D. Webb, J. A. Scott, S. Mehta, J. R. Bend, and D. G. McCormack. 2001. Cytochrome P450 metabolites of arachidonic acid but not cyclooxygenase-2 metabolites contribute to the pulmonary vascular hyporeactivity in rats with acute *Pseudomonas* pneumonia. *J Pharmacol Exp Ther* 297(2):479–88.

Yan, G., S. Chen, B. You, and J. Sun. 2008. Activation of sphingosine kinase-1 mediates induction of endothelial cell proliferation and angiogenesis by epoxyeicosatrienoic acids. *Cardiovasc Res* 78(2):308–14.

Yang, S., S. Wei, A. Pozzi, and J. H. Capdevila. 2009. The arachidonic acid epoxygenase is a component of the signaling mechanisms responsible for VEGF-stimulated angiogenesis. *Arch Biochem Biophys* 489(1–2):82–91.

Yang, W., B. B. Holmes, V. R. Gopal, R. V. Kishore, B. Sangras, X. Y. Yi, J. R. Falck, and W. B. Campbell. 2007. Characterization of 14,15-epoxyeicosatrienoyl-sulfonamides as 14,15-epoxyeicosatrienoic acid agonists: Use for studies of metabolism and ligand binding. *J Pharmacol Exp Ther* 321(3):1023–31.

Yousif, M. H. and I. F. Benter. 2007. Role of cytochrome P450 metabolites of arachidonic acid in regulation of corporal smooth muscle tone in diabetic and older rats. *Vascul Pharmacol* 47(5–6):281–7.

Zeldin, D. C. 2001. Epoxygenase pathways of arachidonic acid metabolism. *J Biol Chem* 276(39):36059–62.

Zeldin, D. C., R. N. DuBois, J. R. Falck, and J. H. Capdevila. 1995a. Molecular cloning, expression and characterization of an endogenous human cytochrome P450 arachidonic acid epoxygenase isoform. *Arch Biochem Biophys* 322(1):76–86.

Zeldin, D. C., J. Foley, J. X. Ma, J. E. Boyle, J. M. S. Pascual, C. R. Moomaw, K. B. Tomer, C. Steenbergen, and S. Wu. 1996. CYP2J subfamily P450s in the lung: Expression, localization, and potential functional significance. *Mol Pharmacol* 50(5):1111–7.

Zeldin, D. C., J. Kobayashi, J. R. Falck, B. S. Winder, B. D. Hammock, J. R. Snapper, and J. H. Capdevila. 1993. Regiofacial and enantiofacial selectivity of epoxyeicosatrienoic acid hydration by cytosolic epoxide hydrolase. *J Biol Chem* 268(9):6402–7.

Zeldin, D. C., J. D. Plitman, J. Kobayashi, R. F. Miller, J. R. Snapper, J. R. Falck, J. L. Szarek, R. M. Philpot, and J. H. Capdevila. 1995b. The rabbit pulmonary cytochrome P450 arachidonic acid metabolic pathway: Characterization and significance. *J Clin Invest* 95(5):2150–60.

Zeldin, D. C., S. Wei, J. R. Falck, B. D. Hammock, J. R. Snapper, and J. H. Capdevila. 1995c. Metabolism of epoxyeicosatrienoic acids by cytosolic epoxide hydrolase: Substrate structural determinants of asymmetric catalysis. *Arch Biochem Biophys* 316(1):443–51.

Zhang, B., H. Cao, and G. N. Rao. 2006. Fibroblast growth factor-2 is a downstream mediator of phosphatidylinositol 3-kinase-Akt signaling in 14,15-epoxyeicosatrienoic acid-induced angiogenesis. *J Biol Chem* 281(2):905–14.

Zhang, W., I. P. Koerner, R. Noppens, M. Grafe, H. J. Tsai, C. Morisseau, A. Luria, B. D. Hammock, J. R. Falck, and N. J. Alkaye. 2007. Soluble epoxide hydrolase: A novel therapeutic target in stroke. *J Cereb Blood Flow Metab* 27(12):1931–40.

Zhang, Y., C. L. Oltman, T. Lu, H. C. Lee, K. C. Dellsperger, and M. VanRollins. 2001. EET homologs potently dilate coronary microvessels and activate BK(Ca) channels. *Am J Physiol Heart Circ Physiol* 280(6):H2430–40.

Zhao, X. and J. D. Imig. 2003. Kidney CYP450 enzymes: Biological actions beyond drug metabolism. *Curr Drug Metab* 4(1):73–84.

Zheng, J., C. G. Plopper, J. Lakritz, D. H. Storms, and B. D. Hammock. 2001. Leukotoxin-diol: A putative toxic mediator involved in acute respiratory distress syndrome. *Am J Respir Cell Mol Biol* 25(4):434–8.

Zhu, D., M. Bousamra, 2nd, D. C. Zeldin, J. R. Falck, M. Townsley, D. R. Harder, R. J. Roman, and E. R. Jacobs. 2000. Epoxyeicosatrienoic acids constrict isolated pressurized rabbit pulmonary arteries. *Am J Physiol Lung Cell Mol Physiol* 278(2):L335–43.

Zimpfer, U., C. Hofmann, S. Dichmann, E. Schopf, and J. Norgauer. 1998. Synthesis, biological effects and pathophysiological implications of the novel arachidonic acid metabolite 5-oxo-eicosatetraenoic acid (review). *Int J Mol Med* 2(2):149–53.

7 The Eosinophil Chemoattractant 5-Oxo-ETE

William S. Powell and Joshua Rokach

CONTENTS

7.1　INTRODUCTION

Arachidonic acid (AA) is converted to a large number of biologically active products by cyclooxygenases, lipoxygenases, and cytochrome P450 enzymes (Funk, 2001). The most important of these are the cyclooxygenase and 5-lipoxygenase (5-LO) pathways, which result in the formation of prostanoids and leukotrienes (LTs). In some cases, different oxygenases cooperate to produce additional products such as the lipoxins (Schwab and Serhan, 2006). 5-LO oxygenates AA at the 5-position to give 5S-hydroperoxy-6E,8Z,11Z,14Z-eicosatetraenoic acid (5-HpETE), which it then cyclizes to LTA$_4$ (Figure 7.1). LTA$_4$ is converted to LTB$_4$ by LTA$_4$ hydrolase and to the cysteinyl-LT LTC$_4$ by LTC$_4$ synthase (Murphy and Gijon, 2007). Alternatively, if relatively large amounts of LTA$_4$ are formed, it decomposes nonenzymatically to biologically inactive 6-*trans* isomers of LTB$_4$. A significant amount of 5-HpETE is normally lost from 5-LO and is reduced by peroxidases such as glutathione peroxidase to 5S-hydroxy-6E,8Z,11Z,14Z-eicosatetraenoic acid (5-HETE) (Rådmark et al., 2007).

LTs are potent proinflammatory mediators and are important in host defense (Serezani et al., 2005) as well as inflammatory diseases such as asthma (Peters-Golden and Henderson, 2007). The neutrophil is a major target for LTB$_4$, which is a potent chemoattractant for these cells. The cysteinyl-LTs, in particular LTD$_4$, are important mediators of asthma, inducing contraction and remodeling of airway smooth muscle, mucous secretion, increased vascular permeability, and T$_{H2}$ cytokine

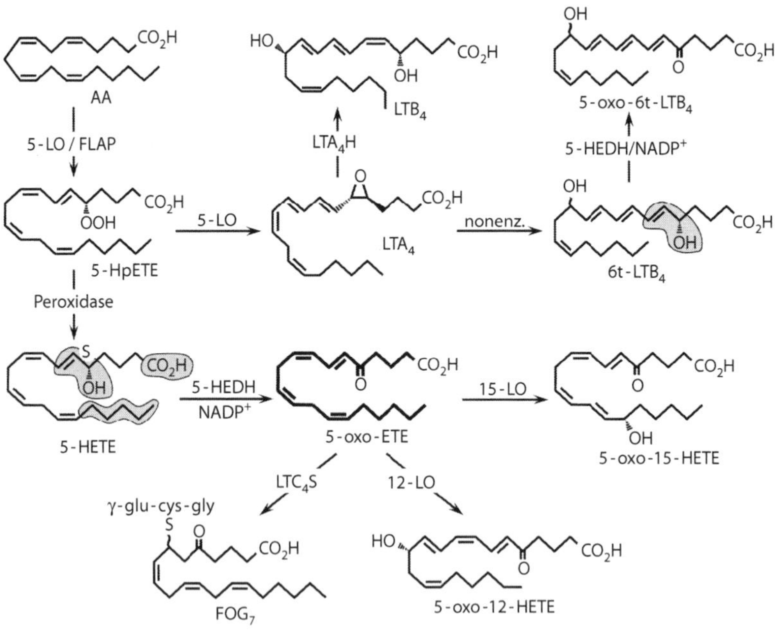

FIGURE 7.1　Biosynthesis and metabolism of 5-oxo-ETE and other 5-LO products. The shaded areas on the structure of 5-HETE indicate the parts of the molecule necessary for metabolism by 5-HEDH. The 5S-hydroxy-δ^6-*trans* region of 6-*trans* isomers of LTB$_4$ is also highlighted for comparison. Not all of the metabolic pathways of 5-oxo-ETE are shown.

production. The biological actions of LTs are mediated principally by three highly selective receptors: the BLT_1 receptor for LTB_4, the $cysLT_1$ receptor for LTD_4, and the $cysLT_2$ receptor for LTC_4 and LTD_4 (Peters-Golden and Henderson, 2007).

5-HETE was known to stimulate neutrophils by a mechanism that did not involve receptors for LTs or certain other lipid mediators (O'Flaherty, 1985; O'Flaherty and Nishihira, 1987; O'Flaherty et al., 1988). However, its potency was lower than would be expected if its actions were mediated by a selective receptor. We subsequently found that 5-HETE is metabolized to a much more potent product by an enzyme in neutrophils (Powell et al., 1992) that we initially identified because of its ability to oxidize 6-*trans* isomers of LTB_4 to 5-oxo metabolites (Powell and Gravelle, 1988) (Figure 7.1). About the same time, Schröder identified a related product that was formed following incubation of AA with soybean lipoxygenase (Schwenk et al., 1992). These two products were identified as 5-oxo-6*E*,8*Z*,11*Z*,14*Z*-eicosatetraenoic acid (5-oxo-ETE) (Powell et al., 1992) and 5-oxo-15*S*-hydroxy-6*E*,8*Z*,11*Z*,13*E*-eicosatetraenoic acid (5-oxo-15-HETE) (Schwenk et al., 1992).

7.2 FORMATION OF 5-OXO-ETE

7.2.1 OXIDATION OF 5-HETE BY 5-HEDH

Enzymatic oxidation of 5-HETE to 5-oxo-ETE was first observed in neutrophil microsomal fractions (Powell et al., 1992). The enzyme responsible for this reaction was named 5-hydroxyeicosanoid dehydrogenase (5-HEDH) and has not yet been cloned or sequenced. 5-HEDH activity is dependent on $NADP^+$, and is found in most types of inflammatory cells, including eosinophils (Powell et al., 1995a), monocytes and circulating lymphocytes (Zhang et al., 1996), tonsillar B cells (Grant et al., 2011a), and monocytes-derived dendritic cells (Zimpfer et al., 2000). It is also present in a variety of cells that do not express significant amounts of 5-LO activity, including platelets (Powell et al., 1999b), endothelial (Erlemann et al., 2006) cells, keratinocytes (Cossette et al., 2008), airway epithelial and smooth muscle cells (Erlemann et al., 2007b), and various tumor cell lines (Grant et al., 2011b).

7.2.1.1 Substrate Selectivity of 5-HEDH

5-HEDH is highly selective for both of its substrates, 5-HETE and $NADP^+$. Although high concentrations of NAD^+ will promote oxidation of 5-HETE to some extent, it is about 10,000 times less potent than $NADP^+$ (Erlemann et al., 2007a). It is also highly selective for its substrate, 5-HETE, as there is little or no metabolism of other regioisomers, including 8-, 9-, 11-, 12-, and 15-HETE (Powell et al., 1992). Similarly, 5R-HETE is a very poor substrate, indicating that it is stereospecific for a 5*S*-hydroxyl group (Powell et al., 1994). Some dihydroxyeicosatetraenoic acids (diHETEs) containing a 5*S*-hydroxyl group followed by a 6-*trans*-double bond are metabolized by 5-HEDH, although at a rate slower than 5-HETE. These include 6-*trans*-LTB_4, 12-*epi*-6-*trans*-LTB_4, and 5*S*,15*S*-diHETE, the latter being converted to 5-oxo-15-HETE. In contrast, LTB_4, which has a 6-*cis* rather than a 6-*trans*-double bond, is not metabolized (Powell et al., 1992). A free carboxyl group is important for metabolism by 5-HEDH, as the methyl ester of 5-HETE is a very poor substrate. The hydrophobic ω-end of the molecule also appears to interact with the enzyme, as substrates with chain lengths of less than 16 carbons are poorly metabolized and hydroxylation of the ω-methyl group blocks metabolism (Patel et al., 2009).

Although AA is clearly the most important substrate for 5-LO, other polyunsaturated fatty acids (PUFA) are also metabolized. For example, sebaleic acid (5Z,8Z-octadecadienoic acid), which is the main PUFA found in secretions of human sebaceous glands, is converted to 5-hydroxy- and 5-oxo-6E,8Z-octadecadienoic acid (5-HODE and 5-oxo-ODE, respectively) by neutrophils stimulated with A23187 and phorbol myristate acetate (PMA) (Cossette et al., 2008). Because this PUFA contains only two double bonds it cannot be converted to LTs, leaving 5-oxo-ODE as the major bioactive 5-LO product. Similarly, the ω9-PUFA Mead acid (5Z,8Z,11Z-eicosatrienoic acid) is metabolized to 5-hydroxy- (5-HETrE) and 5-oxo- (5-oxo-ETrE) 6E,8Z,11Z-eicosatrienoic acid by neutrophils (Patel et al., 2008). Despite the fact that 5-LO converts Mead acid to LTA_3, this intermediate is not metabolized to significant amounts of LTB_3 because LTA_3 inhibits LTA hydrolase (Evans et al., 1985). Finally, the ω3-PUFA 5Z,8Z,11Z,14Z,17Z-eicosapentaenoic acid (EPA) is converted to 5-hydroxy- (5-HEPE) and 5-oxo- (5-oxo-EPE) 6E,8Z,11Z,14Z,17Z-eicosapentaenoic acid, but in this case LTB_5 is also formed (Powell et al., 1995b).

7.2.1.2   Regulation of 5-Oxo-ETE Formation by 5-HEDH

Even though many cell types contain high levels of 5-HEDH activity, they convert 5-HETE to only very modest amounts of 5-oxo-ETE in the absence of stimulation. This is because resting cells contain only low levels of $NADP^+$, which is present principally in its reduced form NADPH, which serves to protect against oxidative stress (Schafer and Buettner, 2001). To permit appreciable synthesis of 5-oxo-ETE, intracellular $NADP^+$ levels must first be raised, as is the case when phagocytic cells undergo the respiratory burst or cells are subjected to oxidative stress. On the other hand, 5-oxo-ETE synthesis is diminished in the presence of glucose, which is converted to glucose-6-phosphate by hexokinase and then oxidized by the pentose phosphate pathway, resulting in the concomitant reduction of $NADP^+$ to NADPH (Erlemann et al., 2004) (Figure 7.2).

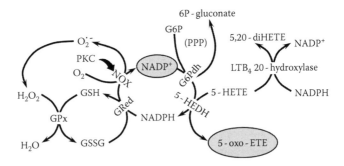

FIGURE 7.2   Regulation of intracellular $NADP^+$ levels and 5-oxo-ETE synthesis by NADPH oxidase (NOX), oxidative stress (through the GSH redox cycle), and the pentose phosphate pathway (PPP). GPx, glutathione peroxidase; GRed, glutathione reductase; G6Pdh, glucose 6-phosphate dehydrogenase; G6P, glucose 6-phosphate; 6P-gluconate, 6-phosphogluconate.

7.2.1.2.1 Respiratory Burst

Phorbol myristate acetate (PMA) strongly stimulates the oxidation of 5-HETE to 5-oxo-ETE (Powell et al., 1994). This effect is mediated by protein kinase C (PKC) and is dependent on the activation of NADPH oxidase (Figure 7.2), as it is blocked by both PKC inhibitors and diphenylene iodonium, an NADPH oxidase inhibitor. NADPH oxidase is activated in phagocytes following the PKC-mediated phosphorylation of p47phox and assembly of the cytosolic components (p47phox, p67phox, and rac) with the membrane-bound NOX2 and p22phox (Bedard and Krause, 2007). This results in a rapid burst of superoxide production, which is dependent on oxidation of NADPH to $NADP^+$, thus providing the cofactor required for the synthesis of 5-oxo-ETE. Under these conditions, there is a dramatic shift in the metabolism of 5-HETE by neutrophils from NADPH-dependent ω-oxidation to $NADP^+$-dependent oxidation to 5-oxo-ETE. Although other types of human phagocytes, including eosinophils and monocytes, do not convert 5-HETE to ω-oxidation products, they respond similarly to neutrophils with increased 5-oxo-ETE production in response to induction of the respiratory burst with PMA or opsonized zymosan (Powell et al., 1995a; Zhang et al., 1996).

7.2.1.2.2 Oxidative Stress Stimulates 5-Oxo-ETE Formation

Cells maintain high levels of GSH and NADPH as a means of protection against oxidative stress. Superoxide dismutase converts superoxide to H_2O_2, which is then reduced to H_2O by glutathione peroxidase. This is accompanied by the oxidation of GSH to GSSG, which is then recycled back to GSH by the NADPH-dependent enzyme glutathione reductase, resulting in the generation of $NADP^+$. Cells exposed to H_2O_2 thus undergo a rapid increase in GSSG levels accompanied by a corresponding rise in $NADP^+$, which dramatically increases their capacity to synthesize 5-oxo-ETE (Figure 7.2). Hydrogen peroxide has been shown to dramatically increase 5-oxo-ETE synthesis in nearly all cell types that contain 5-HEDH with the notable exception of neutrophils (Figure 7.3). The reason for the lack of response of neutrophils is not clear. It does not appear to be due to rapid metabolism of H_2O_2 by other enzymes such as catalase or myeloperoxidase, since H_2O_2 was still inactive in the presence of azide, an inhibitor of these enzyme (Erlemann et al., 2004). Moreover, t-butyl hydroperoxide, which is not a substrate for catalase and a poor substrate for myeloperoxidase, had little effect on 5-oxo-ETE formation by neutrophils. The reason for the resistance of neutrophils to stimulation of 5-oxo-ETE synthesis by H_2O_2 could possibly be due to their relatively low levels of glutathione peroxidase compared to other cells (Pietarinen-Runtti et al., 2000).

Regulation of 5-oxo-ETE synthesis by oxidative stress is particularly apparent in B cells. These cells produce only small amounts of 5-LO products in the presence of a calcium ionophore and AA unless they are exposed to an agent that induces oxidative stress such as H_2O_2 or diamide (Jakobsson et al., 1992). 5-Oxo-ETE synthesis is selectively stimulated in these cells by H_2O_2, which stimulates both the formation of its precursor 5-HETE by 5-LO and the oxidation of 5-HETE to 5-oxo-ETE by 5-HEDH (Grant et al., 2011a).

7.2.1.2.3 5-Oxo-ETE Formation Is Enhanced in Dying Cells

Freshly isolated neutrophils produce relatively small amounts of 5-oxo-ETE compared to LTB_4 and 5-HETE. These cells have a very active ω-oxidation pathway, and

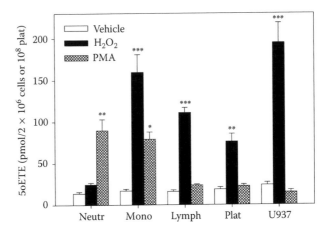

FIGURE 7.3   Regulation of 5-oxo-ETE (5oETE) synthesis by oxidative stress (i.e., H_2O_2) and the respiratory burst (i.e., PMA) in blood cells. The amounts of 5-oxo-ETE synthesized following addition of 5-HETE in the presence of vehicle (open bars), 100 μM H_2O_2 (solid bars), or 100 nM PMA (cross-hatched bars) to neutrophils (Neutr), monocytes (Mono), lymphocytes (Lymph), platelets (Plat), or U937 monocytic cells are shown. *, $p < 0.05$; **, $p < 0.01$; ***, $p < 0.001$. (Data taken from Erlemann, K. R. et al. 2004. *J. Biol. Chem.* 279:40376–84 and are means ± SE.)

convert both 5-HETE and 5-oxo-ETE to their 20-hydroxy metabolites via LTB_4 20-hydroxylase, an NADPH-dependent cytochrome P450 (CYP4F3A). Neutrophils undergo apoptosis when cultured and after 24 h only about 25% are still resistant to annexin V staining (Graham et al., 2009). This is accompanied by a dramatic increase in their ability to synthesize 5-oxo-ETE and a reduction in the ω-oxidation of 5-HETE. These changes appear to be due to oxidative stress associated with neutrophil apoptosis, as they are inhibited by antiapoptotic agents such as GM-CSF and forskolin, as well as by substances that have antioxidant properties, including the NADPH oxidase inhibitor diphenylene iodonium, catalase, and deferoxamine. Apoptotic neutrophils are normally phagocytosed by macrophages following the appearance of phosphatidylserine on the outer surface of the plasma membrane (Serhan and Savill, 2005). However, in conditions of severe inflammation, the capacity of macrophages to remove neutrophils could be exceeded and these cells could proceed to late apoptosis and secondary necrosis (Haslett, 1999; Brown et al., 2006), which could be accompanied by increased production of 5-oxo-ETE. This may be the case in inflammatory airway diseases such as severe asthma and chronic obstructive pulmonary disease, which are characterized by persistently high levels of neutrophils in the lungs (Barnes, 2007). Increased production of 5-oxo-ETE under these circumstances could result in further leukocyte infiltration and prolongation of inflammation.

Increased production of 5-oxo-ETE associated with cell death may not be restricted to neutrophils, as we have observed that exposure of tumor cells to cytotoxic agents also induces this phenomenon (Grant et al., 2011b). 5-Oxo-ETE has been reported to promote the proliferation of a variety of tumor cell lines, and its production by dying tumor cells could potentially promote tumor growth. Another possibility is that 5-oxo-ETE could induce the infiltration of eosinophils and/or neutrophils into the tumor due to its chemoattractant effects. There is evidence for the production of an eosinophil chemoattractant within tumors (Cormier et al., 2006) and 5-oxo-ETE could certainly fulfill this role.

7.2.1.3   Formation of 5-Oxo-ETE by Transcellular Biosynthesis

As noted above, various structural noninflammatory cells that possess little or no 5-LO express 5-HEDH activity. This is reminiscent of the distribution of the enzymes responsible for the formation of LTB_4 and LTC_4 from LTA_4 (i.e., LTA_4 hydrolase and LTC_4 synthase, respectively), which are also more widely distributed than 5-LO. LTs can be synthesized by transcellular biosynthesis with neutrophils serving as donors of LTA_4 and other cells such as epithelial cells and platelets converting it to LTs B_4 and C_4, respectively (Folco and Murphy, 2006). For example, although neutrophils synthesize LTB_4 themselves, the amount synthesized is greater in the presence of epithelial cells. We have recently shown that a similar situation exists for 5-oxo-ETE when neutrophils are stimulated in the presence of prostate epithelial tumor cells, especially after exposure of the tumor cells to oxidative stress (Grant et al., 2011b). Under these conditions tumor cell-associated 5-HEDH is responsible for at least 75% of the 5-oxo-ETE synthesized from AA in the presence of neutrophils.

7.2.2   Formation of 5-Oxo-ETE by Cytochrome P450

It has recently been shown that CYP2S1, a recently discovered cytochrome P450, can convert 5-HpETE and a variety of other fatty acid hydroperoxides to oxo metabolites in the absence of NADPH (Bui et al., 2010). This enzyme is expressed in a variety of tissues, including peripheral leukocytes (Rylander et al., 2001), and could potentially convert 5-HpETE, generated either by 5-LO or during lipid peroxidation, to 5-oxo-ETE. However, the K_m for this reaction (~20 µM) (Bui et al., 2010) is considerably higher than for 5-HEDH (~0.5 µM) (Erlemann et al., 2007a).

7.2.3   Nonenzymatic Generation of 5-Oxo-ETE by Dehydration of 5-HpETE

Hydroperoxy derivatives of PUFA are formed during lipid peroxidation and can be converted to the corresponding oxo products by loss of water, a reaction that is facilitated by the presence of heme compounds (Hamberg, 1975). Although enzymatic synthesis of 5-oxo-ETE by 5-HEDH appears to be the main route for its formation, 5-oxo-ETE can also be generated nonenzymatically by dehydration of 5-HpETE and has been identified following incubation of murine mast cell cytosolic fractions with AA (Bryant et al., 1986). 5-Oxo-ETE can also be generated as a result of lipid peroxidation. Significant amounts of esterified 5-oxo-ETE have been detected on certain phospholipids, in particular arachidonate-containing plasmalogen phospholipids, subjected to oxidative stress (Khaselev and Murphy, 2000). It has also been detected in phospholipid fractions following exposure of red blood cell membranes to *t*-butyl hydroperoxide (Hall and Murphy, 1998a).

5-Oxo-ETE is synthesized from 5-HpETE by a similar mechanism by murine macrophages, which do not contain detectable 5-HEDH activity (Zarini and Murphy, 2003). This process is facilitated by the presence of a cytosolic protein, but is resistant to denaturation by heating and to treatment with trypsin, indicating that it is not an enzymatic reaction, but rather may be due to protein-bound heme.

7.2.4 Formation of 5-Oxo-15-HETE

5-Oxo-15-HETE has similar biological activities to 5-oxo-ETE, but is less potent and efficacious. Its biosynthesis is more complicated than that of 5-oxo-ETE, as an additional step is required to introduce the hydroxyl group in the 15-position. It can be formed either by the 5-HEDH-catalyzed oxidation of 5,15-diHETE (Powell et al., 1992), a product of the combined actions of 5-LO and 15-LO, or by the action of 15-LO on 5-oxo-ETE (Schwenk and Schröder, 1995) (Figure 7.1). Although all the necessary enzymes are present in eosinophils, these cells produce relatively small amounts of 5-oxo-ETE and 5-oxo-15-HETE when incubated with calcium ionophore and AA (Powell et al., 1995a), as the major 5-LO products formed by these cells are cysteinyl-LTs. 5-Oxo-15-HETE could also be synthesized by transcellular biosynthesis by the actions of 5-HEDH and/or 15-LO, both of which are found in epithelial cells (Hunter et al., 1985; Erlemann et al., 2007b), on leukocyte-derived 5-HETE or 5-oxo-ETE, respectively.

7.2.5 5-Oxo-ETE Isomers

The name "5-oxo-ETE" is used here to refer specifically to 5-oxo-6E,8Z,11Z,14Z-eicosatetraenoic acid (i.e., the 8-*cis* isomer), which is the product formed by the oxidation of 5-HETE by 5-HEDH. We have also detected small amounts of the 8-*trans* isomer that were formed during prolonged storage or in incubations with various cell types (Powell et al., 1996). It is presumably formed by the nonenzymatic isomerization of the 8-*cis*-double bond to the more stable *trans* configuration.

A different isomer of 5-oxo-ETE can be formed by the nonenzymatic decomposition of LTA$_4$ following incubation of neutrophils with high concentrations of AA and the calcium ionophore A23187 (Gravel et al., 1993). Unlike 5-oxo-ETE, the conjugated triene chromophore of LTs is retained in this compound, which contains an isolated keto group and was identified as 5-oxo-7E,9E,11Z,14Z-eicosatetraenoic acid. It would appear that this is a minor product and that the 6E,8Z,11Z,14Z isomer (i.e., 5-oxo-ETE) is normally the major 5-oxoeicosanoid formed by neutrophils and other inflammatory cells.

7.3 METABOLISM OF 5-OXO-ETE

5-Oxo-ETE is metabolized by a variety of pathways, most of which result in its biological inactivation. The major pathway for its metabolism in the presence of neutrophils is ω-oxidation, as these cells contain high levels of CYP4F3 (LTB$_4$ 20-hydroxylase), resulting in the formation of 20-hydroxy-5-oxo-ETE (Powell et al., 1996). Murine macrophages also convert 5-oxo-ETE to ω-oxidation products, but in contrast to human neutrophils, 18- and 19-hydroxy metabolites are formed (Hevko et al., 2001). We made a similar observation with respect to the metabolism of LTB_4 by rat neutrophils, which convert this substance to 18- and 19-hydroxyLTB$_4$ rather than 20-hydroxyLTB$_4$, which is the product formed by human neutrophils (Powell and Gravelle, 1990). 5-Oxo-ETE can also be stereospecifically reduced back to its precursor 5-HETE by 5-HEDH, acting in the reverse direction in the presence of NADPH (Erlemann et al., 2007a). Each of these pathways results in a 100-fold reduction in biological potency. Neutrophils also reduce 5-oxo-ETE to its 6,7-dihydro metabolite (5-oxo-8Z,11Z,14Z-eicosatrienoic acid) by the action of the calmodulin-dependent enzyme eicosanoid Δ^6-reductase in neutrophils. This results in a 1000-fold loss in biological activity (Berhane et al., 1998).

Incubation of 5-oxo-ETE with neutrophils results in its incorporation into cellular lipids, primarily triglycerides (O'Flaherty et al., 1998). This process can be blocked by the acyl CoA transferase inhibitor triacsin C. However, it appears that 5-oxo-ETE must first be reduced to 5-HETE prior to esterification into lipids.

5-Oxo-ETE can also be oxygenated by lipoxygenases, including 15-LO, which converts it to the biologically active 5-oxo-15-HETE, as discussed in Section 7.2.4 above. It can also be oxidized by 12-LO to 5-oxo-12-HETE (Figure 7.1), and this is an important product of transcellular metabolism of AA by mixtures of platelets and neutrophils (Powell et al., 1999b). Unlike 5-oxo-15-HETE, 5-oxo-12-HETE has little or no biological activity, and antagonizes 5-oxo-ETE-induced calcium mobilization and cell migration in neutrophils.

Murine macrophages convert 5-oxo-ETE to a glutathione adduct, which was named FOG$_7$ (five-oxo-7-glutathionyl-8,11,14-eicosatrienoic acid) (Bowers et al., 2000). FOG$_7$ is formed by the 1,4 Michael addition of GSH to the 7-position of 5-oxo-ETE, which is catalyzed by LTC$_4$ synthase (Hevko and Murphy, 2002). FOG$_7$ has chemoattractant effects on both eosinophils and neutrophils, and stimulates actin polymerization in these cells, but, unlike 5-oxo-ETE, does not induce calcium mobilization (Bowers et al., 2000). The stereochemistry of the 7-glutathionyl group is not known.

7.4   5-OXO-ETE RECEPTOR

Prior to the discovery of the 5-oxo-ETE pathway, 5-HETE was known to activate neutrophils by a mechanism that was independent of receptors for other lipid mediators (O'Flaherty, 1985; O'Flaherty and Nishihira, 1987; O'Flaherty et al., 1988), suggesting that it acts through a unique selective receptor. However, the rather modest potency of 5-HETE suggested that although a novel receptor might indeed be mediating its response, 5-HETE may not be its primary ligand. The identification of a selective 5-HETE dehydrogenase raised the possibility that its product, 5-oxo-ETE, was the preferred ligand for this receptor, and this was confirmed by studies from both our laboratory (Powell et al., 1993) and that of O'Flaherty (O'Flaherty et al., 1993), showing that 5-oxo-ETE is a potent activator of human neutrophils. Neutrophils exposed to either 5-oxo-ETE or 5-HETE became desensitized and failed to respond to further stimulation by either ligand, but responded normally to other lipid mediators such as LTB$_4$ and platelet-activating factor (PAF). This indicates that 5-oxo-ETE and 5-HETE share the same receptor, of which 5-oxo-ETE is the primary ligand. Furthermore, the actions of 5-oxo-ETE were blocked by pertussis toxin, indicating that its receptor is coupled to G_i and is a member of the G-protein-coupled receptor family (Powell et al., 1996; Czech et al., 1997; O'Flaherty et al., 2000).

Binding studies with radiolabeled 5-oxo-ETE were complicated by the fact that it was rapidly incorporated into cellular lipids, as discussed above. However, O'Flaherty overcame this problem by conducting experiments in the presence of the acyl CoA transferase inhibitor triacsin C (O'Flaherty et al., 1998). In this way it was possible to demonstrate high-affinity selective binding sites for 5-oxo-ETE.

The 5-oxo-ETE receptor was cloned independently by three groups who were screening various orphan G-protein-coupled receptors with libraries containing large numbers of potential ligands. This receptor, originally known as TG1019 (Hosoi et al., 2002), R527 (Jones et al., 2003), and hGPCR48 (Takeda et al., 2002), has been designated as the OXE receptor by the IUPHAR Nomenclature Committee (Brink et al., 2004). It consists of 423 amino acids and is primarily coupled to G_i. RT-PCR analysis indicates that it is highly expressed in human eosinophils > neutrophils > alveolar macrophages (Jones et al., 2003). The most potent ligand for the cloned OXE receptor was found to be 5-oxo-ETE. 5-HpETE (which can decompose to 5-oxo-ETE) was about 100 times less potent, whereas 5-HETE was even less potent. LTs, prostaglandins, 12-HETE, and 15-HETE had no effect on this receptor. The gene encoding the OXE receptor, which maps to 2p21 on chromosome 2 (Hosoi et al., 2002; Jones et al., 2003) and is intronless, is referred to as OXER1.

7.4.1 STRUCTURE–ACTIVITY RELATIONSHIPS FOR ACTIVATION OF THE OXE RECEPTOR

We have conducted fairly extensive structure–activity studies on the 5-oxo-ETE receptor, in which we examined calcium mobilization and other response in both human neutrophils and eosinophils. The OXE receptor is highly selective for 5-oxo-ETE. Activity is lost by the reduction of the 5-oxo group to a hydroxyl group, as 5-HETE is 100 times less potent (Powell et al., 1992). The carboxyl group is also important for the activation of the receptor, as methylation reduces potency by 20-fold (Powell et al., 1996). Reduction of the 6-*trans*-double bond to give 6,7-dihydro-5-oxo-ETE (i.e., 5-oxo-8,11,14-eicosatrienoic acid) results in an even larger 1000-fold reduction in potency (Berhane et al., 1998). Changes in the 8,9-double bond are better tolerated, as replacing the 8-*cis* with the 8-*trans* configuration reduces potency by only sixfold (Powell et al., 1996). We also investigated the activities of a large series of synthetic analogs differing in the numbers of double bonds and the carbon chain length (Patel et al., 2008). The minimum requirements for appreciable activation of the OXE receptor were found to be a chain length of at least 18 carbons and double bonds present in the 6- and 8-positions (Patel et al., 2008). This indicates that the hydrophobic ω-end of the molecule is important for recognition by the OXE receptor. Further evidence for this is the dramatic 100-fold reduction in potency that results from addition of a hydroxyl group in the 20-position (Powell et al., 1996).

At least three polyunsaturated fatty acids in addition to AA are metabolized by neutrophils to products that are potent OXE receptor agonists. Neutrophils convert sebaleic acid (5Z,8Z-octadecadienoic acid) (Cossette et al., 2008) and Mead acid (5Z,8Z,11Z-eicosatrienoic acid) (Patel et al., 2008) to 5-oxo-ODE (5-oxo-6E,8Z-octadecadienoic acid) and 5-oxo-ETrE, respectively. These two 5-oxo fatty acids are approximately equipotent with 5-oxo-ETE in activating neutrophils and eosinophils. EPA (6Z,8Z,11Z,14Z,17Z-eicosapentaenoic acid) is converted to 5-oxo-EPE (5-oxo-6E,8Z,11Z,14Z,17Z-eicosapentaenoic acid), which is about five times less potent than 5-oxo-ETE (Powell et al., 1995b; Patel et al., 2008).

7.5 INTRACELLULAR SIGNALING MECHANISMS OF 5-OXO-ETE

5-Oxo-ETE was initially found to induce Ca^{2+} mobilization in human neutrophils (Powell et al., 1993; O'Flaherty et al., 1993) and subsequently to inhibit cAMP formation in response to addition of forskolin to CHO cells that had been transfected with the OXE receptor (Hosoi et al., 2002) (Figure 7.4). The effects of 5-oxo-ETE are dependent on $G\alpha_{i/o}$ as they can be blocked by pertussis toxin (Norgauer et al., 1996; Powell et al., 1996; O'Flaherty et al., 1996a). Unlike some other chemoattractants, 5-oxo-ETE cannot signal appreciably through $G\alpha_q$ (O'Flaherty et al., 2000; Hosoi et al., 2002).

The stimulatory effects of 5-oxo-ETE on Ca^{2+} mobilization and cell migration were blocked by the phospholipase C (PLC) inhibitor U73122, indicating that these responses were mediated by the PLC-dependent hydrolysis of phosphatidylinositol 4,5-bisphosphate, resulting in the release of inositol trisphosphate from the cell membrane (Hosoi et al., 2005). 5-Oxo-ETE increased the levels of phosphatidylinositol (3,4,5)-trisphosphate in neutrophils, indicating that it also signals through phosphoinositide-3 kinase (PI3K) (Norgauer et al., 1996). Further evidence for signaling via PI3K was found in OXE receptor-transfected CHO cells, in which 5-oxo-ETE-induced phosphorylation of Akt was blocked by the PI3K inhibitor LY294002 (Hosoi et al., 2005). LY294002 also blocked 5-oxo-ETE-induced migration of these cells

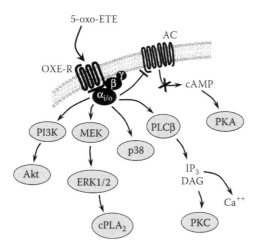

FIGURE 7.4 Signaling pathways for the OXE receptor (OXE-R).

(Hosoi et al., 2005), suggesting that both activation of PI3K and PLC-induced Ca^{2+} mobilization are required for 5-oxo-ETE-induced cell migration in this model.

The phosphorylation of ERK-1 and ERK-2 is also induced by 5-oxo-ETE in a variety of cell types, including neutrophils (O'Flaherty et al., 1996a), eosinophils (O'Flaherty et al., 1996b; Langlois et al., 2009), and PC3 cells (O'Flaherty et al., 2002) as well as OXE receptor-transfected CHO cells (Hosoi et al., 2005). $cPLA_2$ phosphorylation accompanied by AA release is also induced by 5-oxo-ETE in neutrophils (O'Flaherty et al., 1996a), presumably mediated by activation of ERK, which is known to phosphorylate $cPLA_2$ at Ser505 (Gijon and Leslie, 1999). It has also been reported that PKCδ and PKCζ as well as p38 are involved in mediating cellular responses to 5-oxo-ETE (Langlois et al., 2009).

7.6   CELLULAR TARGETS OF 5-OXO-ETE

7.6.1   EOSINOPHILS

7.6.1.1   Eosinophil Migration and Tissue Infiltration

The first indication that 5-oxoeicosanoids activate eosinophils came from Schröder's group, who found that incubation of AA with human eosinophils resulted in the formation of a novel product that they named eosinophil chemotactic lipid (ECL) (Morita et al., 1990). ECL induced a strong chemotactic response in these cells that was subject to homologous desensitization following pretreatment with ECL, but not with either LTB_4 or PAF, indicating that it acted through a distinct receptor. They subsequently showed that incubation of AA with soybean lipoxygenase resulted in the formation of a substance with properties very similar to ECL, which they identified as 5-oxo-15-HETE (Schwenk et al., 1992). 5-Oxo-ETE was subsequently shown to be an even more potent chemotactic agent for eosinophils (Powell et al., 1995a; O'Flaherty et al., 1996b).

The only other eicosanoid with potent chemoattractant effects on eosinophils is PGD_2 (Hirai et al., 2001; Monneret et al., 2001). At low nanomolar concentrations, PGD_2 and 5-oxo-ETE have similar effects, whereas at higher concentrations, 5-oxo-ETE induces a much stronger response (Monneret et al., 2001). Although LTB_4 is a potent chemotactic agent for guinea pig eosinophils (Maghni et al., 1991; Sehmi et al., 1991; Sun et al., 1991), it has only a very weak effect on the migration of human eosinophils (Morita et al., 1989; Sun et al., 1991; Powell et al., 1995a), as is also true for LTD_4 (Powell et al., 1995a). Although cysLTs do appear to be involved in the recruitment of eosinophils into tissues such as the lungs, these effects appear to be mediated by the release of cytokines such as IL-5 rather than by a direct effect on eosinophils (Underwood et al., 1996).

The maximal response to 5-oxo-ETE is about 30 times greater than those to LTs B_4 and D_4 and about 3 times greater than that to PAF, which was previously believed to be the most potent eosinophil chemoattractant among lipid mediators (Powell et al., 1995a). Moreover, low concentrations of 5-oxo-ETE enhance the responsiveness of eosinophils to PAF. O'Flaherty also reported that 5-oxo-ETE is both more potent and more efficacious in inducing eosinophil migration than a variety of other inflammatory mediators, including fMLP, LTB_4, C5a, and PAF (O'Flaherty et al.,

1996b). The potency of 5-oxo-ETE is comparable to those of CC chemokines, being intermediate between eotaxin and RANTES, but the maximal response is about 50% higher than that to eotaxin (Powell et al., 2001). Furthermore, threshold concentrations of both eotaxin and RANTES shift the concentration–response curve to 5-oxo-ETE to the left by a factor of about 10.

The studies referred to above all examined the effects of 5-oxo-ETE on the migration of eosinophils through membranes containing very small pores in Boyden chambers. To mimic more closely *in vivo* tissue infiltration, Laviolette's group examined its effects on migration through a basement membrane Matrigel matrix. They found that 5-oxo-ETE is a potent and efficacious stimulator of eosinophil transmigration through this matrix (Guilbert et al., 1999) and also stimulates transendothelial migration (Dallaire et al., 2003). Although it was slightly less potent than eotaxin, the maximal response to 5-oxo-ETE was about double that to this CC chemokine (Ferland et al., 2001). In contrast, PAF only induced a weak response. In addition to its direct chemotactic effects, the transmigration response to 5-oxo-ETE was found to be mediated by the release of MMP-9 from eosinophils, resulting in the degradation of the matrix, as well as by increased surface expression of urokinase plasminogen activator receptor (uPAR) on eosinophils (Guilbert et al., 1999; Langlois et al., 2006).

5-Oxo-ETE elicits tissue infiltration of eosinophils *in vivo* in humans. Intradermal injection results in the accumulation of these cells in the skin around the injection site, with atopic asthmatic subjects responding more strongly than nonatopic healthy controls (Muro et al., 2003). Intratracheal administration of 5-oxo-ETE to Brown Norway rats elicited pulmonary eosinophilia (Stamatiou et al., 1998). This effect was not mediated by either LTB_4 or PAF receptors, as it was not blocked by selective antagonists of either receptor. LTB_4, PAF (Stamatiou et al., 1998), and PGD_2 (Almishri et al., 2005) also induced pulmonary infiltration of eosinophils in this animal model, but to a slightly lesser extent than 5-oxo-ETE. The eosinophilic response to 5-oxo-ETE in rats cannot be mediated by the OXE receptor, as neither rats nor mice possess an ortholog of this receptor. This suggests that there may be a distinct 5-oxo-ETE receptor in this species.

7.6.1.2   Other Responses Elicited by 5-Oxo-ETE in Eosinophils

5-Oxo-ETE elicits a variety of other responses in eosinophils, some of which are involved in cell movement and tissue infiltration. Among the most rapid responses, which are maximal after only a few seconds, are intracellular calcium mobilization (Schwenk and Schröder, 1995; Powell et al., 1999a) and actin polymerization (Czech et al., 1997; Powell et al., 1999a). Cellular adhesion molecules are also activated, including increased surface expression of CD11b and shedding of L-selectin (Powell et al., 1999a). 5-Oxo-ETE upregulates the leukocyte activation marker CD69 (Urasaki et al., 2001), in contrast to LTs B_4, C_4, D_4, and E_4, which are inactive. PAF and IL-5 also stimulate CD69 expression on eosinophils, but these effects appear to be mediated by the release of 5-oxo-ETE, as they are blocked by 5-LO inhibitors (Urasaki et al., 2001).

LTB_4 also induces many of the responses discussed above, even though it exhibits little chemotactic activity for human eosinophils. It is equipotent with 5-oxo-ETE in inducing calcium mobilization and CD11b expression. However, 5-oxo-ETE induces much stronger actin polymerization and L-selectin responses than both LTB_4 and

PAF (Powell et al., 1999a). Interestingly, PAF elicits a much stronger calcium response than both 5-oxo-ETE and LTB_4, even though it is considerably less potent as an eosinophil chemoattractant than 5-oxo-ETE. It appears that the effectiveness of these lipid mediators as eosinophil chemoattractants correlates much more closely with their abilities to induce actin polymerization than with their capacities to induce calcium mobilization.

Responses involved in pathogen killing, which can also have destructive effects in the lung and other tissues, are also activated by 5-oxo-ETE, but generally to a somewhat lesser extent than responses involved in cell movement. The respiratory burst is triggered by 5-oxo-ETE, with superoxide reaching maximal levels by about 5 min (Czech et al., 1997). 5-Oxo-ETE has only a modest effect on degranulation of otherwise unstimulated eosinophils, but induces a strong response following priming of these cells with GM-CSF, in which case the release of β-glucuronidase, eosinophil peroxidase, and arylsulfatase have all been documented (O'Flaherty et al., 2000). 5-Oxo-ETE also acts synergistically with a variety of other inflammatory mediators, including PAF, C5a, LTB_4, and FMLP, to give dramatically enhanced degranulation responses (O'Flaherty et al., 2000).

7.6.2 Neutrophils

Neutrophils were the first cells in which the biological activities of 5-oxo-ETE were recognized. In general, the responses of neutrophils to 5-oxo-ETE are very similar to those described above for eosinophils. It is a chemoattractant for these cells (Powell et al., 1993) and is a potent inducer of a variety of responses associated with cell movement and tissue infiltration, including actin polymerization (Norgauer et al., 1996; Powell et al., 1997) and calcium mobilization (Powell et al., 1993; O'Flaherty et al., 1993). It also increases the surface expression of the cellular adhesion molecules CD11b and CD11c as well as neutrophil adherence (Powell et al., 1993) and aggregation (O'Flaherty et al., 1994). It is also active *in vivo*, inducing the infiltration of neutrophils into the skin following intradermal injection in humans (Muro et al., 2003).

As is the case for eosinophils, 5-oxo-ETE elicits relatively weak degranulation and superoxide responses in neutrophils, being considerably less potent and efficacious than LTB_4. Unprimed neutrophils release only modest amounts of lysozyme and β-glucuronidase when exposed to 5-oxo-ETE. However, priming with TNFα (O'Flaherty et al., 1993), GM-CSF, or G-CSF (O'Flaherty et al., 1996a) strongly enhanced the maximal response to 5-oxo-ETE to nearly the same level as that observed with LTB_4, and increased its potency to about one-fifth that of LTB_4. Similarly, the respiratory burst is only stimulated modestly (Norgauer et al., 1996) or not at all (O'Flaherty et al., 1996a) by 5-oxo-ETE in untreated neutrophils, in contrast to GM-CSF-primed neutrophils, in which superoxide production is strongly stimulated by 5-oxo-ETE, with a potency approaching that of LTB_4 (O'Flaherty et al., 1996a). 5-Oxo-ETE also strongly enhanced PAF-induced superoxide production (O'Flaherty et al., 1994).

Among eicosanoids, only LTB_4 exceeds the ability of 5-oxo-ETE to activate human neutrophils. Although the efficacies of these two 5-LO products are similar, except with respect to degranulation and superoxide production by unprimed neutrophils,

5-oxo-ETE is about 5–50 times less potent. Because 5-oxo-ETE and LTB$_4$ share the same initial biosynthetic pathway, they would normally be formed at the same time. It would therefore seem likely that under normal circumstances LTB$_4$ would be the most important eicosanoid in regulating neutrophils *in vivo*. However, there may be certain situations in which 5-oxo-ETE could play a role. For example, neutrophils can be selectively desensitized to LTB$_4$ following *in vivo* exposure to this substance (Marleau et al., 1993), raising the possibility that 5-oxo-ETE could play a role in prolonging severe neutrophilic inflammation. The contribution of 5-oxo-ETE could be enhanced by inflammatory cytokines such as GM-CSF and TNFα, which selectively sensitize neutrophils to 5-oxo-ETE as discussed above. Furthermore, 5-oxo-ETE is more resistant to metabolism by neutrophils, which rapidly inactivate LTB$_4$ due to the action of LTB$_4$ 20-hydroxylase (Graham et al., 2009). The formation of 5-oxo-ETE is also selectively enhanced by oxidative stress (Erlemann et al., 2004) and in dying cells (Graham et al., 2009; Grant et al., 2011b), which should favor its formation at inflammatory sites.

7.6.3 Monocytes

5-Oxo-ETE induces monocyte migration, but is less potent and efficacious than fMLP and MCP-1 (Sozzani et al., 1996). However, it acts synergistically with MCP-1 and MCP-3, resulting in considerably enhanced chemotactic responses to these chemokines. 5-Oxo-ETE also stimulates actin polymerization in monocytes, but, in contrast to granulocytes, does not mobilize calcium. Although it does not elicit AA release in monocytes as it does in neutrophils, 5-oxo-ETE augments the response to MCP-1 (Sozzani et al., 1996).

We found that addition of 5-oxo-ETE to eosinophils in the presence of small numbers of monocytes markedly increased eosinophil survival (Stamatiou et al., 2004). A similar effect was observed after the addition of conditioned medium from 5-oxo-ETE-treated monocytes to eosinophils. This effect was blocked by an antibody against GM-CSF. Furthermore, 5-oxo-ETE was a potent stimulator of GM-CSF release from monocytes. The ability of 5-oxo-ETE to induce this response raises the possibility that its actions may be broader than originally thought, because of the wide-ranging effects of this cytokine on inflammatory cells. It is interesting that that GM-CSF, in addition to being released in response to 5-oxo-ETE, also stimulates the synthesis of 5-LO products (DiPersio et al., 1988; Pouliot et al., 1994) and enhances the responsiveness of both eosinophils and neutrophils to 5-oxo-ETE as discussed above (O'Flaherty et al., 1996a, 1996b).

7.6.4 Basophils

Basophils express the OXE receptor and migrate in response to 5-oxo-ETE (Iikura et al., 2005; Sturm et al., 2005). Although they also express receptors for LTB$_4$ and cysLTs, 5-oxo-ETE was the only 5-LO product found to induce migration of these cells, although it is not as potent as eotaxin and slightly less potent than PGD$_2$ (Iikura et al., 2005). It also promotes the transmigration of basophils through a basement membrane Matrigel matrix, but, like other basophil agonists, this requires pretreatment with IL-3 (Suzukawa et al., 2006). 5-Oxo-ETE also induces a shape change in basophils (Sturm et al., 2005) and has modest stimulatory effects on calcium mobilization (Iikura et al., 2005; Sturm et al., 2005) and surface expression of the adhesion molecule CD11b and the basophil activation marker CD203c (ecto-nucleotide pyrophosphatase/phosphodiesterase family member 3) (Monneret et al., 2005). However, it does not elicit degranulation, as evaluated by both histamine release and surface expression of the degranulation marker CD67 (Monneret et al., 2005; Iikura et al., 2005).

7.6.5 AIRWAY SMOOTH MUSCLE

5-Oxo-ETE induces contractile responses in guinea pig airway smooth muscle, which were somewhat lower in amplitude than those elicited by carbamylcholine and the TP receptor agonist U46619 (Mercier et al., 2004). The effect of 5-oxo-ETE was blocked by both indomethacin and the selective TP receptor antagonist SQ-29548 as well as by the PLA_2 inhibitor ONO-RS-082, suggesting that this response was dependent on the release of AA and its conversion to TXA_2 (Mercier et al., 2004). The response to 5-oxo-ETE appears to be mediated by Rho kinase and calcium sensitization of airway smooth muscle cell myofilaments. Interestingly, guinea pig smooth muscle that had been cultured for 3 days responded more strongly to 5-oxo-ETE (EC_{50} 350 nM compared to 900 nM for freshly isolated tissue) and in this case the response was no longer dependent on the release of cyclooxygenase products (Morin and Rousseau, 2007).

In contrast to the contractile effect of 5-oxo-ETE on guinea pig airway smooth muscle, human bronchial airway smooth muscle undergoes relaxation when exposed to this substance (IC_{50}, 400 nM) (Morin et al., 2007). This effect appears to be mediated by the direct activation of large conducting calcium-activated K^+ channels (BK_{Ca}) by 5-oxo-ETE, since it was blocked by iberiotoxin, an inhibitor of this channel. Furthermore, 5-oxo-ETE activated BK_{Ca} channels reconstituted in a lipid bilayer, raising the possibility that in some cases it may act independently of the OXE receptor (Morin et al., 2007).

7.7 POTENTIAL ROLE OF 5-OXO-ETE IN AIRWAY INFLAMMATION

Of all the cell types investigated the OXE receptor is the most highly expressed in eosinophils, suggesting that this cell may be its major target. Furthermore, 5-oxo-ETE is active in humans *in vivo*, inducing eosinophil infiltration into the skin. Although the eosinophil has been regarded as playing a critical role in asthma and other allergic diseases, there has been some debate about its role in asthma since it was reported that anti-IL-5 was not effective in alleviating the symptoms of this disease, while strongly depleting blood and sputum eosinophils (Leckie et al., 2000). However, tissue eosinophils were not examined in this study, and it was subsequently shown that anti-IL-5 depleted eosinophils in the airways only by about 55% and did not affect the degree of bronchial mucosal staining of eosinophil MBP (Flood-Page et al., 2003). Furthermore, a recent study demonstrated that treatment with anti-IL-5 (mepolizumab) resulted in considerable improvements in the symptoms of severe asthmatics (Nair et al., 2009).

Eosinophils have long been thought to contribute to tissue damage in the lungs due to the release of eosinophil peroxidase, eosinophil cationic protein, and major basic protein (Rådinger and Lötvall, 2009). More recent evidence suggests that they may play an important role in promoting airway remodeling that is associated with asthma, in large part due to the release of TGFβ and other cytokines (Ochkur et al., 2007; Venge, 2010). Furthermore, eosinophils, along with basophils and mast cells, are an important source of cysLTs, which are key inflammatory mediators in asthma because of their potent bronchoconstrictor effects and their ability to stimulate mucus production, cytokine release, and airway remodeling (Peters-Golden and Henderson, 2007). 5-Oxo-ETE is also a potent chemoattractant for basophils, which, in addition to having the ability to synthesize cysLTs, may play an important role in asthma because they are a source of both IL-4 and IL-13 (Gibbs, 2005). Drugs designed to prevent the infiltration of eosinophils into the lungs may therefore be useful therapeutic agents in asthma (Jacobsen et al., 2007; Rådinger and Lötvall, 2009).

Because of its potent chemoattractant effects on both eosinophils and basophils (Sections 7.6.1 and 7.6.4), 5-oxo-ETE is an attractive drug target in asthma. Although it is clear that other inflammatory mediators also contribute to eosinophil infiltration, including CC chemokines and PGD$_2$, blocking the effects of 5-oxo-ETE may have some advantages. 5-Oxo-ETE displays very high efficacy when compared to other eosinophil chemoattractants, and stimulates the surface expression of adhesion molecules and the release of MMP-9, which would facilitate eosinophil extravasation and tissue infiltration. Moreover, it acts synergistically with eotaxin and RANTES, as well as PAF, in inducing eosinophil migration. There is also an interesting relationship between 5-oxo-ETE and GM-CSF, which is important for prolonging eosinophil survival in the airways (Park et al., 1998). 5-Oxo-ETE stimulates the release of GM-CSF from monocytes, resulting in increased eosinophil survival, whereas GM-CSF could increase both the synthesis of 5-oxo-ETE through its effects on cPLA$_2$ and 5-LO and its potency, especially with respect to release of its damaging granule contents.

5-Oxo-ETE and PGD$_2$ may play complementary roles in inducing eosinophil infiltration in response to allergen. PGD$_2$ is released very rapidly from mast cells following exposure of asthmatic subjects to antigen (Murray et al., 1986). However, 5-oxo-ETE may be formed later, following the initial infiltration of inflammatory cells, due to the stimulation of its synthesis by the respiratory burst in phagocytes, oxidative stress, and cell death (Section 7.2.1.2). Furthermore, its formation could be enhanced by transcellular biosynthesis, in which leukocyte-derived 5-HETE could be converted to 5-oxo-ETE by 5-HEDH in airway epithelial or smooth muscle cells (cf. Section 7.2.1.3). Thus, it is possible that 5-oxo-ETE could be involved in prolonging inflammation in asthma.

7.8   5-OXO-ETE-RELATED DRUG TARGETS

The actions of 5-oxo-ETE could be blocked by preventing its synthesis either with already existing 5-LO inhibitors or FLAP antagonists or with a selective 5-HEDH inhibitor (Figure 7.5). Alternatively, the activation of the OXE receptor could be

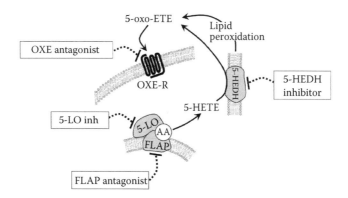

FIGURE 7.5 Possible strategies to block 5-oxo-ETE-induced inflammation. 5-Oxo-ETE synthesis could be inhibited with a 5-LO inhibitor, FLAP antagonist, or 5-HEDH inhibitor, or activation of the OXE receptor could be blocked with a selective antagonist.

blocked with a selective antagonist. The use of 5-LO/FLAP inhibitors would have the advantage of preventing the formation of both 5-oxo-ETE and cysLTs. However, LTB_4 formation would also be blocked, the advantage of which is not as clear because of its important role in host defense against bacterial infection (Bailie et al., 1996; Mancuso et al., 2001). Another possibility would be to selectively target 5-HEDH. We have identified C_{18} and C_{20} 5-hydroxy monoenoic acids that selectively inhibit 5-oxo-ETE synthesis by peripheral blood mononuclear cells (Patel et al., 2009). However, these substances are competitive inhibitors and their potency is somewhat limited (IC_{50}, 2–3 μM). Another problem with inhibiting the enzymatic generation of 5-oxo-ETE is that it can also be formed independently of 5-LO and 5-HEDH by lipid peroxidation (Hall and Murphy, 1998b), although the physiological significance of this pathway has not yet been well defined.

Another strategy would be to target the OXE receptor. The possibility of identifying an OXE receptor antagonist was raised some time ago when we found that 5-oxo-12-HETE, a product formed during coincubations of neutrophils with platelets, was a fairly potent inhibitor (IC_{50}, 0.5 μM) of 5-oxo-ETE-induced calcium mobilization in neutrophils without itself inducing this response (Powell et al., 1999b). 5-Oxo-12-HETE also inhibited 5-oxo-ETE-induced neutrophil migration, but unfortunately displayed weak agonist activity in inducing surface expression of CD11b (Powell and Rokach, unpublished work). More recently, we have identified a series of synthetic compounds that are potent selective OXE receptor antagonists (Powell, Patel, Gore, and Rokach, unpublished results). These compounds are devoid of agonist activity and have no effect on the responses of granulocytes to other related mediators such as LTB_4, while blocking the effects of 5-oxo-ETE on cell migration, calcium mobilization, and other responses. Testing in animal models of asthma is currently underway.

7.9 CONCLUSIONS

5-Oxo-ETE is a 5-LO product that is a chemoattractant for a variety of inflammatory cells, including eosinophils, neutrophils, basophils, and monocytes. It also has effects on other cells, including airway smooth muscle cells, intestinal epithelial cells (volume reduction) (MacLeod et al., 1999), and tumor cells (proliferation) (Ghosh and Myers, 1998; O'Flaherty et al., 2005; Sundaram and Ghosh, 2006). However, its primary target appears to be the eosinophil, for which it is a potent chemoattractant, inducing transendothelial migration and tissue infiltration. 5-Oxo-ETE is synthesized from 5-HETE by a highly selective dehydrogenase, 5-HEDH. Because its synthesis is dependent on $NADP^+$, its synthesis is limited in resting cells, but is strongly stimulated by the respiratory burst in phagocytic cells and by oxidative stress, conditions that exist at the sites of inflammation. The actions of 5-oxo-ETE are mediated by the selective OXE receptor, which is most highly expressed in eosinophils. Because of its potent effects on these cells and their important role in asthma and other allergic diseases, the OXE receptor is an attractive drug target.

ACKNOWLEDGMENTS

This work was supported by grants from the Canadian Institutes of Health Research (WSP; MOP-6254) and the National Heart, Lung, and Blood Institute (Joshua Rokach; Award Number R01HL081873). The Meakins-Christie Laboratories—MUHC-RI are supported in part by a center grant from Le Fonds de la Recherche en Santé du Québec as well as by the J.T. Costello Memorial Research Fund. Joshua Rokach also wishes to acknowledge the National Science Foundation for the AMX-360 (grant number CHE-90-13145) and Bruker 400 MHz (grant number CHE-03-42251) NMR instruments. The content is solely the responsibility of the authors and does not necessarily represent the official views of the National Heart, Lung, and Blood Institute or the National Institutes of Health.

REFERENCES

Almishri, W., C. Cossette, J. Rokach, J.G. Martin, Q. Hamid, and W.S. Powell 2005. Effects of prostaglandin D_2, 15-deoxy-$\Delta^{12,14}$-prostaglandin J_2, and selective DP_1 and DP_2 receptor agonists on pulmonary infiltration of eosinophils in Brown Norway rats. *J. Pharmacol. Exp. Ther.* 313:64–69.

Bailie, M.B., T.J. Standiford, L.L. Laichalk, M.J. Coffey, R. Strieter, and M. Peters-Golden 1996. Leukotriene-deficient mice manifest enhanced lethality from *Klebsiella* pneumonia in association with decreased alveolar macrophage phagocytic and bactericidal activities. *J. Immunol.* 157:5221–24.

Barnes, P.J. 2007. New molecular targets for the treatment of neutrophilic diseases. *J. Allergy Clin. Immunol.* 119:1055–62.

Bedard, K. and K.H. Krause 2007. The NOX family of ROS-generating NADPH oxidases: Physiology and pathophysiology. *Physiol. Rev.* 87:245–313.

Berhane, K., A.A. Ray, S.P. Khanapure, J. Rokach, and W.S. Powell 1998. Calcium/calmodulin-dependent conversion of 5-oxoeicosanoids to 6,7-dihydro metabolites by a cytosolic olefin reductase in human neutrophils. *J. Biol. Chem.* 273:20951–59.

Bowers, R.C., J. Hevko, P.M. Henson, and R.C. Murphy 2000. A novel glutathione containing eicosanoid (FOG_7) chemotactic for human granulocytes. *J. Biol. Chem.* 275:29931–34.

Brink, C., S.E. Dahlen, J. Drazen, J.F. Evans, D.W. Hay, G.E. Rovati, C.N. Serhan, T. Shimizu, and T. Yokomizo 2004. International Union of Pharmacology XLIV. Nomenclature for the oxoeicosanoid receptor. *Pharmacol. Rev.* 56:149–57.

Brown, K.A., S.D. Brain, J.D. Pearson, J.D. Edgeworth, S.M. Lewis, and D.F. Treacher 2006. Neutrophils in development of multiple organ failure in sepsis. *Lancet* 368:157–69.

Bryant, R.W., H.S. She, K.J. Ng, and M.I. Siegel 1986. Modulation of the 5-lipoxygenase activity of MC-9 mast cells: Activation by hydroperoxides. *Prostaglandins* 32:615–27.

Bui, P.H., S. Imaizumi, S.R. Beedanagari, S.T. Reddy, and O. Hankinson 2010. CYP2S1, metabolizes cyclooxygenase—and lipoxygenase—derived eicosanoids. *Drug Metab. Dispos.* 39:180–90.

Cormier, S.A., A.G. Taranova, C. Bedient, T. Nguyen, C. Protheroe, R. Pero, D. Dimina et al. 2006. Pivotal advance: Eosinophil infiltration of solid tumors is an early and persistent inflammatory host response. *J. Leukoc. Biol.* 79:1131–39.

Cossette, C., P. Patel, J.R. Anumolu, S. Sivendran, G.J. Lee, S. Gravel, F.D. Graham et al. 2008. Human neutrophils convert the sebum-derived polyunsaturated fatty acid sebaleic acid to a potent granulocyte chemoattractant. *J. Biol. Chem.* 283:11234–43.

Czech, W., M. Barbisch, K. Tenscher, E. Schopf, J.M. Schröder, and J. Norgauer 1997. Chemotactic 5-oxo-eicosatetraenoic acids induce oxygen radical production, Ca^{2+}-mobilization, and actin reorganization in human eosinophils via a pertussis toxin-sensitive G-protein. *J. Invest. Dermatol.* 108:108–12.

Dallaire, M.J., C. Ferland, N. Page, S. Lavigne, F. Davoine, and M. Laviolette 2003. Endothelial cells modulate eosinophil surface markers and mediator release. *Eur. Respir. J.* 21:918–24.

DiPersio, J.F., P. Billing, R. Williams, and J.C. Gasson 1988. Human granulocyte-macrophage colony-stimulating factor and other cytokines prime human neutrophils for enhanced arachidonic acid release and leukotriene B_4 synthesis. *J. Immunol.* 140:4315–22.

Erlemann, K.R., C. Cossette, G.E. Grant, G.J. Lee, P. Patel, J. Rokach, and W.S. Powell 2007a. Regulation of 5-hydroxyeicosanoid dehydrogenase activity in monocytic cells. *Biochem. J.* 403:157–65.

Erlemann, K.R., C. Cossette, S. Gravel, A. Lesimple, G.J. Lee, G. Saha, J. Rokach, and W.S. Powell 2007b. Airway epithelial cells synthesize the lipid mediator 5-oxo-ETE in response to oxidative stress. *Free Radic. Biol. Med.* 42:654–64.

Erlemann, K.R., C. Cossette, S. Gravel, P.B. Stamatiou, G.J. Lee, J. Rokach, and W.S. Powell 2006. Metabolism of 5-hydroxy-6,8,11,14-eicosatetraenoic acid by human endothelial cells. *Biochem. Biophys. Res. Commun.* 350:151–56.

Erlemann, K.R., J. Rokach, and W.S. Powell 2004. Oxidative stress stimulates the synthesis of the eosinophil chemoattractant 5-oxo-6,8,11,14-eicosatetraenoic acid by inflammatory cells. *J. Biol. Chem.* 279:40376–84.

Evans, J.F., D.J. Nathaniel, R.J. Zamboni, and A.W. Ford-Hutchinson 1985. Leukotriene A_3. A poor substrate but a potent inhibitor of rat and human neutrophil leukotriene A_4 hydrolase. *J. Biol. Chem.* 260:10966–70.

Ferland, C., M. Guilbert, F. Davoine, N. Flamand, J. Chakir, and M. Laviolette 2001. Eotaxin promotes eosinophil transmigration via the activation of the plasminogen-plasmin system. *J. Leukoc. Biol.* 69:772–78.

Flood-Page, P., A. Menzies-Gow, S. Phipps, S. Ying, A. Wangoo, M.S. Ludwig, N. Barnes, D. Robinson, and A.B. Kay 2003. Anti-IL-5 treatment reduces deposition of ECM proteins in the bronchial subepithelial basement membrane of mild atopic asthmatics. *J. Clin. Invest.* 112:1029–36.

Folco, G. and R.C. Murphy 2006. Eicosanoid transcellular biosynthesis: From cell–cell interactions to *in vivo* tissue responses. *Pharmacol. Rev.* 58:375–88.

Funk, C.D. 2001. Prostaglandins and leukotrienes: Advances in eicosanoid biology. *Science* 294:1871–75.

Ghosh, J. and C.E. Myers 1998. Inhibition of arachidonate 5-lipoxygenase triggers massive apoptosis in human prostate cancer cells. *Proc. Natl. Acad. Sci. U. S. A.* 95:13182–87.

Gibbs, B.F. 2005. Human basophils as effectors and immunomodulators of allergic inflammation and innate immunity. *Clin. Exp. Med.* 5:43–49.

Gijon, M.A. and C.C. Leslie 1999. Regulation of arachidonic acid release and cytosolic phospholipase A2 activation. *J. Leukoc. Biol.* 65:330–36.

Graham, F.D., K.R. Erlemann, S. Gravel, J. Rokach, and W.S. Powell 2009. Oxidative stress-induced changes in pyridine nucleotides and chemoattractant 5-lipoxygenase products in aging neutrophils. *Free Radic. Biol. Med.* 47:62–71.

Grant, G.E., S. Gravel, J. Guay, P. Patel, B.D. Mazer, J. Rokach, and W.S. Powell 2011a. 5-Oxo-ETE is a major oxidative stress-induced arachidonate metabolite in B lymphocytes. *Free Radic. Biol. Med.* 50:1297–304.

Grant, G.E., S. Rubino, S. Gravel, X. Wang, P. Patel, J. Rokach, and W.S. Powell 2011b. Enhanced formation of 5-oxo-6,8,11,14-eicosatetraenoic acid by cancer cells in response to oxidative stress, docosahexaenoic acid, and neutrophil-derived 5-hydroxy-6,8,11,14-eicosatetraenoic acid. *Carcinogenesis* 32:822–28.

Gravel, J., J.P. Falgueyret, J. Yergey, L. Trimble, and D. Riendeau 1993. Identification of 5-keto-(7E,9E,11Z,14Z)-eicosatetraenoic acid as a novel nonenzymatic rearrangement product of leukotriene A4. *Arch. Biochem. Biophys.* 306:469–75.

Guilbert, M., C. Ferland, M. Bosse, N. Flamand, S. Lavigne, and M. Laviolette 1999. 5-Oxo-6,8,11,14-eicosatetraenoic acid induces important eosinophil transmigration through basement membrane components: Comparison of normal and asthmatic eosinophils. *Am. J. Respir. Cell. Mol. Biol.* 21:97–104.

Hall, L.M. and R.C. Murphy 1998a. Activation of human polymorphonuclear leukocytes by products derived from the peroxidation of human red blood cell membranes. *Chem. Res. Toxicol.* 11:1024–31.

Hall, L.M. and R.C. Murphy 1998b. Electrospray mass spectrometric analysis of 5-hydroperoxy and 5-hydroxyeicosatetraenoic acids generated by lipid peroxidation of red blood cell ghost phospholipids. *J. Am. Soc. Mass. Spectrom.* 9:527–32.

Hamberg, M. 1975. Decomposition of unsaturated fatty acid hydroperoxides by hemoglobin: Structures of major products of 13L-hydroperoxy-9,11-octadecadienoic acid. *Lipids* 10:87–92.

Haslett, C. 1999. Granulocyte apoptosis and its role in the resolution and control of lung inflammation. *Am. J. Respir. Crit Care Med.* 160:S5–11.

Hevko, J.M., R.C. Bowers, and R.C. Murphy 2001. Synthesis of 5-oxo-6,8,11,14-eicosatetraenoic acid and identification of novel omega-oxidized metabolites in the mouse macrophage. *J. Pharmacol. Exp. Ther.* 296:293–305.

Hevko, J.M. and R.C. Murphy 2002. Formation of murine macrophage-derived 5-oxo-7-glutathionyl-8,11,14-eicosatrienoic acid (FOG$_7$) is catalyzed by leukotriene C$_4$ synthase. *J. Biol. Chem.* 277:7037–43.

Hirai, H., K. Tanaka, O. Yoshie, K. Ogawa, K. Kenmotsu, Y. Takamori, M. Ichimasa et al. Nagata 2001. Prostaglandin D$_2$ selectively induces chemotaxis in T helper type 2 cells, eosinophils, and basophils via seven-transmembrane receptor CRTH2. *J. Exp. Med.* 193:255–61.

Hosoi, T., Y. Koguchi, E. Sugikawa, A. Chikada, K. Ogawa, N. Tsuda, N. Suto, S. Tsunoda, T. Taniguchi, and T. Ohnuki 2002. Identification of a novel eicosanoid receptor coupled to G$_{i/o}$. *J. Biol. Chem.* 277:31459–65.

Hosoi, T., E. Sugikawa, A. Chikada, Y. Koguchi, and T. Ohnuki 2005. TG1019/OXE, a Galpha(i/o)-protein-coupled receptor, mediates 5-oxo-eicosatetraenoic acid-induced chemotaxis. *Biochem. Biophys. Res. Commun.* 334:987–95.

Hunter, J.A., W.E. Finkbeiner, J.A. Nadel, E.J. Goetzl, and M.J. Holtzman 1985. Predominant generation of 15-lipoxygenase metabolites of arachidonic acid by epithelial cells from human trachea. *Proc. Natl. Acad. Sci. U. S. A.* 82:4633–37.

Iikura, M., M. Suzukawa, M. Yamaguchi, T. Sekiya, A. Komiya, C. Yoshimura-Uchiyama, H. Nagase, K. Matsushima, K. Yamamoto, and K. Hirai 2005. 5-Lipoxygenase products regulate basophil functions: 5-Oxo-ETE elicits migration, and leukotriene B(4) induces degranulation. *J. Allergy Clin. Immunol.* 116:578–85.

Jacobsen, E.A., S.I. Ochkur, N.A. Lee, and J.J. Lee, 2007. Eosinophils and asthma. *Curr. Allergy Asthma Rep.* 7:18–26.

Jakobsson, P.J., D. Steinhilber, B. Odlander, O. Rådmark, H.E. Claesson, and B. Samuelsson 1992. On the expression and regulation of 5-lipoxygenase in human lymphocytes. *Proc. Natl. Acad. Sci. U. S. A.* 89:3521–25.

Jones, C.E., S. Holden, L. Tenaillon, U. Bhatia, K. Seuwen, P. Tranter, J. Turner et al. 2003. Expression and characterization of a 5-oxo-6*E*,8*Z*,11*Z*,14*Z*-eicosatetraenoic acid receptor highly expressed on human eosinophils and neutrophils. *Mol. Pharmacol.* 63: 471–77.

Khaselev, N. and R.C. Murphy 2000. Peroxidation of arachidonate containing plasmenyl glycerophosphocholine: Facile oxidation of esterified arachidonate at carbon-5. *Free Radic. Biol. Med.* 29:620–32.

Langlois, A., F. Chouinard, N. Flamand, C. Ferland, M. Rola-Pleszczynski, and M. Laviolette 2009. Crucial implication of protein kinase C (PKC)-δ, PKC-ζ, ERK-1/2, and p38 MAPK in migration of human asthmatic eosinophils. *J. Leukoc. Biol.* 85:656–63.

Langlois, A., C. Ferland, G.M. Tremblay, and M. Laviolette 2006. Montelukast regulates eosinophil protease activity through a leukotriene-independent mechanism. *J. Allergy Clin. Immunol.* 118:113–19.

Leckie, M.J., A. ten Brinke, J. Khan, Z. Diamant, B.J. O'Connor, C.M. Walls, A.K. Mathur et al. 2000. Effects of an interleukin-5 blocking monoclonal antibody on eosinophils, airway hyper-responsiveness, and the late asthmatic response. *Lancet* 356:2144–48.

MacLeod, R.J., P. Lembessis, J.R. Hamilton, and W.S. Powell 1999. 5-Oxo-6,8,11,14-eicosatetraenoic acid stimulates isotonic volume reduction of guinea pig jejunal crypt epithelial cells. *J. Pharmacol. Exp. Ther.* 291:511–16.

Maghni, K., A.J. de Brum-Fernandes, E. Foldes-Filep, P. Gaudry, P. Borgeat, and P. Sirois 1991. Leukotriene B$_4$ receptors on guinea pig alveolar eosinophils. *J. Pharm. Exp. Ther.* 258:784–89.

Mancuso, P., P. Nana-Sinkam, and M. Peters-Golden 2001. Leukotriene B4 augments neutrophil phagocytosis of *Klebsiella pneumoniae*. *Infect. Immun.* 69:2011–16.

Marleau, S., C. Fortin, P.E. Poubelle, and P. Borgeat 1993. *In vivo* desensitization to leukotriene B$_4$ (LTB$_4$) in the rabbit. Inhibition of LTB$_4$-induced neutropenia during intravenous infusion of LTB$_4$. *J. Immunol.* 150:206–13.

Mercier, F., C. Morin, M. Cloutier, S. Proteau, J. Rokach, W.S. Powell, and E. Rousseau 2004. 5-Oxo-ETE regulates tone of guinea pig airway smooth muscle via activation of Ca^{2+} pools and Rho-kinase pathway. *Am. J. Physiol. Lung Cell. Mol. Physiol.* 287: L631–L640.

Monneret, G., R. Boumiza, S. Gravel, C. Cossette, J. Bienvenu, J. Rokach, and W.S. Powell 2005. Effects of prostaglandin D$_2$ and 5-lipoxygenase products on the expression of CD203c and CD11b by basophils. *J. Pharmacol. Exp. Ther.* 312:627–34.

Monneret, G., S. Gravel, M. Diamond, J. Rokach, and W.S. Powell 2001. Prostaglandin D$_2$ is a potent chemoattractant for human eosinophils that acts via a novel DP receptor. *Blood* 98:1942–48.

Morin, C. and E. Rousseau 2007. Effects of 5-oxo-ETE and 14,15-EET on reactivity and Ca^{2+} sensitivity in guinea pig bronchi. *Prostaglandins Other Lipid Mediat.* 82:30–41.

Morin, C., M. Sirois, V. Echave, M.M. Gomes J. Rokach, and E. Rousseau 2007. Relaxing effects of 5-oxo-ETE on human bronchi involve BK Ca channel activation. *Prostaglandins Other Lipid Mediat.* 83:311–19.

Morita, E., J.M. Schröder, and E. Christophers 1989. Differential sensitivities of purified human eosinophils and neutrophils to defined chemotaxins. *Scand. J. Immunol.* 29:709–16.

Morita, E., J.M. Schröder, and E. Christophers 1990. Identification of a novel and highly potent eosinophil chemotactic lipid in human eosinophils treated with arachidonic acid. *J. Immunol.* 144:1893–900.

Muro, S., Q. Hamid, R. Olivenstein, R. Taha, J. Rokach, and W.S. Powell 2003. 5-Oxo-6,8,11,14-eicosatetraenoic acid induces the infiltration of granulocytes into human skin. *J. Allergy Clin. Immunol.* 112:768–74.

Murphy, R.C. and M.A. Gijon 2007. Biosynthesis and metabolism of leukotrienes. *Biochem. J.* 405:379–95.

Murray, J.J., A.B. Tonnel, A.R. Brash, L.J. Roberts, P. Gosset, R. Workman, A. Capron, and J.A. Oates 1986. Release of prostaglandin D_2 into human airways during acute antigen challenge. *N. Engl. J. Med.* 315:800–04.

Nair, P., M.M. Pizzichini, M. Kjarsgaard, M.D. Inman, A. Efthimiadis, E. Pizzichini, F.E. Hargreave, and P.M. O'Byrne 2009. Mepolizumab for prednisone-dependent asthma with sputum eosinophilia. *N. Engl. J. Med.* 360:985–93.

Norgauer, J., M. Barbisch, W. Czech, J. Pareigis, U. Schwenk, and J.M. Schröder 1996. Chemotactic 5-oxo-icosatetraenoic acids activate a unique pattern of neutrophil responses—Analysis of phospholipid metabolism, intracellular Ca^{2+} transients, actin reorganization, superoxide-anion production and receptor up-regulation. *Eur. J. Biochem.* 236:1003–09.

O'Flaherty, J.T. 1985. Neutrophil degranulation: Evidence pertaining to its mediation by the combined effects of leukotriene B4, platelet-activating factor, and 5-HETE. *J. Cell. Physiol.* 122:229–39.

O'Flaherty, J.T., J. Cordes, J. Redman, and M.J. Thomas 1993. 5-Oxo-eicosatetraenoate, a potent human neutrophil stimulus. *Biochem. Biophys. Res. Commun.* 192:129–34.

O'Flaherty, J.T., J.F. Cordes, S.L. Lee, M. Samuel, and M.J. Thomas 1994. Chemical and biological characterization of oxo-eicosatetraenoic acids. *Biochim. Biophys. Acta* 1201:505–15.

O'Flaherty, J.T., D. Jacobson, and J. Redman 1988. Mechanism involved in the mobilization of neutrophil calcium by 5-hydroxyeicosatetraenoate. *J. Immunol.* 140:4323–28.

O'Flaherty, J.T., M. Kuroki, A.B. Nixon, J. Wijkander, E. Yee, S.L. Lee, P.K. Smitherman, R.L. Wykle, and L.W. Daniel 1996a. 5-Oxo-eicosanoids and hematopoietic cytokines cooperate in stimulating neutrophil function and the mitogen-activated protein kinase pathway. *J. Biol. Chem.* 271:17821–28.

O'Flaherty, J.T., M. Kuroki, A.B. Nixon, J. Wijkander, E. Yee, S.L. Lee, P.K. Smitherman, R.L. Wykle, and L.W. Daniel 1996b. 5-Oxo-eicosatetraenoate is a broadly active, eosino-phil- selective stimulus for human granulocytes. *J. Immunol.* 157:336–42.

O'Flaherty, J.T. and J. Nishihira 1987. 5-Hydroxyeicosatetraenoate promotes Ca^{2+} and protein kinase C mobilization in neutrophils. *Biochem. Biophys. Res. Commun.* 148:575–81.

O'Flaherty, J.T., L.C. Rogers, B.A. Chadwell, J.S. Owen, A. Rao, S.D. Cramer, and L.W. Daniel 2002. 5(*S*)-Hydroxy-6,8,11,14-*E,Z,Z,Z*-eicosatetraenoate stimulates PC3 cell signaling and growth by a receptor-dependent mechanism. *Cancer Res.* 62:6817–19.

O'Flaherty, J.T., L.C. Rogers, C.M. Paumi, R.R. Hantgan, L.R. Thomas, C.E. Clay, K. High et al. 2005. 5-Oxo-ETE analogs and the proliferation of cancer cells. *Biochim. Biophys. Acta* 1736:228–36.

O'Flaherty, J.T., J.S. Taylor, and M. Kuroki 2000. The coupling of 5-oxo-eicosanoid receptors to heterotrimeric G proteins. *J. Immunol.* 164:3345–52.

O'Flaherty, J.T., J.S. Taylor, and M.J. Thomas 1998. Receptors for the 5-oxo class of eico-sanoids in neutrophils. *J. Biol. Chem.* 273:32535–41.

Ochkur, S.I., E.A. Jacobsen, C.A. Protheroe, T.L. Biechele, R.S. Pero, M.P. McGarry, H. Wang et al. 2007. Coexpression of IL-5 and eotaxin-2 in mice creates an eosinophil-dependent model of respiratory inflammation with characteristics of severe asthma. *J. Immunol.* 178:7879–89.

Park, C.S., Y.S. Choi, S.Y. Ki, S.H. Moon, S.W. Jeong, S.T. Uh, and Y.H. Kim 1998. Granulocyte macrophage colony-stimulating factor is the main cytokine enhancing survival of eosinophils in asthmatic airways. *Eur. Respir. J.* 12:872–78.

Patel, P., C. Cossette, J.R. Anumolu, S. Gravel, A. Lesimple, O.A. Mamer, J. Rokach, and W.S. Powell 2008. Structural requirements for activation of the 5-oxo-6*E*,8*Z*,11*Z*,14*Z*-eicosatetraenoic acid (5-oxo-ETE) receptor: Identification of a Mead acid metabolite with potent agonist activity. *J. Pharmacol. Exp. Ther.* 325:698–707.

Patel, P., C. Cossette, J.R. Anumolu, K.R. Erlemann, G.E. Grant, J. Rokach, and W.S. Powell 2009. Substrate selectivity of 5-hydroxyeicosanoid dehydrogenase and its inhibition by 5-hydroxy-delta(6)-long-chain fatty acids. *J. Pharmacol. Exp. Ther.* 329:335–41.

Peters-Golden, M. and W.R. Henderson, Jr. 2007. Leukotrienes. *N. Engl. J. Med.* 357:1841–54.

Pietarinen-Runtti, P., E. Lakari, K.O. Raivio, and V.L. Kinnula 2000. Expression of antioxidant enzymes in human inflammatory cells. *Am. J. Physiol. Cell Physiol.* 278:C118–C125.

Pouliot, M., P.P. McDonald, L. Khamzina, P. Borgeat, and S.R. McColl 1994. Granulocyte-macrophage colony-stimulating factor enhances 5-lipoxygenase levels in human polymorphonuclear leukocytes. *J. Immunol.* 152:851–58.

Powell, W.S., S. Ahmed, S. Gravel, and J. Rokach 2001. Eotaxin and RANTES enhance 5-oxo-6,8,11,14-eicosatetraenoic acid-induced eosinophil chemotaxis. *J. Allergy Clin. Immunol.* 107:272–78.

Powell, W.S., D. Chung, and S. Gravel 1995a. 5-Oxo-6,8,11,14-eicosatetraenoic acid is a potent stimulator of human eosinophil migration. *J. Immunol.* 154:4123–32.

Powell, W.S. and F. Gravelle, 1988. Metabolism of 6-*trans* isomers of leukotriene B4 to dihydro products by human polymorphonuclear leukocytes. *J. Biol. Chem.* 263: 2170–77.

Powell, W.S. and F. Gravelle 1990. Metabolism of arachidonic acid by peripheral and elicited rat polymorphonuclear leukocytes. Formation of 18- and 19-oxygenated dihydro metabolites of leukotriene B4. *J. Biol. Chem.* 265:9131–39.

Powell, W.S., F. Gravelle, and S. Gravel 1992. Metabolism of 5(*S*)-hydroxy-6,8,11,14-eicosatetraenoic acid and other 5(*S*)-hydroxyeicosanoids by a specific dehydrogenase in human polymorphonuclear leukocytes. *J. Biol. Chem.* 267:19233–41.

Powell, W.S., F. Gravelle, and S. Gravel 1994. Phorbol myristate acetate stimulates the formation of 5-oxo-6,8,11,14-eicosatetraenoic acid by human neutrophils by activating NADPH oxidase. *J. Biol. Chem.* 269:25373–80.

Powell, W.S., S. Gravel, and F. Gravelle 1995b. Formation of a 5-oxo metabolite of 5,8,11,14,17-eicosapentaenoic acid and its effects on human neutrophils and eosinophils. *J. Lipid Res.* 36:2590–98.

Powell, W.S., S. Gravel, and F. Halwani 1999a. 5-Oxo-6,8,11,14-eicosatetraenoic acid is a potent stimulator of L-selectin shedding, surface expression of CD11b, actin polymerization, and calcium mobilization in human eosinophils. *Am. J. Respir. Cell. Mol. Biol.* 20:163–70.

Powell, W.S., S. Gravel, F. Halwani, C.S. Hii, Z.H. Huang, A.M. Tan, and A. Ferrante 1997. Effects of 5-oxo-6,8,11,14-eicosatetraenoic acid on expression of CD11b, actin polymerization and adherence in human neutrophils. *J. Immunol.* 159:2952–59.

Powell, W.S., S. Gravel, S.P. Khanapure, and J. Rokach 1999b. Biological inactivation of 5-oxo-6,8,11,14-eicosatetraenoic acid by human platelets. *Blood* 93:1086–96.

Powell, W.S., S. Gravel, R.J. MacLeod, E. Mills, and M. Hashefi 1993. Stimulation of human neutrophils by 5-oxo-6,8,11,14- eicosatetraenoic acid by a mechanism independent of the leukotriene B$_4$ receptor. *J. Biol. Chem.* 268:9280–86.

Powell, W.S., R.J. MacLeod, S. Gravel, F. Gravelle, and A. Bhakar 1996. Metabolism and biologic effects of 5-oxoeicosanoids on human neutrophils. *J. Immunol.* 156:336–42.

Rådinger, M. and J. Lötvall 2009. Eosinophil progenitors in allergy and asthma—Do they matter? *Pharmacol. Ther.* 121:174–84.

Rådmark, O., O. Werz, D. Steinhilber, and B. Samuelsson 2007. 5-Lipoxygenase: Regulation of expression and enzyme activity. *Trends Biochem. Sci.* 32:332–41.

Rylander, T., E.P. Neve, M. Ingelman-Sundberg, and M. Oscarson 2001. Identification and tissue distribution of the novel human cytochrome P450 2S1 (CYP2S1). *Biochem. Biophys. Res. Commun.* 281:529–35.

Schafer, F.Q. and G.R. Buettner 2001. Redox environment of the cell as viewed through the redox state of the glutathione disulfide/glutathione couple. *Free Radic. Biol. Med.* 30:1191–212.

Schwab, J.M. and C.N. Serhan 2006. Lipoxins and new lipid mediators in the resolution of inflammation. *Curr. Opin. Pharmacol.* 6:414–20.

Schwenk, U., E. Morita, R. Engel, and J.M. Schröder 1992. Identification of 5-oxo-15-hydroxy-6,8,11,13-eicosatetraenoic acid as a novel and potent human eosinophil chemotactic eicosanoid. *J. Biol. Chem.* 267:12482–88.

Schwenk, U. and J.M. Schröder 1995. 5-Oxo-eicosanoids are potent eosinophil chemotactic factors—Functional characterization and structural requirements. *J. Biol. Chem.* 270: 15029–36.

Sehmi, R., O. Cromwell, G.W. Taylor, and A.B. Kay 1991. Identification of guinea pig eosinophil chemotactic factor of anaphylaxis as leukotriene B$_4$ and 8(S),15(S)-dihydroxy-5,9,11,13(Z,E,Z,E)-eicosatetraenoic acid. *J. Immunol.* 147:2276–83.

Serezani, C.H., D.M. Aronoff, S. Jancar, P. Mancuso, and M. Peters-Golden 2005. Leukotrienes enhance the bactericidal activity of alveolar macrophages against *Klebsiella pneumoniae* through the activation of NADPH oxidase. *Blood* 106:1067–75.

Serhan, C.N. and J. Savill 2005. Resolution of inflammation: The beginning programs the end. *Nat. Immunol.* 6:1191–97.

Sozzani, S., D. Zhou, M. Locati, S. Bernasconi, W. Luini, A. Mantovani, and J.T. O'Flaherty 1996. Stimulating properties of 5-oxo-eicosanoids for human monocytes: Synergism with monocyte chemotactic protein-1 and -3. *J. Immunol.* 157:4664–71.

Stamatiou, P., Q. Hamid, R. Taha, W. Yu, T.B. Issekutz, J. Rokach, S.P. Khanapure, and W.S. Powell 1998. 5-Oxo-ETE induces pulmonary eosinophilia in an integrin-dependent manner in Brown Norway rats. *J. Clin. Invest.* 102:2165–72.

Stamatiou, P.B., C.C. Chan, G. Monneret, D. Ethier, J. Rokach, and W.S. Powell 2004. 5-Oxo-6,8,11,14-eicosatetraenoic acid stimulates the release of the eosinophil survival factor granulocyte-macrophage colony stimulating factor from monocytes. *J. Biol. Chem.* 279:28159–64.

Sturm, G.J., R. Schuligoi, E.M. Sturm, J.F. Royer, D. Lang-Loidolt, H. Stammberger, R. Amann, B.A. Peskar, and A. Heinemann 2005. 5-Oxo-6,8,11,14-eicosatetraenoic acid is a potent chemoattractant for human basophils. *J. Allergy Clin. Immunol.* 116:1014–19.

Sun, F.F., N.J. Crittenden, C.I. Czuk, B.M. Taylor, B.K. Stout, and H.G. Johnson 1991. Biochemical and functional differences between eosinophils from animal species and man. *J. Leukoc. Biol.* 50:140–50.

Sundaram, S. and J. Ghosh 2006. Expression of 5-oxoETE receptor in prostate cancer cells: Critical role in survival. *Biochem. Biophys. Res. Commun.* 339:93–98.

Suzukawa, M., A. Komiya, M. Iikura, H. Nagase, C. Yoshimura-Uchiyama, H. Yamada, H. Kawasaki et al. 2006. Trans-basement membrane migration of human basophils: Role of matrix metalloproteinase-9. *Int. Immunol.* 18:1575–83.

Takeda, S., A. Yamamoto, and T. Haga 2002. Identification of a G protein-coupled receptor for 5-oxo-eicosatetraenoic acid. *Biomed. Res. Tokyo* 23:101–08.

Underwood, D.C., R.R. Osborn, S.J. Newsholme, T.J. Torphy, and D.W. Hay 1996. Persistent airway eosinophilia after leukotriene (LT) D_4 administration in the guinea pig: Modulation by the LTD4 receptor antagonist, pranlukast, or an interleukin-5 monoclonal antibody. *Am. J. Respir. Crit. Care Med.* 154:850–57.

Urasaki, T., J. Takasaki, T. Nagasawa, and H. Ninomiya 2001. Pivotal role of 5-lipoxygenase in the activation of human eosinophils: Platelet-activating factor and interleukin-5 induce CD69 on eosinophils through the 5-lipoxygenase pathway. *J. Leukoc. Biol.* 69:105–12.

Venge, P. 2010. The eosinophil and airway remodelling in asthma. *Clin. Respir. J.* 4 Suppl 1:15–19.

Zarini, S. and R.C. Murphy 2003. Biosynthesis of 5-oxo-6,8,11,14-eicosatetraenoic acid from 5-hydroperoxyeicosatetraenoic acid in the murine macrophage. *J. Biol. Chem.* 278:11190–96.

Zhang, Y., A. Styhler, and W.S. Powell 1996. Synthesis of 5-oxo-6,8,11,14-eicosatetraenoic acid by human monocytes and lymphocytes. *J. Leukoc. Biol.* 59:847–54.

Zimpfer, U., S. Dichmann, C.C. Termeer, J.C. Simon, J.M. Schroder, and J. Norgauer 2000. Human dendritic cells are a physiological source of the chemotactic arachidonic acid metabolite 5-oxo-eicosatetraenoic acid. *Inflamm. Res.* 49:633–38.

8 Cyclooxygenases and Prostaglandin Receptors

Jitesh P. Iyer, Jitendra Anant Sattigeri, and Sunanda Ghosh Dastidar

CONTENTS

8.1 INTRODUCTION

There is strong evidence that arachidonic acid (AA) metabolites play an important proinflammatory role in inflammatory diseases of the airways such as asthma and chronic obstructive pulmonary disease (COPD) (Wenzel et al., 1997; Henderson, 1994; O'Byrne, 1997). AA is metabolized by two different families of enzymes, namely cyclooxygenase (COX) and lipoxygenase (LO/LOX), leading to the generation of prostanoids and leukotrienes. The contribution of the COX family of enzymes, together with the prostanoids, has been well documented in the pathophysiology of several inflammatory diseases, including COPD. Current treatment for COPD relies mainly on long-acting bronchodilators or inhaled corticosteroids. Both of these have little impact in controlling the inflammatory component of this disease, thereby creating a large unmet medical need in this area. The objective of this chapter is to examine the contribution of the COX pathway in the pathophysiology of respiratory conditions such as COPD and asthma. Drug discovery effort around COX and prostanoid receptor with special reference to prostaglandin D_2 receptor will be explored.

8.2 PROSTANOIDS

In Latin, the term "eicosa" stands for "twenty." The term "eicosanoids" hence refers to derivatives of long-chain fatty acids having 20 carbon atoms in the chain, such as AA. The oxygenation of such poly-unsaturated, long-chain fatty acids leads to the generation of multiple eicosanoids. Along with their precursors, eicosanoids are found in most species of the plant and animal kingdom. Prostanoids are a subclass of eicosanoids comprising of the prostaglandins (PGs), prostacyclin, and thromboxanes.

It was in the 1930s that the significance of lipids as signaling intermediates in human biochemistry was discovered. Slow-reacting substance of anaphylaxis (SRS-A), a mixture of leukotrienes such as leukotriene C_4 (LTC_4), leukotriene D_4 (LTD_4), and leukotriene E_4 (LTE_4) (now known as cysteinyl leukotrienes), provided the earliest indication of such a role for lipids (Feldberg et al., 1938). Around the same time, prostaglandins were found to have vasodilatory actions (von Euler, 1936). Probably the most important discovery that paved the path for our current understanding of lipid signaling was that AA served as the source of both prostaglandins as well as leukotrienes, providing the first instance of a single lipid being responsible for diverse cellular outcomes, depending on cell type and the signaling network of the target cell.

AA can undergo two fates within cells, usually dependent on the cell type. It could get stereospecifically oxygenated through the cyclic prostaglandin synthesis pathway, mediated by COX, forming prostanoids. AA can also be a substrate for lipoxygenase to generate leukotrienes. The biology of lipoxygenase and leukotrienes are covered in more detail in Chapters 4 and 5.

All prostanoids are synthesized by the action of COX on AA (**1**, Figure 8.1). There are three isoforms of COX, namely COX-1, COX-2 (Carey et al., 2003), and COX-3 (Shaftel et al., 2003), a splice variant of the *COX-1* gene. Prostanoids comprise the following prostaglandins (PG): PGD_2 (**3**), PGE_2 (**4**), $PGF_{2\alpha}$ (**5**), PGG_2 (**6**), thromboxane A_2 (**7**, TxA_2), and prostacyclin (**8**, PGI_2). Though derived from a common precursor, unlike prostaglandins, thromboxanes have six-carbon heterocyclic oxane.

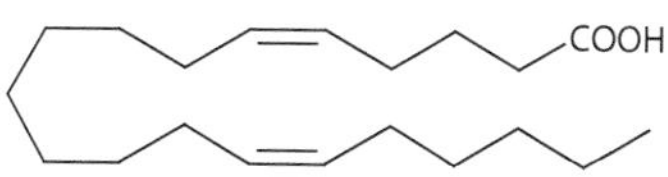

1, Arachidonic acid

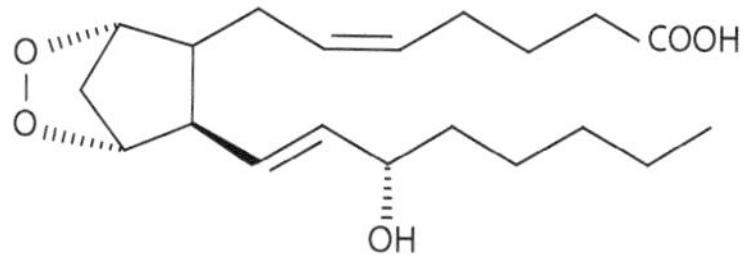

2, Prostaglandin H_2 (PGH_2)

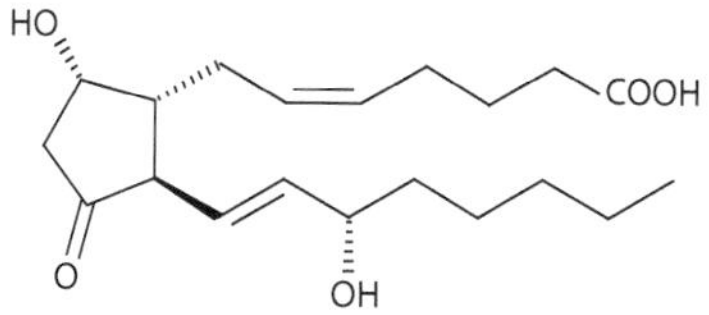

3, Prostaglandin D_2 (PGD_2)

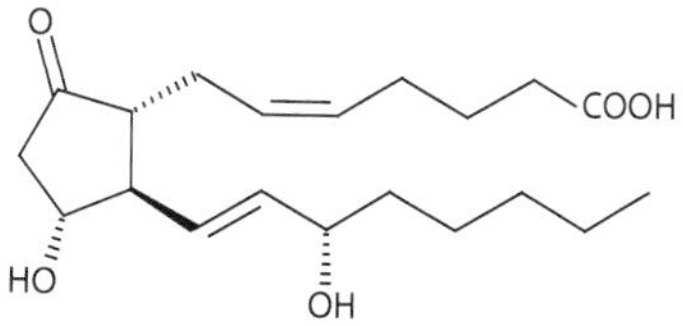

4, Prostaglandin E_2 (PGE_2)

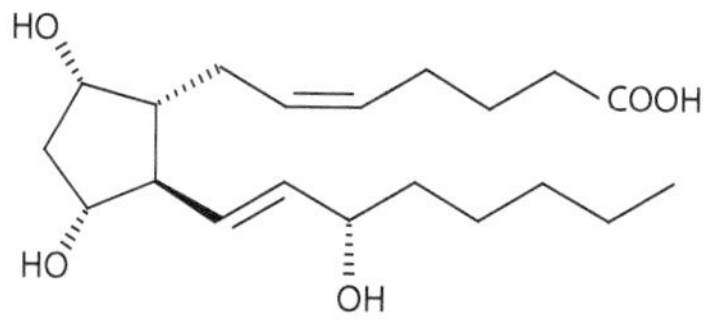

5, Prostaglandin $F_{2\alpha}$ ($PGF_{2\alpha}$)

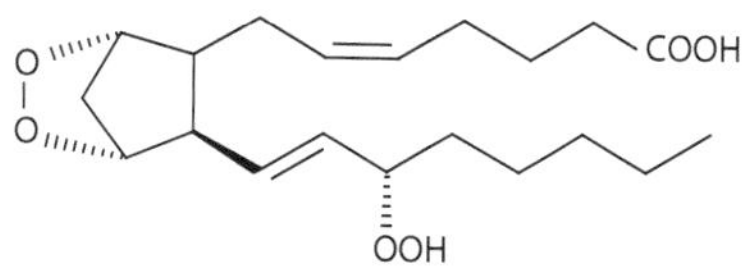

6, Prostaglandin G_2 (PGG_2)

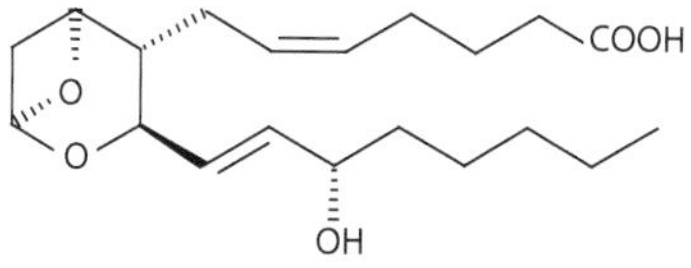

7, Thromboxane A_2 (TxA_2)

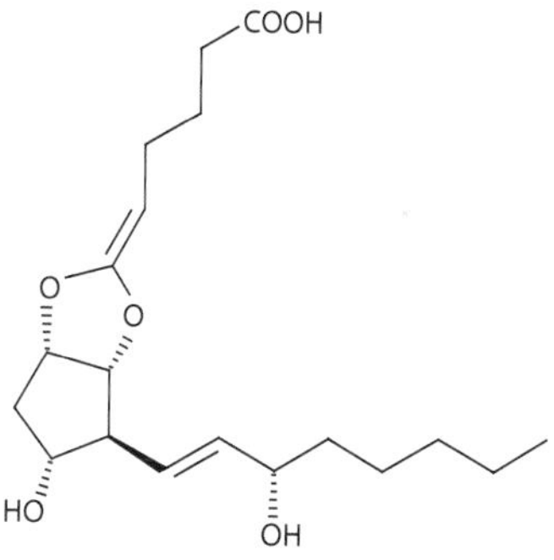

8, Prostacyclin (PGI_2)

FIGURE 8.1 Structure of various prostanoids.

Prostanoids are synthesized by specific enzymes, which confer stereospecificity and chirality on every functional group. Prostanoids thus formed are immediately released outside the cell, and elicit their action locally, near the site of production. Although cyclooxygenases are present in most cells, the type(s) of prostanoid(s) produced by a particular cell type generally reflects expression of specific synthases within the cell. For example, mast cells, which express abundant hematopoietic PGD_2 synthase, predominantly generate PGD_2 (Schuligoi et al., 2009) and macrophages, which express PGE_2 synthase, produce mainly PGE_2 (Graf et al., 1999; Iyer et al., 2009). Most prostanoids are extremely unstable with a half-life of less than

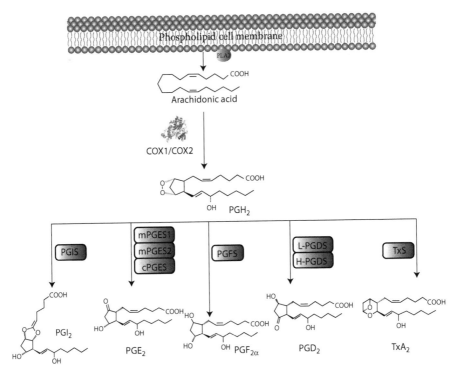

FIGURE 8.2 Biosynthetic pathway of prostanoids. Arachidonic acid is released from the cell membrane by the action of cytosolic phospholipase A_2 ($cPLA_2$) and is the precursor for all prostanoids. COX1/COX2 mediates the conversion of arachidonic acid to PGH_2, which is converted by specific enzymes to the respective prostaglandins.

60 s in blood. Their main function is that of serving as key mediators and modulators in both physiological as well as pathological conditions. The biosynthetic pathway of prostanoids is summarized in Figure 8.2.

8.3 PROSTANOID RECEPTORS

Membrane receptors for various prostanoids have been identified, through which they elicit their effects on target cells. All prostanoid receptors are G-protein-coupled rhodopsin-type receptors with seven transmembrane domains, each encoded by a different gene. Phylogenetically, all the prostanoid receptors have been found to originate from the primitive PGE_2 receptor. Gene duplication of functionally related PGE_2 subtypes led to the evolution of the other prostaglandin and thromboxane receptors (Narumiya and FitzGerald, 2001).

Overall homology among the various prostanoid receptors varies between 20% and 30%. These receptors though are conserved between species, with interspecies homology for the same receptor subtype being around 70–80%.

The accepted receptor nomenclature is the name of the most potent agonist followed by the term "prostanoid" (Coleman et al., 1994) (Table 8.1). Hence, thromboxane

TABLE 8.1
Prostanoid Receptor Subtypes

Ligand	Receptor Subtype	G-Protein Involved	Second Messenger	Physiological Role
PGD_2	DP_1	G_s	Increase in cAMP	Mediator of allergic asthma
	DP_2 (CRTH2)	G_i	Inhibition of cAMP	Mediator of allergic asthma
			Increase in Ca^{2+}	
PGE_2	EP_1	Not known	Increase in Ca^{2+}	Colon carcinogenesis
	EP_2	G_s	Increase in cAMP	Ovulation, fertilization
				Salt-sensitive hypertension
	EP_3	G_i, G_s, G_q	Decrease in cAMP	Mediator of febrile
			Activation of Rho	responses to pyrogen
			Increase in Ca^{2+}	Urinary concentration
	EP_4	G_s	Increase in cAMP	Closure of ductus arteriosus
				Bone resorption
$PGF_{2\alpha}$	FP	G_q	Increase in Ca^{2+}	Induction of parturition
PGI_2	IP	G_s, G_q	Increase in cAMP	Antithrombotic
			Increase in Ca^{2+}	Nociception
TxA_2	TP	G_q, G_i, G_s	Increase in Ca^{2+}	Hemostasis
			Decrease in cAMP	Lung hyperplasia and
			Increase in cAMP	airway remodeling
				Pulmonary hypertension

receptor is designated the thromboxane prostanoid receptor or the TP receptor. Similarly, the PGE_2 receptor is the EP receptor, the PGD_2 receptor is the DP receptor, and the $PGF_{2\alpha}$ receptor is the FP receptor. Single-gene receptors have been identified for PGI_2, $PGF_{2\alpha}$, and TxA_2, whereas there are two receptors for PGD_2 (DP_1 and DP_2), while four receptors have been identified for PGE_2 (EP_{1-4}). Although each prostanoid has its own cognate receptor(s), an extensive amount of cross-talk between them has been reported.

8.3.1 Receptors for PGE_2: EP Receptors

PGE_2 is an important and physiologically relevant prostaglandin in airway disease. Synthesis of PGE_2 is governed by three synthases. Cytosolic prostaglandin E_2 synthase (cPGES) is localized in the cytosol while mitochondrial prostaglandin E_2 synthase-1 and -2 (mPGES-1 and mPGES-2) are localized in the mitochondria. Four receptors for PGE_2 have been identified till date, namely EP_1, EP_2, EP_3, and EP_4 (Coleman et al., 1994; Norel et al., 2004). As different stress stimuli can evoke the release of PGE_2 from different cell types, it is proposed that PGE_2 and its receptors integrate multiple stress stimuli for producing adaptive responses.

- EP_1 receptor is expressed in a variety of tissues, including the kidney, lung, dorsal root ganglia, and sensory neuron (Funk et al., 1993). Within the kidney, the EP_1 receptor is expressed at high levels in the cortical, outer medullary, and inner medullary collecting duct.

- EP_2 receptors are expressed in airway and vascular smooth muscle, epithelial cells, inflammatory cells such as mast cells and alveolar macrophages, dorsal root ganglia, and sensory neurons (Coleman et al., 1994; Ratcliffe et al., 2007).
- EP_3 receptors are distributed in various organs and tissues, including airway epithelial cells and vascular smooth muscle. They exist in three alternatively spliced isoforms (Narumiya et al., 1999; Norel et al., 2004; Matsuoka and Narumiya, 2007).
- EP_4 receptors have been localized on human alveolar macrophages (Ratcliffe et al., 2007), dorsal root ganglia, and airway smooth muscle (ASM) (Clarke et al., 2005).

8.3.2 RECEPTORS FOR PGD_2: DP_1 AND CRTH2/DP_2

The synthesis of PGD_2 is governed by two enzymes, the lipocalin-type PGD_2 synthase (L-PGDS) found mainly in the heart, brain, and adipose tissue and the hematopoietic PGD_2 synthase (H-PGDS) expressed in mast cells, macrophages, dendritic cells, and T_H2 cells. PGD_2 is generated by mast cells on challenge by allergens. It is produced extensively in various allergic conditions. PGD_2 and/or its metabolites bind to multiple receptors, expressed both on the cell surface as well as intracellularly. The DP_1 receptor is the first identified receptor of PGD_2 and is widely found in the airway and vascular smooth muscles, airway epithelium, platelets, and nervous tissue (Coleman et al., 1994; Matsuoka et al., 2000). PGD_2 also activates the chemoattractant receptor-homologous molecule expressed on T_{H2} lymphocytes (CRTH2) (DP_2) receptor. This receptor is mainly expressed on the T_H2 cells, eosinophils (Abe et al., 1999), and ASMs (Parameswaran et al., 2007). The role of DP_1 and CRTH2 receptors is explained in Figure 8.3. PGD_2 also binds to the TP receptor as well as the nuclear receptor peroxisome proliferator-activated receptor (PPAR-γ). Hence the actions of PGD_2 as well as its metabolites (which include PGJ_2 and $9\alpha,11\beta$-PGF_2) are mediated by the above-mentioned three receptors, two belonging to the 7-transmembrane G-protein-coupled receptor (GPCR) family (DP/DP_1 receptor and the CRTH2/DP_2 receptor) and PPAR-γ, a nuclear receptor for metabolites of PGD_2.

When mice deficient in DP receptors ($DP^{-/-}$) were examined for airway inflammation, a marked reduction in airway obstruction and hypersensitivity was observed, suggesting the role of this receptor in mediating the PGD_2 responses in allergy (Matsuoka et al., 2000). Activated mast cells contribute to asthmatic pulmonary inflammation mediated by PGD_2, leading to its use as a biomarker for mast cell activation in asthma (Figure 8.4). The DP_1 receptor is the most studied of the PGD_2 receptors and mediates its action via G_s-induced elevation of cyclic adenosine monophosphate (cAMP) levels. The ligands for the DP_1 receptor include PGD_2 and its metabolites PGF_2, Δ^{12}-PGJ_2, 15d-PGJ_2, and $9\alpha,11\beta$-PGF_2, associated with vasodilatation, bronchodilatation, and antiaggregation of platelets. However, $9\alpha,11\beta$-PGF_2 also binds to the thromboxane receptor TP with high affinity. TP receptor is a

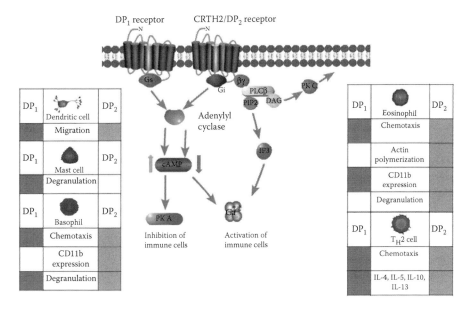

FIGURE 8.3 Role of DP_1 and CRTH2 receptors. PGD_2 elicits its action through two receptors, DP_1 and DP_2. DP_1 is a G_s-mediated GPCR, and causes increase in cAMP, leading to an inhibition of immune cells. DP_2 mediates its action through G_i, leading to Ca^{2+} release and subsequent activation of immune cells.

G_q-coupled receptor and causes bronchoconstriction, which often supersedes the bronchodilatory effects mediated by the DP_1 receptor.

DP_2/CRTH2 receptor, initially identified as an orphan receptor on T_{H2} cells, basophils, eosinophils, and monocytes, mediates its effects by G_i-independent increase in intracellular calcium levels and reduction of intracellular cAMP levels, leading to further proinflammatory cascades. CRTH2 bears no structural relation to the prostanoid DP_1 receptor and belongs to the family of chemoattractant receptors, bearing homology to fMLP and C5a receptors. CRTH2 binds to PGD_2 and its metabolites, including DK-PGD_2, $\Delta12$-PGD_2, 15d-PGD_2, $\Delta12$-PGJ_2, 15d-PGJ_2, and $9\alpha,11\beta$-PGF_2. Other ligands for this receptor include PGF_2 and TxA_2 metabolite 11-dehydro-TxB_2. The activation of CRTH2 receptor has been substantially implicated in the role of PGD_2 in asthma as well as allergic rhinitis, whereas single nucleotide polymorphisms (SNPs) on the CRTH2 gene have been linked to severity of asthma.

PPAR-γ is a transcription factor directly activated by ligands (Lemberger et al., 1996). PPAR-γ is activated by the PGD_2 metabolites PGJ_2, $\Delta12$-PGJ_2, and 15d-PGJ_2, negatively regulating inflammation by its actions on T cells, monocytes, dendritic cells, and mast cells. Table 8.2 compares the essential properties of DP_1, CRTH2, and TP receptor while Figure 8.5 summarizes the various PGD_2 receptors and their ligands.

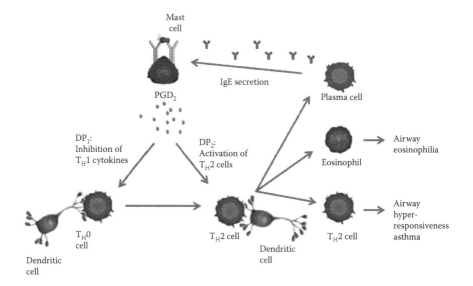

FIGURE 8.4 A combined action of DP_1 and CRTH2 (DP_2) leads to the activation of T_H2 cells. PGD_2 is secreted by mast cells by IgE cross-linking. The secreted PGD_2 can act via two receptors. Engagement of the DP_1 receptor leads to the inhibition of cytokines responsible for T_H1 phenotype, whereas engagement of DP_2/CRTH2 receptors leads to the activation of T_H2 cells. This favors T_H2 polarization, leading to a plethora of pathological effects, such as asthma, airway hyperreactivity, airway eosinophilia, and production of IgE-secreting plasma B cells. Secreted IgE interacts with mast cells and facilitates antigen presentation, exacerbating the allergic response.

8.3.3 TP Receptor

The human TP receptor exists in two spliced isoforms, TPα and TPβ (Raychowdhury et al., 1994), with mRNA for both detected in smooth muscle cells from human bronchi (Capra et al., 2003). TP receptors are also expressed on vascular smooth muscle, blood platelets, and myofibroblasts (Coleman et al., 1994). TxA_2 has two major activities—it causes vasoconstriction and acts as a mitogen for the vasculature. This effect is mediated by both isoforms of the TP receptor, which is spatially regulated. Only the TPα isoform is present in platelets whereas both isoforms are present in the vascular smooth muscle cells.

8.3.4 FP Receptor

Two spliced variants of the FP receptor are known—FP_A and FP_B (Abramovitz et al., 1994; Pierce et al., 1997). The receptor acts via coupling to the G_q subunit and activation of phospholipase C (PLC). It is implicated in mammalian reproduction (Sugimoto et al., 1997), kidney function (Breyer and Breyer, 2001; Comporti et al., 2008), and regulation of intraocular pressure (Akaishi et al., 2009; Costagliola et al., 2009; Ishida et al., 2006; Woodward et al., 2004; Al-Jazzaf et al., 2003; Weinreb et al., 2002; Linden and Alm, 1999). The FP receptor has not been found on any

TABLE 8.2

Comparison of DP$_1$, DP$_2$, and TP Receptors

Receptor	Ligands (Endogenous)	Synthetic Agonists	Synthetic Antagonists	Physiological Role
DP$_1$	PGD$_2$	BW245C	BWA868C MK0524, S5751	Bronchodilatation, vasodilatation, inhibition of platelet aggregation, decreases dendritic cell cytokine secretion
DP$_2$/CRTH2	PGD$_2$, Δ^{12}PGD$_2$, Δ^{12}PGJ$_2$, 15-deoxy-$\Delta^{12,14}$PGJ$_2$, 15-deoxy-$\Delta^{12,14}$PGD$_2$, 11-dehydro-TxB$_2$, DK-PGD$_2$	Indomethacin	Ramatroban	Chemotaxis, activation of T$_H$2 cells, eosinophils, and basophils
TP	TxA$_2$, 9α11βPGF$_2$	U44612	Ramatroban GR32191, SQ29558, ICI192605	Bronchoconstriction, vasoconstriction, platelet aggregation

immune cell population (Hata and Breyer, 2004; Tilley et al., 2001), though low-level expression has been found in lungs (Coleman et al., 1994). No role for this receptor has been proposed in inflammation or allergy.

8.3.5 IP RECEPTORS

They are mainly expressed on vascular smooth muscle and platelets (Oliva and Nicosia, 1987). The role of IP receptors has been proven in pain and inflammation.

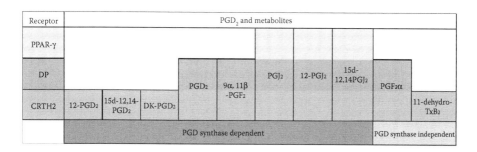

FIGURE 8.5 PGD$_2$ receptors and ligands. Various PGD$_2$ receptors and their ligands, derived from PGD$_2$ in mechanisms dependent or independent on PGD synthase.

Murata et al. (1997) showed that animals null for the IP receptor were completely devoid of any response in classic pain and inflammation models such as acetic acid-induced writhing and carrageenan-induced paw edema. PGI_2 is known to mimic the responses of aspirin in suppressing acute inflammation and this study could nail the activity to be mediated by the IP receptor. But other PGs are also involved in mediated pain, hyperalgesia, and allodynia. Dorsal root ganglia express receptors for IP, EP_1, EP_3, and EP_4. The contribution of PGE_2 in mediating pain is under evaluation.

In spite of prostaglandin receptors being classified based on agonist specificity, the lipids are promiscuous and cross-react with each of the receptors to a varying extent. In other words, a particular prostaglandin may bind to different receptors and elicit different responses. For example, PGE_2 elicits opposing responses in alveolar macrophages and ASM cells, depending on whether the signal is mediated through the EP_3 or EP_4 receptor subtypes. Similarly, there are four prostaglandin ligands for the DP_1 receptor and PGF_2 binds to both DP_1 and TP receptors in addition to the FP receptor. But there is an underlying order behind such complexity. Each particular response is elicited by a unique downstream signaling cascade, mediated by one of the three transducer G-protein-based mechanisms. Based on the mechanism of the coupling of the receptor with the effector via transducer G protein, prostanoid receptors can be grouped as relaxant, inhibitory, or contractile receptors.

1. Relaxant receptors:
   EP_2, EP_4, IP, and DP_1 receptors activate adenylyl cyclase via G_s. This leads to increased intracellular cAMP levels, which in turn activates specific protein kinases. These receptors are termed "relaxant" receptors. They decrease intracellular calcium levels by phosphorylating calcium pumps.
2. Inhibitory receptors:
   EP_3 receptors mediate their action via the G_i subunit, causing an increase in intracellular calcium (Irie et al., 1994) as well as a decrease in cAMP. The CRTH2 receptor, which is unrelated to the other prostanoid receptors, is a member of the fMLP receptor superfamily. This receptor too couples through a G_i-type G protein and leads to inhibition of cAMP synthesis and increase in intracellular calcium in a variety of cell types.
3. Contractile receptors:
   In contrast, EP_1, FP, and TP receptors activate phosphatidylinositol metabolism, mobilize calcium from intracellular stores via $InsP_3$, and lead to an increase in intracellular free calcium. Hence, these receptors are known as "contractile" receptors.

The fact that a particular prostanoid may have diverse and often opposing effects, mediated by different receptors, has a strong bearing on drug discovery. Earlier, inhibiting synthesis or downstream activity of a particular prostanoid was thought to be sufficient to provide relief to symptoms associated with it. For example, PGE_2 is strongly implied in inflammation. But simply inhibiting PGE_2 activity, earlier thought of as a good anti-inflammatory strategy, may not be such a good idea as PGE_2 is now known to elicit both pro- as well as anti-inflammatory responses via different receptors. EP_3 receptor mediates pro-inflammatory effects of PGE_2 whereas

EP_2 and EP_4 mediate anti-inflammatory effects. A nonspecific inhibitor of PGE_2 may lead to a decrease in EP_3-mediated inflammation, but at the same time, also inhibit the EP_2 and EP_4-mediated protection. Hence, a better strategy would be to develop receptor antagonists to particular prostaglandin receptors than to develop inhibitors of PG synthesis or activity. In the above scenario, the ideal molecule would be a selective EP_3 inhibitor that does not inhibit EP_2 or EP_4. Again, assuming that multiple symptoms of a particular disease are completely mediated by a single prostaglandin receptor too would be naive. Hence, developing a highly specific PG receptor antagonist may not again yield an effective drug. The strategy therefore, would be to develop a molecule that selectively inhibits two or more relevant prostaglandin while sparing all the others, especially those whose effects help in disease resolution. It is only when we consider this fact that the enormity and complexity of such an approach dawn. Developing a specific inhibitor to a single receptor is a great task by itself; developing inhibitors to a hand-picked set of receptors while sparing others is going to be a far greater one. In addition, if the two receptors in question are structurally quite divergent, such as the DP_1 and CRTH2 receptors, another level of complexity gets introduced. Thus, a multipronged approach seems to be the only way forward, to offer any hope for success. Researchers in academia and pharmaceutical companies have taken up this challenge, and have achieved some degree of success too, which will be discussed in detail in Section 8.4.

8.3.6 ISOPROSTANES

Isoprostanes are the least studied yet important players of the prostaglandin pathway. Isoprostanes are isomers of the prostaglandins that are produced *in vivo* by free radical-mediated peroxidation of poly-unsaturated fatty acids. Isoprostanes were first characterized *in vivo* in the early 1990s (Morrow et al., 1990) when it was discovered that reactive oxygen and nitrogen species can react with the unsaturated bonds of AA, to form four different classes of isoprostane regioisomers—D_2-, E_2-, F_2-isoprostanes, and isothromboxanes. Peroxidation of eicosapentaenoic acid generates F_3-isoprostanes, α-linoleic acid, and γ-linoleic acids, which leads to E_1- and F_1-isoprostanes, while peroxidation of docosahexaenoic acid leads to eight classes of D_4- and E_4-isoprostanes (Burke et al., 2000; Reich et al., 2000; Parchmann and Mueller, 1998; Nourooz-Zadeh et al., 1997; Rokach et al., 1997). COX has an indirect contribution to isoprostane production, mediated by increase in reactive oxygen species in the endothelium, vascular smooth muscle cells, monocytes, macrophages, and platelets (Watkins et al., 1999a,b). The role of F_2-isoprostanes and 8-iso-$PGF_{2\alpha}$ has been implicated in airway disease (Janssen, 2001). Their unique mechanism of formation lends them specific structural features that differentiate isoprostanes from other free radical-generated products.

Isoprostanes lend themselves beautifully to being biomarkers of oxidative stress. 8-Iso-$PGF_{2\alpha}$ is generated in substantial amounts in otherwise normal persons exposed to cigarette smoke (CS), allergen, hyperoxia, or ozone and after ventilated ischemia, in patients of COPD, in interstitial lung disease, in cystic fibrosis, in acute respiratory distress, and in severe respiratory failure. Elevated isoprostanoid level has also been associated with a variety of cardiovascular conditions, ischemia–reperfusion injury, angina, coronary heart disease, ischemic stroke, hypercholesterolemia, atherosclerosis, and preeclampsia.

In addition, isoprostanes are known to act as ligands for TP receptors. Their significance in drug discovery arises from the fact that they may compensate for effects of thromboxane synthesis inhibition by COX inhibitors as they are produced independent of COX and elicit response through these receptors.

8.4 COX AND PROSTAGLANDIN RECEPTORS AS DRUG TARGETS IN AIRWAY DISEASE

The following sections give an overview on efforts made by various groups toward the validation of COX and PG receptors as drug targets, and efforts made toward discovering candidate molecules acting on them.

8.4.1 COX INHIBITORS IN AIRWAY DISEASE

Regulation of COX is of critical importance in prostaglandin production. The constitutive isoform of COX (COX-1) serves a homeostatic role whereas the inducible isoform (COX-2), normally absent from cells, is transcriptionally upregulated by proinflammatory cytokines and growth factors (Goppelt-Struebe, 1995). COX-2 is expressed in response to proinflammatory stimuli, and it plays a role in the pathophysiology of asthma. Several studies have shown that ASM cells can express COX-2 and release PGE_2, PGI_2, $PGF_{2\alpha}$, and PGD_2 in response to proinflammatory cytokines (Lazzeri et al., 2001; Pang et al., 1998). The relative contribution of the individual COX products on airway inflammation depends ultimately on the presence of their respective receptors on target tissues (McKay and Sharma, 2002). Recent studies have suggested an important homeostatic role for this enzyme. Hence, expression of COX-2 in human airway epithelium occurs in the upper and lower airways, is widespread in airway epithelial and airway resident inflammatory cells in the absence of overt airway inflammation, and is detectable in cultured human airway epithelial cells in the absence of inflammatory cytokine stimulation (Watkins et al., 1999a,b).

Cyclooxygenase metabolites have diverse effects in the lung and are known to modify airway tone, as well as inflammatory responses. Studies have shown that mice deficient in either COX-1 or COX-2 have increased lung inflammation following allergen challenge. Moreover, selective COX-1 and COX-2 inhibitors increase T_{H2} cell-mediated allergic inflammation and airway hyperresponsiveness in mice (Carey et al., 2003). Expression of COX-2 mRNA in response to inflammatory cytokines has been well documented, and *in vitro* studies have demonstrated expression and activity of COX-2 in response to cytokine stimulation in human airway epithelium (Mitchell et al., 1994) and smooth muscle cells (Vigano et al., 1997). Furthermore, it has been shown that endogenous nitric oxide (NO) is an important regulator of COX-2 activity in the human airway epithelium *in vitro* (Watkins et al., 1997). Airborne allergens often lead to an accumulation of eosinophils, lymphocytes (predominantly CD4 type), mast cells, and macrophages, resulting in an inflammatory reaction in the mucosa. Neutrophil numbers can increase during an exacerbation.

The release of mediators from these inflammatory cells has been proposed to contribute directly or indirectly to changes in airway structure and function. Important structural changes of inflamed airways include epithelial cell shedding, basement membrane thickening, goblet cell hyperplasia (increase in cell number), and hypertrophy (increase in cell size), as well as an increase in ASM content. These structural changes consequently form the basis for airway remodeling, a phenomenon believed to have profound consequences for airway function (James, 1997). Bronchial vascular remodeling, with an increase in size and number of blood vessels as well as vascular hyperemia, has been proposed as a contributing factor in airway wall remodeling in patients with chronic asthma and COPD.

The status of COX expression was studied in asthma patients, but results obtained were conflicting. Levels of prostanoids in bronchoalveolar lavage fluid (BALF) are increased in asthma, and studies have found enhanced expression of both COX-1 and COX-2 in the airways of asthmatics. In similar experiments, Demoly et al. (1997) found that neither COX-1 nor COX-2 is upregulated in stable asthma, Sousa et al. (1997) found enhanced expression of COX-2, but not COX-1, in asthmatic airways, while Taha et al. (2000) found increased immunoreactivity of both COX-1 and COX-2 in induced sputum cells from asthmatics. Recent reports also imply that COX-2 may be involved in controlling cough reflex sensitivity in asthma because cough is a major symptom of asthma. A nonspecific COX inhibitor, indomethacin (Fujimura et al., 1995), and a potent, specific COX-2 inhibitor, etodolac, has been shown to attenuate airway cough reflex sensitivity to inhaled capsaicin in the airways of patients with asthma, indicating a possible role of COX-2 in airway cough reflex sensitivity in asthmatic airways with chronic eosinophilic inflammation (Ishiura et al., 2009).

Aspirin and other nonsteroidal anti-inflammatory drugs (NSAIDs) inhibit both COX-1 and COX-2 but differ in their inhibitory potency. The therapeutic and adverse effects associated with NSAID therapy are thought to be mediated by modulation of prostaglandin synthesis. It has been suggested that the anti-inflammatory and therapeutic effects of NSAIDs are due to inhibition of COX-2 whereas the adverse effects of NSAIDs are due to inhibition of COX-1 (Mitchell et al., 1993). Aspirin-sensitive asthma, which affects about 10% of adult asthmatics, is characterized by the precipitation of asthma and rhinitis after the ingestion of aspirin and most NSAIDs (Quiralte et al., 1996; Szczeklik and Stevenson, 1999). Asthma attacks usually occur within 3 h following ingestion of aspirin or NSAIDs, are often severe, and in some cases, may be life-threatening (Picado et al., 1989). Precipitation of asthma attacks by aspirin and NSAIDs was thought to be due to the pharmacological action of these drugs, namely inhibition of COX enzymes in the lung, rather than an allergic response to the drug (Szczeklik, 1990), leading to shunting of AA down the 5-lipoxygenase pathway with resultant increased production of leukotrienes, which are potent bronchoconstrictors and induce mucus secretion, airway edema, and eosinophil influx (Busse, 1996; Sampson, 1996; Sanak and Sampson, 1999). It has been suggested that PGE_2 inhibits the 5-lipoxygenase pathway, and that COX inhibition causes removal of the suppressive effect of PGE_2 with a resultant rise in leukotrienes (Sestini et al., 1996) (Figure 8.6). The significance of the protective role of PGE_2 is evident in cases of aspirin-intolerant asthma, wherein inhibition of COX-1-derived PGE_2 leads to mast cell activation and leukotriene-mediated bronchial spasms (Dahlen et al.,

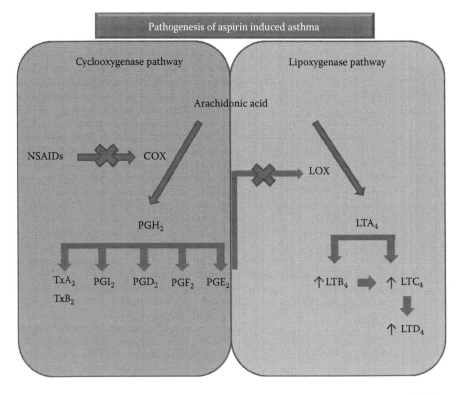

FIGURE 8.6 Pathogenesis of aspirin-induced asthma. NSAIDs such as aspirin inhibit the COX pathway, thereby shunting the arachidonic acid down the lipoxygenase pathway and increasing levels of proinflammatory leukotrienes.

2002). Leukotriene antagonists such as montelukast were found to be protective in this situation (Christie et al., 1991).

Cigarette smoking is a risk factor in the pathogenesis of COPD, which is characterized by abnormal inflammatory responses in the lungs (Moodie et al., 2004; Yang et al., 2009). Acrolein, a known toxin in tobacco smoke, has been shown to induce COX-2 and prostaglandin production in endothelial cells (Park et al., 2007). In addition, acrolein and crotonaldehyde have been shown to elicit IL-8 release in pulmonary cells (Moretto et al., 2009). Thus, exposure to CS may induce COX-2, PGs, or IL-8 production, and this may contribute to airway inflammation in smokers (Martey et al., 2005). ASM is considered as an end-response effector mediating regional differences in ventilation by contraction in response to various proinflammatory mediators and exogenous substances under homeostatic or pathological conditions (Hirst et al., 2004). Although COX-2 has been shown to regulate airway inflammatory responses (Howarth et al., 2004), the mechanisms of intracellular signaling pathways involved in CS-induced COX-2 expression in human tracheal smooth muscle cells are not completely defined. Therefore, CS plays a potential role in the regulation of the expression of inflammatory genes, such as COX-2, and thereby promotes inflammatory responses.

There is an extensive literature on COX inhibitors for pain and arthritis that are widely used in the clinics and is beyond the scope of this chapter. However, there is only limited data available on COX inhibitors for respiratory diseases. There has been no report of a COX inhibitor being developed to treat airway inflammation.

8.4.2 Prostaglandin Receptor Antagonists in Airway Disease

Prostanoids have multiple mechanisms of action and exert both proinflammatory and anti-inflammatory actions. These actions are often exerted in a context-dependent manner, and require context-dependent regulation of signaling to have the desired therapeutic effect. Designing such molecules would be a challenge worth investing in for the field of drug discovery.

8.4.2.1 PGE_2 Receptors

Often considered a potent proinflammatory mediator, PGE_2 is involved in the pathogenesis of several inflammatory conditions. The activity of PGE_2 is mediated by four receptors, EP_1–EP_4. But inhibition of PGE_2 by NSAIDs in diseases such as asthma often increased the severity of the attack (Szczeklik and Stevenson, 2003). The reason for this was initially thought to be the diversion of AA metabolism from COX to the 5-lipoxygenase pathway, causing enhanced production of leukotrienes. But later studies showed that PGE_2 can exhibit a range of antiallergic effects, primary among which are the following:

- Inhibition of bronchoconstriction and airway hyperresponsiveness (Pavord et al., 1993; Gauvreau et al., 1999; Hartert et al., 2000)
- Reduction in the release of mast cell mediators of inflammation (Gomi et al., 2000; Wang and Lau, 2006)
- Reduction in the recruitment of eosinophils to inflamed tissue (Chan et al., 2000; Hartert et al., 2000)
- Increase in IgE levels

These protective effects of PGE_2 have been attributed to the action on the EP_3 receptor (Gomi et al., 2000; Kunikata et al., 2005; Wang and Lau, 2006). It was suggested that the site of action of PGE_2 via EP_3 receptors was on mast cells, reducing secretion of inflammatory mediators, as well as on airway epithelial cells, by down-regulating the expression of relevant genes. EP_3-mediated suppression of allergic responses have been recorded in mouse models of other allergic syndromes such as allergic conjunctivitis (Ueta et al., 2009) and contact hypersensitivity (Honda et al., 2009). It was hence realized that simply blocking all the actions of PGE_2 would be detrimental and the ideal drug should block the proinflammatory actions of PGE_2 while sparing or enhancing EP_3-mediated anti-inflammatory actions. Based on these studies, EP_3 receptor agonist, in conjunction with DP antagonist, was proposed as a potential treatment strategy for various allergies, including asthma.

The progress in the development of an EP_3 receptor agonist has been slow, given the challenges associated with such a multipronged approach. No dedicated industrial effort seems to have been made toward developing an EP_3 agonist, targeting allergic airway diseases. Till date, five molecules targeting other PGE_2 receptors have been launched in the market, four of which are EP_1 agonists and are for the indication of erectile dysfunction, with the fifth being an EP_2 agonist for peptic ulcer. Another nine compounds in various phases of clinical trials target a host of conditions pertaining to the reproductive system, kidney, eye, and vasculature and target the various EP receptors.

8.4.2.2 PGI_2

The major role of PGI_2 in the body is to prevent platelet aggregation, and it is secreted by the vascular endothelial cells. It is known to be released during acute allergic inflammation or anaphylaxis, and has a relaxant effect on ASMs (Nagao et al., 2003) and suppresses synthesis of leukotrienes (Takahashi et al., 2002). When mice lacking the gene for IP receptor were subjected to airway inflammation, there was enhancement of T_{H2} cytokine levels, increased airway eosinophilia, and increased BALF protein levels. In splenocytes isolated from these mice, there was increased IL4 production on antigen stimulation *in vitro*. PGI_2 has been shown to enhance the production of IL10 (Jaffar et al., 2002), a known suppressor of T_{H2} responses, providing a probable mechanism of the immunomodulatory effects. Promising as the results may seem, testing of PGI_2 analogs/receptor agonists in airway inflammation did not yield the expected results (Nagai, 2008), indicating the necessity for a multipronged approach while developing prostaglandin modulators for complex conditions such as allergies. There are only two molecules being developed targeting the IP receptor and both are for pulmonary hypertension.

8.4.2.3 TxA_2

TxA_2 has been implicated in allergic inflammations and is known to cause bronchoconstriction (Dogne et al., 2002; Rolin et al., 2006; Allen et al., 2006). It has been shown to function through distinct mechanisms, causing opposite effects in allergies of the lung and that of the skin (Kabashima et al., 2003), playing a pathological role in asthma and a protective one in skin inflammations. More studies need to be performed to fully evaluate the role of TxA_2 in allergies before a call can be taken to develop agonists or antagonists of the TP receptor as a treatment for allergies. An advantage of developing TP antagonists would be the blockade of isoprostane-induced effects, which cannot be inhibited even by COX inhibitors like aspirin, but would be blocked by TP antagonists. But as the case of COX-2 selective inhibitors have shown, the perturbation of the delicate TxA_2/PGI_2 balance must be duly considered.

Ramatroban (Bay u3405, Bayer, **9**, Figure 8.7) and seratrodast (AA-2414, Takeda, **10**, Figure 8.7) are the only drugs that have been launched as TxA_2 receptor antagonists for the treatment of asthma. Seratrodast is a long-acting thromboxane A_2 and prostaglandin endoperoxide receptor antagonist, and was approved in Japan for the treatment of asthma in 1995. In a Phase II, randomized, double-blind, parallel-group, placebo-controlled 15-center study of seratrodast in patients with mild to moderate asthma, 183 patients received either placebo or 80/120 mg seratrodast for 8 weeks. Seratrodast was found to be orally bioavailable with an apparent volume of distribution of 8.5 mL/h/kg and oral clearance of 43.3 mL/kg, respectively. Seratrodast at a

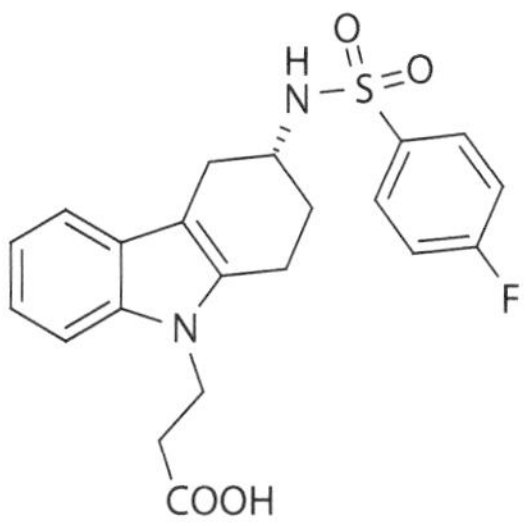

9, Ramatroban 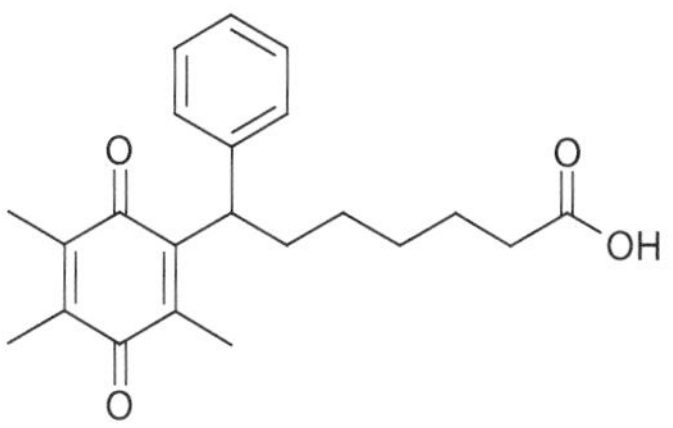   **10**, Seratrodast

FIGURE 8.7   Ramatroban and seratrodast are the first TP/CRTH2 antagonists to be clinically launched.

dose of 120 mg daily produced an increase in forced expiratory volume in 1 s (FEV1) from baseline that was linearly correlated with its plasma concentrations. A lower percentage of predicted FEV1 (i.e., more severe obstruction) was associated with higher slopes, and greater increases in FEV1 (Samara et al., 1997). A second multicenter, double-blind, randomized, placebo-controlled study was conducted on 45 patients with mild to moderate asthma. They received 40 mg seratrodast or placebo for 6 weeks during which pulmonary function, sputum production, and mucociliary function were assessed. Changes in forced expiratory volume (FEV) and peak expiratory flow (PEF) were similar among the two patient groups, but there were significant reductions in diurnal variation of PEF ($p = 0.034$), frequency of daytime asthma symptoms ($p = 0.030$), and daytime supplemental use of β_2-agonist ($p = 0.032$) in the seratrodast group. For sputum analysis, seratrodast treatment decreased the amount of sputum ($p = 0.005$), dynamic viscosity ($p = 0.007$), and albumin concentration ($p = 0.028$). Nasal clearance time was shortened in the seratrodast group at week 4 ($p = 0.031$) and week 6 ($p = 0.025$) as compared with the placebo group. This led to the conclusion that inhibiting TxA$_2$ receptor decreases the viscosity of airway secretions, increases mucociliary clearance, and improves pulmonary function (Tamaoki et al., 2000).

Ramatroban was initially developed as a TP antagonist but its activity is now attributed to competitive inhibition of the CRTH2 receptor. It was confirmed to suppress the increase of nasal mucosa vascular permeability and sneezing in a nasal allergy model. At a dose of 150 mg/day for 4 weeks, it showed an improvement of over 70% for nasal obstruction. It also has an antiasthmatic effect because it blocks bronchoconstriction, hyperresponsiveness of the airways, and infiltration of inflammation cells. In another clinical study, 75 mg of ramatroban or placebo were administered orally, twice a day for 2 weeks each in a crossover design. Bronchial hyperresponsiveness was measured by the astograph method. The D_{min} value of 0.533 U (GSEM 1.675) after the ramatroban treatment was significantly greater than that of 0.135 U (GSEM 1.969) after the placebo treatment ($p = 0.0139$). There were no safety concerns in either treatment group (Aizawa et al., 1996). Ramatroban has now been launched for the treatment of perennial allergic rhinitis, after being approved in Japan.

Other research focusing on the TxA_2 receptor is for indications other than asthma/COPD.

8.4.2.4 PGD_2

Perhaps the most significant prostaglandin, as far as asthma and airway diseases are concerned, is PGD_2. There is clinical evidence that PGD_2 is detected in BALF from patients of asthma. PGD_2 is the major prostaglandin produced by eosinophils and mast cells, but not by basophils. It is also known to cause constriction of human bronchial smooth muscles *in vitro* (Wenzel et al., 1989). Mice lacking the DP_1 receptor failed to develop airway hyperreactivity (AHR)-like symptoms and showed only marginal eosinophil infiltration when sensitized and challenged with aerosolized antigen whereas such a treatment caused AHR in wild-type mice (Matsuoka et al., 2000). Conversely, overproduction of PGD_2 in mice overexpressing the PGD_2 synthetase gene caused an increase in T_H2 cytokine levels and enhanced accumulation of eosinophils and lymphocytes in lungs (Fujitani et al., 2002).

The DP_1 receptor is known to be present on monocytes, the precursor to macrophages. Inhibition of this receptor is hence predicted to inhibit macrophage chemotaxis and recruitment of proinflammatory cells at the site of inflammation, resulting in reduced pathological consequences. SNP analysis of the PGD_2 gene locus (PTGDR) in humans revealed that populations with higher incidence of asthma correlated with a different haplotype, which corresponded to a different promoter activity for this gene (Oguma et al., 2004). One fact that clearly emerges out of the many studies is the relevance of CRTH2-mediated eosinophil mobilization (Heinemann et al., 2003; Almishri et al., 2005; Shiraishi et al., 2005). This fact assumes significance in the light of the argument that CRTH2 receptors are upregulated in eosinophils/T_H2 cells and PGD_2 in airways of asthma patients, as well as those suffering from asthma, pollen sensitization, or atopic dermatitis.

A point must be made here with respect to the effect of COX inhibition on PGD_2. It would be expected that aspirin and related COX-1/COX-2 inhibitors would decrease levels of PGD_2, but paradoxically, especially in patients of aspirin-intolerant asthma, PGD_2 levels were enhanced on treatment with aspirin (Bochenek et al., 2003). There is still lack of clarity on this unexpected action of aspirin, though it has been proposed that the reason might be the lesser sensitivity of COX-2 vis-a-vis COX-1 to inhibition by aspirin, leading to high induction of COX-2 in aspirin-intolerant asthmatics (Sousa et al., 1997). This points to the possibility that selectively inhibiting PGD_2 may lead to a completely different pharmacological profile than a nonspecific COX inhibition, a situation that reminds one of the infamous coxib episode (Halpern, 2005; Levesque et al., 2005; Liew and Krum, 2005; Sanghi et al., 2006; Brophy et al., 2007; Moodley, 2008; Ray et al., 2009). Going by the experience of how selective COX-2 inhibitors completely altered the PGI_2/thromboxane ratio with grave cardiovascular consequences (Grosser et al., 2006), it would be wise to study this pathway thoroughly while embarking on selective inhibitors targeting PGD_2 production or activity. Though the role of the PGD_2 receptors in asthma is established, the exact strategy to be followed—a selective CRTH2 antagonist, a nonselective DP_1/CRTH2 dual antagonist, or a CRTH2 antagonist in conjunction with EP_3 agonist activity—is still open to debate and experiment.

Pharmaceutical industries have realized the immense market potential DP receptor antagonists hold and much research has been focused on the same. The subsequent section briefs on molecules targeting these receptors that have either made it to the market or reached an advanced stage of development. Merck Frosst launched laropiprant, a DP_1 antagonist (Sturino et al., 2007), to counter the facial flushing induced by niacin, rather than act as a therapeutic agent by itself. Recent clinical studies involving laropiprant did not show efficacy in treating asthma or allergic rhinitis, as the molecule exhibited no improvement in therapy over either placebo (when administered singly) or montelukast (when administered in combination with montelukast) (Philip et al., 2009). Though laropiprant was launched in the United Kingdom, it failed to get FDA approval in the United States.

Oxagen has a few molecules taken for Phase II studies but only a single one, OC459 (aka ODC9101), gave promising results, with the company declaring it safe, orally active, and suitable for once-a-day dosing. It cleared Phase IIa in 2009 with improvement of 7.4% over placebo, and a Phase IIb study of longer duration and higher dose of the molecule is now underway. This clinical study replicated data obtained in animals that blockade of CRTH2 receptor can be demonstrated in humans. Whether such a blockade is clinically efficacious in asthma or not will only be revealed in the next couple of years when the results of ongoing clinical trials are declared.

In a related study, antiallergic efficacy of a small-molecule CRTH2 antagonist TM30089 was described. It displayed nanomolar affinity, high selectivity of over 1000-fold over DP_1 receptors, and high antagonistic potency (pA_2: 9.15 ± 0.11) on mouse CRTH2 but lacked affinity to TP, C3a and C5a receptors, COX-1, and COX-2. TM30089 also inhibited asthma pathology *in vivo* by reducing peribronchial eosinophilia and mucus cell hyperplasia. This data suggested that CRTH2 antagonism alone may be effective in mouse allergic airway inflammation (Uller et al., 2007).

Most companies that are developing drugs in this class have focused on developing selective CRTH2 inhibitors. But there exists a sequential CRTH2 role of DP_1 and CRTH2 receptor in initiating and maintaining allergic rhinitis, and superior clinical results might be achieved if both the DP receptors are concurrently blocked. Amgen is pursuing this strategy, and AMG853 is reported to have progressed to Phase II clinical trials. A dual DP_1/CRTH2 inhibitor hence may be a very promising therapeutic approach.

8.5 CHEMISTRY STRATEGIES FOR DEVELOPMENT OF DP RECEPTOR ANTAGONIST

In the following sections, a perspective of approaches taken toward identification of DP_1/CRTH2 antagonists will be provided. Drug discovery efforts to develop a DP receptor antagonist can be grouped in two distinct phases. In the first phase of research to identify small-molecule DP_1 or CRTH2 antagonists, efforts were focused toward the synthesis of prostaglandin analog and their derivatives. These structural modifications, as seen in BW A868C and S5751 (**11 and 12**, Figure 8.8), highlight the extent to which PG structure has been subjected to exploration and modifications. BW A868C is constructed around a hydantoin scaffold (Giles et al., 1989) and binds tightly to DP_1 (K_d: 1.45 nM). On the other hand, S5751 is constructed around a

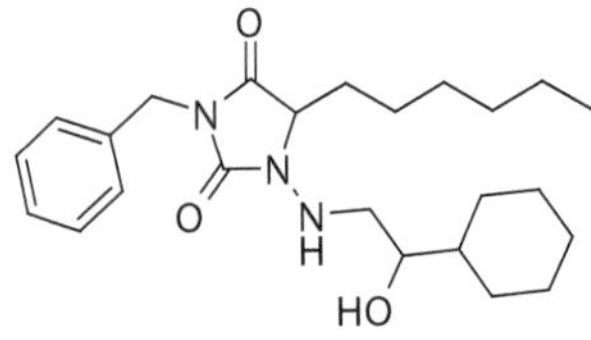

11, BW A868C
$DP_1 K_d = 1.45$ nM

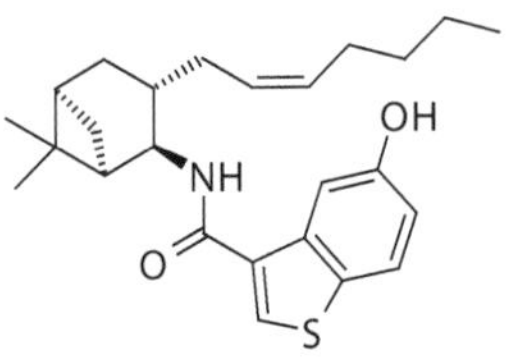

12, S-5751
$DP_1 IC_{50} = 1.9$ nM

FIGURE 8.8   Prostaglandin analogs.

bicyclo[3.1.1]heptane skeleton (Arimura et al., 2001; Mitsumori et al., 2003) and also shows tight binding to DP_1 (IC_{50}: 1.9 nM) in [^{3}H] PGD_2 binding assay. S5751 was orally active in models of allergy and inflammation in guinea pig.

In the second phase, research was focused on nonprostanoid structures. Developing an inhibitor with the correct pharmacological profile, based on a novel nonprostanoid structure, has proved to be very challenging, and the following section will discuss the efforts in this direction.

The initial approach toward the development of a DP antagonist was a ligand-based one. Work by scientists at Merck Frosst and Bayer laid the foundation for the identification of nonprostanoid structures that were potent and selective antagonist of DP_1 as well as CRTH2. HTS screening played a very important role in the identification of lead structures for chemical modification.

8.5.1   INDOMETHACIN: LEAD STRUCTURE FOR CRTH2 ANTAGONIST EXPLORATION

Several NSAIDs were screened in [^{3}H] PGD_2 binding assay. Indomethacin (Figure 8.9) was found to have high binding affinity (K_i: 25 nM) for CRTH2 (Sawyer et al., 2002; Hirai et al., 2002). However, indomethacin was found to be a full agonist at CRTH2 and decreased forskolin-stimulated intracellular cAMP with EC_{50} of 14.9 nM. Though indomethacin exhibited agonistic activity at CRTH2, it provided a starting point to arrive at potent and selective CRTH2 antagonists.

8.5.1.1   Modification of Indomethacin to CRTH2 Antagonist

Indomethacin (**13**, Figure 8.9) has been subjected to several modifications by different groups. Replacing the N-carboxamide group with an N-sulfonamide group resulted in a compound (**14**, Figure 8.9) that was found to be potent and selective for CRTH2 in a binding experiment (K_i: 29 nM) (Armer et al., 2005). This compound was found to potently inhibit PGD_2-stimulated Ca^{2+} influx in intact Chinese hamster ovary (CHO) cells expressing CRTH2 (IC_{50}: 26 nM), indicating that a small structural change had converted indomethacin into a competitive antagonist.

8.5.1.2   Modification of Indomethacin to DP_1 Antagonist

Although indomethacin was shown to have a weak binding to the DP_1 receptor (K_i: 10 μm) (Hirai et al., 2002; Stubbs et al., 2002), Torisu et al. at Ono Pharmaceuticals

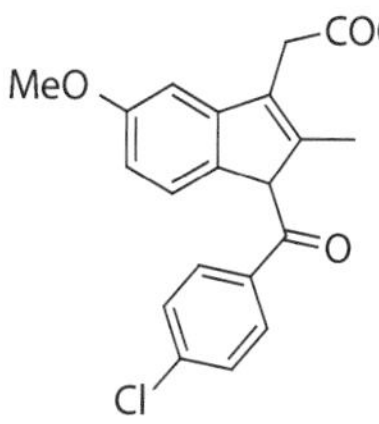

13, Indomethacin
K_i mCRTH2 = 25 nM
K_i mDP$_1$ > 10 µM

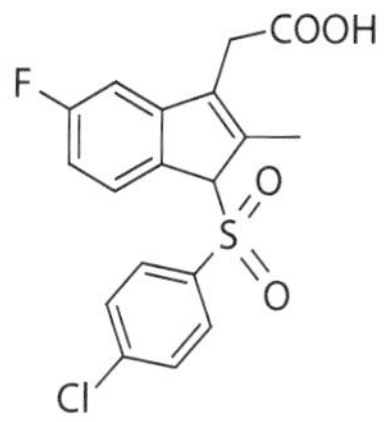

14
K_i mCRTH2 = 29 nM; IC$_{50}$ mCRTH2 = 26 nM

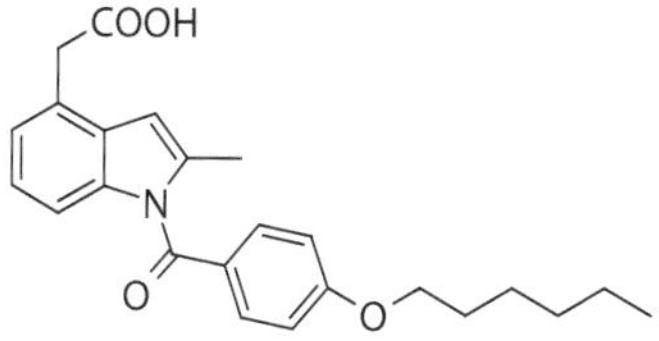

15
K_i mDP$_1$ = 8.5 nM; IC$_{50}$ mDP$_1$ = 20 nM

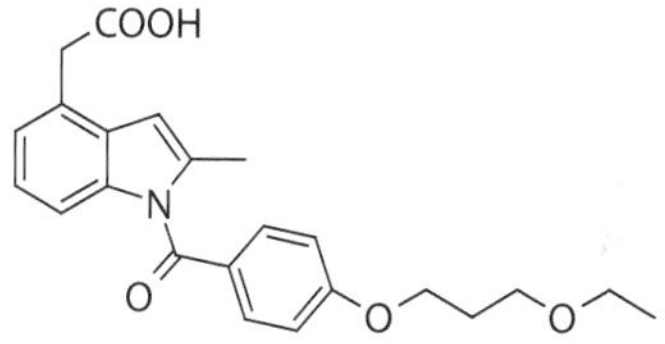

16
K_i mDP$_1$ = 4.3 nM; IC$_{50}$ mDP$_1$ = 23 nM

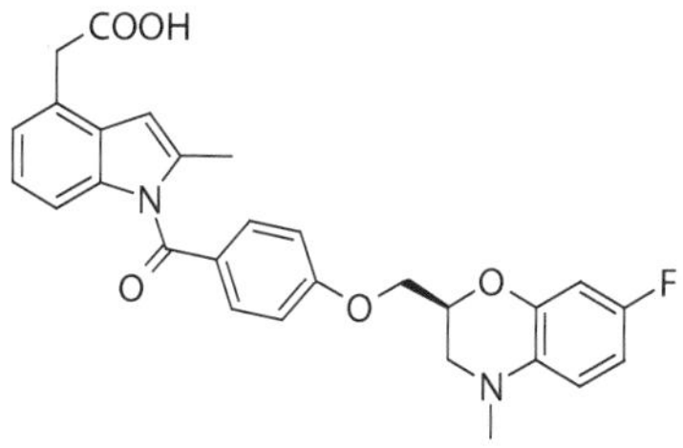

17
K_i mDP$_1$ = 23 nM; K_i hDP$_1$ = 5.3 nM;
IC$_{50}$ hDP$_1$ = 0.8 nM

FIGURE 8.9 Indomethacin as a lead structure for the identification of DP$_1$ and DP$_2$ antagonists.

made several modifications on the indole-3-acetic acid skeleton of indomethacin with a view to induce DP$_1$ receptor antagonist activity. Interestingly and more so unexpectedly, transformation of the indole-4-acetic acid scaffold introduced potent DP$_1$ antagonist activity against both murine and human DP$_1$ receptor (**15**, **16**, and **17**, Figure 8.9). All these compounds exhibited high-affinity binding to murine DP$_1$ receptor (K_i: 8.5, 4.3, and 23 nM, respectively). These compounds also exhibited potent inhibition of PGD$_2$-stimulated cAMP formation in CHO cell lines stably expressing the DP_1 receptor. While compounds **15** and **16** were reported to inhibit cAMP formation in murine DP_1 receptor with an IC_{50} of 20 and 23 nM, respectively, compound **17** exhibited potent inhibition of cAMP formation in human DP_1 receptor with an IC_{50} of 0.8 nM. Further, compound **17**, having a benzoxazine substitution, when taken *in vivo* studies, showed 60% inhibition of PGD_2-induced increase of vascular permeability at an oral dose of 0.3 mg/kg (Torisu et al., 2004a,b).

8.5.2 RAMATROBAN AS LEAD STRUCTURE

Following the report that ramatroban exhibited CRTH2 antagonistic activity along with the TP receptor affinity, several groups focused on the modification of the ramatroban scaffold. The tetrahydrocarbazol-9-yl propanoic acid scaffold of ramatroban was subjected to several modifications with a view to improve CRTH2 affinity and dial out TP affinity. Ulven et al. at 7TM Pharma demonstrated that

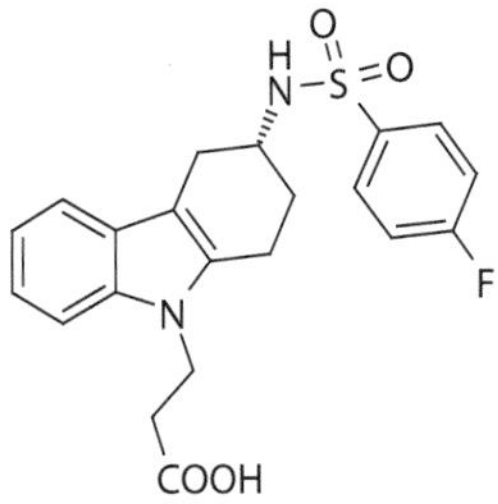

9

K_i hCRTH2 = 4.3 nM; IC_{50} hCRTH2 = 29 nM
K_i hTP = 4.5 nM; IC_{50} hTP = 9.6 nM

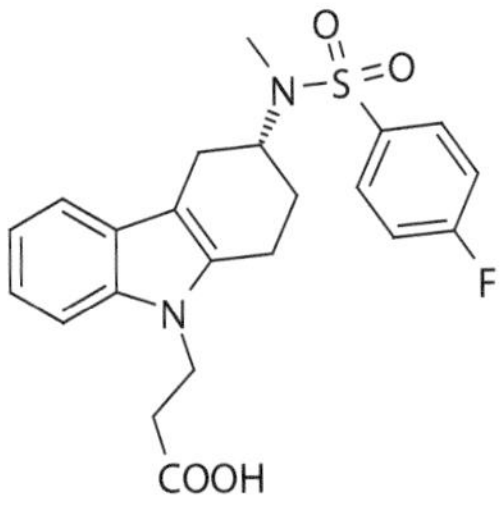

18

K_i hCRTH2 = 1.9 nM; IC_{50} hCRTH2 = 29 nM
K_i hTP = 3000 nM; IC_{50} hTP = 4600 nM

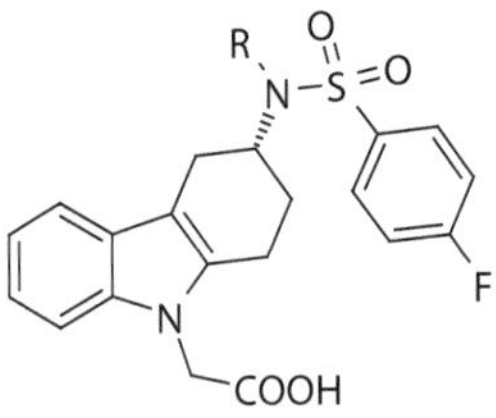

19; R = H

K_i hCRTH2 = 0.51 nM; IC_{50} hCRTH2 = 3.8 nM
K_i hTP = 540 nM; IC_{50} hTP = 1700 nM

20; R = Me

K_i hCRTH2 = 0.6 nM; IC_{50} hCRTH2 = 1.2 nM
K_i hTP >10,000 nM; IC_{50} hTP >10,000 nM

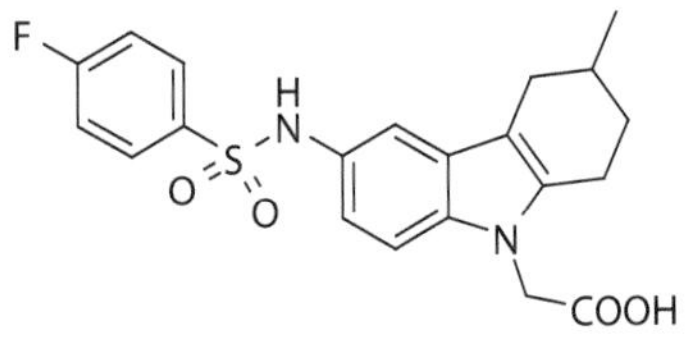

21

K_i hCRTH2 = 13 nM; IC_{50} hCRTH2 = 9.7 nM

FIGURE 8.10 Ramatroban as a starting point for CRTH2 antagonist exploration.

minor structural modifications like N-methylation and shortening of the acidic chain of ramatroban by one carbon (**18**, **19**, and **20**, Figure 8.10) could drastically change the ramatroban biological profile. N-alkylation resulted in compound **18**, which had comparable CRTH2 activity but exhibited significant loss of TP activity. Compound **18** exhibited >1000-fold selectivity over TP. Shortening of the acid chain length by one carbon not only led to improved binding to and inhibition of CRTH2 but also resulted in significant loss of binding to TP as seen in compound **19**. Further, combination of N-alkylation and shortening of acid chain resulted in compound **20**, which not only exhibited further improvement in CRTH2 affinity (Ki: 0.6 nM) and antagonistic activity (IC_{50}:1.2 nM) in an PGD_2-induced inositol phosphate or cAMP formation study but also resulted in complete abolition of affinity for and functional activity mediated through TP receptor (selectivity > 10,000-fold) (Ulven and Kostenis, 2005). Robarge et al. at Athersys Inc. synthesized isosteric ramatroban analogs. By just reversing the scaffold, they found that the binding affinity for CRTH2 remained unaltered. Further modification involving shortening of the acidic chain led to the discovery of compound **21** (Figure 8.10), which was a very potent CRTH2 antagonist (inhibition of PGD_2-mediated receptor activation in a fluorescence assay) and >400-fold selective over TP (Robarge et al., 2005).

8.5.3 FENCLOFENAC AS LEAD STRUCTURE

While trying to identify nonprostanoid small-molecule antagonists, Sawyer et al. at Merck Frosst found that NSAID fenclofenac also exhibited moderate binding affinity (K_i ~3–4 μM) at both DP_1 and CRTH2 (Sawyer et al., 2002). This discovery prompted several modifications on the phenylacetic acid scaffold fenclofenac.

Pfizer filed a patent mentioning a biaryl thioether as having a 40-fold selectivity over DP_1 for CRTH2 (Bauer, 2002). Notable success in this field was the discovery of Amira's AM211, which is a potent DP antagonist and has entered into clinical trials. Though the specific structure for this compound has not yet been disclosed, it is clear from the patent data that it belongs to the class of phenylacetic acid series (**23**, representative structure **24**) (Figure 8.11). Astra Zeneca has patented biaryl acetic acid derivatives (pIC_{50}: 8.2 for CRTH2) and biaryl ether acetic acid derivatives (pIC_{50}: 9.0) for CRTH2 (Bonnert et al., 2008). Realizing the immense potential for developing a cure for the more debilitating condition of COPD, Astra Zeneca has initiated three Phase II studies for AZD1981, one for asthma and two for COPD, with results still awaited at this point of time. Astra Zeneca's AZD5985 and AZD8075 are other candidates in Phase I. Phenoxyacetic acid derivative AZ11805131 was found to be efficacious in murine cigarette smoke model of COPD with a long half-life and high metabolic stability (Bonnert et al., 2009).

8.5.4 OTHER SCAFFOLDS

8.5.4.1 Tetrahydrocarbazoles

Screening of the Merck compound collection led to the identification of a tetrahydrocarbazolyl hit **25** (Figure 8.12), which exhibited potent binding affinity at both DP and TP (K_i: 11 and 1.7 nM, respectively). Medicinal chemistry efforts to optimize

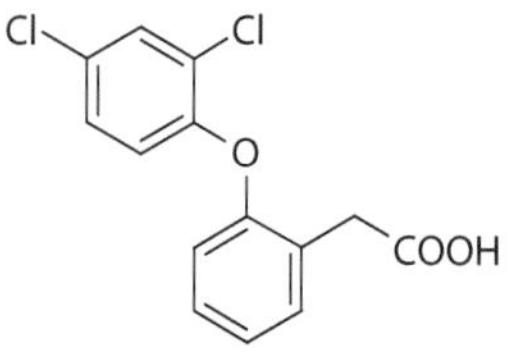

22, Fenclofenac
K_i hDP$_1$ = 3.2 μM
K_i hCRTH2 = 4.9 μM

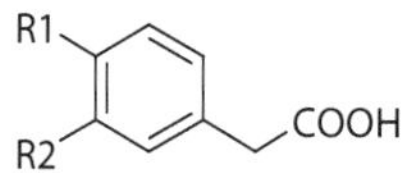

23

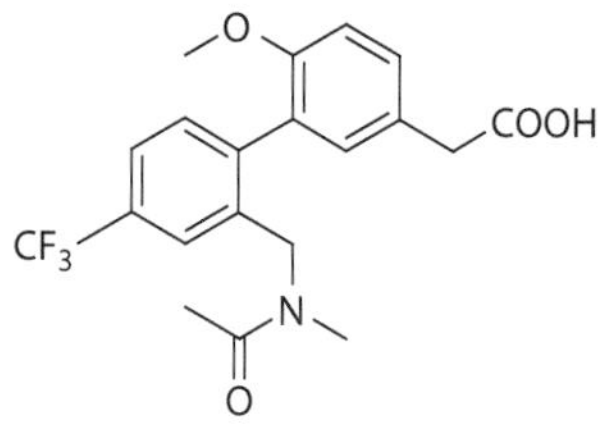

24

FIGURE 8.11 Fenclofenac as a lead structure.

this hit led to a tetrahydrocyclopenta[*b*]indol-3-acetic acid scaffold as in **26**, **27**, and **28**. Compound **28** (also known as MK 0524, laropiprant) exhibited potent binding affinity (K_i: 0.57 nM) and potently inhibited accumulation of cAMP in platelet-rich plasma challenged with PGD_2 (IC_{50}: 4.0 nM). Compound **28** further exhibited ~200-fold functional selectivity over TP. The chirality of the acetic acid side chain was found to be very critical (Sturino et al., 2006, 2007).

8.5.4.2 Tetrahydroquinolines

Tetrahydroquinolines dominate the new class of nonacidic CRTH2 antagonists, which provide the immediate benefit of a potent molecule that is able to cross the blood–brain barrier without having to resort to a carboxylic acid prodrug approach. Millennium, Pfizer, and Kyowa Hakko Kogyo have reported molecules having potency in the low nanomolar range (K-117: 5.5 nM, K-604: 11 nM) with minimal cross-reactivity with TP and DP_1 receptors, up to a concentration of 1 μM (Mimura et al., 2005).

The stage is set for the launch of a PG receptor antagonist for asthma/COPD/hypersensitivity as there are 13 molecules in various stages of clinical trials involving nine of the top pharmaceutical companies of the world, with a further two having preclinical research programs running. Table 8.3 lists the various drugs targeting PG and its receptors that have entered at least Phase I clinical trials. Future chemistry studies will focus on developing expanded libraries for detailed quantitative structure–activity relationship, molecular modeling studies, and future drug development based on the crystal structures of the respective enzymes. We strongly feel that molecular modeling studies can improve our understanding of the molecular interactions and the structural factors responsible for selectivity of drugs, and, therefore, hasten the much needed research.

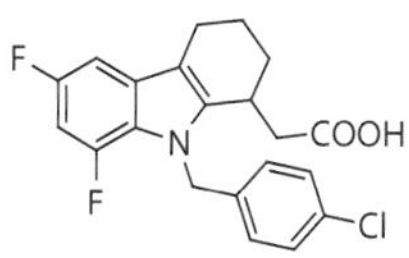

25

$K_i DP_1$ = 11 nM

$K_i TP$ = 1.7 nM

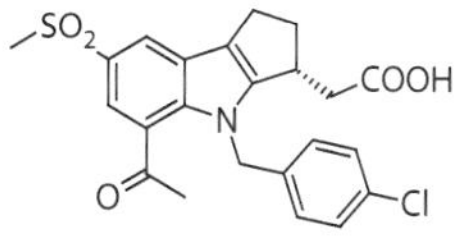

26

$K_i DP_1$ = 2.6 nM $K_i TP$ = 1200 nM

$IC_{50} DP_1$ (washed platelets, cAMP inhibition) = 2.0 nM

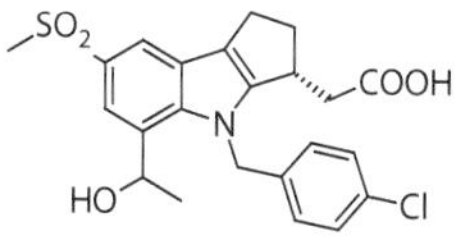

27

$K_i DP_1$ = 1.8 nM $K_i TP$ = 7100 nM

$IC_{50} DP_1$ (washed platelets, cAMP inhibition) = 2.0 nM

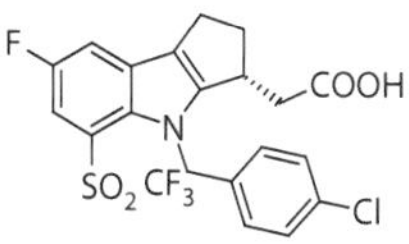

28, MK-0524; Laropirant

$K_i DP_1$ = 0.57 nM $K_i TP$ = 2.95 nM

$IC_{50} DP_1$ (washed platelets, cAMP inhibition) = 0.09 nM

FIGURE 8.12   Tetrahydrocarbazolyl analogs.

8.6   EXPERT OPINION

Prostaglandins are lipid derivatives that are critical in maintaining homoeostasis in our body. With time, our understanding of these complex set of molecules has increased substantially, leading to exploitation of its biology to develop drugs. Aspirin, which works by interfering with the formation of PGs, has been a highly successful anti-inflammatory agent, spurring further research in this field. One of the major issues with aspirin was the syndrome of aspirin-induced asthma, the cause of which was found to be the COX-1 inhibitory action of aspirin. Hence, it was suggested that specifically inhibiting COX-2 while sparing COX-1 could be a solution to this issue. In support of this argument, it was found that drugs that are more potent inhibitors of COX-1 than COX-2, such as nimesulide and meloxicam, are better tolerated by aspirin-sensitive asthmatic patients (Bianco et al., 1993; Kosnik et al., 1998). The most frenetic activity was seen in the 1990s and early years of the twenty-first century, when selective COX-2 inhibitors were launched with much promise. After becoming blockbuster drugs, these so-called "wonder" pills met an inglorious end to their life with the discovery that they cause an almost twofold increase in the incidence of acute myocardial infarction and sudden cardiac death in patients (Caldwell et al., 2006), severe enough to justify their withdrawal from the market (Halpern, 2005; Levesque et al., 2005; Liew and Krum, 2005; Sanghi et al., 2006; Brophy et al., 2007; Moodley,

TABLE 8.3

Status of Drugs Affecting PG or Its Receptors for Treatment of Asthma

Drug (Brand)	Highest Phase (Originator)	Mechanism of Action	Route	Therapeutic Area
Seratrodast (Bronica)	Launched (Takeda)	Prostaglandin receptor antagonist, Thromboxane A_2 receptor antagonist	PO	Ischemic heart disease, obstructive airways disease, respiratory tract disorders, thromboses
Treprostinil (Remodulin, Tyvaso, Uniprost)	Launched Aradigm Corporation, GlaxoSmithKline, Pfizer	Prostacyclin agonist	Inhalation, IV, SC, PO	Cancer, heart failure, respiratory tract disorders, skin disorders, transplant rejection, vascular disorders
AZD 1981 (Astra Zeneca)	Phase II	Prostaglandin D_2 receptor antagonist	PO	Obstructive airways disease, respiratory tract disorders
AM 211 (Amira Pharmaceuticals)	Phase I	Prostaglandin D_2 receptor antagonist	PO	Immunological disorders, inflammation, obstructive airways disease, respiratory tract disorders
AZD 5985 (Astra Zeneca)	Phase I	Prostaglandin D_2 receptor antagonists	PO	Obstructive airways disease, respiratory tract disorders
AZD 8075 (Astra Zeneca)	Phase I	Prostaglandin D_2 receptor antagonist	PO	Obstructive airways disease, respiratory tract disorders
ONO 1301 (Ono Pharmaceuticals)	Preclinical	Prostacyclin agonist, prostaglandin agonist, thromboxane synthase inhibitor	Parenteral, PO	Genitourinary disorders, ischemic heart disease, respiratory tract disorders, thromboses

Source: Data from Adis R&D Insight. http://bi.adisinsight.com.

2008; Ray et al., 2009). This incident raised pertinent questions on the very logic of interfering with a pathway that is so critical to various facets of human biology, and about which we still cannot claim to have an exhaustive knowledge of. As discussed earlier, COX-1 and COX-2 control the synthesis of five different prostaglandins, which mediate their actions through a multitude of receptors that have either complementary or compensatory functions among each other. The pathway shows a high level of interplay, and interfering with one of the players may potentially affect the others adversely. Consequently, bringing about selective inhibition of any component of the prostaglandin pathway should be accompanied by thorough investigations on its effect on all related prostaglandins.

Drug companies have been working toward developing a selective inhibitor of a particular PG synthase or a selective PG receptor antagonist on the premise that the target syndrome is tied to a single prostaglandin. To expect that a complex disease state like asthma/respiratory inflammation is uniquely controlled by a single prostaglandin acting through a single receptor would not be a practical assumption to have. A case in point is the role of DP_1 and CRTH2 receptors in asthma and allergic rhinitis. There is sufficient pharmacological and genetic evidence that both DP_1 and CRTH2 receptors play a crucial role in mediating such allergic diseases that are resistant to current therapy. Here, DP_1 and CRTH2 behave complementary to each other, in initiating and maintaining the disease. Hence, a dual DP_1/CRTH2 antagonist selective over all other prostaglandin receptors appears to be an attractive target for drug development.

There exists a difference between the earlier approach and this one. By targeting downstream receptors instead of the upstream rate-limiting enzyme, we are making an attempt not to imbalance multiple prostaglandins. Again, we are not interfering with the synthesis or release of PGs, but only controlling its effector functions on specific targets, making sure the others PGs are unaffected. Ramatroban, a known inhibitor of CRTH2 and TxA_2, is in clinical use in Japan. To date, there has been no report of any serious adverse effects associated with the molecule. This works as a proof of the concept that inhibiting PGD_2 receptors may prove to be an effective strategy. Nevertheless, the safety profile would need to be evaluated for each such molecule to rule out potentially fatal prostaglandin imbalances, before a final call on the efficacy and safety of the candidate molecule can be taken. Lessons learnt from earlier COX-2 inhibitor trials will help us design clinical trials where the effect of the test molecule on the balance between the effector functions of other prostaglandins in addition to the one being targeted can be monitored early. This could help in minimizing unexpected adverse effects and achieving a better success rate. Evaluating each candidate on an individual case-to-case basis for each specific disease seems to be a better way forward if we want to realize the dream of tapping the immense potential that the prostaglandin pathway holds for disease therapy.

REFERENCES

Abe, H., T. Takeshita, K. Nagata, T. Arita, Y. Endo, T. Fujita, H. Takayama, M. Kubo, and K. Sugamura. 1999. Molecular cloning, chromosome mapping and characterization of the mouse CRTH2 gene, a putative member of the leukocyte chemoattractant receptor family. *Gene* 227(1):71–7.

Abramovitz, M., Y. Boie, T. Nguyen, T. H. Rushmore, M. A. Bayne, K. M. Metters, D. M. Slipetz, and R. Grygorczyk. 1994. Cloning and expression of a cDNA for the human prostanoid FP receptor. *J Biol Chem.* 269(4):2632–6.

Aizawa, H., M. Shigyo, H. Nogami, T. Hirose, and N. Hara. 1996. BAY u3405, a thromboxane A_2 antagonist, reduces bronchial hyperresponsiveness in asthmatics. *Chest* 109(2):338–42.

Akaishi, T., N. Odani-Kawabata, N. Ishida, and M. Nakamura. 2009. Ocular hypotensive effects of anti-glaucoma agents in mice. *J Ocul Pharmacol Ther.* 25(5):401–8.

Al-Jazzaf, A. M., L. DeSantis, and P. A. Netland. 2003. Travoprost: A potent ocular hypotensive agent. *Drugs Today (Barc)*. 39(1):61–74.

Allen, I. C., J. M. Hartney, T. M. Coffman, R. B. Penn, J. Wess, and B. H. Koller. 2006. Thromboxane A_2 induces airway constriction through an M_3 muscarinic acetylcholine receptor-dependent mechanism. *Am J Physiol Lung Cell Mol Physiol*. 290(3):L526–33.

Almishri, W., C. Cossette, J. Rokach, J. G. Martin, Q. Hamid, and W. S. Powell. 2005. Effects of prostaglandin D_2, 15-deoxy-d12,14-prostaglandin J_2, and selective DP_1 and DP_2 receptor agonists on pulmonary infiltration of eosinophils in Brown Norway rats. *J Pharmacol Exp Ther*. 313(1):64–9.

Arimura, A., K. Yasui, J. Kishino, F. Asanuma, H. Hasegawa, S. Kakudo, M. Ohtani, and H. Arita. 2001. Prevention of allergic inflammation by a novel prostaglandin receptor antagonist, S-5751. *J Pharmacol Exp Ther*. 298(2):411–9.

Armer, R. E., M. R. Ashton, E. A. Boyd, C. J. Brennan, F. A. Brookfield, L. Gazi, S. L. Gyles et al. 2005. Indole-3-acetic acid antagonists of the prostaglandin D_2 receptor CRTH2. *J Med Chem*. 48:6174–7.

Bauer, P. H. A., R. P. G. Gladue, B. G. Li, K. S. Neote, and J. R. Zhang. 2002. Methods for the identification of compounds useful for the treatment of disease states mediated by prostaglandin D_2, edited by E. P. Office, Pfizer Inc., EP1170594A2.

Bianco, S., M. Robuschi, G. Petrigni, M. Scuri, M. G. Pieroni, R. M. Refini, A. Vaghi, and P. S. Sestini. 1993. Efficacy and tolerability of nimesulide in asthmatic patients intolerant to aspirin. *Drugs* 46(Suppl 1):115–20.

Bochenek, G., K. Nagraba, E. Nizankowska, and A. Szczeklik. 2003. A controlled study of 9a,11b-PGF_2 (a prostaglandin D_2 metabolite) in plasma and urine of patients with bronchial asthma and healthy controls after aspirin challenge. *J Allergy Clin Immunol* 111(4):743–9.

Bonnert, R. V., A. Patel, and S. Thom. 2008. Substituted diphenylethers, amines, -sulfides and -methanes for the treatment of respiratory disease. Astra Zeneca, WO/2005/018529.

Bonnert, R., V., T. J. Luker, A. Patel, and A. Rigby. 2009. Novel compounds 951: A biphenyloxypropanoic acid as CRTH2 modulator and intermediates. edited by W. I. P. Organization, Astra Zeneca, WO/2009/004379.

Breyer, M. D. and R. M. Breyer. 2001. G protein-coupled prostanoid receptors and the kidney. *Annu Rev Physiol*. 63:579–605.

Brophy, J. M., L. E. Levesque, and B. Zhang. 2007. The coronary risk of cyclo-oxygenase-2 inhibitors in patients with a previous myocardial infarction. *Heart* 93(2):189–94.

Burke, A., J. A. Lawson, E. A. Meagher, J. Rokach, and G. A. FitzGerald. 2000. Specific analysis in plasma and urine of 2,3-dinor-5,6-dihydro-isoprostane F(2alpha)-III, a metabolite of isoprostane F_{2a}-III and an oxidation product of g-linolenic acid. *J Biol Chem*. 275(4):2499–504.

Busse, W. W. 1996. The role of leukotrienes in asthma and allergic rhinitis. *Clin Exp Allergy* 26(8):868–79.

Caldwell, B., S. Aldington, M. Weatherall, P. Shirtcliffe, and R. Beasley. 2006. Risk of cardiovascular events and celecoxib: A systematic review and meta-analysis. *J R Soc Med*. 99(3):132–40.

Capra, V., A. Habib, M. R. Accomazzo, S. Ravasi, S. Citro, S. Levy-Toledano, S. Nicosia, and G. E. Rovati. 2003. Thromboxane prostanoid receptor in human airway smooth muscle cells: A relevant role in proliferation. *Eur J Pharmacol*. 474(2–3):149–59.

Carey, M. A., D. R. Germolec, R. Langenbach, and D. C. Zeldin. 2003. Cyclooxygenase enzymes in allergic inflammation and asthma. *Prostaglandins Leukot Essent Fatty Acids* 69(2–3):157–62.

Chan, C. L., R. L. Jones, and H. Y. Lau. 2000. Characterization of prostanoid receptors mediating inhibition of histamine release from anti-IgE-activated rat peritoneal mast cells. *Br J Pharmacol*. 129(3):589–97.

Christie, P. E., C. M. Smith, and T. H. Lee. 1991. The potent and selective sulfidopeptide leukotriene antagonist, SK&F 104353, inhibits aspirin-induced asthma. *Am Rev Respir Dis.* 144(4):957–8.

Clarke, D. L., M. G. Belvisi, S. J. Smith, E. Hardaker, M. H. Yacoub, K. K. Meja, R. Newton, D. M. Slater, and M. A. Giembycz. 2005. Prostanoid receptor expression by human airway smooth muscle cells and regulation of the secretion of granulocyte colony-stimulating factor. *Am J Physiol Lung Cell Mol Physiol.* 288(2):L238–50.

Coleman, R. A., W. L. Smith, and S. Narumiya. 1994. International Union of Pharmacology classification of prostanoid receptors: Properties, distribution, and structure of the receptors and their subtypes. *Pharmacol Rev.* 46(2):205–29.

Comporti, M., C. Signorini, B. Arezzini, D. Vecchio, B. Monaco, and C. Gardi. 2008. F_2-isoprostanes are not just markers of oxidative stress. *Free Radic Biol Med.* 44(3):247–56.

Costagliola, C., R. dell'Omo, M. R. Romano, M. Rinaldi, L. Zeppa, and F. Parmeggiani. 2009. Pharmacotherapy of intraocular pressure: Part I. Parasympathomimetic, sympathomimetic and sympatholytics. *Expert Opin Pharmacother.* 10(16):2663–77.

Dahlen, S. E., K. Malmstrom, E. Nizankowska, B. Dahlen, P. Kuna, M. Kowalski, W. R. Lumry et al. 2002. Improvement of aspirin-intolerant asthma by montelukast, a leukotriene antagonist: A randomized, double-blind, placebo-controlled trial. *Am J Respir Crit Care Med.* 165(1):9–14.

Demoly, P., D. Jaffuel, N. Lequeux, B. Weksler, C. Creminon, F. B. Michel, P. Godard, and J. Bousquet. 1997. Prostaglandin H synthase 1 and 2 immunoreactivities in the bronchial mucosa of asthmatics. *Am J Respir Crit Care Med.* 155(2):670–5.

Dogne, J. M., X. de Leval, P. Benoit, J. Delarge, and B. Masereel. 2002. Thromboxane A_2 inhibition: Therapeutic potential in bronchial asthma. *Am J Respir Med.* 1(1):11–7.

Feldberg, W., H. F. Holden, and C. H. Kellaway. 1938. The formation of lysocithin and of a muscle-stimulating substance by snake venoms. *J Physiol.* 94(2):232–48.

Fujimura, M., Y. Kamio, K. Kasahara, T. Bando, T. Hashimoto, and T. Matsuda. 1995. Prostanoids and cough response to capsaicin in asthma and chronic bronchitis. *Eur Respir J.* 8(9):1499–05.

Fujitani, Y., Y. Kanaoka, K. Aritake, N. Uodome, K. Okazaki-Hatake, and Y. Urade. 2002. Pronounced eosinophilic lung inflammation and Th2 cytokine release in human lipocalin-type prostaglandin D synthase transgenic mice. *J Immunol.* 168(1):443–9.

Funk, C. D., L. Furci, G. A. FitzGerald, R. Grygorczyk, C. Rochette, M. A. Bayne, M. Abramovitz, M. Adam, and K. M. Metters. 1993. Cloning and expression of a cDNA for the human prostaglandin E receptor EP1 subtype. *J Biol Chem.* 268(35):26767–72.

Gauvreau, G. M., R. M. Watson, and P. M. O'Byrne. 1999. Protective effects of inhaled PGE_2 on allergen-induced airway responses and airway inflammation. *Am J Respir Crit Care Med.* 159(1):31–6.

Giles, H., P. Leff, M. L. Bolofo, M. G. Kelly, and A. D. Robertson. 1989. The classification of prostaglandin DP-receptors in platelets and vasculature using BW A868C, a novel, selective and potent competitive antagonist. *Br J Pharmacol.* 96(2):291–300.

Gomi, K., F. G. Zhu, and J. S. Marshall. 2000. Prostaglandin E_2 selectively enhances the IgE-mediated production of IL-6 and granulocyte-macrophage colony-stimulating factor by mast cells through an EP1/EP3-dependent mechanism. *J Immunol.* 165(11):6545–52.

Goppelt-Struebe, M. 1995. Regulation of prostaglandin endoperoxide synthase (cyclooxygenase) isozyme expression. *Prostaglandins Leukot Essent Fatty Acids* 52(4):213–22.

Graf, B. A., D. A. Nazarenko, M. A. Borrello, L. J. Roberts, J. D. Morrow, J. Palis, and R. P. Phipps. 1999. Biphenotypic B/macrophage cells express COX-1 and up-regulate COX-2 expression and prostaglandin E_2 production in response to pro-inflammatory signals. *Eur J Immunol.* 29(11):3793–803.

Grosser, T., S. Fries, and G. A. FitzGerald. 2006. Biological basis for the cardiovascular consequences of COX-2 inhibition: Therapeutic challenges and opportunities. *J Clin Invest.* 116(1):4–15.

Halpern, G. M. 2005. COX-2 inhibitors: A story of greed, deception and death. *Inflammopharmacology* 13:419–25.

Hartert, T. V., R. T. Dworski, B. G. Mellen, J. A. Oates, J. J. Murray, and J. R. Sheller. 2000. Prostaglandin E$_2$ decreases allergen-stimulated release of prostaglandin D$_2$ in airways of subjects with asthma. *Am J Respir Crit Care Med.* 162(2 Pt 1):637–40.

Hata, A. N. and R. M. Breyer. 2004. Pharmacology and signaling of prostaglandin receptors: Multiple roles in inflammation and immune modulation. *Pharmacol Ther.* 103(2):147–66.

Heinemann, A., R. Schuligoi, I. SabroeA. Hartnell, and B. A. Peskar. 2003. d12-Prostaglandin J$_2$, a plasma metabolite of prostaglandin D$_2$, causes eosinophil mobilization from the bone marrow and primes eosinophils for chemotaxis. *J Immunol.* 170(9):4752–8.

Henderson, W. R., Jr. 1994. Role of leukotrienes in asthma. *Ann Allergy.* 72(3):272–8.

Hirai, H., K. Tanaka, S. Takano, M. Ichimasa, M. Nakamura, and K. Nagata. 2002. Cutting edge: Agonistic effect of indomethacin on a prostaglandin D$_2$ receptor, CRTH2. *J Immunol.* 168(3):981–5.

Hirst, S. J., J. G. Martin, J. V. Bonacci , V. Chan, E. D. Fixman, Q. A. Hamid, B. Herszberg et al. 2004. Proliferative aspects of airway smooth muscle. *J Allergy Clin Immunol.* 114(2 Suppl):S2–17.

Honda, T., T. Matsuoka, M. Ueta, K. Kabashima, Y. Miyachi, and S. Narumiya. 2009. Prostaglandin E$_2$-EP$_3$ signaling suppresses skin inflammation in murine contact hypersensitivity. *J Allergy Clin Immunol.* 124(4):809–18 e2.

Howarth, P. H., A. J. Knox, Y. Amrani, O. Tliba, R. A. Panettieri, Jr., and M. Johnson. 2004. Synthetic responses in airway smooth muscle. *J Allergy Clin Immunol.* 114 (2 Suppl): S32–50.

Irie, A., E. Segi, Y. Sugimoto, A. Ichikawa, and M. Negishi. 1994. Mouse prostaglandin E receptor EP3 subtype mediates calcium signals via Gi in cDNA-transfected Chinese hamster ovary cells. *Biochem Biophys Res Commun.* 204(1):303–9.

Ishida, N., N. Odani-Kawabata, A. Shimazaki, and H. Hara. 2006. Prostanoids in the therapy of glaucoma. *Cardiovasc Drug Rev.* 24(1):1–10.

Ishiura, Y., M. Fujimura, H. Yamamoto T. Ishiguro, N. Ohkura, and S. Myou. 2009. COX-2 inhibition attenuates cough reflex sensitivity to inhaled capsaicin in patients with asthma. *J Investig Allergol Clin Immunol.* 19(5):370–4.

Iyer, J. P., P. K. Srivastava, R. Dev, S. G. Dastidar, and A. Ray. 2009. Prostaglandin E$_2$ synthase inhibition as a therapeutic target. *Expert Opin Ther Targets.* 13(7):849–65.

Jaffar, Z., K. S. Wan, and K. Roberts. 2002. A key role for prostaglandin I$_2$ in limiting lung mucosal Th2, but not Th1, responses to inhaled allergen. *J Immunol.* 169(10):5997–04.

James, A. L. 1997. Relationship between airway wall thickness and airway hyperresponsiveness. In *Airway Wall Remodelling in Asthma*, A. Stewart (ed.), London: CRC.

Janssen, L. J. 2001. Isoprostanes: An overview and putative roles in pulmonary pathophysiology. *Am J Physiol Lung Cell Mol Physiol.* 280(6):L1067–82.

Kabashima, K., T. Murata, H. Tanaka, T. Matsuoka, D. Sakata, N. Yoshida, K. Katagiri et al. 2003. Thromboxane A$_2$ modulates interaction of dendritic cells and T cells and regulates acquired immunity. *Nat Immunol.* 4(7): 694–701.

Kosnik, M., E. Music, F. Matjaz, and S. Suskovic. 1998. Relative safety of meloxicam in NSAID-intolerant patients. *Allergy* 53(12):1231–3.

Kunikata, T., H. Yamane, E. Segi, T. Matsuoka, Y. Sugimoto, S. Tanaka, H. Tanaka, H. Nagai, A. Ichikawa, and S. Narumiya. 2005. Suppression of allergic inflammation by the prostaglandin E receptor subtype EP3. *Nat Immunol.* 6(5):524–31.

Lazzeri, N., M. G. Belvisi, H. J. Patel, M. H. Yacoub, K. F. Chung, and J. A. Mitchell. 2001. Effects of prostaglandin E_2 and cAMP elevating drugs on GM-CSF release by cultured human airway smooth muscle cells. Relevance to asthma therapy. *Am J Respir Cell Mol Biol.* 24(1):44–8.

Lemberger, T., B. Desvergne, and W. Wahli. 1996. Peroxisome proliferator-activated receptors: A nuclear receptor signaling pathway in lipid physiology. *Annu Rev Cell Dev Biol.* 12:335–63.

Levesque, L. E., J. M. Brophy, and B. Zhang. 2005. The risk for myocardial infarction with cyclooxygenase-2 inhibitors: A population study of elderly adults. *Ann Intern Med.* 142(7):481–9.

Liew, D. and H. Krum. 2005. The cardiovascular safety of celecoxib. *Future Cardiol.* 1(6):709–22.

Linden, C. and A. Alm. 1999. Prostaglandin analogues in the treatment of glaucoma. *Drugs Aging* 14(5):387–98.

Martey, C. A., C. J. Baglole, T. A. Gasiewicz, P. J. Sime, and R. P. Phipps. 2005. The aryl hydrocarbon receptor is a regulator of cigarette smoke induction of the cyclooxygenase and prostaglandin pathways in human lung fibroblasts. *Am J Physiol Lung Cell Mol Physiol.* 289(3):L391–9.

Matsuoka, T., M. Hirata, H. Tanaka, Y. Takahashi, T. Murata, K. Kabashima, Y. Sugimoto et al. 2000. Prostaglandin D_2 as a mediator of allergic asthma. *Science* 287(5460):2013–7.

Matsuoka, T. and S. Narumiya. 2007. Prostaglandin receptor signaling in disease. *Scientific World Journal* 7:1329–47.

McKay, S. and H. S. Sharma. 2002. Autocrine regulation of asthmatic airway inflammation: Role of airway smooth muscle. *Respir Res.* 3:11.

Mimura, H., T. Ikemura, O. Kotera, M. Sawada, S. Tashiro, E. Fuse, K. Ueno et al. 2005. Inhibitory effect of the 4-aminotetrahydroquinoline derivatives, selective chemoattractant receptor-homologous molecule expressed on T helper 2 cell antagonists, on eosinophil migration induced by prostaglandin D_2. *J Pharmacol Exp Ther.* 314(1):244–51.

Mitchell, J. A., P. Akarasereenont, C. Thiemermann, R. J. Flower, and J. R. Vane. 1993. Selectivity of nonsteroidal antiinflammatory drugs as inhibitors of constitutive and inducible cyclooxygenase. *Proc Natl Acad Sci USA* 90(24):11693–7.

Mitchell, J. A., M. G. Belvisi, P. Akarasereenont, R. A. Robbins, O. J. Kwon, J. Croxtall, P. J. Barnes, and J. R. Vane. 1994. Induction of cyclo-oxygenase-2 by cytokines in human pulmonary epithelial cells: Regulation by dexamethasone. *Br J Pharmacol.* 113(3):1008–14.

Mitsumori, S., T. Tsuri, T. Honma, Y. Hiramatsu, T. Okada, H. Hashizume, S. Kida et al. 2003. Synthesis and biological activity of various derivatives of a novel class of potent, selective, and orally active prostaglandin D2 receptor antagonists. 2. 6,6-Dimethylbicyclo[3.1.1] heptane derivatives. *J Med Chem.* 46(12):2446–55.

Moodie, F. M., J. A. Marwick, C. S. Anderson, P. Szulakowski, S. K. Biswas, M. R. Bauter, I. Kilty, and I. Rahman. 2004. Oxidative stress and cigarette smoke alter chromatin remodeling but differentially regulate NF-kB activation and proinflammatory cytokine release in alveolar epithelial cells. *FASEB J.* 18(15):1897–9.

Moodley, I. 2008. Review of the cardiovascular safety of COXIBs compared to NSAIDS. *Cardiovasc J Afr.* 19(2):102–7.

Moretto, N., F. Facchinetti, T. Southworth, M. Civelli, D. Singh, and R. Patacchini. 2009. Alpha,beta-unsaturated aldehydes contained in cigarette smoke elicit IL-8 release in pulmonary cells through mitogen-activated protein kinases. *Am J Physiol Lung Cell Mol Physiol.* 296(5):L839–48.

Morrow, J. D., K. E. Hill, R. F. Burk, T. M. Nammour, K. F. Badr, and L. J. Roberts, 2nd. 1990. A series of prostaglandin F_2-like compounds are produced *in vivo* in humans by a non-cyclooxygenase, free radical-catalyzed mechanism. *Proc Natl Acad Sci USA* 87(23):9383–7.

Murata, T., F. Ushikubi, T. Matsuoka , M. Hirata, A. Yamasaki, Y. Sugimoto, A. Ichikawa et al. 1997. Altered pain perception and inflammatory response in mice lacking prostacyclin receptor. *Nature* 388(6643):678–82.

Nagai, H. 2008. Prostaglandin as a target molecule for pharmacotherapy of allergic inflammatory diseases. *Allergol Int.* 57(3):187–96.

Nagao, K., H. Tanaka, M. Komai, T. Masuda, S. Narumiya, and H. Nagai. 2003. Role of prostaglandin I$_2$ in airway remodeling induced by repeated allergen challenge in mice. *Am J Respir Cell Mol Biol.* 29(3 pt 1):314–20.

Narumiya, S. and G. A. FitzGerald. 2001. Genetic and pharmacological analysis of prostanoid receptor function. *J Clin Invest.* 108(1):25–30.

Narumiya, S., Y. Sugimoto, and F. Ushikubi. 1999. Prostanoid receptors: Structures, properties, and functions. *Physiol Rev.* 79(4):1193–226.

Norel, X., V. de Montpreville, and C. Brink. 2004. Vasoconstriction induced by activation of EP1 and EP3 receptors in human lung: Effects of ONO-AE-248, ONO-DI-004, ONO-8711 or ONO-8713. *Prostaglandins Other Lipid Mediat.* 74(1–4):101–12.

Nourooz-Zadeh, J., B. Halliwell, and E. E. Anggard. 1997. Evidence for the formation of F$_3$-isoprostanes during peroxidation of eicosapentaenoic acid. *Biochem Biophys Res Commun.* 236(2):467–72.

O'Byrne, P. M. 1997. Leukotrienes in the pathogenesis of asthma. *Chest* 111(2 suppl):27S–34S.

Oguma, T., L. J. Palmer, E. Birben, L. A. Sonna, K. Asano, and C. M. Lilly. 2004. Role of prostanoid DP receptor variants in susceptibility to asthma. *N Engl J Med.* 351(17):1752–63.

Oliva, D. and S. Nicosia. 1987. PGI$_2$-receptors and molecular mechanisms in platelets and vasculature: State of the art. *Pharmacol Res Commun.* 19(11):735–65.

Pang, L., A. Pitt, D. Petkova, and A. J. Knox. 1998. The COX-1/COX-2 balance in asthma. *Clin Exp Allergy* 28(9):1050–8.

Parameswaran, K., K. Radford, A. Fanat, J. Stephen, C. Bonnans, B. D. Levy, L. J. Janssen, and P. G. Cox. 2007. Modulation of human airway smooth muscle migration by lipid mediators and Th-2 cytokines. *Am J Respir Cell Mol Biol.* 37(2):240–7.

Parchmann, S. and M. J. Mueller. 1998. Evidence for the formation of dinor isoprostanes E$_1$ from alpha-linolenic acid in plants. *J Biol Chem.* 273(49):32650–5.

Park, Y. S., J. Kim, Y. Misonou, R. Takamiya, M. Takahashi, M. R. Freeman, and N. Taniguchi. 2007. Acrolein induces cyclooxygenase-2 and prostaglandin production in human umbilical vein endothelial cells: Roles of p38 MAP kinase. *Arterioscler Thromb Vasc Biol.* 27(6):1319–25.

Pavord, I. D., C. S. Wong, J. Williams, and A. E. Tattersfield. 1993. Effect of inhaled prostaglandin E$_2$ on allergen-induced asthma. *Am Rev Respir Dis.* 148(1):87–90.

Philip, G., J. van Adelsberg, T. Loeys, N. Liu, P. Wong, E. Lai, S. B. Dass, and T. F. Reiss. 2009. Clinical studies of the DP$_1$ antagonist laropiprant in asthma and allergic rhinitis. *J Allergy Clin Immunol.* 124(5):942–8 e1–9.

Picado, C., J. A. Castillo, J. M. Montserrat, and A. Agusti-Vidal. 1989. Aspirin-intolerance as a precipitating factor of life-threatening attacks of asthma requiring mechanical ventilation. *Eur Respir J.* 2(2):127–9.

Pierce, K. L., T. J. Bailey, P. B. Hoyer, D. W. Gil, D. F. Woodward, and J. W. Regan. 1997. Cloning of a carboxyl-terminal isoform of the prostanoid FP receptor. *J Biol Chem.* 272(2):883–7.

Quiralte, J., C. Blanco, R. Castillo, J. Delgado, and T. Carrillo. 1996. Intolerance to nonsteroidal antiinflammatory drugs: Results of controlled drug challenges in 98 patients. *J Allergy Clin Immunol.* 98(3):678–85.

Ratcliffe, M. J., A. Walding, P. A. Shelton, A. Flaherty, and I. G. Dougall. 2007. Activation of EP4 and EP2 receptors inhibits TNF-a release from human alveolar macrophages. *Eur Respir J.* 29(5):986–94.

Ray, W. A., C. Varas-Lorenzo, C. P. Chung, J. Castellsague, K. T. Murray, C. M. Stein, J. R. Daugherty, P. G. Arbogast, and L. A. Garcia-Rodriguez. 2009. Cardiovascular risks of nonsteroidal antiinflammatory drugs in patients after hospitalization for serious coronary heart disease. *Circ Cardiovasc Qual Outcomes.* 2(3):155–63.

Raychowdhury, M. K., M. Yukawa, L. J. Collins, S. H. McGrail, K. C. Kent, and J. A. Ware. 1994. Alternative splicing produces a divergent cytoplasmic tail in the human endothelial thromboxane A_2 receptor. *J Biol Chem.* 269(30):19256–61.

Reich, E. E., W. E. Zackert, C. J. Brame, Y. Chen, L. J. Roberts, 2nd, D. L. Hachey, T. J. Montine, and J. D. Morrow. 2000. Formation of novel D-ring and E-ring isoprostanelike compounds (D_4/E_4-neuroprostanes) *in vivo* from docosahexaenoic acid. *Biochemistry* 39(9):2376–83.

Robarge, M. J., D. C. Bom, L. N. Tumey, N. Varga, E. Gleason, D. Silver, J. Song et al. 2005. Isosteric ramatroban analogs: Selective and potent CRTH-2 antagonists. *Bioorg Med Chem Lett.* 15(6):1749–53.

Rokach, J., S. P. Khanapure, S. W. Hwang, M. Adiyaman, J. A. Lawson, and G. A. FitzGerald. 1997. The isoprostanes: A perspective. *Prostaglandins* 54(6):823–51.

Rolin, S., B. Masereel, and J. M. Dogne. 2006. Prostanoids as pharmacological targets in COPD and asthma. *Eur J Pharmacol.* 533(1–3):89–100.

Samara, E., G. Cao, C. Locke, G. R. Granneman, R. Dean, and A. Killian. 1997. Population analysis of the pharmacokinetics and pharmacodynamics of seratrodast in patients with mild to moderate asthma. *Clin Pharmacol Ther.* 62(4):426–35.

Sampson, A. P. 1996. The leukotrienes: Mediators of chronic inflammation in asthma. *Clin Exp Allergy* 26(9):995–1004.

Sanak, M. and A. P. Sampson. 1999. Biosynthesis of cysteinyl-leucotrienes in aspirin-intolerant asthma. *Clin Exp Allergy.* 29(3):306–13.

Sanghi, S., E. J. MacLaughlin, C. W. Jewell, S. Chaffer, P. J. Naus, L. E. Watson, and D. E. Dostal. 2006. Cyclooxygenase-2 inhibitors: A painful lesson. *Cardiovasc Hematol Disord Drug Targets* 6(2):85–100.

Sawyer, N., E. Cauchon, A. Chateauneuf, R. P. Cruz, D. W. Nicholson, K. M. Metters, G. P. O'Neill, and F. G. Gervais. 2002. Molecular pharmacology of the human prostaglandin D_2 receptor, CRTH2. *Br J Pharmacol.* 137(8):1163–72.

Schuligoi, R., M. Sedej, M. Waldhoer, A. Vukoja, E. M. Sturm, I. T. Lippe, B. A. Peskar, and A. Heinemann. 2009. Prostaglandin H_2 induces the migration of human eosinophils through the chemoattractant receptor homologous molecule of Th2 cells, CRTH2. *J Leukoc Biol.* 85(1):136–45.

Sestini, P., L. Armetti, G. Gambaro, M. G. Pieroni, R. M. Refini, A. Sala, A. Vaghi, G. C. Folco, S. Bianco, and M. Robuschi. 1996. Inhaled PGE_2 prevents aspirin-induced bronchoconstriction and urinary LTE_4 excretion in aspirin-sensitive asthma. *Am J Respir Crit Care Med.* 153(2):572–5.

Shaftel, S. S., J. A. Olschowka, S. D. Hurley, A. H. Moore, and M. K. O'Banion. 2003. COX-3: A splice variant of cyclooxygenase-1 in mouse neural tissue and cells. *Brain Res Mol Brain Res.* 119(2):213–5.

Shiraishi, Y., K. Asano, T. Nakajima, T. Oguma, Y. Suzuki, T. Shiomi, K. Sayama et al. 2005. Prostaglandin D_2-induced eosinophilic airway inflammation is mediated by CRTH2 receptor. *J Pharmacol Exp Ther.* 312(3):954–60.

Sousa, A., R. Pfister, P. E. Christie, S. J. Lane, S. M. Nasser, M. Schmitz-Schumann, and T. H. Lee. 1997. Enhanced expression of cyclo-oxygenase isoenzyme 2 (COX-2) in asthmatic airways and its cellular distribution in aspirin-sensitive asthma. *Thorax* 52(11):940–5.

Stubbs, V. E., P. Schratl, A. Hartnell, T. J. Williams, B. A. Peskar, A. Heinemann, and I. Sabroe. 2002. Indomethacin causes prostaglandin D_2-like and eotaxin-like selective responses in eosinophils and basophils. *J Biol Chem.* 277(29):26012–20.

Sturino, C. F., N. Lachance, M. Boyd, C. Berthelette, M. Labelle, L. Li, B. Roy et al. 2006. Identification of an indole series of prostaglandin D$_2$ receptor antagonists. *Bioorg Med Chem Lett.* 16(11):3043–8.

Sturino, C. F., G. O'Neill, N. Lachance, M. Boyd, C. Berthelette, M. Labelle, L. Li et al. 2007. Discovery of a potent and selective prostaglandin D$_2$ receptor antagonist, [(3R)-4-(4-chloro-benzyl)-7-fluoro-5-(methylsulfonyl)-1,2,3,4-tetrahydrocy clopenta[b]indol-3-yl]-acetic acid (MK-0524). *J Med Chem.* 50(4):794–6.

Sugimoto, Y., A. Yamasaki, E. Segi, K. Tsuboi, Y. Aze, T. Nishimura, H. Oida et al. 1997. Failure of parturition in mice lacking the prostaglandin F receptor. *Science* 277(5326):681–3.

Szczeklik, A. 1990. The cyclooxygenase theory of aspirin-induced asthma. *Eur Respir J.* 3(5):588–93.

Szczeklik, A. and D. D. Stevenson. 1999. Aspirin-induced asthma: Advances in pathogenesis and management. *J Allergy Clin Immunol.* 104(1):5–13.

Szczeklik, A. and D. D. Stevenson. 2003. Aspirin-induced asthma: Advances in pathogenesis, diagnosis, and management. *J Allergy Clin Immunol.* 111(5):913–21quiz 922.

Taha, R., R. Olivenstein, T. Utsumi, P. Ernst, P. J. Barnes, I. W. Rodger, and A. Giaid. 2000. Prostaglandin H synthase 2 expression in airway cells from patients with asthma and chronic obstructive pulmonary disease. *Am J Respir Crit Care Med.* 161(2 pt 1):636–40.

Takahashi, Y., S. Tokuoka, T. Masuda, Y. Hirano, M. Nagao, H. Tanaka, N. Inagaki, S. Narumiya, and H. Nagai. 2002. Augmentation of allergic inflammation in prostanoid IP receptor deficient mice. *Br J Pharmacol.* 137(3):315–22.

Tamaoki, J., M. Kondo, J. Nakata, Y. Nagano, K. Isono, and A. Nagai. 2000. Effect of a thromboxane A$_2$ antagonist on sputum production and its physicochemical properties in patients with mild to moderate asthma. *Chest* 118(1):73–9.

Tilley, S. L., T. M. Coffman, and B. H. Koller. 2001. Mixed messages: Modulation of inflammation and immune responses by prostaglandins and thromboxanes. *J Clin Invest.* 108(1):15–23.

Torisu, K., K. Kobayashi, M. Iwahashi, H. Egashira, Y. Nakai, Y. Okada, F. Nanbu, S. Ohuchida, H. Nakai, and M. Toda. 2004a. Discovery of new chemical leads for prostaglandin D$_2$ receptor antagonists. *Bioorg Med Chem Lett.* 14(17):4557–62.

Torisu, K., K. Kobayashi, M. Iwahashi, Y. Nakai, T. Onoda, T. Nagase, I. Sugimoto et al. 2004b. Discovery of orally active prostaglandin D$_2$ receptor antagonists. *Bioorg Med Chem Lett.* 14(19):4891–5.

Ueta, M., T. Matsuoka, S. Narumiya, and S. Kinoshita. 2009. Prostaglandin E receptor subtype EP3 in conjunctival epithelium regulates late-phase reaction of experimental allergic conjunctivitis. *J Allergy Clin Immunol.* 123(2):466–71.

Uller, L., J. M. Mathiesen, L. Alenmyr, M. Korsgren, T. Ulven, T. Hogberg, G. Andersson, C. G. Persson, and E. Kostenis. 2007. Antagonism of the prostaglandin D$_2$ receptor CRTH2 attenuates asthma pathology in mouse eosinophilic airway inflammation. *Respir Res.* 8:16.

Ulven, T. and E. Kostenis. 2005. Minor structural modifications convert the dual TP/CRTH2 antagonist ramatroban into a highly selective and potent CRTH2 antagonist. *J Med Chem.* 48(4):897–900.

Vigano, T., A. Habib, A. Hernandez, A. Bonazzi, D. Boraschi, M. Lebret, E. Cassina, J. Maclouf, A. Sala, and G. Folco. 1997. Cyclooxygenase-2 and synthesis of PGE$_2$ in human bronchial smooth-muscle cells. *Am J Respir Crit Care Med.* 155(3):864–8.

von Euler, U. S. 1936. On the specific vaso-dilating and plain muscle stimulating substances from accessory genital glands in man and certain animals (prostaglandin and vesiglandin). *J Physiol.* 88(2):213–34.

Wang, X. S. and H. Y. Lau. 2006. Prostaglandin E potentiates the immunologically stimulated histamine release from human peripheral blood-derived mast cells through EP1/EP3 receptors. *Allergy* 61(4):503–6.

Watkins, D. N., M. J. Garlepp, and P. J. Thompson. 1997. Regulation of the inducible cyclooxygenase pathway in human cultured airway epithelial (A549) cells by nitric oxide. *Br J Pharmacol.* 121(7):1482–8.

Watkins, D. N., D. J. Peroni, J. C. Lenzo, D. A. Knight, M. J. Garlepp, and P. J. Thompson. 1999a. Expression and localization of COX-2 in human airways and cultured airway epithelial cells. *Eur Respir J.* 13(5):999–1007.

Watkins, M. T., G. M. Patton, H. M. Soler, H. Albadawi, D. E. Humphries, J. E. Evans, and H. Kadowaki. 1999b. Synthesis of 8-epi-prostaglandin F_{2a} by human endothelial cells: Role of prostaglandin H_2 synthase. *Biochem J.* 344(pt 3):747–54.

Weinreb, R. N., C. B. Toris, B. T. Gabelt, J. D. Lindsey, and P. L. Kaufman. 2002. Effects of prostaglandins on the aqueous humor outflow pathways. *Surv Ophthalmol.* 47 Suppl 1:S53–64.

Wenzel, S. E., S. J. Szefler, D. Y. Leung, S. I. Sloan, M. D. Rex, and R. J. Martin. 1997. Bronchoscopic evaluation of severe asthma. Persistent inflammation associated with high dose glucocorticoids. *Am J Respir Crit Care Med.* 156(3 pt 1):737–43.

Wenzel, S. E., J. Y. Westcott, H. R. Smith, and G. L. Larsen. 1989. Spectrum of prostanoid release after bronchoalveolar allergen challenge in atopic asthmatics and in control groups. An alteration in the ratio of bronchoconstrictive to bronchoprotective mediators. *Am Rev Respir Dis.* 139:450–7.

Woodward, D. F., R. L. Phelps, A. H. Krauss, A. Weber, B. Short, J. Chen, Y. Liang, and L. A. Wheeler. 2004. Bimatoprost: A novel antiglaucoma agent. *Cardiovasc Drug Rev.* 22(2):103–20.

Yang, C. M., I. T. Lee, C. C. Lin, Y. L. Yang, S. F. Luo, Y. R. Kou, and L. D. Hsiao. 2009. Cigarette smoke extract induces COX-2 expression via a PKCa/c-Src/EGFR, PDGFR/PI3K/Akt/NF-kB pathway and p300 in tracheal smooth muscle cells. *Am J Physiol Lung Cell Mol Physiol.* 297(5):L892–902.

9 Proresolution Mediators of Inflammation in Airway Diseases
Resolvins Pave New Directions

Bruce D. Levy and Charles N. Serhan

CONTENTS

9.1 INTRODUCTION

Inflammation is a physiological response to injury, infection, or noxious stimuli, such as allergens. In health, acute inflammation is self-limited with complete restoration of tissue homeostasis, a process defined as catabasis (Serhan 2007). If an acute inflammatory response is unrestrained or fails to resolve, then pathology results, including several common lung diseases, such as acute respiratory distress syndrome (ARDS), asthma, chronic obstructive pulmonary disease (COPD), and cystic fibrosis. While there is a complex array of proinflammatory mediators that promote the onset of inflammation, the identification of several distinct families of proresolution lipid mediators indicates that inflammation resolution is a similarly complex and tightly orchestrated process in healthy subjects.

Inflammation resolution is an active and dynamic host response with points of regulatory control (Bannenberg et al. 2005, Schwab et al. 2007). At sites of inflammation, leukocyte trafficking, survival, functional responses to soluble and particulate stimuli, and clearance of pathogens, apoptotic cells, and debris are all highly regulated (reviewed in Serhan 2010). Several cellular and molecular mechanisms can limit the acute inflammatory response, including lipid mediator class switching (Levy et al. 2001). During acute inflammation, early-phase prostaglandin (PG) E_2 and PGD_2 can decrease neutrophil leukotriene (LT) generation and increase the expression of 15-lipoxygenase (15-LO) to switch LO-derived arachidonic acid (C20:4) metabolism to biosynthesis of LXs (Levy et al. 2001) that are "stop" signals for neutrophil transmigration and activation as well as "go" signals for macrophage clearance of apoptotic neutrophils. Combined, these actions actively promote resolution by blocking further PMN entry and activation as well as promoting clearance of these inflammatory cells and their corpses (apoptotic and necrotic).

Distinct macrophage subsets play important roles in promoting the resolution of acute inflammation via phagocytic clearance of apoptotic leukocytes (Hou et al. 2010, Navarro et al. 2010). Apoptotic inflammatory cells also serve a proresolving role by sequestering and scavenging extracellular chemokines via upregulation of cysteine–cysteine (CC) chemokine receptor 5 (CCR5) (Ariel et al. 2006). Neutrophil clearance from mucosal tissues can also proceed by transmigration from the apical epithelial surface into the lumen via an epithelial CD55-mediated process (Louis et al. 2005) (reviewed in Uller et al. 2006). Mucosal epithelium can also express antimicrobial peptides and LXs enhance host defense at mucosal surfaces by inducing epithelial bactericidal/permeability-increasing protein (BPI) (Canny et al. 2002). On tissue resident cells, proresolving mediators also display antiangiogenic and antifibrotic actions (Cezar-de-Mello et al. 2008, Sodin-Semrl et al. 2000). These findings are relevant to chronic inflammatory diseases of the lungs as both angiogenesis and fibrosis are recognized features of airway remodeling in many diseases of airway inflammation (Belvisi 2009). There is now increasing evidence that defined signal transduction pathways relay cell-type-specific proresolution signals that impact a diverse range of cellular events that are fundamental to resolution (Gilroy et al. 2004, Lawrence et al. 2005, Leitch et al. 2010, Levy et al. 2005, Perretti et al. 2002, Rossi et al. 2006). Of translational relevance, defects in counterregulatory signaling pathways can contribute to chronic inflammation in disease states, including severe asthma (Levy et al. 2005, Planagumà et al. 2008). This chapter illustrates how uncovering the natural endogenous mediator for host resolution programs is providing abundant new targets for drug discovery to control pathological airway inflammation.

9.2 BASIC BIOLOGY OF PRORESOLUTION MEDIATORS

9.2.1 C20:4: LIPOXINS

Lipoxins (LXs) were the first described endogenous anti-inflammatory and proresolution lipid mediators (reviewed in Serhan 2007). These compounds are enzymatically derived from arachidonic acid (C20:4) (Serhan et al. 1984), but are structurally and functionally distinct from other eicosanoids, such as prostaglandins (PGs), thromboxane (TX), and leukotrienes (LTs). LXs are formed during cell–cell interactions via bidirectional, transcellular biosynthesis (Fiore and Serhan 1990). In the mucosa, 15-lipoxygenase (15-LO) is a key enzyme for LX generation and is expressed by many cells in the airway, including epithelial cells, cytokine-primed macrophages, and eosinophils. 15-LO converts C20:4 to 15(S)-hydroperoxyeicosatetraenoic acid (15(S)-H(p)ETE) that can be further converted to LXs by leukocyte 5-LO (Levy et al. 1993). LX biosynthesis is enhanced in the lung when infiltrating leukocytes interact with epithelia in inflamed airways. LXs can also be generated in the vasculature during interactions between leukocyte 5-LO and platelet 12-LO, as the human platelet 12-LO is an LX synthase (Fiore and Serhan 1990).

LXs are potent inhibitors of polymorphonuclear leukocyte (PMN) and eosinophil locomotion and transmigration, expression of adhesion molecules, and cell activation, including phospholipase D, azurophilic granule degranulation, and superoxide anion generation (reviewed in Serhan, 2007). LXs display cell-type-specific actions that are also notable for promoting monocyte adhesion and chemotaxis and macrophage engulfment of apoptotic PMNs, as well as neuromodulatory actions at capsaicin-sensitive sensory nerves to dampen inflammatory pain (Svensson et al. 2007).

Generation of epimeric LXs was first identified in the presence of aspirin [acetylsalicylic acid (ASA)], leading to their naming as aspirin-triggered LXs (ATL). Acetylation of COX-2 by ASA blocks PG synthesis, but the acetylated enzyme remains catalytically active, converting C20:4 to 15(R)-hydroxyeicosatetraenoic acid (15(R)-HETE). 5-LO can then utilize 15(R)-HETE for subsequent transformation to 15-epimer-LXs (Claria and Serhan 1995). ASA is not required for 15-epi-LX biosynthesis, as 15(R)-HETE can also be generated by cytochrome P450 metabolism of C20:4 (Claria et al. 1996). Of interest, atorvastatin and pioglitazone increase 15-epimer-LX production in rat myocardium (Birnbaum et al. 2006) and statins can trigger 15-epi-LX production during interactions between airway epithelial cells and PMNs *in vitro* and *in vivo* during airway inflammation (Planaguama et al., 2010).

LXs and ATLs are rapidly formed and rapidly inactivated. The enzyme 15-hydroxyprostaglandin dehydrogenase (15-PGDH) metabolizes LXs by dehydration to convert 5(S),6(R),15(S)-trihydroxy-7E,9E,11Z,13E-eicosatetraenoic acid (lipoxin A$_4$, LXA$_4$) into 15-oxo-LXA$_4$ or LXB$_4$ into 5-oxo-LXB$_4$ (Serhan et al. 1993). This inactivates the LXs, as the oxo-LXs no longer display counterregulatory bioactions. 15-Oxo-LXA$_4$ is further metabolized by eicosanoid oxidoreductase, which specifically reduces the double bond adjacent to the ketone, and then again by 15-PGDH to a 13,14-dihydro-LXA$_4$. LX metabolism is stereospecific, so 15-epi-LXs are metabolized less efficiently than LXs, increasing their biological half-life approximately twofold and thereby enhancing their ability to evoke bioactions (Serhan et al. 1995).

Of note, this biosynthetic paradigm for C20:4-derived, ASA-triggered counterregulatory lipid mediators is also observed with the omega-3 polyunsaturated fatty acids C20:5 (eicosapentaenoic acid) and C22:6 (docosahexaenoic acid) (see below).

LXA$_4$ and 5(S),6(R),15(R)-trihydroxy-7E,9E,11Z,13E-eicosatetraenoic acid (15-epi-lipoxin A$_4$, 15-epi-LXA$_4$) can interact with a specific receptor [termed ALX/FPR2 (RefSeq nucleotide NM_001462; protein NP_001453)], which is a seven-transmembrane, G-protein-coupled receptor expressed on several cell types, including human PMNs, eosinophils, monocytes/macrophages, enterocytes, synovial fibroblasts, and airway epithelium (Chiang et al. 2006). LXA$_4$ binds to ALX/FPR2 with high affinity (K_d = 1.7 nM) (Chiang et al. 2000). Of interest, ALX/FPR2 was the first receptor described that can bind to both peptide and lipid ligands (Serhan et al. 1994). This paradigm is now appreciated to be a more generalized phenomenon (Chiang et al. 2006). Another example includes CMKLR1 (RefSeq nucleotide NM_004072; protein NP_004063) that interacts with both chemerin and resolvin E1 (see below) (Arita et al. 2005a). In PMNs, ALX/FPR2 signals in part via polyisoprenyl phosphate remodeling (Levy et al. 1999) and inhibition of leukocyte-specific protein (LSP-1) phosphorylation, a downstream regulator of the p38-MAPK cascade (Ohira et al. 2004).

Diverse cytokines, such as IL-13 and IFN-γ, induce ALX/FPR2 expression (Gronert et al. 1998). Lung ALX/FPR2 expression is induced *in vivo* in a murine model of allergic airway inflammation (Levy et al. 2002). The contribution of ALX/FPR2 to counterregulatory signaling was demonstrated *in vivo* using transgenic mice that express human ALX/FPR2 directed by a component of the myeloid CD11b promoter (Devchand et al. 2003). Of note, these human ALX/FPR2-transgenic mice are protected from the development of allergic airway inflammation with markedly decreased eosinophil activation and tissue accumulation (Levy et al. 2002). ALX/FPR2-deficient mice were prepared and display ligand-specific effects on leukocyte responses and experimental inflammation (Dufton et al. 2010). Transgenic mice have facilitated the study of the ALX/FPR2 receptor signaling properties *in vitro* and *in vivo* in a wide range of experimental models of inflammation. In addition to ALX/FPR2, LXs can also interact with CysLT1 (RefSeq nucleotide NM_006639; protein NP_006630) receptors to antagonize CysLT binding (see below) (Gronert et al. 2001). These findings emphasize the notion that LXs can serve as agonists (via ALX) to promote resolution and as antagonists (at CysLT1) to block prophlogistic signaling.

9.2.2 RESOLVINS

9.2.2.1 C22:6: D-Series Resolvins

In addition to C20:4-derived LXs, omega-3 fatty acids can also be enzymatically transformed to anti-inflammatory and proresolving mediators, including resolvins (*resolution phase interaction products*) (Serhan et al. 2000, 2002), protectins, and maresins. Resolvins are classified by their parent fatty acid, namely D-series resolvins from docosahexaenoic acid (DHA, C22:6) or E- series resolvins from eicosapentaenoic acid (EPA, C20:5). Generation of these compounds can be increased by dietary modification and in mice transgenic for the *fat-1* gene [GeneID: 178291 (NC_003282.5)], which encodes a prokaryote omega-3 fatty acid desaturase (Connor et al. 2007, Hudert et al. 2006).

DHA is an omega-3 fatty acid found abundantly in neural and mucosal tissues and serves as a substrate for enzymatic conversion to anti-inflammatory and proresolving mediators (Figure 9.1). In addition to tissue stores, DHA is delivered during the early stages of an acute inflammatory response in albumin-rich edematous fluid (Kasuga et al. 2008). DHA can be metabolized via COX-2 to generate the D-series resolvins (e.g., Resolvin D1–D4) (Hong et al. 2003). These D-series resolvins can inhibit PMN infiltration and migration and have been found to be protective in murine models of ischemia/reperfusion (Duffy et al. 2008, Serhan et al. 2006, Sun et al. 2007). Stereochemical conformation of RvD1 and RvD2 have been assigned as 7*S*,8*R*,17*S*,-trihydroxy-4*Z*,9*E*,11*E*,13*Z*,15*E*, 19*Z*-DHA and 7*S*,16*R*,17*S*-trihydroxy-docosa-4*Z*,8*E*,10*Z*,12*E*,14*E*,19*Z*-hexaenoic

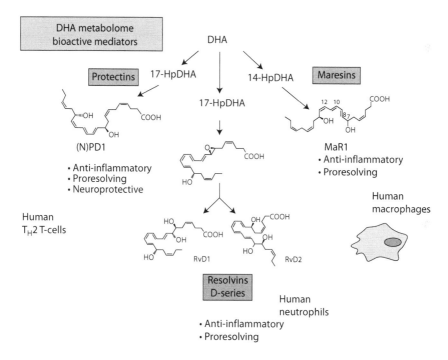

FIGURE 9.1   DHA metabolome: bioactive mediators. The polyunsaturated fatty acid DHA (*n*–3, C22:6) is enzymatically converted to proresolving mediators in a cell-type- and context-specific manner. Protectin D1 is generated from DHA by lipoxygenase activity, proceeding via a 17-H(peroxy)-DHA intermediate. PD1 potently inhibits activated T cells, and when generated in neural tissues, the compound is named neuroprotectin D1 (NPD1) in recognition of its potent neuroprotective actions. The anti-inflammatory and proresolving D-series resolvins (RvD1 and RvD2) are also derived from the 17-H(peroxy)-DHA intermediate, and inhibit neutrophil activation and tissue accumulation. Maresins are the newest members of the DHA metabolome. Maresin 1 (7*R*,14*R*-dihydroxy-4*Z*,8*E*,10*Z*,12*Z*,16*Z*,19*Z*-docosahexaenoic acid, MaR1) is generated by activated macrophages and displays proresolving actions.

acid, respectively (Spite et al. 2009, Sun et al. 2007). The determination of the complete stereochemistry of RvD3 and RvD4 is the subject of ongoing research. Similar to 15-epi LXs, enantiomers of the resolvins can be generated in the presence of ASA (Serhan 2007). Aspirin-triggered resolvins of the D-series (AT-RvD1–D4) are formed from DHA by acetylated COX-2 to generate 17(R)-H(peroxy)DHA that can be further transformed to two epoxide intermediates; 7(S)-H(peroxy),17(R)-HDHA and 4(S)-H(peroxy),17(R)-HDHA for subsequent conversion to the bioactive mediators AT-RvD1–D4. All four of these ASA-triggered D-series resolvins stop PMN infiltration *in vivo* (Serhan 2007). The stereochemical assignment for AT-RvD1 is 7S,8R,17R-trihydroxy-4Z,9E,11E,13Z,15E,19Z-DHA (AT-RvD1) (Spite et al. 2009, Sun et al. 2007). Determination of the complete stereochemistry of AT-RvD3 and AT-RvD4 is also the subject of ongoing research. Only limited information currently exists on D-series resolvin receptors; however, resolvin D1 can serve as a ligand at ALX/FPR2 and GPR32 (RefSeq nucleotide NM_001506; protein NP_001497) receptors (Krishnamoorthy et al. 2010).

9.2.2.2   C22:6: Protectins

Another family of DHA-derived proresolving mediators is the protectins. These compounds can be distinguished from D-series resolvins by the presence of a conjugated triene double bond (Serhan et al. 2006). Protectin D1 (10R,17S-dihydroxy-docosa-4Z,7Z,11E,13E,15Z,19Z-hexaenoic acid, PD1) is the lead member of this class (Serhan et al. 2006). When produced by neural tissues, they are termed neuroprotectins to affix the site of origin and have many anti-inflammatory actions (Lukiw et al. 2005, Mukherjee et al. 2004).

9.2.2.3   C22:6: Maresins

The newest family of anti-inflammatory and proresolving mediators is the maresins (named for macrophage mediators in resolving inflammation). Maresins are 7,14 dihydroxy-containing products that are generated by activated macrophages (Serhan et al. 2009). DHA is delivered to inflamed or injured tissue by plasma exudation (Kasuga et al. 2008) and can be enzymatically converted by macrophages to maresins to decrease the acute inflammatory response (Serhan et al. 2009). These novel proresolving compounds also block PMN trafficking and stimulate macrophage clearance of apoptotic PMNs (Serhan et al. 2009). Roles for maresins in lung biology are areas of active investigation. The presence of large numbers of alveolar macrophages in the lung and their critical role in tissue catabasis and host defense suggest important functions for maresins in the regulation of airway inflammation.

9.2.2.4   C20:5: E-Series Resolvins

The first molecular family of resolvins derived from EPA was identified by unbiased lipidomic analyses of resolving exudates, namely the E-series resolvins (Serhan et al. 2000, 2002). Two members of the E-series resolvins have been characterized and are named resolvin E1 (RvE1) and resolvin E2 (RvE2). Both are generated in mammalian tissues *in vivo* (Serhan et al. 2000, 2002). In the vasculature, transcellular synthesis of RvE1 proceeds in the presence of aspirin with the transformation of C20:5 to 18R-HEPE

(18*R*-hydroxyeicosapentaenoic acid) by aspirin-acetylated COX-2 in endothelial cells. 18*R*-HEPE is subsequently converted by leukocyte 5-LOX to RvE1 via a 5(6) epoxide-containing biosynthetic intermediate (Arita et al. 2005a, Serhan et al. 2000). RvE1 levels are increased in humans taking aspirin and/or following dietary EPA supplementation (Arita et al. 2005a). The stereochemical assignment for RvE1 is 5*S*,12*R*,18*R*-trihydroxyeicosa-6*Z*,8*E*,10*E*,14*Z*,16-EPA (Arita et al. 2005a, Serhan et al. 2000). Functionally, RvE1's actions are highly stereoselective both *in vivo* and *in vitro* (Arita et al., 2005a,b). RvE2 (5*S*,18(*R/S*)-dihydroxy-8*Z*,11*Z*,14*Z*,16*E*-eicosapentaenoic acid) is a structurally distinct member of the E-series resolvins, yet it displays similar bioactivities as RvE1 (Tjonahen et al. 2006). Administration of RvE1 and RvE2 in nanogram amounts exhibits additive cytoprotective effects (Tjonahen et al. 2006).

Resolvin E1 has potent anti-inflammatory/proresolution effects in a number of murine models of inflammation and is bioactive in very low concentrations (nanomolar to picomolar) *in vivo* and *in vitro* (Serhan et al. 2000). RvE1 serves as a ligand for the G-protein-coupled receptor CMKLR1 that is expressed on myeloid cells (Arita et al. 2005a). CMKLR1 was earlier identified as a receptor for the peptide mediator chemerin (Wittamer et al. 2003). RvE1 is also an agonist for this receptor conveying cell-type-specific counterregulatory actions. RvE1–CMKLR1 interactions attenuate NF-κB activation in response to proinflammatory cytokine signaling (Arita et al. 2005a). Of interest, CMKLR1 null mice have increased inflammation in a murine model of peritonitis, demonstrating the importance of counterregulatory signaling via this receptor (Cash et al. 2008). RvE1 can also act as a receptor-level antagonist at BLT1 (RefSeq nucleotide NM_181657; protein NP_858043), a receptor for the proinflammatory lipid mediator LTB_4 (Arita et al. 2007). Together, these findings indicate that as an agonist for CMKLR1 and antagonist for BLT-1, RvE1 can mediate proresolving actions. When given together, RvE1 and RvE2's protective actions are additive at low doses, suggesting distinct receptors for RvE2 and RvE1 (Tjonahen et al. 2006). A cognate receptor for RvE2 has yet to be identified, but is the subject of ongoing research (Tjonahen et al. 2006).

9.3   EVIDENCE FOR A ROLE IN AIRWAY INFLAMMATION

Chronic diseases of airway inflammation are common conditions without curative therapy and lead to excess morbidity. Examples include asthma, cystic fibrosis, and chronic obstructive pulmonary disease. We have become accustomed to thinking of inflammatory diseases as conditions precipitated by an overabundance of provocative stimuli. In contrast, an emerging body of evidence supports a new view that these conditions can also result from defective counterregulatory signaling. Several lines of evidence support important roles for endogenous proresolving lipid mediators in healthy tissue responses to limit airway inflammation (Table 9.1).

Specialized proresolving mediators are generated in the lung during a broad range of respiratory illnesses characterized by inflammation (Lee et al. 1990) (Figure 9.2). Lipoxins are present in asthmatic lung (Lee et al. 1990) and can prevent LT-mediated smooth muscle migration (Parameswaran et al. 2007) and bronchoconstriction

TABLE 9.1
Proresolving Mediators in Airway Inflammation Mediator

	Basic Biology	Preclinical	Clinical

Lipoxins/15-epi-lipoxins	Inhibit PMNs:
	Chemotaxis
	Transmigration
	Degranulation
	O_2^- generation
	PLD
	LSP-1
		Murine:
		Allergic airway
		  inflammation
		Acute lung injury (acid)
		Interstitial lung disease
		  (bleomycin)
		Lung second organ
		  injury (I/R)
		Allergic edema
			Asthma
			Interstitial lung disease
			Sarcoidosis
			Pneumonia
			Cystic fibrosis
			Lung transplant
			  rejection
			Exudative pleural
			  effusion

	Inhibit eosinophil:
	Chemotaxis
	Stimulate monocyte:
	Adhesion
	Transmigration
	Stimulate
	  macrophage:
	Phagocytosis of
	  apoptotic cells
		Guinea pig:
		Airway tissues
		Human:
		Bronchi
		Nasal polyps	

D-series resolvins/
  AT-resolvins
	Inhibit PMNs:
	Transmigration
		Murine:
		RvD1—acute lung
		  injury (I/R)
		RvD2—sepsis	

	Inhibit capsaicin- sensitive nerves: Inflammatory pain		
Protectins	Inhibit T cells: Migration	Murine: Allergic airway inflammation	Asthma

E-series resolvins	Inhibit PMNs:
	Chemotaxis
	Retention on mucosal
	  surfaces (CD55)
		Murine:
		Allergic airway
		  inflammation
		Acute lung injury
		Bacterial pneumonia	

	Inhibit dendritic cells: Cytokine release Migration Inhibit capsaicin- sensitive nerves: Inflammatory pain		

(Christie et al. 1992). For leukocytes, LXs are potent "stop" signals for PMN (limiting their entry into inflamed tissues) and "go" signals for monocyte locomotion and macrophage phagocytosis of apoptotic PMNs (Serhan et al. 2007). By blocking further PMN or eosinophil entry and enhancing macrophage-mediated clearance of inflamed tissues, LXs promote resolution of lung inflammation.

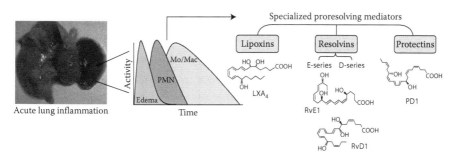

FIGURE 9.2 Generation of specialized proresolving mediators during acute lung inflammation. During acute lung inflammation in response to pathogens, injury, or noxious stimuli (including allergen), leukocytes interact with tissue resident cells to form biosynthetic circuits for specialized mediators. These mediators display both anti-inflammatory and proresolving actions that restrain the lung inflammation and promote tissue catabasis. Several distinct families of specialized proresolving mediators have been elucidated, including lipoxins (LXA_4), E-series resolvins (RvE1), D-series resolvins (RvD1), and protectins (PD1) (the structure of a representative member of each family is shown).

Experimental animal models of asthma have provided several important insights into the impact of LX signaling on the regulation of allergic airway inflammation and hyperresponsiveness. Using chicken ovalbumin as an allergen, mice can be sensitized and then aerosol-challenged to direct the allergic responses to the airway. In this setting, LX analogs that resist metabolic inactivation display potent inhibition of the development of airway inflammation and hyperresponsiveness to methacholine (Levy et al. 2002, 2007). Both eosinophil and T-cell accumulation are dampened by the LX analog and levels of the important T_H2 cytokines, interleukin-5 (IL-5) and IL-13, are markedly decreased (Levy et al. 2002, 2007). LX analogs can be administered by intravenous, intraperitoneal, and oral routes and display an IC_{50} of ~0.05 mg/kg (Levy et al. 2002, 2007). In a distinct murine model of asthma based on a clinically relevant allergen for human asthma, cockroach antigen can be used to sensitize and airway-challenge mice to generate a robust peribronchial allergic inflammatory infiltrate and airway hyperresponsiveness (Levy et al. 2007). LX analogs are similarly potent regulators of this model (Levy et al. 2007). LXs and their stable analogs also prevent PMN-mediated tissue injury (reactive oxygen species), enhance bronchial epithelial cell proliferation in response to acid injury, and block TNF-α signaling for the release of proinflammatory cytokines from epithelial cells (Bonnans et al. 2006a, Fukunaga et al. 2005).

When animals are no longer exposed to allergen, there is a natural resolution of the allergic airway inflammation that is notable for increased endogenous LXA_4 generation in the airways (Haworth et al. 2008). Of interest, when animals are given a bioactive LX analog at peak inflammation, resolution of the allergic airway responses is accelerated, in part by inhibition of IL-17 (Haworth et al. 2008). In several experimental model systems, disruption of LX biosynthesis is resolution "toxic" for acute inflammation (Fukunaga et al. 2005, Schwab et al. 2007), consistent with pivotal roles for LXs in the natural restitution of airway homeostasis.

RvE1 displays potent *in vivo* regulation of mucosal inflammation at many levels of the aerodigestive tract. In a rabbit model of periodontitis, RvE1 both dampens inflammation and promotes restoration of periodontal tissues, including bone (Hasturk et al. 2006, 2007). RvE1 also markedly decreases the inflammatory sequellae of 2,4,6-trinitrobenzene sulfonic acid-induced colitis (Arita et al. 2005b) and can facilitate the resolution of allergic airway inflammation (Haworth et al. 2008). At mucosal surfaces, RvE1 enhances clearance of PMN from the apical surfaces of epithelial cells (Campbell et al. 2007). In comparison with other agents that are clinically available to decrease inflammation, RvE1 proved log-orders more potent than dexamethasone or aspirin in a murine dorsal air pouch model of dermal wounding and inflammation. In nanogram quantities, RvE1 decreases leukocyte infiltration by 50–70%, whereas the IC_{50} for dexamethasone and aspirin in this model are in the microgram and milligram range, respectively (Arita et al. 2005a, Serhan et al. 2000). RvE1 also displays important actions on structural cell functional responses, including the facilitation of wound healing by epithelial cells (Gronert et al. 2005) and augmentation of host defense via increased intestinal alkaline phosphatase expression and activity to detoxify endotoxin (Campbell et al. 2010). Thus, RvE1 displays characteristics *in vitro* and *in vivo* of a mediator for catabasis.

PD1 is generated in both human and murine lung. In a model of OVA-induced allergic airway inflammation, PD1 administered in nanogram quantities decreases eosinophil trafficking into the lung and dampens airway hyperresponsiveness, two of the clinical hallmarks of asthma. When PD1 was given at the peak of inflammation, it also accelerated the resolution of allergic airway inflammation (Levy et al. 2007). In a head-to-head comparison between ATLa and RvE1 in acute, self-limited murine peritonitis, PD1 displays the lowest ψ_{max} (maximum number of infiltrating leukocytes) and T_{max} (the time taken to resolve) and the shortest resolution interval [the time taken for maximal leukocyte infiltration to reduce by 50% (R_i)], indicative of PD1's potent bioactions (Bannenberg et al. 2005).

9.4   DRUG DISCOVERY EFFORT AROUND THE TARGET FOR AIRWAY DISEASE

With the discovery of LXs, resolvins, and protectins as mediators of anti-inflammation and resolution, potential lead compounds were uncovered for the development of novel therapeutic strategies. While there have been many agents developed to block inflammatory mediator pathways, the LXs, resolvins, and protectins are natural agonists for cognate receptors that initiate an array of signaling pathways for catabasis. If successfully developed as new drugs, chemical mimetics as agonists of natural resolution pathways would represent a new genus of therapeutic agents (Serhan and Chiang. 2008).

9.4.1   *In Vitro Biology*

Proresolving mediators can serve as agonists at specific receptors. The signaling properties of these mediators are emerging from *in vitro* studies of leukocytes and structural cells of the airway. With multiple receptors to transduce the potent proresolving actions of these compounds, the responses are cell-type and tissue specific. The transcription factor NF-κB is a critical regulator of innate immune responses. A wide range of proinflammatory cytokines, chemokines, adhesion molecules, and enzymes are regulated by the NF-κB pathway (Barnes and Karin 1997). In addition, NF-κB repression has been identified as a major inducer of granulocyte apoptosis (Fujihara et al. 2002, Ward et al. 1999), an integral part of the resolution of inflammation that is linked to the pathogenesis of chronic inflammatory lung diseases, including asthma and chronic obstructive pulmonary disease (COPD) (Barnes and Karin 1997, Hart et al. 1998). Hence, NF-κB signaling pathways represent a therapeutic target to manipulate apoptotic programs and consequently facilitate resolution (Gilroy et al. 2004, Rossi et al. 2007, Sawatzky et al. 2006). Several lines of evidence indicate that LXs and resolvins can modulate NF-κB signaling. For instance, RvE1 can serve as an agonist at CMKLR1 to block TNF-α-induced activation of NF-κB in a concentration-dependent manner (Arita et al. 2005b). RvE1 can also attenuate proinflammatory LTB$_4$-mediated BLT1 signaling by decreased activation of NF-κB (Arita et al. 2007). In primary osteoclast cultures derived from murine bone marrow, RvE1 acts via BLT1 receptors to inhibit the nuclear translocation of the p50 subunit of NF-κB (Herrera et al. 2008). In a murine model of pneumonia, RvE1 inhibits the translocation and activation of NF-κB (p65) in lung tissue homogenates (Seki et al. 2010). Collectively, these findings point to NF-κB as an important downstream site of action for resolvins to modulate inflammatory responses and signal for resolution.

Signal transduction pathways via the phosphatidylinositol 3-kinase (PI3-K) and the MAPK family member extracellular signal-regulated kinase (ERK) also have critical roles in proinflammatory intracellular signaling events (Barnes 2008). PI3-Ks are a family of enzymes that generate phosphatidylinositol-3,4,5-trisphosphate (PtdIns(3,4,5)P$_3$), a lipid second messenger critical for the localization and function of the downstream serine/threonine kinase, Akt (protein kinase B) (reviewed in Vanhaesebroeck et al. 2001). Intracellular signaling via the PI3-K/Akt pathway plays an important role in regulating many cellular processes, especially granulocyte migration, activation, and survival (Bonnans et al. 2006b, Cantley 2002, Hawkins et al. 2006). Similarly, the ERK 1/2 pathway represents a point of convergence for multiple signaling events, thereby regulating important cellular functions and fate of human neutrophils (Downey et al. 1998, Nolan et al. 1999, Sawatzky et al. 2006). RvE1–CMKLR1 interactions initiate the activation of PI3-K resulting in phosphorylation of Akt and ribosomal S6 protein via mTOR signaling (Ohira et al. 2010). RvE1 activation leads to ERK phosphorylation that is inhibited by the pan-selective PI3-K inhibitor wortmannin and a specific ERK 1/2 inhibitor PD98059 (Ohira et al. 2010). *In vitro*, RvE1 exerts direct protective actions on cardiomyocytes by regulating apoptotic programs in cells exposed to hypoxia or hypoxia/reoxygenation (Keyes et al. 2010). It does so by increasing phosphorylation of Akt, ERK1/2, and eNOS and attenuating the levels of key proapoptotic proteins, including caspase-3 and Bax. Of interest, RvE1 attenuates inflammatory pain via a glutamate ionotropic receptor mechanism in spinal dorsal horn neurons through inhibition of the ERK signaling pathway (Xu et al. 2010). Thus, resolvins can regulate these pivotal signal transduction pathways that coordinate cell activation in inflammation.

LXA$_4$ also regulates cell activation by inhibiting polyisoprenyl phosphate remodeling (Levy et al. 1999). In freshly isolated PMNs, presqualene diphosphate (PSDP) is present in cell membranes. Upon cell activation, PSDP is rapidly converted to its monophosphate, presqualene monophosphate (PSMP) (Levy et al. 1997). This remodeling facilitates cellular responses because PSDP, but not PSMP, blocks important signaling enzymes, namely PI3-K and phospholipase D (PLD) (Bonnans et al. 2006b, Levy et al. 1999, 2005b), as well as O$_2^-$ generation from human PMNs. LX stable analogs dramatically block PMN PSDP remodeling in response to proinflammatory agonists (Levy et al. 1999). LX-mediated inhibition of PSDP remodeling is linked to its anti-inflammatory effects on PMN functional responses, including O$_2^-$ generation Levy et al. 1999, 2005b. Polyisoprenyl diphosphate phosphatase 1 (PDP1) (originally identified as CSS2α and PPAPDC2) serves as a pivotal phosphatase for PSDP remodeling to PSMP (Carlo et al. 2009, Fukunaga et al. 2006) and represents an intracellular signaling target for LXs to block PSDP remodeling. In addition to PDSP remodeling, LX stable analogs can also regulate PMN activation by decreasing agonist-initiated phosphorylation of leukocyte-specific protein 1 (Ohira et al. 2004) and several other components of the p38-MAPK pathway (MAPK kinase 3/MAPK kinase 6, p38-MAPK, MAPK-activated protein kinase-2) (Ohira et al. 2004). Inhibition of the p38-MAPK cascade by LXs is likely responsible, in part, for their regulation of PMN chemotaxis.

9.4.2 PRECLINICAL EVIDENCE

Because LXA$_4$ is rapidly metabolized, LX stable analogs have been prepared for preclinical studies to probe *in vivo* mechanisms for LX signaling and provide more potent pharmacological actions than the native compound (Petasis et al. 2005, Serhan et al. 1995). The first generation of LXA$_4$ analogs were designed to resist metabolic inactivation at carbon 15 and displayed potent bioactivity *in vivo* in several murine model systems of acute inflammation (Petasis et al. 2005). To block the first metabolic step for LXA$_4$, structural analogs were initially designed to target carbon 15 by either addition of a methyl group or substitution of the alkyl tail of the molecule with a bulky phenoxyl group. To prevent dehydrogenation of LXs *in vivo*, a halide was added to the para position of the phenoxyl group (Clish et al. 1999). An example of a bioactive first-generation LX analog is 15-epi-16-(*para*-fluorophenoxy)-LXA$_4$ (also known as ATLa in the literature) that has been investigated in several settings and found to be efficacious *in vivo* in many models of inflammation. With further recognition of *in vivo* LX metabolism, a second generation of stable LXA$_4$ analogs, the 3-oxa-LXA$_4$ analogs, were introduced to resist *in vivo* β-oxidation (Guilford and Parkinson 2005). These 3-oxa-LXA$_4$ analogs also display similar anti-inflammatory and proresolving properties *in vivo* in several murine models of acute inflammation. Both first- and second-generation analogs of LX and 15-epi-LX administered to OVA-sensitized mice prior to OVA aerosol challenge block the development of allergic airway inflammation at doses as low as 1 μg (~0.05 mg/kg) (Levy et al. 2002).

RvE1 is metabolically inactivated in a rapid and regiospecific manner. The first step in its inactivation is a regiospecific conversion at carbon 18 to generate 18-oxo-RvE1 (Arita et al. 2006). RvE1 inactivation is complex and shows species-, tissue-, and cell-type-specific pathways (Arita et al. 2006). Rapid inactivation of RvE1 proceeds in a regulated manner to provide local control for cells and tissues in catabasis. In addition, elucidation of its metabolic pathways has informed the design of structural analogs of RvE1 that resist inactivation (Arita et al. 2006). The first stable RvE1 analog has been prepared by chemical modification at the ω-end of RvE1 to block metabolic inactivation (Arita et al. 2006). This RvE1 analog, 19-(*p*-fluorophenoxy)-RvE1 methyl ester, displays protective actions for acute inflammation in a murine experimental model of periodontitis and resists ω-oxidation and rapid dehydrogenation (Arita et al. 2006).

PD1 can serve as a potent regulator of allergic airway inflammation and broncho-provocation (Levy et al. 2007). In nanogram quantities, PD1 prevents the development of airway hyperreactivity and markedly inhibits infiltration of the lung eosinophils and T cells as well as mucus metaplasia (Levy et al. 2007). Of even more relevance for clinical asthma, when PD1 is given as a treatment after allergic airway inflammation is established, resolution of this experimental model of asthma is accelerated (Levy et al. 2007). Similar tissue-protective actions for PD1 are observed in renal tissues. After ischemia–reperfusion injury, murine kidneys generate PD1 (Duffield et al. 2006) and display renoprotective actions.

9.4.3 CLINICAL EVIDENCE

Asthma is a prototypical example of chronic airway inflammation. Despite several available therapies to dampen inflammation, bronchodilate, and prevent broncho-constriction, asthma still has a high morbidity in many parts of the world (Barnes 2004, Busse and Lemanske 2001). Of note, no therapies are available to completely resolve asthma.

LXA_4 (0.4–2.8 ng/mL) is found in bronchoalveolar lavage fluid (BALF) from individuals with lung diseases, including asthma (Lee et al. 1990). LXA_4 has also been detected in the pleural fluid of patients with systemic and pulmonary inflammation (Levy et al. 2001). LXs block LTD_4-mediated constriction of isolated lung strips *in vitro* (Dahlen et al. 1987) and nebulized LXA_4 can protect asthmatic subjects during airway provocation challenges with inhaled LTC_4 (Christie et al. 1992), indicating that LXA_4 displays protective actions in human asthmatic airways. In this study, there were no deleterious side effects on normal physiology, including blood pressure. To date, this proof of principle study remains the only reported human clinical trial of LXs (Christie et al. 1992).

Lung biopsies and bronchoalveolar lavage fluids from most patients with uncontrolled or severe asthma are notable for increased leukocyte accumulation, in particular eosinophils and PMNs (Fahy et al. 1995, Wenzel et al. 1997). Because LXs are such potent regulators of granulocyte activation and trafficking, these pathological findings suggest a defect in LX formation or action. To determine LX biosynthetic capacity in asthma, peripheral venous whole blood was obtained from individuals with severe asthma and mild to moderate asthma was exposed to the divalent cation ionophore A23187 at concentrations known to activate phospholipase A_2 and 5-LO (Levy et al. 2005a). The results uncovered a dramatic difference in the amounts of LXs generated *in vitro* (Levy et al. 2005a). Blood from individuals with mild to moderate asthma displayed a marked increase in LX generation (~2 ng LXA_4/mL) compared to healthy subjects (not detectable). This increase in LX biosynthesis in mild asthma was not shared by samples of activated blood from subjects with severe asthma that had significantly lower LX biosynthetic capacity (~0.5 ng LXA_4/mL). To identify the enzyme responsible for decreased LX formation, phospholipase A_2, 5-LO and 15-LO activities were measured by quantitation of LTs and C20:4 monooxygenation products. In contrast to decrements in LXs, 5-LO-derived products, including LTB_4 and CysLTs, were increased two- and four-fold, respectively, in activated whole blood from severe asthma subjects (Levy et al. 2005a). 5-HETE was also increased approximately twofold, while 15-HETE levels were decreased approximately threefold in the severe asthma-activated whole blood, indicative of increased 5-LO and decreased 15-LO activity. Evidence for the potent airway actions of these eicosanoids was also confirmed, as CysLTs levels were inversely correlated and LXA_4 levels positively correlated with lung function (as measured by the forced expiratory volume in 1 s (FEV_1) (percent predicted values). In addition, levels of these mediators were also determined in nonstimulated whole blood (Levy et al. 2005a). Circulating LXA_4 levels were lower in severe asthma and normalized values of LXA_4 to CysLTs in nonstimulated whole blood was also correlated with airflow obstruction. Together, these findings were consistent with a fundamental difference in arachidonic acid metabolism in whole blood from individuals with asthma of different severity, and point to 15-LO as a pivotal regulator of LX formation.

To determine whether these changes in whole blood were reflective of asthmatic airways, BALFs were obtained from severe and nonsevere asthma subjects. Compared to healthy subjects, 5-LO-derived CysLTs and 15-LO-derived 15-HETE levels were increased in individuals with asthma, independent of severity (Planaguma et al. 2008). Of interest, levels of LXA_4 were again markedly decreased in severe asthma BALFs (mean: 11.2 pg LXA_4/mL BALF) relative to nonsevere asthma (mean: 150.1 pg LXA_4/mL BALF). In addition, similar to whole blood, a decreased ratio of BALF LXA_4 to CysLTs sharply distinguished severe from not severe asthma. These findings extend to the respiratory tract the earlier findings of dysregulated LX biosynthesis in severe asthma whole blood.

Unlike CysLTs or 15-HETE, LX biosynthesis requires the actions of more than one LO. Transcellular LX biosynthesis occurs most efficiently during cell–cell interactions and decreases when direct cell contact is prevented (Mayadas et al. 1996). 5-LO activity is consistently increased in asthma, and successful therapeutics have been developed to inhibit 5-LO or CysLT signaling (Drazen et al. 1999). In contrast, 15-LO is a key enzyme capable of both initiating LX biosynthesis as well as converting 5-LO-derived LTA_4 to LXs (Serhan 2007). No therapies are currently available to modulate 15-LO activity. Increased expression of 15-LO in experimental animal systems and human subjects attenuates inflammation, in part via increased LX formation (Serhan et al. 2003, Wittwer et al. 2007).

Altered eicosanoid metabolism in severe asthma is related in part to changes in expression of LX biosynthetic genes (Planaguma et al. 2008). Materials obtained from whole blood, endobronchial biopsy (EBB) (large airways), and BAL cells (peripheral airways) were analyzed by quantitative PCR for 5-LO and both 15-LO isoforms to explore LX biosynthetic gene expression in different anatomical compartments in addition to asthma severity. Of note, expression of 15-LO-1 mRNA was decreased in BALF cells in severe asthma (Planaguma et al. 2008). In this report, BALF levels of LXA_4 were decreased 15-fold in severe asthma. Yet, changes in the expression of 15-LO-1 and 15-LO-2 combined in BALF cells revealed only a sixfold decrease, indicating the involvement of posttranscriptional factors to the decreased LXA_4 generation in severe asthma airways. Unlike 15-HETE, LX biosynthesis requires both 15-LO and 5-LO activity. 5-LO-derived CysLTs in BALF were increased in both severe and not severe asthma subjects, correlating with results from 5-LO gene expression. Because biosynthetic intermediates are shared for LTs and LXs, there is a reciprocal relationship between LT and LX formation that can be influenced by redox state as well as several other factors (Fiore and Serhan 1990). Thus, altered LT and LX generation in severe asthma appears to relate to changes in eicosanoid biosynthetic gene expression, cell–cell interactions, and redox state; any one of which can disrupt LX biosynthesis without also concurrently decreasing 15-HETE or LT production.

Patients with severe asthma differ from nonsevere cohorts by exposure to larger amounts of systemic and inhaled corticosteroids (Levy et al. 2005, Moore et al. 2007). Relationships between corticosteroids and either 15-LO expression or LXA_4 have not been identified in severe asthma. While not seen in the respiratory compartments sampled or with oral corticosteroids, there was a significant correlation in nonsevere asthma whole blood between inhaled corticosteroids and decreased 15-LO-1 and 15-LO-2 expression (Planaguma et al. 2008), a relationship potentially explained by corticosteroid-induced decreases in viable circulating leukocytes, in particular eosinophils and cytokine-primed monocytes that can express 15-LO (Kuhn et al. 1999). Together, these findings raise the possibility that corticosteroid therapy may adversely impact LX biosynthesis. Of interest, corticosteroids can also potentiate this counterregulatory signaling pathway by increasing ALX/FPR2 and annexin A1 expression (Hashimoto et al. 2007, Perretti and D' Acquisto 2009).

In addition to 5-LO and 15-LOs, cyclooxygenase-2 (COX-2) is also pivotal to the resolution of airway injury and inflammation, in part via PG-mediated induction of 15-LO-1 and LX formation (Fukunaga et al. 2005, Levy et al. 2001). Of note, COX-2 expression was down-regulated in all samples from subjects with severe asthma (Planaguma et al. 2008). Because COX-2 expression is sensitive to glucocorticoids, an indirect effect of steroids on LX production by decreased COX-2 expression is also possible.

Induced sputum from individuals with severe asthma also contains lower concentrations of LXA_4 than do samples from mild or moderate asthmatic subjects (Vachier et al. 2005), indicating similar changes in LX production in both upper and lower airways. Many individuals with severe asthma are aspirin-sensitive (Moore et al. 2007) and similar to the larger severe asthma cohort, activated whole blood from carefully phenotyped aspirin-intolerant asthmatics displays relative decrements in LXA_4 biosynthesis (Sanak et al. 2000). Recently, lower levels of LXA_4 were also associated with exercise-induced bronchoconstriction in asthma (Tahan et al. 2008), raising the possibility that these changes in C20:4 metabolism may also be operative during asthma exacerbations. Together, analyses of patient-derived materials from cohorts with asthma from diverse cultures on three continents have all confirmed a defect in LX generation in uncontrolled and severe asthma in adults and children (Celik et al. 2007, Levy et al. 2005, Planaguma et al. 2008, Vachier et al. 2005, Wu et al. 2010).

Alterations in arachidonic acid metabolism have also been observed in other chronic inflammatory airway diseases. Low levels of LXs and a defect in LX signaling have been linked to excess PMN-rich inflammation in the airways of patients with cystic fibrosis (Karp et al. 2004). In addition, lower LX levels are related to persistent inflammation during the progression of scleroderma lung disease (SLD) (Kowal-Bielecka et al. 2005). In SLD, BALF LTB_4 and LTC_4 levels are increased disproportionately to LXA_4 levels that are decreased (Kowal-Bielecka et al. 2003). Similar to severe asthma, overproduction of LTB_4 with underproduction of LXs in SLD would lead to a proinflammatory airway typical of fibrosing alveolitis.

Together, these recent results suggest that uncontrolled and more severe variants of asthma and other diseases associated with airway inflammation may result from a defect in LX-mediated counterregulatory signaling that is present in whole blood and large and small airways. The genesis of the decreased LX generation is likely multifactorial with some of the defect attributable to changes in gene expression and some reflective of posttranscriptional regulation, perhaps by microenvironmental factors in the lung or current therapeutic agents. A defect in the generation of LXs in severe asthma would lead to unregulated airway inflammation and relatively unopposed LT-mediated bronchoconstriction, findings characteristic of the disease.

Epidemiological evidence supports a beneficial influence of diets with increased amounts of cold water fish, enriched in omega-3 fatty acids, on the development of asthma (Schwartz and Weiss 1994). Recent results in murine model systems of asthma have identified potent actions for RvE1 in blocking the development of experimental asthma and facilitating its resolution in part by increasing endogenous LXA_4 generation (Haworth et al. 2008). RvE1 is present in human tissues (Arita et al. 2005a). Clinical results for this mediator in asthma are likely to be obtained in the near future.

Airway mucosal tissues contain substantial amounts of DHA (Freedman et al. 2004). Protectin D1 is a 15-LO-derived product of DHA that is generated in human asthma and displays potent actions—reducing both airway inflammation and hyperresponsiveness in experimental asthma (Levy et al. 2007). Of note, protectin D1 and its biosynthetic precursor 17(*S*)-HDHA are present in exhaled breath condensates from volunteer human subjects and levels of protectin D1 are lower during asthma exacerbations (Levy et al. 2007), suggesting, together with the LX decrements in severe asthma, an overall decrease in the capacity to generate 15-LO-derived counterregulatory lipid mediators during uncontrolled asthma.

9.5   EXPERT OPINION

Several common diseases, such as asthma, are characterized by airway inflammation and still have important unmet clinical needs. Currently, the principal anti-inflammatory agents used clinically are corticosteroids, but in most patients, they lead to serious side effects with chronic use (Barnes 2004). No therapies, including corticosteroids or biologics, such as anti-IgE or anti-IL-5, are available to modify the evolution of airway diseases and cure these conditions. New therapeutic approaches are needed.

An understanding of natural proresolving mediators enlisted by healthy tissues to resolve inflammation and injury represents a fundamentally new approach to uncovering mechanisms to control host innate and adaptive immunity. The polyunsaturated fatty acid-derived proresolving mediators reviewed in this chapter are natural small molecules that, along with their bioactive, metabolically stable structural analogs, offer promise as a new genus of therapeutic agents for asthma and other chronic inflammatory conditions of the airway.

Proof of this concept has been provided by LXA_4 and RvE1 mimetics. For example, the first-generation analog ATLa has a rapid oral absorption and a relatively short half-life in plasma, cleared within 15 min after intravenous injection in the mouse (Clish et al. 1999). Additional metabolic studies identified modification of ATLa *in vivo* by β-oxidation in plasma (Guilford and Parkinson 2005). To improve its *in vivo* half-life, second-generation 3-oxa-LXA_4 analogs were developed (Guilford and Parkinson 2005) and their properties to dampen inflammation were tested in a murine model of allergic airway inflammation. First- and second-generation LXA_4 analogs were compared to a cysteinyl leukotriene (CysLT1) receptor antagonist (montelukast), commonly used in asthma treatment, and distinct mechanisms of actions were identified between these compounds (Levy et al. 2007). The 3-oxo-LXA_4 analog (ZK-994) was comparable to ATLa in inhibiting eosinophil and lymphocyte trafficking to the lungs. The regulation of T_H2 cytokines and lymphocyte trafficking was less potent with montelukast than the LX analogs. Cockroach allergen is a common clinical allergen trigger in human asthma (Busse and Lemanske 2001) and the LX analogs, given either intraperitoneally or in the drinking water, potently inhibited allergic airway inflammation and hyperresponsiveness to methacholine (Levy et al. 2007). In addition, CCL5, CCL11 (eotaxin), and CCL2 (MCP-1) release were blocked by the LX analogs. The second-generation 3-oxo-LXA_4 analog also increased lung IFN-γ, a cytokine that can antagonize T_H2 immune responses (Barnes 2008) and accelerate the resolution of airway inflammation in this model (Haworth et al. 2008).

Recognition of increased generation of prophlogistic LTs in airway inflammation led to development of therapeutic agents for asthma that modify CysLT formation or sites of action (Hui and Funk 2002, Barnes 2004). CysLTs signal via several receptors, including CysLT1, and montelukast and other CysLT1 antagonists were developed to block CysLT1-mediated airway responses (Barnes 2004). Targeting this receptor provides selectivity, but the presence of multiple LT receptors has limited the utility and therefore the clinical acceptance of these agents by many health care providers in asthma management (Hui and Funk 2002). Select patients, such as those with aspirin-exacerbated respiratory disease, markedly overproduce LTs and may derive particular clinical benefits from CysLT1 blockade or the 5-LO inhibitor zileuton. In practice, many asthmatics without this variant do not enjoy these same therapeutic benefits (Smith 1998).

9.5.1 RESOLUTION TOXICITY

A comprehensive view of the actions of LT biosynthetic enzymes is required for drug development, as these same enzymes, including 5-LO, can participate in the generation of proresolving mediators, such as LXs and resolvins. In animal systems, inhibitors of LO that block LX, resolvin, or protectin generation can be resolution "toxic," delaying the clearance of leukocytes from inflamed tissue (Schwab et al. 2007). In addition, leukotriene A$_4$ hydrolase (LTA$_4$H) converts the 5-LO-derived LTA$_4$ into LTB$_4$. Of interest, LTA$_4$H inhibition both decreases LTB$_4$ formation and increases LXA$_4$ production to decrease PMN accumulation in tissue inflammation (Rao et al. 2007). These findings with LTA$_4$H inhibition suggest preferential conversion of LTA$_4$ to LXA$_4$ over LTC$_4$ at sites of acute PMN-enriched inflammation. Thus, either targeting LTA$_4$H or combining an LX analog with an LT biosynthetic enzyme inhibitor would block LT and enhance LX-mediated signaling. Together, data on the pharmacokinetic properties and proresolving actions of LXs, resolvins, and protectins and their analogs support their continued clinical development as new therapeutic leads for asthma.

The identification of several classes of proresolving mediators for airway inflammation has provided an improved understanding of the natural resolution of airway inflammation and host mechanisms for lung catabasis. However, many unanswered questions persist related to the roles of these chemical mediators in physiological and pathophysiological processes. First, bioactive proresolving mediators are generated from C20:4, C20:5, and C22:6, but how do these mediators relate to each other in terms of the timing of their formation? Second, do these distinct classes of mediators employ shared or distinct sites of action? Third, are there specific signaling checkpoints that are critical to regulating their cell-type-specific actions?

Another consideration is the impact of these proresolving mediators and their bioactive analogs on host defense. The natural resolution of tissue injury or inflammation is not associated with an increase in susceptibility to infections. Thus, proresolving mechanisms would not be predicted to be immunosuppressive during physiological responses. Toward this point, LXA$_4$ and RvE1 can actually enhance host defense. For example, in gastrointestinal mucosa, LXA$_4$ induces the expression of bactericidal/permeability-increasing protein and increases bacterial killing by mucosal epithelial cells. In addition, RvE1 protects from bacterial pneumonia by increasing bacterial clearance and reducing pathogen-associated inflammation (Seki et al. 2010) and RvE1 also can induce intestinal alkaline phosphatase expression and activity to detoxify lipopolysaccharide during the resolution of an infectious challenge (Campbell et al. 2010). Moreover, adversely affecting the capacity of the host to generate LXs can threaten viability secondary to exacerbation of pathogen-mediated inflammation (Machado et al. 2006). Much more work is needed to understand the timing and contributions of proresolving mediators to host defense.

Lastly, the findings of defective generation of LXs and PD1 during severe and uncontrolled asthma and related conditions of airway inflammation emphasize the importance of both pro- and anti-inflammatory signals in orchestrating a self-limited response to provocative stimuli. Mechanisms underlying aberrant LX generation in airway disease appear to involve both transcriptional and posttranscriptional events, but are still being explored. Moreover, *in vitro* studies have raised the important possibility that corticosteroids may influence the biosynthesis of proresolving mediators and expression of their receptors. It is crucial to develop a better understanding of the endogenous and pharmacological factors that impact LX, resolvin, protectin, and maresin formation and their respective role(s) in airway inflammation in physiological responses and disease.

9.6   SUMMARY AND CONCLUSIONS

In summary, several lines of evidence support the existence and fundamental roles for specific mediators to actively promote the resolution of inflammation. This chapter has focused on select mediators enzymatically generated by injured and inflamed host tissues from the polyunsaturated fatty acids C20:4, C20:5, and C22:6. Early during acute inflammation in healthy airways, biosynthetic and signaling circuits are constructed to restrain the inflammatory responses and promote their eventual resolution. These molecular circuits lead to cell-type- and tissue-specific responses for anti-inflammation and resolution. Select proresolving mediators have been detected in human airways in health and disease and defects in their production exist during severe or uncontrolled airway inflammation. When administered in animals with airway inflammation, proresolving mediators or their stable analogs display potent protective actions. Only limited information on these compounds is currently available from clinical trials in human disease (Christie et al. 1992). Nonetheless, the promise of this genus of proresolving agonists as disease-modifying agents has led to the establishment of several drug development programs. With no available medical therapy to cure lung diseases that are characterized by chronic airway inflammation, such as asthma, there is a substantial unmet clinical need that serves as a powerful impetus to motivate scientists to develop a more thorough understanding of these common and highly morbid conditions.

9.7   DISCLOSURES

Bruce D. Levy is supported by National Institutes of Health grants HL068669, AI068084, and HL090927. Charles N. Serhan is supported by National Institutes of Health grants GM378765 and DK074448. The content is solely the responsibility of the authors and does not necessarily reflect the official views of NHLBI, NIAID, NIGMS, NIDDK, or the National Institutes of Health.

The authors are inventors on patents assigned to Brigham and Women's Hospital and Partners HealthCare on the composition of matter, uses, and clinical development of anti-inflammatory and proresolving lipid mediators. These are licensed for clinical development. Charles N. Serhan retains founder stock in Resolvyx Pharmaceuticals.

REFERENCES

Ariel, A., G. Fredman, Y.P. Sun et al. 2006. Apoptotic neutrophils and T cells sequester chemokines during immune response resolution through modulation of CCR5 expression. *Nat Immunol* 7:1209–15.

Arita, M., F. Bianchini, J. Aliberti et al. 2005a. Stereochemical assignment, antiinflammatory properties, and receptor for the omega-3 lipid mediator resolvin E1. *J Exp Med* 201:713–22.

Arita, M., S.F. Oh, T. Chonan et al. 2006. Metabolic inactivation of resolvin E1 and stabiliza-
    tion of its anti-inflammatory actions. *J Biol Chem* 281:22847–54.
Arita, M., T. Ohira, Y.P. Sun et al. 2007. Resolvin E1 selectively interacts with leukotriene B4
    receptor BLT1 and ChemR23 to regulate inflammation. *J Immunol* 178:3912–7.
Arita, M., M. Yoshida, S. Hong et al. 2005b. Resolvin E1, an endogenous lipid mediator
    derived from omega-3 eicosapentaenoic acid, protects against 2,4,6-trinitrobenzene
    sulfonic acid-induced colitis. *Proc Natl Acad Sci USA* 102:7671–6.
Bannenberg, G.L., N. Chiang, A. Ariel et al. 2005. Molecular circuits of resolution: Formation
    and actions of resolvins and protectins. *J Immunol* 174:4345–55.
Barnes, P.J. 2004. New drugs for asthma. *Nat Rev Drug Discov* 3:831–44.
Barnes, P.J. 2008. Frontrunners in novel pharmacotherapy of COPD. *Curr Opin Pharmacol*
    8:300–07.
Barnes, P.J. 2008. Immunology of asthma and chronic obstructive pulmonary disease. *Nat Rev
    Immunol* 8:183–92.
Barnes, P.J. and M. Karin. 1997. Nuclear factor-kappaB: A pivotal transcription factor in
    chronic inflammatory diseases. *N Engl J Med* 336:1066–71.
Belvisi, M.G. and J.A. Mitchell. 2009. Targeting PPAR receptors in the airway for the treat-
    ment of inflammatory lung disease. *Br J Pharmacol* 158:994–1003.
Birnbaum, Y., Y. Ye, Y. Lin et al. 2006. Augmentation of myocardial production of
    15-epi-lipoxin-a4 by pioglitazone and atorvastatin in the rat. *Circulation* 114:929–35.
Bonnans, C., K. Fukunaga, R. Keledjian et al. 2006a. Regulation of phosphatidylinositol
    3-kinase by polyisoprenyl phosphates in neutrophil-mediated tissue injury. *J Exp Med*
    203:857–63.
Bonnans, C., K. Fukunaga, M.A. Levy et al. 2006b. Lipoxin A(4) regulates bronchial epithelial
    cell responses to acid injury. *Am J Pathol* 168:1064–72.
Busse, W.W., Jr. and R.F. Lemanske. 2001. Asthma. *N Engl J Med* 344:350–62.
Campbell, E.L., N.A. Louis, S.E. Tomassetti et al. 2007. Resolvin E1 promotes mucosal sur-
    face clearance of neutrophils: A new paradigm for inflammatory resolution. *FASEB J*
    21:3162–70.
Campbell, E.L., C.F. MacManus, D.J. Kominsky et al. 2010. Resolvin E1-induced intestinal
    alkaline phosphatase promotes resolution of inflammation through LPS detoxification.
    Proc Nat Acad Sci USA 107:14298–303.
Canny, G., O. Levy, G.T. Furuta et al. 2002. Lipid mediator-induced expression of bacteri-
    cidal/permeability-increasing protein (BPI) in human mucosal epithelia. *Proc Natl Acad
    Sci USA* 99:3902–07.
Cantley, L.C. 2002. The phosphoinositide 3-kinase pathway. *Science* 296:1655–57.
Carlo, T., N.A. Petasis, and B.D. Levy. 2009. Activation of polyisoprenyl diphosphate phos-
    phatase 1 remodels cellular presqualene diphosphate. *Biochemistry* 48:2997–3004.
Cash, J.L., R. Hart, A. Russ et al. 2008. Synthetic chemerin-derived peptides suppress inflam-
    mation through ChemR23. *J Exp Med* 205:767–75.
Celik, G.E., F.O. Erkekol, Z. Misirligil et al. 2007. Lipoxin A4 levels in asthma: Relation with
    disease severity and aspirin sensitivity. *Clin Exp Aller* 37:1494–501.
Cezar-de-Mello, P.F., A.M. Vieira, V. Nascimento-Silva et al. 2008. ATL-1, an analogue of
    aspirin-triggered lipoxin A4, is a potent inhibitor of several steps in angiogenesis
    induced by vascular endothelial growth factor. *Br J Pharmacol* 153:956–65.
Chiang, N., I.M. Fierro, K. Gronert et al. 2000. Activation of lipoxin A(4) receptors by aspirin-
    triggered lipoxins and select peptides evokes ligand-specific responses in inflammation.
    J Exp Med 191:1197–208.
Chiang, N., C.N. Serhan, S.E. Dahlen et al. 2006. The lipoxin receptor ALX: Potent ligand-
    specific and stereoselective actions *in vivo*. *Pharmacol Rev* 58:463–87.
Christie, P.E., B.W. Spur, and T.H. Lee. 1992. The effects of lipoxin A4 on airway responses
    in asthmatic subjects. *Am Rev Respir Dis* 145:1281–4.

Claria, J., M.H. Lee, and C.N. Serhan. 1996. Aspirin-triggered lipoxins (15-epi-LX) are generated by the human lung adenocarcinoma cell line (A549)-neutrophil interactions and are potent inhibitors of cell proliferation. *Mol Med* 2:583–96.

Claria, J. and C.N. Serhan. 1995. Aspirin triggers previously undescribed bioactive eicosanoids by human endothelial cell-leukocyte interactions. *Proc Natl Acad Sci USA* 92: 9475–9.

Clish, C.B., J.A. O'Brien, K. Gronert et al. 1999. Local and systemic delivery of a stable aspirin-triggered lipoxin prevents neutrophil recruitment *in vivo*. *Proc Natl Acad Sci USA* 96:8247–52.

Connor, K.M., J.P. SanGiovanni, C. Lofqvist et al. 2007. Increased dietary intake of omega-3-polyunsaturated fatty acids reduces pathological retinal angiogenesis. *Nat Med* 13:868–73.

Dahlen, S.E., J. Raud, C.N. Serhan et al. 1987. Biological activities of lipoxin A include lung strip contraction and dilation of arterioles in vivo. *Acta Physiol Scand* 130: 643–7.

Devchand, P.R., M. Arita, S. Hong et al. 2003. Human ALX receptor regulates neutrophil recruitment in transgenic mice: Roles in inflammation and host defense. *FASEB J* 17:652–9.

Downey, G.P., J.R. Butler, H. Tapper et al. 1998. Importance of MEK in neutrophil microbicidal responsiveness. *J Immunol* 160:434–43.

Drazen, J.M., E. Israel, and P.M. O'Byrne 1999. Treatment of asthma with drugs modifying the leukotriene pathway. *N Engl J Med* 340:197–206.

Duffield, J.S., S. Hong, V.S. Vaidya et al. 2006. Resolvin D series and protectin D1 mitigate acute kidney injury. *J Immunol* 177:5902–11.

Duffy, D., S.M. Sparshott, C.P. Yang et al. 2008. Transgenic CD4 T cells (DO11.10) are destroyed in MHC-compatible hosts by NK cells and CD8 T cells. *J Immunol* 180:747–53.

Dufton, N., R. Hannon, V. Brancaleone et al. 2010. Anti-inflammatory role of the murine formyl-peptide receptor 2: Ligand-specific effects on leukocyte responses and experimental inflammation. *J Immunol* 184:2611–9.

Fahy, J.V., K.W. Kim, J. Liu et al. 1995. Prominent neutrophilic inflammation in sputum from subjects with asthma exacerbation. *J Allergy Clin Immunol* 95:843–52.

Fiore, S. and C.N. Serhan. 1990. Formation of lipoxins and leukotrienes during receptor-mediated interactions of human platelets and recombinant human granulocyte/macrophage colony-stimulating factor-primed neutrophils. *J Exp Med* 172:1451–7.

Freedman, S.D., P.G. Blanco, M.M. Zaman et al. 2004. Association of cystic fibrosis with abnormalities in fatty acid metabolism. (see comment). *New Eng J Med* 350:560–9.

Fujihara, S., C. Ward, I. Dransfield et al. 2002. Inhibition of nuclear factor-kappaB activation un-masks the ability of TNF-alpha to induce human eosinophil apoptosis. *Eur J Immunol* 32:457–66.

Fukunaga, K., M. Arita, M. Takahashi et al. 2006. Identification and functional characterization of a presqualene diphosphate phosphatase. *J Biol Chem* 281:9490–7.

Fukunaga, K., P. Kohli, C. Bonnans et al. 2005. Cyclooxygenase 2 plays a pivotal role in the resolution of acute lung injury. *J Immunol* 174:5033–9.

Gilroy, D.W., T. Lawrence, M. Perretti et al. 2004. Inflammatory resolution: New opportunities for drug discovery. *Nat Rev Drug Dis* 3:401–16.

Gronert K., A. Gewirtz, J.L. Madara et al. 1998. Identification of a human enterocyte lipoxin A4 receptor that is regulated by interleukin (IL)-13 and interferon gamma and inhibits tumor necrosis factor alpha-induced IL-8 release. *J Exp Med* 187:1285–94.

Gronert, K., N. Maheshwari, N. Khan et al. 2005. A role for the mouse 12/15-lipoxygenase pathway in promoting epithelial wound healing and host defense. *J Biol Chem* 280:15267–78.

Gronert, K., T. Martinsson-Niskanen, S. Ravasi et al. 2001. Selectivity of recombinant human leukotriene D(4), leukotriene B(4), and lipoxin A(4) receptors with aspirin-triggered 15-epi-LXA(4) and regulation of vascular and inflammatory responses. *Am J Pathol* 158:3–9.

Guilford, W.J. and J.F. Parkinson. 2005. Second-generation beta-oxidation resistant 3-oxa-lipoxin A4 analogs. *Prostaglandins Leukot Essent Fatty Acids* 73:245–50.

Hart, L.A., V.L. Krishnan, I.M. Adcock et al. 1998. Activation and localization of transcription factor, nuclear factor-kappaB, in asthma. *Am J Respir Crit Care Med* 158:1585–92.

Hashimoto, A., Y. Murakami, H. Kitasato et al. 2007. Glucocorticoids co-interact with lipoxin A4 via lipoxin A4 receptor (ALX) up-regulation. *Biomed Pharmacother* 61:81–5.

Hasturk, H., A. Kantarci, E. Goguet-Surmenian et al. 2007. Resolvin E1 regulates inflammation at the cellular and tissue level and restores tissue homeostasis *in vivo*. *J Immunol* 179:7021–9.

Hasturk, H., A. Kantarci, T. Ohira et al. 2006. RvE1 protects from local inflammation and osteoclast-mediated bone destruction in periodontitis. *FASEB J* 20:401–03.

Hawkins, P.T., K.E. Anderson, K. Davidson et al. 2006. Signalling through Class I PI3Ks in mammalian cells. *Biochem Soc Trans* 34:647–62.

Haworth, O., M. Cernadas, R. Yang et al. 2008. Resolvin E1 regulates interleukin 23, interferon-gamma and lipoxin A4 to promote the resolution of allergic airway inflammation. *Nat Immunol* 9:873–9.

Herrera, B.S., T. Ohira, L. Gao et al. 2008. An endogenous regulator of inflammation, resolvin E1, modulates osteoclast differentiation and bone resorption. *Br J Pharmacol* 155:1214–23.

Hong, S., K. Gronert, P.R. Devchand et al. 2003. Novel docosatrienes and 17S-resolvins generated from docosahexaenoic acid in murine brain, human blood, and glial cells. Autacoids in anti-inflammation. *J Biol Chem* 278:14677–87.

Hou, T.Z., J. Bystrom, J.P. Sherlock et al. 2010. A distinct subset of podoplanin (gp38) expressing F4/80+ macrophages mediate phagocytosis and are induced following zymosan peritonitis. *FEBS Lett* 584:3955–61.

Hudert, C.A., K.H, Weylandt, Y. Lu et al. 2006. Transgenic mice rich in endogenous omega-3 fatty acids are protected from colitis. *Proc Natl Acad Sci USA* 103:11276–81.

Hui, Y. and C.D. Funk 2002. Cysteinyl leukotriene receptors. *Biochem Pharmacol* 64: 1549–57.

Karp, C.L., L.M. Flick, K.W. Park et al. 2004. Defective lipoxin-mediated anti-inflammatory activity in the cystic fibrosis airway. *Nat Immunol* 5:388–92.

Kasuga, K., R. Yang, T.F. Porter et al. 2008. Rapid appearance of resolvin precursors in inflammatory exudates: Novel mechanisms in resolution. *J Immunol* 181:8677–87.

Keyes, K.T., Y. Ye, Y. Lin et al. 2010. Resolvin E1 protects the rat heart against reperfusion injury. *Am J Physiol Heart Circ Physiol* 299: H153–64.

Kowal-Bielecka, O., O. Distler, K. Kowal et al. 2003. Elevated levels of leukotriene B4 and leukotriene E4 in bronchoalveolar lavage fluid from patients with scleroderma lung disease. *Arthritis Rheum* 48:1639–46.

Kowal-Bielecka, O., K. Kowal, O. Distler et al. 2005. Cyclooxygenase- and lipoxygenase-derived eicosanoids in bronchoalveolar lavage fluid from patients with scleroderma lung disease: An imbalance between proinflammatory and antiinflammatory lipid mediators. *Arthritis Rheum* 52:3783–91.

Krishnamoorthy S., A. Recchiuti, N. Chiang et al. 2010. Resolvin D1 binds human phagocytes with evidence for proresolving receptors. *Proc Natl Acad Sci USA.* 107:1660–5.

Kuhn, H., D. Heydeck, R. Brinckman et al. 1999. Regulation of cellular 15-lipoxygenase activity on pretranslational, translational, and posttranslational levels. *Lipids* 34 (Suppl):S273–9.

Lawrence, T., M. Bebien, G.Y. Liu et al. 2005. IKKalpha limits macrophage NF-kappaB activation and contributes to the resolution of inflammation. *Nature* 434:1138–43.

Lee, T.H., A.E. Crea, V. Gant et al. 1990. Identification of lipoxin A4 and its relationship to the sulfidopeptide leukotrienes C4, D4, and E4 in the bronchoalveolar lavage fluids obtained from patients with selected pulmonary diseases. *Am Rev Respir Dis* 141:1453–8.

Leitch, A.E., N.A. Riley, T.A. Sheldrake et al. 2010. The cyclin-dependent kinase inhibitor R-roscovitine down-regulates Mcl-1 to override pro-inflammatory signalling and drive neutrophil apoptosis. *Eur J Immunol* 40:1127–38.

Levy, B.D., C. Bonnans, E.S. Silverman et al. 2005a. Diminished lipoxin biosynthesis in severe asthma. *Am J Resp Crit Care Med* 172:824–30.

Levy, B.D., C.B. Clish, B. Schmidt et al. 2001. Lipid mediator class switching during acute inflammation: Signals in resolution. *Nat Immunol* 2:612–9.

Levy, B.D., G.T. De Sanctis, P.R. Devchand et al. 2002. Multi-pronged inhibition of airway hyper-responsiveness and inflammation by lipoxin A(4). *Nat Med* 8:1018–23.

Levy B.D, V.V. Fokin, J.M. Clark et al. 1999. Polyisoprenyl phosphate (PIPP) signaling regulates phospholipase D activity: A 'stop' signaling switch for aspirin-triggered lipoxin A4. *FASEB J* 13:903–11.

Levy, B.D., L. Hickey, A.J. Morris et al. 2005b. Novel polyisoprenyl phosphates block phospholipase D and human neutrophil activation *in vitro* and murine peritoneal inflammation *in vivo*. *Br J Pharmacol* 146:344–51.

Levy, B.D., P. Kohli, K. Gotlinger et al. 2007. Protectin D1 is generated in asthma and dampens airway inflammation and hyperresponsiveness. *J Immunol* 178:496–502.

Levy, B.D., N.W. Lukacs, A.A. Berlin et al. 2007. Lipoxin A4 stable analogs reduce allergic airway responses via mechanisms distinct from CysLT1 receptor antagonism. *FASEB J* 21:3877–84.

Levy, B.D., N.A. Petasis, and C.N. Serhan. 1997. Polyisoprenyl phosphates in intracellular signalling. *Nature* 389:985–90.

Levy, B.D., M. Romano, H.A. Chapman et al. 1993. Human alveolar macrophages have 15-lipoxygenase and generate 15(S)-hydroxy-5,8,11-*cis*-13-*trans*-eicosatetraenoic acid and lipoxins. *J Clin Invest* 92:1572–9.

Louis, N.A., K.E. Hamilton, T. Kong et al. 2005. HIF-dependent induction of apical CD55 coordinates epithelial clearance of neutrophils. *FASEB J* 19:950–9.

Lukiw, W.J., J.G. Cui, V.L. Marcheselli et al. 2005. A role for docosahexaenoic acid-derived neuroprotectin D1 in neural cell survival and Alzheimer disease. *J Clin Invest* 115:2774–83.

Machado, F.S., J.E. Johndrow, L. Esper et al. 2006. Anti-inflammatory actions of lipoxin A4 and aspirin-triggered lipoxin are SOCS-2 dependent. *Nat Med* 12:330–4.

Mayadas, T.N., D.L. Mendrick, H.R. Brady et al. 1996. Acute passive anti-glomerular basement membrane nephritis in P-selectin-deficient mice. *Kidney Int* 49:1342–9.

Moore, W.C., E.R. Bleecker, D. Curran-Everett et al. 2007. Characterization of the severe asthma phenotype by the National Heart, Lung, and Blood Institute's Severe Asthma Research Program. *J Aller Clin Immunol* 119:405–13.

Mukherjee, P.K., V.L. Marcheselli, C.N. Serhan et al. 2004. Neuroprotectin D1: A docosahexaenoic acid-derived docosatriene protects human retinal pigment epithelial cells from oxidative stress. *Proc Natl Acad Sci USA* 101:8491–6.

Navarro-Xavier, R.A., J. Newson, V.L. Silveira et al. 2010. A new strategy for the identification of novel molecules with targeted proresolution of inflammation properties. *J Immunol* 184:1516–25.

Nolan, B., A. Duffy, L. Paquin et al. 1999. Mitogen-activated protein kinases signal inhibition of apoptosis in lipopolysaccharide-stimulated neutrophils. *Surgery* 126:406–12.

Ohira, T., M. Arita, K. Omori et al. 2010. Resolvin E1 receptor activation signals phosphorylation and phagocytosis. *J Biol Chem* 285:3451–61.

Ohira, T., G. Bannenberg, M. Arita et al. 2004. A stable aspirin-triggered lipoxin A4 analog blocks phosphorylation of leukocyte-specific protein 1 in human neutrophils. *J Immunol* 173:2091–8.

Parameswaran, K., K. Radford, A. Fanat et al. 2007. Modulation of human airway smooth muscle migration by lipid mediators and Th-2 cytokines. *Am J Respir Cell Mol Biol* 37:240–7.

Perretti, M., N. Chiang, M. La et al. 2002. Endogenous lipid- and peptide-derived anti-inflammatory pathways generated with glucocorticoid and aspirin treatment activate the lipoxin A4 receptor. *Nat Med* 8:1296–1302.

Perretti, M. and F. D'Acquisto. 2009. Annexin A1 and glucocorticoids as effectors of the resolution of inflammation. *Nat Rev Immunol* 9:62–70.

Petasis, N.A., I. Akritopoulou-Zanze, V.V. Fokin et al. 2005. Design, synthesis and bioactions of novel stable mimetics of lipoxins and aspirin-triggered lipoxins. *Prostaglandins Leukot Essent Fatty Acids* 73:301–21.

Planaguma, A., S. Kazani, G. Marigowda et al. 2008. Airway lipoxin A4 generation and lipoxin A4 receptor expression are decreased in severe asthma. *Am J Respir Crit Care Med* 178:574–82.

Planagumà A., M.A. Pfeffer, G. Rubin et al. 2010. Lovastatin decreases acute mucosal inflammation via 15-epi-lipoxin A4. *Mucosal Immunol.* 3:270–9.

Rao, N.L., P.J. Dunford, X. Xue et al. 2007. Anti-inflammatory activity of a potent, selective leukotriene A4 hydrolase inhibitor in comparison with the 5-lipoxygenase inhibitor zileuton. *J Pharmacol Exp Ther* 321:1154–60.

Rossi, A.G., J.M. Hallett, D.A. Sawatzky et al. 2007. Modulation of granulocyte apoptosis can influence the resolution of inflammation. *Biochem Soc Trans* 35:288–91.

Rossi, A.G., D.A. Sawatzky, A. Walker et al. 2006. Cyclin-dependent kinase inhibitors enhance the resolution of inflammation by promoting inflammatory cell apoptosis. *Nat Med* 12:1056–64.

Sanak, M., B.D. Levy, C.B. Clish et al. 2000. Aspirin-tolerant asthmatics generate more lipoxins than aspirin-intolerant asthmatics. *Eur Respir J* 16:44–9.

Sawatzky, D.A., D.A. Willoughby, P.R. Colville-Nash et al. 2006. The involvement of the apoptosis-modulating proteins ERK 1/2, Bcl-xL and Bax in the resolution of acute inflammation *in vivo*. *Am J Pathol* 168:33–41.

Schwab, J.M., N. Chiang, M. Arita et al. 2007. Resolvin E1 and protectin D1 activate inflammation-resolution programmes. *Nature* 447:869–74.

Schwartz, J. and S.T. Weiss 1994. The relationship of dietary fish intake to level of pulmonary function in the first National Health and Nutrition Survey (NHANES I). *Eur Respir J* 7:1821–4.

Seki, H., K. Fukunaga, M. Arita et al. 2010. The anti-inflammatory and proresolving mediator resolvin E1 protects mice from bacterial pneumonia and acute lung injury. *J Immunol* 184:836–43.

Serhan, C.N. 2007. Resolution phase of inflammation: Novel endogenous anti-inflammatory and proresolving lipid mediators and pathways. *Ann Rev Immunol* 25:101–37.

Serhan, C.N. 2010. Novel lipid mediators and resolution mechanisms in acute inflammation: To resolve or not? *Am J Pathol* 177:1576–91.

Serhan, C.N., S.D. Brain, C.D. Buckley et al. 2007. Resolution of inflammation: State of the art, definitions and terms. *FASEB J* 21:325–32.

Serhan, C.N. and N. Chiang. 2008. Endogenous pro-resolving and anti-inflammatory lipid mediators: A new pharmacologic genus. *Br J Pharmacol* 153 Suppl 1:S200–15.

Serhan, C.N., C.B. Clish, J. Brannon et al. 2000. Novel functional sets of lipid-derived mediators with antiinflammatory actions generated from omega-3 fatty acids via cyclooxygenase 2-nonsteroidal antiinflammatory drugs and transcellular processing. *J Exp Med* 192:1197–204.

Serhan, C.N., S. Fiore, D.A. Brezinski et al. 1993. Lipoxin A4 metabolism by differentiated HL-60 cells and human monocytes: Conversion to novel 15-oxo and dihydro products. *Biochemistry* 32:6313–9.

Serhan, C.N., S. Fiore, and B.D. Levy 1994. Cell-cell interactions in lipoxin generation and characterization of lipoxin A4 receptors. *Ann New York Acad Sci* 744:166–80.

Serhan, C.N., K. Gotlinger, S. Hong et al. 2006. Anti-inflammatory actions of neuroprotectin D1/protectin D1 and its natural stereoisomers: Assignments of dihydroxy-containing docosatrienes. (erratum appears in *J Immunol* Mar 15; 176(6):3843). *J Immunol* 176:1848–59.

Serhan, C.N., M. Hamberg, and B. Samuelsson. 1984. Lipoxins: Novel series of biologically active compounds formed from arachidonic acid in human leukocytes. *Proc Natl Acad Sci USA* 81:5335–9.

Serhan, C.N., S. Hong, K. Gronert et al. 2002. Resolvins: A family of bioactive products of omega-3 fatty acid transformation circuits initiated by aspirin treatment that counter proinflammation signals. *J Exp Med* 196:1025–37.

Serhan, C.N., A. Jain, S. Marleau et al. 2003. Reduced inflammation and tissue damage in transgenic rabbits overexpressing 15-lipoxygenase and endogenous anti-inflammatory lipid mediators. *J Immunol* 171:6856–65.

Serhan, C.N., J.F. Maddox, N.A. Petasis et al. 1995. Design of lipoxin A4 stable analogs that block transmigration and adhesion of human neutrophils. *Biochemistry* 34:14609–15.

Serhan, C.N., R. Yang, K. Martinod et al. 2009. Maresins: Novel macrophage mediators with potent antiinflammatory and proresolving actions. *J Exp Med* 206:15–23.

Smith, L.J. 1998. A risk-benefit assessment of antileukotrienes in asthma. *Drug Saf* 19: 205–18.

Sodin-Semrl, S., B. Taddeo, D. Tseng et al. 2000. Lipoxin A4 inhibits IL-1 beta-induced IL-6, IL-8, and matrix metalloproteinase-3 production in human synovial fibroblasts and enhances synthesis of tissue inhibitors of metalloproteinases. *J Immunol* 164:2660–6.

Spite, M., L.V. Norling, L. Summers et al. 2009. Resolvin D2 is a potent regulator of leukocytes and controls microbial sepsis. *Nature* 461:1287–91.

Sun, Y.P., S.F. Oh, J. Uddin et al. 2007. Resolvin D1 and its aspirin-triggered 17R epimer. Stereochemical assignments, anti-inflammatory properties, and enzymatic inactivation. *J Biol Chem* 282:9323–34.

Svensson, C.I., M. Zattoni, and C.N. Serhan. 2007. Lipoxins and aspirin-triggered lipoxin inhibit inflammatory pain processing. *J Exp Med* 204:245–52.

Tahan, F., R. Saraymen, and H. Gumus. 2008. The role of lipoxin A4 in exercise-induced bronchoconstriction in asthma. *J Asthma* 45:161–4.

Tjonahen, E., S.F. Oh, J. Siegelman et al. 2006. Resolvin E2: Identification and anti-inflammatory actions: Pivotal role of human 5-lipoxygenase in resolvin E series biosynthesis (see comment). *Chem Biol* 13:1193–202.

Uller, L., C.G. Persson, and J.S. Erjefalt 2006. Resolution of airway disease: Removal of inflammatory cells through apoptosis, egression or both? *Trends Pharmacol Sci* 27: 461–6.

Vachier, I., C. Bonnans, C. Chavis et al. 2005. Severe asthma is associated with a loss of LX4, an endogenous anti-inflammatory compound. *J Aller Clin Immunol* 115:55–60.

Vanhaesebroeck, B., S.J. Leevers, K. Ahmadi et al. 2001. Synthesis and function of 3-phosphorylated inositol lipids. *Annu Rev Biochem* 70:535–602.

Ward, C., E.R. Chilvers, M.F. Lawson et al. 1999. NF-kappaB activation is a critical regulator of human granulocyte apoptosis in vitro. *J Biol Chem* 274:4309–18.

Wenzel, S.E., S.J. Szefler, D.Y. Leung et al. 1997. Bronchoscopic evaluation of severe asthma. Persistent inflammation associated with high dose glucocorticoids. *Am J Respir Crit Care Med* 156:737–43.

Wittamer, V., J.D. Franssen, M. Vulcano et al. 2003. Specific recruitment of antigen-presenting cells by chemerin, a novel processed ligand from human inflammatory fluids. *J Exp Med* 198:977–85.

Wittwer, J., M. Bayer, A. Mosandl et al. 2007. The c.-292C > T promoter polymorphism increases reticulocyte-type 15-lipoxygenase-1 activity and could be atheroprotective. *Clin Chem Lab Med* 45:487–92.

Wu, S.H., P.L. Yin, Y.M. Zhang et al. 2010. Reversed changes of lipoxin A4 and leukotrienes in children with asthma in different severity degree. *Pediatric Pulmonol* 45:333–40.

Xu, Z.Z., L. Zhang, T. Liu et al. 2010. Resolvins RvE1 and RvD1 attenuate inflammatory pain via central and peripheral actions. *Nat Med* 16:592–7.

10 Platelet-Activating Factor Antagonist Therapy for Airway Disorders

Punit Kumar Srivastava, Kent E. Pinkerton, and Rishabh Dev

CONTENTS

10.1 INTRODUCTION

Asthma and chronic obstructive pulmonary disease (COPD) are widely recognized as chronic airway inflammatory lung diseases characterized by cellular infiltration, airway remodeling, bronchoconstriction, hypermucous secretion, and declined lung function. Phospholipase A2 (PLA_2) enzymes are central mediators of inflammatory responses. Within the subfamily of PLA_2, cytosolic PLA_2 begins the conversion of

239

membrane phospholipids into platelet-activating factor (PAF; 1-*O*-alkyl-2-acetyl-*sn*-glycero-3-phosphorylcholine). PAF, also known as acetyl-glyceryl-ether-phosphoryl-choline (AGEPC), is a proinflammatory phospholipid involved in the pathophysiology of inflammation and an airway hyperreactivity mediator released by a wide variety of inflammatory immune cells, such as eosinophils, basophils, endothelial cells, neu-trophils, monocytes, macrophages, and mastocytes (Barnes et al., 1998; Kay, 1983).

PAF promotes platelet activation and regulates the functions of various immune and inflammatory cells. It influences the release of histamine from basophils, lead-ing to degranulation of cells; chemotaxis of eosinophils and their transmigration across the endothelial cells; and the production of reactive oxygen species (ROS) and leukotriene C4 (LTC_4), thus inducing bronchoconstriction of the airways. PAF also stimulates delayed-type responses, such as hyperalgesia and eosinophil infiltration. In secondary responses, PAF promotes the release of IL-6 from basophils and increases the expression of matrix metalloproteinases (MMPs) from bronchial epi-thelial cells (Kasperska-Zajac et al., 2007).

PAF mediates its action by binding to a G-protein-coupled membrane receptor (GPCR) called the PAF receptor (PAFR). Expression of functional PAFR has been shown in various immune cells, including platelets, eosinophils, monocytes, and neutrophils (Hwang, 1990). PAFR mRNA levels have been found to be increased in the lungs of asthmatic patients compared to normal subjects (Kishimoto et al., 1997).

Several PAFR antagonist therapies have been attempted in the past with remark-able improvement in animal models of asthma, but none were successful in the clinic. In this chapter, the basic biology of the PAFR and its function will be reviewed along with the reasons why PAFR antagonist therapy has been hindered in the past. This is followed by proposed future development strategies for PAF antagonist ther-apy, including using PAF antagonist drugs in combination.

10.2   LIPID PATHWAY AND PRODUCTION OF PAF

PAF is produced by several cell types activated by specific stimuli, including neutro-phils, monocytes, platelets, and endothelial cells. Endothelial cells respond to mechanical stress, anoxia, ischemia, and oxidative stress, as well as a variety of hor-mones and vasoactive mediators by producing PAF and eicosanoids as mediators of inflammation. The synthesis of PAF is initiated by the PLA_2 metabolic pathway that cleaves membrane phospholipids, yielding lyso-PAF, which is further acetylated by an acetyltransferase to form biologically active PAF (i.e., acetyl-PAF) (Figure 10.1).

Acetyltransferase is induced by cytokines, such as TNF-α and IL-1α, and can cause further synthesis of PAF. In particular, endothelial cells activated by specific stimuli produce acetyl-PAF as the predominant molecular species of PAF.

10.2.1   BIOSYNTHESIS OF MEDIATORS

Generally, the metabolic route responsible for PAF biosynthesis involves the remod-eling of membrane lipid constituents (remodeling pathway) or *de novo* synthesis (*de novo* pathway). The remodeling pathway is mainly activated during inflamma-tion and other hypersensitivity responses, whereas the *de novo* pathway is a source

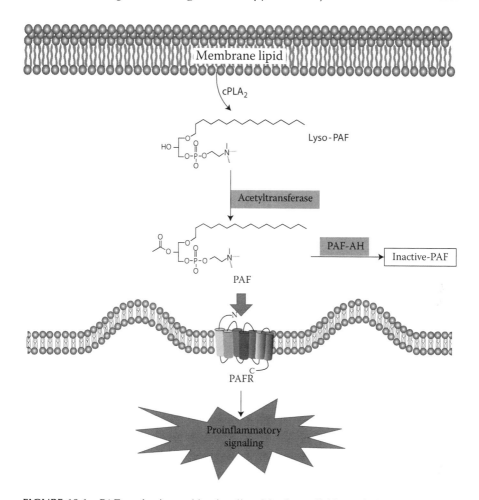

FIGURE 10.1 PAF production and its signaling. Membrane lipids are hydrolyzed by cytosolic phospholipase A2 (cPLA$_2$) enzymes to Lyso-PAFs, which are inactive molecules. They are acted upon by acetyltransferases to produce active PAF, which binds to its receptor GPCR and transduces proinflammatory signals downstream such as hyperresponsiveness and airway remodeling. Once the PAF molecule binds to its receptor, it is immediately hydrolyzed by PAF-AH to inactive PAF.

of PAF under physiological conditions. This chapter provides an up-to-date view of PAF biosynthesis by endothelial cells and, in particular, recent knowledge about the role of CoA-independent PAF-dependent transacetylase in the biosynthesis of acyl analogs of PAF by stimulated endothelial cells.

PAF is degraded by hydrolysis of the acetyl group to produce the biologically inactive lyso-PAF. This reaction is catalyzed by PAF-specific acetyl hydrolase, (PAF-AH), which is present abundantly in plasma and intracellularly in several inflammatory cells. Inherited deficiency of this enzyme has been associated with asthma severity in children (Miwa et al., 1988). PAF-AH is expressed and secreted primarily by cells of the hematopoietic system and has been reported to play essential roles in inflammation, vascular disorders, anaphylaxis, and bowel diseases (Prescott et al., 2000). PAF-AH is a Ca^{2+}-independent enzyme implicated in atherosclerosis (Davis et al., 2008). It is mostly bound to low-density lipoprotein (LDL) particles and to a lesser extent to high-density lipoprotein (HDL) and very low-density lipoprotein (VLDL) particles and it is mainly produced by monocytes, macrophages, platelets, and mast cells (Davis et al., 2008). An accumulating body of evidence supports the concept that PAF-AH is a novel risk marker for cardiovascular disease (Tsoukatos et al., 2008). As far as its action is concerned, it cleaves short-chain acyl chains at the *sn*-2 position of phospholipids, such as oxidized phospholipids and PAF. From this point of view, PAF-AH could be characterized as an antiatherogenic enzyme. However, its action goes hand in hand with the production of inflammatory molecules, such as lysophosphatidylcholine and oxidized nonesterified fatty acids, which are proatherogenic (Davis et al., 2008; Stafforini, 2009).

10.2.2 SYNTHESIS IN ANIMALS AND HUMANS

It has been known for many years that mouse and human plasma express vastly different levels of PAF-AH activity. Recently, a study was conducted in which the authors surveyed plasma PAF-AH levels in various mammalian species and found that basal PAF-AH activity was significantly lower (10-fold) in human compared to mouse plasma (Prescott et al., 2000). A number of mechanisms could potentially account for this observation, including variations in transcriptional and/or translational activation, protein modification, stability, and differences in the efficiency of substrate hydrolysis by each ortholog. It was previously reported that key roles for members of the Sp family of transcription factors are in the transcriptional regulation of human and mouse PAF-AH (Cao et al., 1998). Results suggested similar, albeit not identical, transcriptional regulation of both genes (Cao et al., 1998; Wu et al., 2004). These observations suggest that mechanisms other than promoter-mediated regulatory events likely account for the observed differences in plasma activity levels. We found evidence indicating that residues located at N- and C-terminal regions play key roles in the determination of protein expression levels. Studies unveil novel information related to the mechanisms that control expression of plasma PAF-AH, a PLA_2 with important, but incompletely characterized, roles in physiological and pathological inflammation.

10.3 BIOLOGICAL SIGNIFICANCE OF PAF

PAF is a potent phospholipid activator and mediator of many leukocyte functions, including platelet aggregation, inflammation, and anaphylaxis. Besides PAF's ability to activate platelets, it also modulates the activity of several inflammatory and immune cells, such as eosinophils, basophils, and neutrophils, leading to a cascade of proinflammatory responses such as chemotaxis, infiltration and transmigration of cells across the epithelial barrier, and release of inflammatory cytokines and proteases. It was also observed that when PAF was injected intradermally, it elicits immediate weal-and-flare reaction and a delayed type of hypersensitivity characterized by erythema, hyperalgesia, and cell infiltrations (Kasperska-Zajac et al., 2007).

10.3.1 CELL-SPECIFIC RESPONSES

PAF has a proinflammatory role in airway diseases, and it has been found that NF-κB activity is necessary for airway smooth muscle cell (ASMC) proliferation. In a recent study, administration of PAF (a peak response at 100 nM) resulted in a rise in the cell proliferation index (PI), the PCNA-positive rate in the ASMCs, and an increase in NF-κB cDNA. The addition of 20 mM *N*-acetylcysteine (NAC) completely blocked the above responses. The levels of NF-κB had concomitant increases upon PAF administration as seen by mobility shift assay, and the levels were inhibited with NAC treatment (Ying-Fang et al., 2007).

It was also observed that allergen exposure during provocation test in patients with bronchial asthma causes the release of PAF (Beer et al., 1995). A significant increase in PAF was also observed during allergen-induced bronchoconstriction (Chan-Yeung et al., 1991). Increased level of PAF was also seen in BAL fluid of asthmatic patients as compared to normal individuals (Stenton et al., 1990), with concomitant increase in levels of PAF-AH enzyme.

Measurement of PAF in biological fluids is difficult because it is immediately captured by the surrounding inflammatory cells or rapidly inactivated to lyso-PAF. Therefore, most of the biological effects of PAF have been studied using exogenously administered PAF. On the basis of experimental models of asthma, it has been shown that PAF administration induces similar pathophysiology as seen in asthma, such as bronchoconstriction, bronchial hyperresponsiveness, impaired gas-exchange capabilities, and increase in PAFR levels (Cuss et al., 1986; Chung et al., 1989; Louis and Radermecker, 1996; Adamus et al., 1990; Hsieh, 1991).

10.4 PAF RECEPTOR AND DISTRIBUTION

PAF elicits its biological responses through a receptor (PAFR), which is a membrane GPCR. Upon binding with its ligand, PAF transmits signals downstream through GTP-coupled GTPase enzyme. The PAFR has structural characteristics of the rhodopsin (MIM 180380) gene family and binds to PAF. PAFRs are G-protein linked, and interaction with PAF initiates intracellular signaling mechanisms resulting in increased calcium and ionositol triphosphate levels, intracellular phospholipid turnover, and expression of regulatory genes (Nakamura et al., 1993; Izumi and Shimizu, 1995; Mazer et al., 1991; Schulam et al., 1991). PAFR couples with various phospholipases, such as PLA_2, C, and D, and with several kinases, including mitogen-activated protein kinase, phosphatidylinositol-3 kinase, and tyrosine kinases. Thus, it exerts pleiotropic effects (Ishii and Shimizu 2000). Nevertheless, the existence of an alternate pathway has been implicated because nonspecific binding of [³H]-PAF also leads to intercellular signaling (Valone et al., 1982).

10.4.1 IDENTIFICATION OF RECEPTOR

Overexpression data indicate that a high level of transgenic PAFR mRNA exists in the trachea (Ishii et al., 1996). This suggests that airway smooth muscle cells overexpressing PAFR play a major role in airway hyperresponsiveness. In airway hyperresponsiveness, PAF and PAFR gene expression may affect airway structure, specifically the lamina propria and airway smooth muscle architecture (Nagase et al., 1999). Recently, a transgenic mouse model overexpressing the PAFR gene has been established, and the mice exhibit hyperresponsiveness to both methacholine and PAF challenge. Administration of atropine significantly blocked PAF-induced hyperresponsiveness in the mice (Nagase et al., 2002). PAFR overexpression did not change the binding capacity of the muscarinic receptors, suggesting that PAFR gene expression is involved in airway hyperresponsiveness to methacholine challenge through muscarinic receptor functionally, but may not be related structurally (Nagase et al., 1998).

10.4.2 POLYMORPHISM IN PAFR GENE

Since the PAFR system is important in the physiology of the reproductive, cardiovascular, respiratory, and central nervous systems and in the pathophysiology of inflammation and thromboembolic diseases, functional variants in PAFR may act as predisposing factors for these diseases or as modifiers of the disease phenotypes and therapeutic responses (Prescott et al., 2000). PAFR is located on chromosome 1 (location: 1p35-p34.3), and mRNA translates a 342-amino-acid GPCR protein. Fukunaga et al. (2001) have identified a novel DNA variant in the human PAFR gene that substitutes an aspartic acid for an alanine residue at position 224 (A224D) in the putative third cytoplasmic loop. This mutation was observed in the Japanese population at an allele frequency of 7.8%. Chinese hamster ovary (CHO) cells overexpressing PAFR (A224D) showed significant reduction in calcium mobilization, inositol phosphate production, and inhibition of adenyl cyclase and chemotaxis, suggesting that the variant receptor produced naturally by occurring mutation exhibits impaired coupling of G-proteins; it may be a basis for interindividual variation in PAF-related physiological responses, disease predisposition, and drug responsiveness (Fukunaga et al., 2001).

10.5 ANIMAL MODELS AND PAFR ANTAGONISM

10.5.1 ROLE OF MEDIATORS IN AIRWAY INFLAMMATION

Several investigators have studied the pathophysiological mechanism underlying PAF-associated airway hyperresponsiveness (AHR). In a guinea pig study, animals were transnasally treated with PAF or vehicle twice a week with simultaneous administration of histamine or methacholine by inhalation. Eosinophil infiltration of bronchoalveolar lavage (BAL) was significantly increased in PAF-treated guinea pigs compared to those treated with vehicle (Ishiura et al., 2005). In same study, bronchial responsiveness to inhaled histamine, but not methacholine, was significantly decreased by PAF treatment.

In another study, the role of endothelin, PAF, and thromboxane A2 in AHR to carbachol induced by ovalbumin (OVA) sensitization and challenge in Balb/c mice was investigated. OVA sensitization and challenge induced significant AHR to carbachol in actively sensitized and challenged mice. Treatment of these mice with the PAF antagonist CV-3988 (10 µg/kg i.v.) completely abolished OVA-induced AHR to carbachol. Treatment of sensitized mice with the TXA2 antagonist L-654,664 (1 mg/kg, i.v.) partially blocked the induction of AHR in OVA-challenged mice. These results suggest an important role for endothelin, PAF, and thromboxane A2 in AHR in mice actively sensitized and challenged with OVA (Richter et al., 2007).

10.5.2 OTHER INFLAMMATORY MODELS

Goblet cell metaplasia is an important morphological feature in the airways of patients with chronic airway diseases. The role of PAF has been investigated by cigarette smoke-induced metaplasia *in vivo* in guinea pig animal model. Cigarette smoke exposure significantly increased the goblet cell number, with increase in eosinophil, neutrophil, and PAF levels in bronchoalveolar lavage. Treatment with the PAF antagonist E-6123 significantly inhibited the goblet cell number, and neutrophil and eosinophil counts, showing a possible role of PAF in goblet cell metaplasia (Komori et al., 2001).

10.5.3 TRANSGENIC AND KNOCKOUT ANIMAL DATA

Bronchial asthma is a complex disease of the lung characterized by reversible airway obstruction, chronic airway inflammation, and AHR to nonspecific stimuli. The progression of airway inflammation involves several cell types, including $CD4^+$ Th2 cells, eosinophils, and mast cells (Wills-Karp, 1999). Eosinophils and mast cells are the origin of PAF, as shown in *in vitro* studies. In fact, PAF is reported to be a potent chemotactic factor for eosinophils and their degranulation *in vitro*. PAFR-deficient (PAFr$^{-/-}$) and overexpressing PAFR mice demonstrate that PAF plays a major role in anaphylaxis and acute injury in the lung, suggesting a role for PAF in early-phase allergic and inflammatory responses (Ishii et al., 1998; Nagase et al., 1999). It has been shown in a recent study that mice deficient in PAFR in asthma models exhibit reduced AHR to cholinergic stimulation. However, PAFR-deficient mice develop an eosinophilic response comparable to wild-type mice (Ishii et al., 2004). PAF-induced bronchial responsiveness and airway eosinophilic accumulation has been shown in guinea pig models as well (Ishiura et al., 2005).

10.5.4 ROLE OF PAF-AH AND AN APPROACH TO INHIBIT PAF FUNCTIONS

PAF action is limited in asthma by the administration of recombinant PAF-AH to elevate the endogenous plasma level of the enzyme. Recombinant PAF-AH is able to block PAF-induced inflammation *in vivo*, suggesting that PAF-AH might be useful in anti-inflammation therapy (Tjoelker et al., 1995). In a mouse model of OVA-induced and metacholine-challenged airway hyperreactivity, it was found that exogenously administered recombinant PAF-AH reduced airway inflammation (mucus release and eosinophil infiltration) and airway hyperreactivity of allergen-induced asthma, suggesting that such therapy can be extended to humans in order to reduce allergic airway inflammation (Henderson et al., 2000).

The effect of intravenously injected human recombinant PAF-AH (rPAF-AH) on the dual phase of asthmatic response in atopic subjects with mild asthma has been evaluated in a randomized, double-blind, placebo-controlled study and it was found that treatment did not significantly inhibit eosinophil counts, although there was a trend of reduction in sputum neutrophil numbers, concluding that anti-PAF agent rPAF-AH does not significantly inhibit allergen-induced dual-phase asthmatic responses (Henig et al., 2000).

10.6   DRUG DISCOVERY AND SYNTHETIC PAF ANTAGONISTS

The complex nature of the inflammatory process in bronchial asthma demands development of therapies targeting different steps of the inflammatory cascade. Specific and potent PAFR antagonists and human recombinant PAF-AH have emerged, and their acute as well as long-term effects in asthma have been evaluated.

10.6.1   Various PAF Antagonists Evaluated in Clinical Trials

Table 10.1 shows the list of various PAFR antagonists tested in human clinical trials in various models of airway inflammation and allergy, and their chemical structures have been given in Figure 10.2.

Several PAFR antagonists have been developed and tested successfully in various animal models of asthma. So far, the clinical use of PAF antagonists fails to show any benefit. However, PAF might play a greater role in certain subsets of asthmatic patients (Henig et al., 2000). The protective effect of PAFR antagonists has been evaluated in several PAF-induced airway responses. First-generation PAFR antagonists, such as Ginkgolide mixture (BN-52063), were tested and found to attenuate PAF-induced (injected through epidermal route) platelet aggregations and bronchoconstriction but not neutrophilic infiltration in *ex vivo* studies. The low efficacy was thought to be due to poor bioavailability.

Second-generation PAFR antagonists, such as UK-74505 and WEB-2086 (Apafant), completely inhibit PAF-induced platelet aggregation, airway bronchoconstriction, and neutropenia in *ex vivo* studies. BN-52010 demonstrates significant inhibition of PAF-induced neutrophilia and bronchoconstriction. These studies

TABLE 10.1
PAFR Antagonists Explored in Clinical Trials in Asthma and Allergy

PAFR Antagonist	Field of Investigation
BN-52063	AHR and allergen challenge studies
BN-52021	AHR and allergen challenge studies
RP-59227	Allergen challenge studies
SR-27417A (Foropafant)	AHR, asthma, and allergen challenge studies
UK-74505	AHR and allergen challenge studies
UK-80067 (Modipafant)	AHR and asthma challenge studies
WEB-2086 (Apafant)	AHR, asthma, and allergen challenge studies
Y-24180	AHR challenge studies

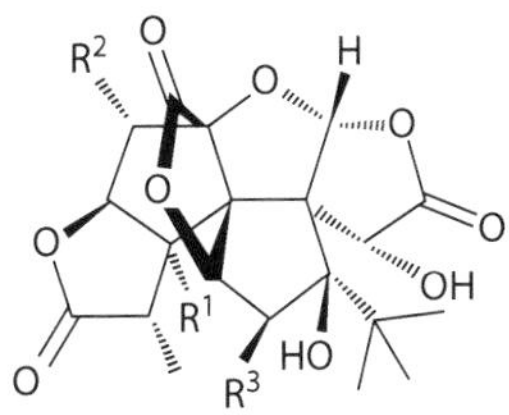

FIGURE 10.2 Chemical structures of PAFR antagonists.

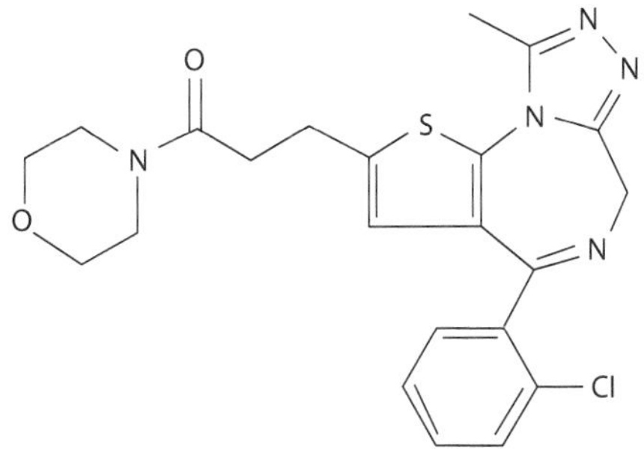

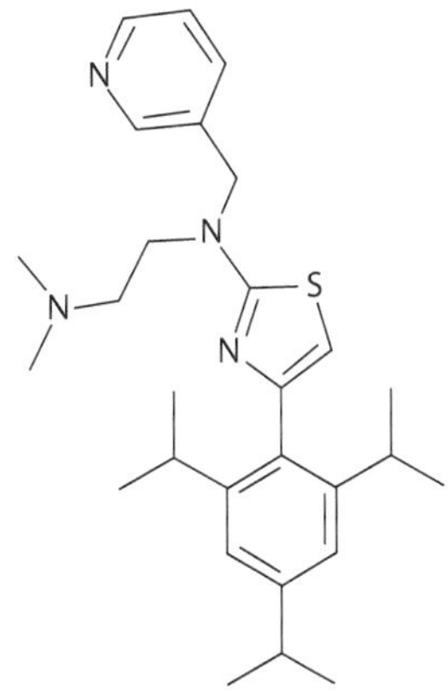

demonstrated that PAFR antagonists can achieve therapeutic concentration in blood and airway lumen. Unfortunately, randomized controlled trials have not clearly proven the clinical benefit of WEB-2086 for asthma (Spence et al., 1994; Kuitert et al., 1995). For example, there was no significant difference between placebo and Modipafant (one 50 mg capsule twice daily) effects on lung function (PEF and FEV1), rescue bronchodilator usage, symptom score, or airway responsiveness in patients with moderate asthma, suggesting that PAF may not be an important mediator in human asthma (Kuitert et al., 1995). Similarly, UK-74, 505, another specific PAF antagonist, had no effect on the early asthmatic response (EAR) and the late asthmatic response (LAR) or allergen-induced airway hyperresponsiveness (Kuitert et al., 1993). The efficacy of PAF antagonist SR27417A in mild asthma patients has been evaluated and it was found to be effective in inhibiting systemic cellular and pulmonary effects after PAF challenge (Gomez et al., 1998). But it failed to show promise as a complementary drug that could be used for the treatment of an acute asthma exacerbation as no significant changes in FEV1, respiratory system resistance, alveolar–arterial pressure difference for oxygen, or arterial oxygen tension were observed (Gomez et al., 1999).

Recently, a potent PAFR antagonist Foropafant (SR-27417) showed significant results in mild asthmatic patients. Compared to placebo, Foropafant, when administered orally, significantly inhibited PAF-induced neutropenia in addition to platelet aggregation. Foropafant also increased respiratory system resistance and alveolar–arterial oxygen pressure difference for oxygen (PA-a, O_2) and arterial oxygen tension (Pa-O_2). Foropafant had a modest inhibitory effect on LAR and showed no effects upon EAR, allergen-induced airway responsiveness, or baseline lung measurements (Evans et al., 1997).

Clinical trials performed to evaluate the therapeutic potential of recombinant PAF-AH in asthma have been disappointing. Administered intravenously at a dose of 1 mg/kg, human recombinant PAF-AH demonstrated no significant effect upon the allergen-induced dual-phase asthmatic response (Henig et al., 2000).

10.6.2 PAF Inhibitors in Combinations

Table 10.2 shows various dual PAF antagonists in combination with other inflammatory antagonists. Rupatadine is a dual PAF and histamine H1 receptor antagonist that originated from Uriach (information obtained from Adis Data Information BV © 2010). The compound has high potency in both of its antagonist roles and has prolonged activity after oral administration. Rupatadine is an antiallergic agent that acts through antagonism of PAF and histamine receptors and was launched for the treatment of seasonal and perennial allergic rhinitis and chronic idiopathic urticaria in patients aged 12 years and older. In the treatment of allergic rhinitis, Rupatadine is administered orally, once daily. Rupatadine has linear pharmacokinetics in healthy volunteers and has demonstrated efficacy in clinical trials in both seasonal and perennial allergic rhinitis. It is comparable to established antihistamines in efficacy. The launched Kyowa Hakko drug, Olopatadine, also has these properties. Rupatadine has a similar chemical structure to Schering-Plough's antihistamine, Loratadine. Rupatadine is well tolerated in healthy volunteers; oral doses of 10 or 20 mg do not affect psychomotor performance or mood. Animal studies have indicated that Rupatadine lacks sedative effects and has a favorable cardiovascular profile compared with astemizole and terfenadine. Uriach has stated that Rupatadine lacks cardiotoxic potential, as no QT segment prolongation has been recorded in preclinical or clinical studies. Coadministration of Rupatadine and inhibitors of cytochrome P450 CYP3A4 should be avoided.

Saiboku-to (TJ-96) is a Chinese herbal medicine that is marketed in Japan by Tsumura for the treatment of asthma and respiratory tract disorders. Saiboku-to is a blend of 10 herbal compounds that inhibits IgE-mediated release of histamine, the release of PAF, and down-regulation of glucocorticoid and 4-adrenergic receptors. It is also thought to stimulate ciliary motility and sodium absorption and appears to increase nitric oxide (NO) generation in airway epithelial cells. Increased airway response in asthmatic patients may be due to diminished production of epithelium-derived relaxing factors, such as NO, so stimulation of epithelial NO production is a desirable property in an antiasthmatic agent. Saiboku-to also inhibits the production of IL-3 and IL-4 by mononuclear cells. Japanese researchers at Kinki University in Osaka consider Saiboku-to as "one of a new class of antiasthmatic drugs." In a randomized, double-blind, crossover trial, 33 patients with mild-to-moderate allergic

TABLE 10.2
Dual PAFR Antagonists (p.o.) Tested in Clinical Trials

Drug	Phase	Indication
Rupatadine	Launched	Airway disorders, skin diseases
Saiboku-to	Launched	Airway diseases
CPL-7075	Phase II	Airway disease, pain, rheumatoid arthritis, neurological disorders

asthma received placebo or Saiboku-to 2.5 g/day by oral administration for 4 weeks. Saiboku-to lowered symptom scores and decreased bronchial hyperresponsiveness to methacholine. It also decreased eosinophil cationic protein levels and eosinophil counts in blood and sputum.

CPL-7075 (formerly IP 751) is a nonpsychoactive cannabinoid being developed by Cervelo Pharmaceuticals for the treatment of neuropathic pain. The compound was originally discovered by the University of Massachusetts Medical School and the initial development was carried out by Indevus Pharmaceuticals (later Endo Pharmaceuticals Solutions). CPL-7075 is ajulemic acid, a 1,1-dimethyl heptyl derivative of the carboxylic metabolite of tetrahydrocannabinol, and has been shown to suppress inflammatory cytokines, including interleukin-19 and MMPs, through a peroxisome proliferator-activated receptor (PPAR) gamma-mediated mechanism. Preclinical studies have suggested that CPL-7075 may lack the gastrointestinal ulceration associated with nonsteroidal anti-inflammatory drugs (NSAIDs) and the cardiovascular effects seen with cyclooxygenase-2 (COX-2) inhibitors. Indevus has pursued interstitial cystitis as a lead indication but also believes that CPL-7075 has a broad potential to treat pain and inflammatory conditions, such as arthritis, postoperative pain, and musculoskeletal injuries. The agent may also be useful in treating noninflammatory conditions such as headache and neuropathic pain.

CPL-7075 underwent preclinical development for inflammation associated with arthritis and asthma, in addition to spasticity associated with multiple sclerosis. However, no recent development has been reported for these indications. A phase II trial of CPL-7075 for neuropathic pain has been completed in Germany, and this is the indication that is now being pursued by Cervelo.

10.6.3   FUTURE APPROACHES FOR PAF-TARGETED THERAPY

Though the role of PAF is well established in bronchial asthma through *in vitro* and animal studies, clinical efficacy failure raises concern whether PAFR antagonist therapy alone will provide sufficient relief to the asthmatic patients. Based on the clinical data obtained with some potent PAFR antagonists, it can be concluded that PAF alone may not play a major role in the development of asthmatic inflammation. This is because the pulmonary effects of PAF are mediated through various pathways. Most of the time, PAFR antagonists are unable to inhibit systemic and pulmonary effects raised by PAF mediators but rather are able to inhibit secondary inflammatory mediators. Interestingly, it has been suggested that dual PAF and histamine H1 receptor antagonists may appear more efficacious than single mediator antagonists in allergic diseases with potential therapeutic use in asthma treatment (Billah et al., 1991, 1992).

10.7   CONCLUSION

PAF mediates several proinflammatory responses in the body, including lung cells, and various studies have established its role in the pathogenesis of airway disorders. Therefore, it has been postulated that PAF antagonism may act as a novel anti-inflammatory strategy. Several attempts have been made to design and synthesize PAF receptor antagonists, which have been very successful in animal models of asthma and airway hyperresponsiveness. Though their clinical efficacy in humans is yet to be established, dual PAFR antagonists in combination with other receptor antagonists like H1 receptor have proven to be a great success in animal models of asthma as well as in human clinical efficacy trials. We recommend that the development of dual antagonists of PAFR with other inflammatory receptor will have a potential in therapeutic intervention of airway disorders.

ACKNOWLEDGMENTS

The authors acknowledge Dr. Suzette Smiley-Jewell, University of California, Davis for providing invaluable suggestions to improve the chapter.

REFERENCES

Adamus, W.S., H.O. Heuer, S. Meade, and J.C. Schilling 1990. Inhibitory effects of the new PAF acether antagonist WEB-2086 on pharmacologic changes induced by PAF inhalation in human beings. *Clin Pharmacol Ther* 47:456–62.

Barnes, P.J., K.F. Chung, and C.P. Page 1998. Inflammatory mediators of asthma: An update. *Pharmacol Rev* 50:515–96.

Beer, J.H., B. Wuthrich, and A. Von-Felten 1995. Allergen exposure in acute asthma causes the release of platelet-activating factor (PAF) as demonstrated by the densitization of platelets to PAF. *Int Arch Allergy Immunol* 106:291–96.

Billah, M., M. Egan, R.W. Ganguly, M.J. Green, W. Kreutner, J.J. Piwinski, M.I. Seigel, F.J. Villoni, and J.K. Wong 1991. Discovery and preliminary pharmacology of Sch 37370, a dual antagonist of PAF and histamine. *Lipids* 26:1172–74.

Billah, M.M., H.G. Gilchrest, S.P. Eckel, C.A. Granzao, P.J. Lawton, E. Radwanski, M.D. Brannan, M.B. Attrine, and J.D. Christopher 1992. Differential plasma duration of antiplatelet-activating factor and antihistamine activities of oral Sch 37370 in humans. *Clin Pharmacol Ther* 52:151–59.

Cao, Y., D.M. Stafforini, G.A. Zimmerman, T.M. McIntyre, and S.M. Prescott 1998. Expression of plasma platelet activating acetohydrolase is transcriptionaly regulated by mediators of inflammation *J Biol Chem* 273:4012–20.

Chan-Yeung, M.S., H. Lam, H. Chan, K. Tse, and H. Salari 1991. The release of platelet-activating factor into plasma during allergen-induced brocnhoconstriction. *J Allergy Clin Immunol* 87:657–73.

Chung, K.F., F.M. Cuss, and P.J. Barnes 1989. Platelet activating factor: Effects on bronchomotor tone and bronchial responsiveness in human beings. *Allergy Proc* 10:333–37.

Cuss, F.M., C.M. Dixon and P. J. Barnes 1986. Effects of inhaled platelet activating factor on pulmonary function and bronchial responsiveness in man. *Lancet* 26:189–92.

Davis, B., G. Koster, L.J. Donet, M. Scigelova, G. Woffendin, J.M. Ward, A. Smith et al. 2008. Electrospray ionization mass spectrometry identifies substrates and products of lipoprotein-associated phospholipase A2 in oxidized human low density protein. *J Biol Chem* 283:6428–37.

Evans, D.J., P.J. Barnes, M. Cluzel, and O'Connor 1997. Effect of a potent platelet-activating factor antagonist, SR27417A, on allergen-induced asthmatic responses. *Am J Respir Crit Care Med* 156:11–16.

Fukunaga, K., S. Ishii, K. Asano, T. Yokomizo, T. Shiomi, T. Shimizu, and K. Yamaguchi 2001. Single nucleotide polymorphism of human platelet-activating factor receptor impairs G-protein activation. *J Biol Chem* 276:43025–30.

Gomez, F.P., R.M. Marrades, R. Lglesia, J. Roca, J.A. Barbera, K.F. Churg, and R.M. Rodriguez-Roisin 1999. Gas exchange response to a PAF receptor antagonist, SR 27417A, in acute asthma: A pilot study. *Eur Respir J* 14:622–26.

Gomez, F.P., J. Roca, J.A. Barbera, K.F. Churg, V.I. Peinadovi, and R.M. Rodriguez-Roisin 1998. Effect of platelet-activating factor (PAF) antagonist SR 27417A, on PAF-induced gas exchange abnormalities in mild asthma. *Eur Respir J* 11:835–39.

Henderson, W.R., J. Lu, K.M. Poole, G.N. Dietsch, and E.Y. Chi 2000. Recombinant human platelet-activating factor acetylhydrolase inhibits airway inflammation and hyperreactivity in mouse asthma model. *J Immunol* 164:3360–67.

Henig, N.R., M.L. Aitken, M.C. Liu, A.S. Yu, and W.R. Henderson Jr. 2000. Effect of recombinant human platelet-activating factor acetylhydrolase on allergen-induced asthmatic responses. *Am J Respir Crit Care Med* 162:523–27.

Hsieh, K.H. 1991. Effects of PAF antagonist BN52021, on the PAF-, methacholine-, and allergen-induced bronchoconstriction in asthmatic children. *Chest* 99:877–82.

Hwang, S.B. 1990. Specific receptors of platelet-activating factor, receptor heterogeneity, and signal transduction mechanisms. *J Lipid Mediat* 2:123–58.

Ishii, S. and T. Shimizu 2000. Platelet-activating factor (PAF) receptor and genetically engineered PAF receptor mutant mice. *Prog Lipid Res* 39:41–82.

Ishii, S., T. Kuwaki, T. Nagase, K. Maki, F. Tashiro, S. Sunaga, and W.H. Cao 1998. Impaired anaphylactic responses but intact sensitivity to endotoxin in mice lacking a platelet-activating factor receptor. *J Exp Med* (1998) 187:1779–88.

Ishii, S., Y. Matsuda, M. Nakamura, J. Miazaki, K. Kume, T. Izumi, and T. Shimizu 1996. A murine platelet-activating factor receptor gene: Cloning, chromosomal localization and up-regulation of expression by lipopolysaccharide in peritoneal resident macrophages. *Biochem J* 1:671–78.

Ishii, S., T. Nagase, H. Shindou, I. Waga, K. Kume, T. Izumi, and T. Shimizu 2004. Platelet-activating factor receptor develops airway hyperresponsiveness independently of airway inflammation in a murine asthma model. *J Immunol* 172:7095–02.

Ishiura, Y., M. Fujimura, K. Nobata, Y. Oribe, M. Abo, S. Myou, and A. Nonamura 2005. *In-vivo* PAF-induced airway eosinophil accumulation reduces bronchial responsiveness to inhaled histamine. *Prostagland Other Lipid Med* 75:1–12.

Izumi, T. and T. Shimizu 1995. Platelet-activating factor receptor: Gene expression and signal transduction. *Biochim Biophys Acta* 1259:317–33.

Kasperska-Zajac, A., Z. Brzoza, and B. Rogala 2007. Platelet activator factor as a mediator and therapeutic approach in bronchial asthma. *Inflamm* 31:112–20.

Kay, A.B. 1983. Mediators of hypersensitivity and inflammatory cells in the pathogenesis of bronchial asthma. *Eur J Respir Dis Sullp* 129:1–44.

Kishimoto, S., W. Shimadzu, T. Izumi, T. Shimizu, H. Sagara, T. Fukuda, S. Makino, T. Suginara, and K. Waku 1997. Comparison of platelet-activating factor receptor mRNA levels in peripheral blood eosinophils from normal subjects and atopic asthmatic patients. *Int Arch Allergy Immunol* 14:60–63.

Komori, M., H. Inoue, S. Makino, H. Koto, S. Fukuyama, H. Aizawa, and N. Hara 2001. PAF mediate cigarette smoke-induced goblet cell metaplasia in guinea pig airways. *Am J Physiol Lung Cell Mol Physiol* 280:436–41.

Kuitert, L.M., R.M. Angus, N.C. Barnes, P.J. Barnes, M.F. Bone, K.F. Churg, A.J. Fairfax, T.W. Higenbotham, B.J. O'Connor, and B. Piotrowska 1995. Effect of a novel potent platelet-activating factor antagonist, modipafant, in clinical asthma. *Am J Respir Crit Care Med* 151:1331–35.

Kuitert, L.M., K.P. Hui, S. Uthayarkumar, W. Burk, A.C. Narland, S. Uden, and N.C. Barnes 1993. Effect of the platelet-activating factor antagonist UK-74,505 on the early and late response to allergen. *Am Rev Respir Dis* 147:82–86.

Louis, R.E. and M.F. Radermecker 1996. Acute bronchial obstruction following inhalation of PAF in asthmatic and normal subjects: Comparison with methacholine. *Eur Respir J* 9:1414–20.

Mazer, B., J. Domenico, H. Sawami, and E.W. Geltand 1991. Platelet-activating factor induces an increase in intracellular calcium and expression of regulatory genes in human B lymphoblastoid cells. *J Immunol* 146:1914–20.

Miwa, M., T. Miyake, T. Yamanaka, J. Sugatani, Y. Suzuki, S. Sakata, Y. Araki, and M. Matsumoto 1988. Characterization of serum platelet-activating factor (PAF) acetylhydrolase. Correlation between deficiency of serum PAF acetylhydrolase and respiratory symptoms in asthmatic children. *J Clin Invest* 82:1983–91.

Nagase, T., S. Ishii, K. Kume, N. Uozumi, T. Izumi, Y. Ouchi, and T. Shimizu 1999. Platelet-activating factor mediates acid-induced lung injury in genetically engineered mice. *J Clin Invest* 104:1071–76.

Nagase, T., S. Ishii, H. Shindou, H. Takizawa, Y. Ouchi, and T. Shimizu 2002. Airway hyper-responsiveness in transgenic mice overexpressing platelet activating factor receptor is mediated by an atropine-sensitive pathway. *Am J Respir Crit Care Med* 165:200–05.

Nagase, T., H. Kurihara, Y. Kurihara, T. Aoki, Y. Fukuchi, Y. Yazaki, and Y. Ouchi 1998. Airway hyperresponsiveness to methacholine in mutant mice deficient in endothelin-1. *Am J Respir Crit Care Med* 157:560–64.

Nakamura, M., Z. Honda, T. Matsumoto, M. Noma, and T. Shimizu 1993. Isolation and prop-erties of platelet-activating factor receptor cDNAs. *J Lipid Mediat* 6:163–68.

Prescott, S.M., G.A. Zimmerman, D.M. Stafforini, and T.M. Prescott 2000. Platelet activating factors and lipid mediators. *Annu Rev Biochem* 69:419–45.

Richter, M., S. Cloutier, and P. Sirois 2007. Endothelin, PAF and thromboxane A2 in allergic pulmonary hyperreactivity in mice. *Prost Leuk Ess Fatty Acids* 76:299–08.

Schulam, P.G., A. Kuruvilla, G. Putcha, L. Mangas, J. Franklin-Jhonson, and W.T. Shearer 1991. Platelet-activating factor induces phospholipid turnover, calcium flux, arachidonic acid liberation, eicosanoid generation, and oncogene expression in a human B cell line. *J Immunol* 146:1642–48.

Spence, D.P., S.L. Johnston, P.M. Calverley, P. Dhilon, S.C. Higgins, E. Ranhamaden, S. Turner, A. Winning, J. Vinter, and S. T. Holgate 1994. The effect of the orally active platelet-activating factor antagonist WEB 2086 in the treatment of asthma. *Am J Respir Crit Care Med* 149:1142–48.

Stafforini, D.M. 2009. Biology of platelet-activating factor acetohydrolase (PAF-AH, lipopro-tein associated phospholipase A2). *Cadiovas Drugs Ther* 33:73–83.

Stenton, S.C., E.N. Court, W.P. Kingston, P. Goadby, C.A. Kelly, M. Duddridge, C. Ward, D.J. Hendride, and E.H. Waters 1990. Platelet-activating factor in bronchoalveolar lavage fluid from asthmatic subjects. *Eur Respir J* 3:408–13.

Tjoelker, L.W., C. Wilder, C. Eberhardt, D.M. Stefforini, G. Dietsch, B. Schimpf, S. Hooper, H. Letrong, L.S. Cousens, and G.A. Zimmerman 1995. Anti-inflammatory properties of a platelet-activating factor acetylhydrolase. *Nature* 374:549–53.

Tsoukatos, D.C., I. Brochériou, V. Moussis, C.P. Panopoulou, E.D. Christofidou, S. Koussissis, S. Sismanidiss, E. Ninio, and S. Siminelakis 2008. Platelet-activating factor acetylhy-drolase and transacetylase activities in human aorta and mammary artery. *J Lipid Res* 49:2240–49.

Valone, F.H., E. Coles, V.R. Reinhold, and E.J. Goetzl 1982. Specific binding of phospholipid platelet-activating factor by human platelets. *J Immunol* 129:1637–41.

Wills-Karp, M. 1999. Immunologic basis of antigen-induced airway hyperresponsiveness. *Annu Rev Immunol* 17:255–81.

Wu, X., G.A. Zimmerman, G.M. Prescott, and D.M. Stafforini 2004. The p38 MAPK pathway mediates transcriptional activation of the plasma platelet activating factor gene in mac-rophages stimulated with lipo-polysaccharide. *J Biol Chem* 279:36158–65.

Ying-Fang, S., H. Jing-Fang, L. Huan-Zhang, and Q. Hao-Weu 2007. Effect of platelet-activating factor on cell proliferation & NF-KB activation in airway smooth muscle cells in rats. *Ind J Med Res* 126:139–45.

11 Lysophosphatidic Acid in Airway Inflammation

Yutong Zhao and Viswanathan Natarajan

CONTENTS

11.1 INTRODUCTION

Phospholipids, sphingolipids, and cholesterol are major constituents of all biological membranes in mammalian cells. While the central role of phospho-, sphingo-, and neutral-lipids in biological processes has been well defined, only recently have the physiological and pathological effects of lysolipids in cell functions been described. Among the lysophospholipids, lysophosphatidic acid (LPA) has the simplest structure, consisting of a long-chain fatty acid- or alkyl- or alk-1-enyl- moiety linked to *sn*-1 carbon of the glycerol or of a polyunsaturated fatty acid(s) linked to *sn*-2 and a phosphate group at *sn*-3 position of the glycerol backbone. Unlike many lysophospholipids, LPA is more water soluble and is present in nM to µM concentrations in plasma and tissues. LPA plays a critical role in *de novo* biosynthesis of phospholipids; additionally, it is a serum-derived growth factor and is involved in many cellular processes such as proliferation, migration, cytokine/chemokines secretion, platelet aggregation, smooth muscle cell contraction, and neurite retraction. These cellular effects are mediated through G-protein-coupled receptors (GPCRs) specific to LPA, which are localized on the plasma membrane. At least seven lysophosphatidic acid receptors (LPA-Rs) (LPA_{1-7}) have been cloned and partially characterized, and in the past two decades several studies related to physiological actions of LPA via LPA-Rs have been carried out. Among the LPA-Rs, at least one putative LPA-R responds to both LPA and sphingosine-1-phosphate (S1P) (Murakami et al., 2008). In addition to the signaling that occurs via G-protein-coupled LPA-Rs, peroxisome proliferators-activator receptor-γ (PPARγ) has been identified as an intracellular receptor for LPA.

In contrast to efforts to understand LPA action through LPA-Rs, mechanisms of LPA generation have been addressed more recently. At least three potential pathways of LPA generation have been identified in mammalian tissues and serum. *De novo* biosynthesis of LPA is mediated by acyltransferases that catalyze acylation of *sn*-glycero-3-phosphate or dihydroxyacetone phosphate. Additionally, phosphorylation of monoacylglycerol (MAG) by acylglycerol kinase (AGK) can also generate LPA intracellularly. In recent years, two phospholipases, lysophospholipase D (lyso PLD) or autotaxin (ATX) and phosphatidic acid-specific phospholipase $A_1\alpha$, have been identified as playing a major role in LPA production in plasma/serum and platelets, respectively. Our understanding of the physiological and pathological roles of LPA and LPA-Rs has been advanced with analyses of the gene knockout mice for most of the LPA-Rs and partial knockdown of ATX.

The purpose of this review is to focus on the functions of LPA signaling in the airway under normal and pathological conditions such as asthma, idiopathic pulmonary fibrosis, and other airway inflammatory diseases.

11.2 METABOLISM OF LPA

11.2.1 TISSUE DISTRIBUTION OF LPA

LPA is a naturally occurring bioactive lipid and is present in all mammalian tissues and cells. This includes plasma where LPA levels range from 100 nM to 1 μM, and serum where concentrations can exceed 1 μM and be as high as 10 μM. The low LPA levels in plasma are regulated by synthesis, degradation, and uptake by tissues and circulating cells. In the plasma, LPA is bound to either albumin or gelsolin and levels of LPA increase under pathological conditions such as breast or ovarian cancer. These increases result from activation of platelets and circulating monocytes/polymorphonuclear leukocytes due to enhanced biosynthesis. In bronchoalveolar lavage (BAL) fluids, LPA levels are altered in asthma and sepsis-induced lung injury. Mechanisms that regulate low LPA levels in plasma or BAL are not yet fully defined.

11.2.2 MOLECULAR SPECIES OF LPA

LPA, one of the simplest glycerophospholipids, has a mixture of long-chain fatty acids linked to *sn*-1 or *sn*-2 position on the glycerol backbone. In addition to fatty acids, LPA containing long-chain alkyl- or alk-1-enyl- moieties on the *sn*-1 position of the glycerol backbone have been detected in biological samples. Molecular species of LPA containing both saturated (16:0 and 18:0) and unsaturated fatty acids (16:1, 18:1, 18:2, 20:4, and 22:6) have been reported in serum, plasma, activated platelets, saliva, and other biological fluids such as BAL fluid. It is likely that saturated and monounsaturated fatty acids are predominantly linked to *sn*-1 and polyunsaturated fatty acids to *sn*-2 position on the glycerol backbone. This differential distribution of saturated/monounsaturated and polyunsaturated fatty acids in LPA suggests (1) the involvement of both phospholipase A_1 (PLA_1) and A_2 (PLA_2) types of activities in the *de novo* production of LPA from phosphatidic acid and (2) the specificity of fatty acyl CoA acyltransferases in transferring saturated and monounsaturated fatty acyl CoA's to *sn*-1 position of glycerol-3-phosphate or dihydroxyacetone phosphate. The different molecular species of LPA seem to exhibit varying biological activities, most likely due to differential activation of LPA-Rs. Interestingly, BAL fluids from segmental allergen-challenged asthmatics exhibited significantly higher proportion of 22:6 molecular species of LPA as compared to 16:0 LPA from nonsegmental asthmatics that closely resemble the LPA molecular species in circulation. Structures of four major molecular species of naturally occurring LPA are shown in Figure 11.1.

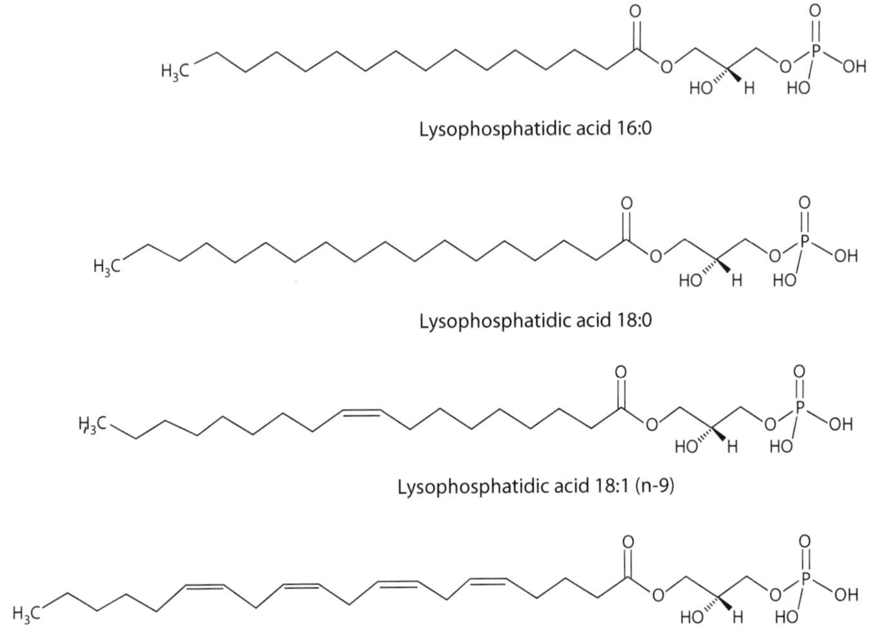

Lysophosphatidic acid 16:0

Lysophosphatidic acid 18:0

Lysophosphatidic acid 18:1 (n-9)

Lysophosphatidic acid 20:4 (n-6)

FIGURE 11.1 Molecular species of naturally occurring LPA in mammalian cells.

11.2.3 BIOSYNTHESIS OF LPA

In mammalian tissues and cells, intracellular or extracellular LPA generation involves at least two major and two minor pathways. The four pathways involved in LPA generation are depicted in Figure 11.2.

11.2.3.1 Intracellular Generation of LPA: Role of Phospholipase D and PA-Specific Phospholipase A

In the first minor pathway, acylation of glycerol-3-phosphate or dihydroxyacetone phosphate by long-chain CoA (16:0, 18:0, or 18:1) produces predominantly 1-acylglycerol-3-phosphate (LPA) or 1-acyl dihydroxyacetone-3-phosphate that is subsequently reduced by a dehydrogenase to 1-acylglycerol-3-phosphate. The second minor pathway involves phosphorylation of MAG to LPA by AGK, wherein MAG is derived by the action of either lipase(s) on diacylglycerol (DAG) (Bektas et al., 2005; Spiegel and Milstien, 2005) or lipid phosphatases on LPA (Furneisen and Carman, 2000; Hiroyama and Takenawa, 1998; Pilquil et al., 2001). However, the major pathway of intracellular LPA production is via phosphatidic acid (PA) generated by phospholipase D (PLD) or phosphorylation of DAG catalyzed by DAG kinase (Aoki, 2004; Ito et al., 1996) and subsequent conversion of PA to LPA by PLA_1 or A_2. The role of the PLD/PA pathway in LPA generation has been demonstrated in the SK-OV-3 ovarian cancer cell line, where incubation with 1-butanol, which diverts PA

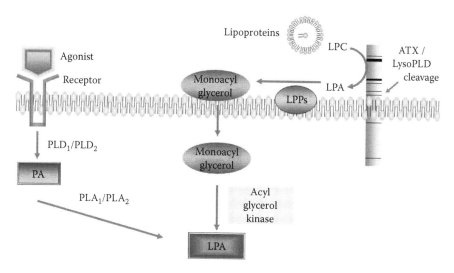

FIGURE 11.2 Mechanisms of LPA generation. Intracellular and extracellular LPA generation is dependent on several enzymes, such as phospholipase D, phospholipase A1 or A2, acylglycerol kinase, and ATX/lyso PLD.

formed by PLD to phosphatidylbutanol, caused a significant reduction in constitutively as well as inducible LPA formation (Luquain et al., 2003). Recent studies have identified that a genetic defect in the mPA-PLA$_1\alpha$/LIPH gene is linked to an inherited form of hair loss and a hair growth defect through its LPA-producing activity (Ali et al., 2007; Kazantseva et al., 2006).

11.2.3.2 Extracellular Generation of LPA: Role of Autotaxin/ Lysophospholipase D

The ability of plasma to generate LPA from lysophosphatidylcholine (LPC) by a plasma enzyme called lyso PLD was reported about three decades ago by Tokumura et al. (1986). Since then, partial amino acid sequence of lyso PLD purified from human plasma (Tokumura et al., 2002) and fetal calf serum (Umezu-Goto et al., 2002) was coincident with that of ATX, also known as nucleotide pyrophosphatase/ phosphodiesterase 2 (NPP2). ATX/lyso PLD does not hydrolyze phosphatidylcholine (PC), phosphatidylethanolamine, or phosphatidylserine (PS) and is highly selective for LPC, lysophosphatidylethanolamine (LPE), or lysophosphatidylserine (LPS) (Aoki et al., 2008; Okudaira et al., 2010; Tokumura et al., 1986; Umezu-Goto et al., 2002). LPC is produced from PC associated with lipoproteins by lecithin-cholesterol acyltransferase (LCAT) and PLA$_1$ (Aoki, 2004; Aoki et al., 2002). Very little data are available on the role of ATX levels in airway inflammation and lung diseases.

11.2.3.3 Cyclic Phosphatidic Acid Generation by Autotaxin/ Lysophospholipase D

ATX/lyso PLD, in addition to LPA generation from LPC, also produced cyclic phosphatidic acid (cPA) that has a dioxaphospholane ring at the *sn*-2 and *sn*-3 positions of the glycerol backbone (Fujiwara, 2008; Fujiwara et al., 2003; Gendaszewska-Darmach, 2008; Kobayashi et al., 1999). cPA is present in biological fluids, including human serum (Kobayashi et al., 1999), with 16:0 as the most abundant molecular species. The levels of cPA in human serum are ~0.1 µM, which is less than one-tenth of LPA concentration (1–5 µM) (Eichholtz et al., 1993). The biological activity of cPA is distinct from that of LPA as it exhibits antimitogenic, antimetastatic, and platelet inhibitory properties (Fujiwara, 2008). Intravenous administration of a stable analog of cPA, carba-cPA, increased tidal volume and respiratory frequency, resulting in elevated total ventilation in rats (Hotta et al., 2006), while LPA has a hypertensive action on systemic blood pressure (Tokumura, 2002; Tokumura et al., 1995). The biological targets of cPA are yet to be identified; however, it activates LPA_{1-4} with a higher EC_{50} concentration than LPA (Baker et al., 2006).

11.2.3.4 Catabolism of LPA

LPA levels are very low in tissues and biological fluids under normal physiological conditions, and at least three pathways have been characterized for LPA degradation. In the first pathway, LPA is converted to MAG by phosphatases that belong to PA phosphatase type 2 (PAP-2) or lipid phosphate phosphatases (LPPs). Three major isoforms of LPPs, LPP_{1-3}, have been cloned and characterized in mammalian cells (Brindley, 2004; Kai et al., 1997; Leung et al., 1998; Roberts et al., 1998). *In vitro*, a variety of lipid phosphates, including LPA, PA, S1P, and ceramide-1-phosphate, serve as substrates for LPP_{1-3}; however, LPA is the preferred substrate for LPP_1, compared to LPP_2 and LPP_3 (Brindley, 2004). LPPs are expressed in human lung and human bronchial epithelial cells (Zhao et al., 2005) as evidenced by real-time RT-PCR and Western blotting with specific antibodies. Further, in human bronchial epithelial cells, exogenously added [^{3}H]oleoyl LPA was hydrolyzed to [^{3}H]MAG and [^{3}H]oleic acid and overexpression of LPP-1 wild type increased LPA hydrolysis 2–3-fold over control (Zhao et al., 2005). As LPP-1 is an ecto-enzyme, one of its major roles may be to modulate LPA levels in circulation and thereby regulate LPA signaling via LPA-Rs. In the second pathway, LPA is converted to PA by an LPA acyltransferase (Tang et al., 2006; West et al., 1997) that has not been well defined in the lung and airway. The third pathway involves lysophospholipases that can hydrolyze LPA to fatty acid and glycerol-3-phosphate, and at least two distinct lysophospholipases with specificity toward LPC and LPA have been reported (Baker and Chang, 1999; Thompson and Clark, 1994; Wang and Dennis, 1999). The relative contributions of these pathways in regulating LPA accumulation in the airway need to be determined.

11.2.3.5 LPA Measurement in Tissues and Biological Samples

The measurement of LPA levels in tissues and biological samples has been carried out using several different methods. Most of these procedures require extraction of total lipids from biological samples using nonacidic extraction techniques developed by Folch et al. (1957) or Bligh and Dyer (1959). However, acidic extraction procedures with strong acids resulted in elevated levels of LPA and other lysophospholipids due to breakdown of alk-1-eny phospholipids (Georas et al., 2007; Tokumura et al., 2009). LPA-like activity was determined using bioassays such as voltage clamping of

Xenopus oocytes (Tigyi and Miledi, 1992) while measurement of LPA-derived fatty acid methyl esters by gas chromatography (GC) provided a semiquantitative analysis of LPA levels in tissues (Tokumura et al., 1986). Immunoassay(s) suffers from lack of antibody specificity (Chen et al., 2000) while radioenzymatic assays require the use of radiolabeled LPA (Saulnier-Blache et al., 2000). As this technique requires a large sample size, time-consuming isolation procedures of LPA by thin-layer chromatography followed by GC of the fatty acid methyl esters or the use of radiolabeled LPA were utilized. Moreover, the use of mass spectrometry (MS) in recent times has revolutionized quantification of LPA measurements. While the analysis of LPA by flow injection coupled directly to MS is useful (Shan et al., 2008), ion suppression effects form major phospholipid species, and other matrix components affected the sensitivity and reliability of this technique. Coupling of chromatographic separation of LPA from other lipids has considerably improved detection limits and sensitivity and currently, liquid chromatography (LC/MS) (Baker et al., 2001) and liquid chromatography tandem mass (Georas et al., 2007; Scherer et al., 2009; Shan et al., 2008; Tokumura et al., 2009) provide accurate, selective, and precise quantitation of LPA and molecular species of LPA in biological samples and fluids.

11.3   LPA RECEPTORS

Molecular cloning of LPA receptors led researchers to further investigate the role and mechanisms of LPA-induced biological function. So far, seven cell surface LPA receptors have been cloned and characterized. Nuclear receptor PPARγ also binds to LPA (McIntyre et al., 2003). Except for PPARγ, all seven LPA receptors are GPCRs (Fukushima and Chun, 2001; Lin et al., 2010; Zhao and Natarajan, 2009). Among them, LPA_{1-3} have been well studied under various biological, physiological, and pathological conditions. Targeting of LPA receptors by developing LPA receptor agonists and antagonists should provide new therapies in treating cancer and inflammatory diseases.

11.3.1   MOLECULAR CLONING OF LPA RECEPTORS

Hecht et al. (1996), for the first time, showed that LPA was a ligand for ventricular zone-1 (vzg-1) receptor. Vzg-1 expression was necessary and sufficient in LPA-induced multiple effects such as DNA synthesis, stress fiber formation, and serum response element activation. Later, vzg-1 was named as endothelial differentiation gene (EDG)-2, since it has a high homology to GPCRs. Based on recent nomenclature, vzg-1/EDG2 was renamed as LPA_1. $EDG4/LPA_2$ was identified by searching the GenBank™ for LPA receptor homologs (An et al., 1998). Cloning of $EDG7/LPA_3$ was carried out by a PCR procedure based on comparison of amino acid sequences of $EDG2/LPA_1$ and $EDG4/LPA_2$ (Bandoh et al., 1999). Recently, four additional LPA receptors have been cloned that do not strictly belong to the EDG family. To date, seven cell surface LPA receptors, LPA_{1-7}, have been cloned and described in mammals, and based on sequence homology, LPA_1, LPA_2, and LPA_3 belong to the EDG family. $LPA_1/EDG2$, $LPA_2/EDG4$, and $LPA_3/EDG7$ share ~50% sequence homology, while LPA_4 /GPR23/ P2Y9, LPA_5/GPR92, LPA_6/GPR87, and LPA_7/P2Y5 are structurally distinct from the EDG family and share <40% homology with conventional LPA_{1-3}.

11.3.2   LPA Signaling through LPA Receptors are G-Protein-Coupled

G-proteins are important signal-transducing molecules coupled to receptors on the cell surface. Heterotrimeric G-proteins are composed of $G\alpha$, $G\beta$, and $G\gamma$. Exchange between GDP and GTP on $G\alpha$ subunit, by several factors, regulate $G\alpha$-mediated signaling. $G\alpha$ subunit includes $G\alpha_s$, $G\alpha_i$, $G\alpha_{q/11}$, and $G\alpha_{12/13}$. The biological effects of LPA are mediated by ligation to specific LPA receptors that are coupled to G-protein families, the $G\alpha_s$, $G\alpha_i$, $G\alpha_q$, and $G\alpha_{12/13}$ (Fukushima and Chun, 2001; Lin et al., 2010; Zhao and Natarajan, 2009). Different LPA receptors couple to different $G\alpha$ subunits. LPA_1 and LPA_2 interact with $G\alpha_i$, $G\alpha_q$, and $G\alpha_{12/13}$; LPA_3 interacts with $G\alpha_i$ and $G\alpha_q$ but not $G\alpha_{12/13}$. LPA_4 appears to couple with all the G-proteins, and LPA_5/GPR92 is coupled to $G\alpha_s$, $G\alpha_{12/13}$, and $G\alpha_q$. LPA_7 functions dependent on $G\alpha_s$ and $G\alpha_{13}$ (Figure 11.3). Tissue expression, intracellular signaling, targets, and *in vivo* functions of LPA and LPA receptors are listed in Table 11.1.

11.4   LPA AND LPA RECEPTORS IN ASTHMA

Asthma is a chronic airway inflammatory disease characterized by airway inflammation, hyperresponsiveness, and airflow obstruction. Asthma causes recurring periods of wheezing, chest tightness, shortness of breath, and coughing. Several cells in the lung are involved in the pathogenesis of asthma, including airway epithelial, smooth muscle, dendritic, eosinophils, and T_H2 cells. Induction of T_H2-associated cytokines, such as IL-4, IL-5, IL-10, and IL-13, and lipid mediators, such

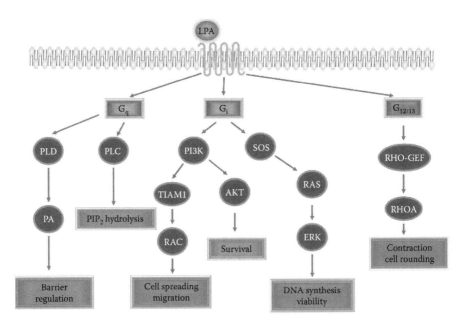

FIGURE 11.3   LPA signaling via LPA receptors. LPA-induced intracellular signal transduction and biological responses through G-protein-coupled LPA receptors.

TABLE 11.1
LPA Targets and *In Vivo* Functions

Receptor Type	Tissue Expression	G-Protein Coupling	Intracellular Targets	*In Vivo* Functions
LPA$_1$/EDG2/vzg1	Brain, lung, testis	G$_i$, G$_q$, G$_{12/13}$	Akt, cAMP, MAPK, GTPases, PLD	Brain development, neuropathy pain, lung fibrosis, tumor metastasis
LPA$_2$/EDG4	Hemopoietic cells, spleen, thymus, intestine	G$_i$, G$_q$, G$_{12/13}$	Akt, MAPK, cAMP, Rho, PLC, [Ca^{2+}]$_i$ PLD	Inhibition of apoptosis, attenuation of eosinophil infiltration into lungs
LPA$_3$/EDG7	Kidney, lung, uterus, ovary	G$_i$, G$_q$	cAMP, ERK, PLC, [Ca^{2+}]$_i$	Embryo spacing, inflammation
LPA$_4$/p2Y9/GPR23	Ovary	G$_i$, G$_q$, G$_{12/13}$, G$_s$	cAMP, [Ca^{2+}]$_i$	Suppression of cell migration and tumor invasion
LPA$_5$/GPR92	Intestine, spleen, neurons	Gq, G12/13, Gs	cAMP, [Ca^{2+}]$_i$	Activation of platelets
LPA$_6$/GPR87	Hair follicles	G$_{12/13}$	cAMP	Wooly hair with hypotrichosis
PPARγ	Ubiquitous	—	—	Metabolism and differentiation

as prostaglandins and LPA, contribute to the pathogenesis of asthma. Here, we will discuss the role of LPA and LPA receptor in the pathogenesis of asthma.

11.4.1 LPA Levels in BAL Fluids and Plasma from Asthmatics

Georas et al. (2007) were the first to demonstrate that LPA is constitutively present in human BAL fluids. Using LC-MS/MS, ~50 nM LPA were detected in baseline, and levels increased about threefold after 18 h of segmental allergen challenge of allergic subjects. Polyunsaturated LPA species, such as 22:6LPA, 20:4LPA, 20:3LPA, 22:4LPA, and 20:2LPA, were the most increased followed by unsaturated 16:1LPA and saturated 18:0LPA. LPA levels are also detectable in normal mouse BAL fluids. In *Schistosoma mansoni* soluble egg-induced allergic murine model, LPA levels in BAL fluids increased ~2.8-fold (Zhao et al., 2009), consistent with findings from human samples.

Recently, LPA levels in plasma from normal healthy control and subjects of asthma were measured by LC-MS/MS (Duff et al., 2010). In contrast to increases in unsaturated LPA levels in BAL fluids, 18:0 LPA in asthmatic subjects was significantly higher than in normal control. The source of LPA in plasma and BAL fluids is not clear. It generates in cells and releases to biological fluids or generates in biological fluids by the action of ATX/lyso PLD, which hydrolyzes LPC to LPA. It is not likely that increases in LPA levels in BAL fluids are through leakage from plasma given the different nature of LPA species in these two biological fluids.

11.4.2   EFFECTS OF LPA ON AIRWAY SMOOTH MUSCLE CELLS

Airway smooth muscle cells contribute to the pathogenesis of asthma through (1) causing airflow obstruction and (2) inducing cytokine release. A study using RT-PCR reported that LPA_{1-3} mRNA are expressed in isolated human airway smooth muscle cells (Ammit et al., 2001). LPA induces airway smooth muscle cell contraction. Isolated tracheal rings from rabbits and cats were stimulated with LPA, and LPA increased the contractile response to the muscarinic agonist methacholine (Toews et al., 1997). Also, LPA decreased the relaxation response to the β-adrenergic-agonist adenylate cyclase activator forskolin (Toews et al., 1997). However, the effect of LPA on contraction is indirect and LPA alone had no effect on airway smooth muscle cell contraction (Toews et al., 1997). LPA induces actin reorganization through $G\alpha_{i-2}$ and $G\alpha_q$ proteins. LPA stimulated cAMP accumulation, PI hydrolysis, and activation of Rho in airway smooth muscle cells and the detailed mechanisms by which LPA regulates the contraction of airway smooth muscle cell will be determined in a future study.

LPA also stimulates airway smooth muscle cell proliferation. LPA induced a concentration-dependent stimulation of [³H]thymidine incorporation into airway smooth muscle cellular DNA (Cerutis et al., 1997). The role of EGF receptor (EGFR) in LPA-induced airway smooth muscle cell proliferation was demonstrated. LPA-induced increases in EGFR binding activity were caused by an increase in EGFR protein expression. EGF alone induced mitogenesis and simultaneously stimulation with EGF and LPA exhibited a significantly synergistic activation of mitogenesis (Cerutis et al., 1997). However, this result was different with other cell types, such as human bronchial epithelial cells, where LPA had an effect on tyrosine phosphorylation and transactivation of EGFR in human airway smooth muscle cells (Cerutis et al., 1997). Meanwhile, β2-adrenergic receptor agonist and cAMP-mediated EPAC pathway inhibits LPA-induced human airway smooth muscle cell proliferation (Kassel et al., 2008).

Although there is no direct evidence to show that LPA induces cytokine release in airway smooth muscle cells, LPA activates transcriptional factors, including NF-κB, AP-1, CREB, NFAT, and SRE complex, in human airway smooth muscle cells (Ediger et al., 2003). AP-1 and NF-κB are known to regulate a diverse group of cytokine production, and their activation by LPA suggests that LPA may regulate cytokine production and release in airway smooth muscle cells. In aortic smooth muscle cells, LPA stimulates IL-6 expression via an LPA_1-regulated PKC and p38MAPK alpha-mediated pathways (Hao et al., 2010). Human airway smooth muscle cells, like other smooth muscle cells, secrete IL-6 and other cytokines in response to stimuli. LPA may regulate airway inflammation through cytokine release from airway smooth muscle cells.

11.4.3   EFFECTS OF LPA ON DENDRITIC CELLS, EOSINOPHILS, AND T_H2 CELLS

Dendritic cells function as antigen-presenting cells to process antigen and present it on the cell surface. Activation of dendritic cells has been considered to be one of the major risk factors for developing asthma. Both immature dendritic cells isolated from bone marrow and LPS-matured dendritic cells express LPA_1, LPA_4, and LPA_5, while immature dendritic cells express higher levels of LPA_3 (Chan et al., 2007). Human immature and mature dendritic cells express the same levels of LPA_{1-3} (Panther et al., 2002). Effects of LPA on immature and mature dendritic cells are significantly different. In immature dendritic cells, LPA induced increases in intracellular Ca^{2+}, actin polymerization, and chemotaxis (Panther et al., 2002). Only unsaturated LPA species induced chemotaxis of immature dendritic cells. Mature mouse dendritic cells had no response to LPA treatment (Chan et al., 2007). The effect of unsaturated LPA on dendritic cell migration is dependent on LPA_3 as evidenced by using isolated immature dendritic cells from mice lacking LPA_3 (Chan et al., 2007). In LPS-matured myeloid dendritic cells, LPA treatment enhanced IL-4 secretion. Also, LPA inhibited IL-12 and TNF-α secretion and enhanced IL-10 secretion from mature dendritic cells (Panther et al., 2002).

Inflammatory cell infiltration into alveolar spaces is a primary marker of airway inflammation. Eosinophils are major inflammatory cells in allergic response and play a central role in the development of late asthmatic reaction. The toxins from eosinophil granules damage the lining of the airway cells, including epithelial cells. RT-PCR analysis shows that human eosinophils express LPA_1 and LPA_3 (Idzko et al., 2004). Eosinophil chemotaxis assay with Boyden chambers showed that LPA has chemotactic activity for eosinophils and reaches maximal reaction at 10 μM. LPA also stimulates superoxide production, increases in intracellular Ca^{2+}, and actin polymerization in eosinophils (Idzko et al., 2004). An *in vivo* experiment showed that after 6 h of LPA inhalation, eosinophil infiltration was detected in BAL fluids from guinea pigs (Hashimoto et al., 2003). A Rho kinase inhibitor, Y-27632, reduced LPA-increased eosinophil cell counts and superoxide production in eosinophils (Hashimoto et al., 2003), suggesting that LPA-induced eosinophil migration may be through Rho kinase-mediated superoxide production in guinea pigs. However, intratracheal administration of LPA to mice did not increase eosinophils accumulation in BAL fluids (Cummings et al., 2004; Zhao et al., 2009). Increasing LPA levels in BAL fluids from human asthmatic subjects does not appear to be a dominant chemoattractant for eosinophils (Georas et al., 2007).

T-helper lymphocyte consists of two terminally differentiated T_H1 and T_H2 cells. Both T_H1 and T_H2 cells express LPA_{1-3}, evidenced by flow cytometry and immunoblotting analysis (Wang et al., 2004). T_H2 cells express higher LPA_1 levels compared to T_H1 cells (Wang et al., 2004). T_H2 cells have been shown to play a central role in allergic asthma in mice and humans. Allergic asthma patients have a greater percentage of T_H2 cells, compared with healthy subjects. LPA increased intracellular Ca^{2+} and chemotaxis of T_H2 cells in a pertussis toxin-sensitive and PI3K-independent manner (Wang et al., 2004). Also, LPA augmented IL-13 secretion under conditions of submaximal T-cell activation and enhanced transcriptional activation of the IL-13 promoter in T_H2 cells (Rubenfeld et al., 2006). IL-13 was identified as a critical regulator of the allergic response. These studies suggest a role of LPA in T_H2 cell chemotaxis and T_H2-type cytokine release.

11.4.4 DUAL EFFECTS OF LPA ON EPITHELIAL CELLS

Airway epithelium is the first contact for inhaled stimuli and has host defense effects in airway inflammatory diseases via its physical barrier and modulation of inflammation. Airway epithelial cells express LPA_{1-3} (Wang et al., 2003). Analysis of total RNA for mRNA expression of LPA receptors by real-time RT-PCR revealed that expression of $LPA_2 > LPA_4 > LPA_1 \geq LPA_3$ in mouse tracheal epithelial cells (Zhao et al., 2009). The production of thymic stromal lymphopoietic (TSLP) and chemokines CCL20 by epithelial cells is critical for the development of allergic airway inflammation by maturation of airway dendritic cells as well as increases in myeloid dendritic cells and T-cell accumulation in airways. LPA treatment of mouse tracheal epithelial cells stimulated TSLP and CCL20 expression via CARMA3-mediated NF-κB activation (Medoff et al., 2009).

However, an anti-inflammatory effect of LPA has been evidenced by induction of IL-13 decoy receptor, IL-13Rα2, and prostaglandin E_2 (PGE_2) by LPA in human bronchial epithelial cells. IL-13Rα2 has higher binding efficiency to IL-13, compared to IL-13Rα1, a receptor for IL-13, and blocks IL-13-mediated signaling and biological functions. LPA induced IL-13Rα2 expression and release in human bronchial epithelial cells through PLD- and JNK-mediated pathway. Pretreatment with LPA attenuated IL-13-induced phosphorylation of STAT6 and cytokine releases, suggesting an anti-inflammatory effect of LPA via attenuation of IL-13-induced inflammation (Zhao et al., 2007b). Furthermore, LPA induced cyclooxygenase-2 (COX-2) and PGE_2 release in human bronchial epithelial cells (He et al., 2008). Although different from results seen in other tissues, in the lung, COX-2 and PGE_2 play an anti-inflammatory role (Sugiura et al., 2007; Suman et al., 2000). The pro- and anti-inflammatory effects of LPA may depend on its target cells.

11.4.5 LPA RECEPTOR KNOCKOUT MURINE MODELS OF ASTHMA

LPA receptors 1–3 expressed in most types of lung cells, including airway epithelial and smooth muscle cells. Chen et al. (2006) showed that LPA_3 was not detectable in both mature and immature dendritic cells; however, Panther et al. (2002) detected LPA_{1-3} in both mature and immature dendritic cells by RT-PCR. Eosinophils express LPA_1 and LPA_3, but not LPA_2 (Idzko et al., 2004). LPA receptors in knockout mice were used to investigate the role of LPA and LPA receptors in a murine model of asthma. *Schistosoma mansoni* egg sensitization and challenge induced eosinophil infiltration into airways and mucus production. LPA_1 heterozygous mice reduce mucus production but not eosinophil infiltration. LPA_2 heterozygous mice reduce both eosinophil infiltration and mucus production (Zhao et al., 2009). These results suggest a proinflammatory role of LPA receptor in an allergic model.

11.5 LPA AND LPA RECEPTORS IN CHRONIC OBSTRUCTIVE PULMONARY DISEASE

Chronic obstructive pulmonary disease (COPD) is characterized by airway inflammation, and progressive destruction of lung parenchyma, a process that in most cases is initiated by noxious gases, air pollutants, including particulate matter and cigarette smoke. Several mechanisms are involved in the development of the disease: influx of inflammatory cells into the lung (leading to chronic inflammation of the airways), imbalance between proteolytic and antiproteolytic enzymes, and oxidant imbalance. Further, recent studies suggest a role for ceramide-mediated apoptosis as a key event in the pathogenesis of COPD (Petrache et al., 2005). However, very little is known regarding the role of LPA and LPA-Rs in the pathogenesis of COPD. Here, the potential role of LPA and LPA receptors in mucus generation and the effects of LPA and cigarette smoke on airway epithelial cells-mediated signaling and inflammation will be discussed.

11.5.1 Effect of LPA on Mucus Generation

Increased mucus production in the respiratory tract, a common symptom of airway inflammation, and airway goblet cell hyperplasia or hypertrophy, is a pathological characteristic of COPD. There is no report to support the effect of LPA on goblet cell hyperplasia and mucin production; however, periodic acid-Schiff (PAS) staining showed reduced goblet cell hyperplasia in LPA_1 and LPA_2 knockout mice (Zhao et al., 2009). Although these results suggest that LPA and LPA receptors may contribute to goblet cell hyperplasia in COPD, a more detailed analysis of LPA and LPA-Rs in lung tissues/BAL fluids from control subjects and COPD patients is necessary to better understand the pathogenesis.

11.5.2 LPA and Cigarette Smoke Induce EGFR Transactivation in Airway Epithelial Cells

Cigarette smoke is a risk factor for COPD. Several studies have shown that cigarette smoke induces mucin production through EGF receptor transactivation in human airway epithelial cells (Cortijo et al., 2011; Khan et al., 2008; Shao et al., 2004). Similar to cigarette smoke inhalation, LPA induces EGF receptor transactivation in human airway epithelial cells (Zhao et al., 2006) that was attenuated by metalloprotease inhibitor, GM-6001. Further, EGF receptor-specific inhibitor, AG-1478, blocked IL-8 production by cigarette smoke and LPA in human airway epithelial cells (Cortijo et al., 2011; Shao et al., 2004; Zhao et al., 2006). These studies indicate that EGF receptor transactivation may play a central role in both cigarette smoke and LPA-mediated airway inflammation.

11.6 LPA AND LPA RECEPTORS IN PULMONARY FIBROSIS

Pulmonary fibrosis is an inflammatory lung disorder characterized by abnormal formation of fibrous connective tissue in the lungs. Fibroblast proliferation and migration and differentiation of fibroblast are keys to the pathogenesis of fibrosis. The recent studies demonstrate that LPA and LPA receptors are involved in the pathogenesis of pulmonary fibrosis via stimulation of fibroblast migration, cytokine releases, and differentiation of fibroblast.

11.6.1 LPA Levels in BAL Fluids in Pulmonary Fibrosis

Bleomycin has been widely used in mice to model pulmonary fibrosis to investigate its pathogenesis and for the evaluation of potential therapies. LPA levels in BAL fluids significantly increased after 5–10 days of bleomycin challenge (Tager et al., 2008). In idiopathic pulmonary fibrosis subjects, LPA levels in BAL fluids significantly increased compared to normal healthy control (Tager et al., 2008). The source of LPA in BAL fluids has not been demonstrated.

11.6.2 Effects of LPA on Fibroblasts

LPA_1 is a major receptor in mouse pulmonary fibroblast. LPA treatment of human fetal lung fibroblast stimulated mitogenesis (Tager et al., 2008). Combined with growth factors, EGF, bFGF, or PDGF-BB stimulated synergistic mitogenic responses. LPA also stimulates fibroblast proliferation in NIH 3T3 and Rat-1 fibroblast (Malcolm et al., 1996). However, LPA_1 knockout mice did not show changes in fibroblast proliferation in bleomycin-induced BAL fluid, which contains ~200 nM of LPA, suggesting bleomycin-induced BAL fluids-mediated lung fibroblast proliferation through the LPA/LPA_1 pathway (Tager et al., 2008). The bleomycin-induced BAL fluids-mediated lung fibroblast proliferation may not be involved with LPA or may depend on other LPA receptors. However, BAL fluids from bleomycin-induced mice stimulated chemotaxis of fibroblast, while fibroblasts isolated from LPA_1 knockout mice show reduced chemotaxis in response to BAL fluids from bleomycin-induced mice, suggesting that LPA in BAL fluids contributes to chemotaxis of fibroblast (Tager et al., 2008). The effect of LPA on fibroblast migration was determined by using isolated fibroblasts from wild-type mice. LPA treatment stimulates fibroblast chemotaxis and reaches maximal effect at 10 nM. Inhibition of $G\alpha_i$-protein by pertussis toxin blocked LPA-induced fibroblast chemotaxis, suggesting that $G\alpha_i$-coupled LPA receptor is involved in the effect of LPA (Tager et al., 2008). These results are consistent with findings from other sources of fibroblast. LPA induces Rat2 fibroblast migration and overexpression of LPP_1 decreases fibroblast migration in response to LPA (Pilquil et al., 2006). Similar to the effect on airway smooth muscle cells, LPA treatment of human fetal lung fibroblasts augmented contraction via a pertussis toxin-insensitive, phospholipase C-mediated pathway (Mio et al., 2002). Furthermore, LPA induces contraction of murine embryonic fibroblast cells (Emmert et al., 2004).

11.6.3 Effects of LPA on TGFβ1 Signaling

TGFβ1 is a multifunctional cytokine that regulates proliferation, differentiation, wound healing, and tissue repair. TGFβ1 levels in plasma and BAL fluids elevate in pulmonary fibrosis and $\alpha_v\beta_6$ integrin-dependent TGFβ1 activation plays a central role in the pathogenesis of pulmonary fibrosis. Recent studies show that LPA enhances TGFβ1 activation in lung epithelial cells and airway smooth muscle cells. LPA induces $\alpha_v\beta_6$ integrin-dependent TGFβ1 activation via LPA_2, RhoA, and Rho kinase in human epithelial cells (Xu et al., 2009). LPA treatment increased the expression of the type-1 inhibitor of plasminogen activator (PAI-1), which is an important physiological regulator of cell motility and is regulated by TGFβ. Neutralizing antibodies to $\alpha_v\beta_6$ and TGFβ blocked the effect of LPA on PAI-1 expression, suggesting that LPA activates TGFβ pathway (Xu et al., 2009). Further, RhoA or Rho kinase inhibitor blocked LPA-induced $\alpha_v\beta_6$-dependent TGFβ1 activation (Xu et al., 2009). Similarly, in airway smooth muscle cells, LPA activates TGFβ via the $\alpha_v\beta_5$ pathway (Tatler et al., 2010). The role of LPA receptor in TGFβ-induced biological functions was determined. TGFβ stimulates procollagen type I α, smooth muscle actin, and fibronectin gene expression in both fibroblasts isolated from wild-type and LPA_1 knockout mice (Tager et al., 2008). TGFβ induces biological function independent of LPA receptor.

11.6.4   LPA Receptor Knockout Murine Models of Fibrosis

Bleomycin challenge of mice induces fibrosis evidenced by the accumulation of collagen of FSP1-positive fibroblasts, inflammatory cells infiltration, and vascular leak (Tager et al., 2008). LPA_1 knockout mice significantly protect inflammation and injury by bleomycin (Tager et al., 2008). After 21 days of bleomycin challenge, there was no mortality of LPA_1 knockout mice, while mortality of wild-type mice was 50%. LPA_1 knockout mice displayed reduced accumulation of collagen and FSP1-positive fibroblasts, as well as total protein, compared to bleomycin-challenged wild-type mice. Bleomycin challenge-increased inflammatory cell infiltration is independent of LPA_1. LPA_1 knockout mice show no significant changes in inflammatory cells, such as neutrophils, CD3[+], and CD8[+] T cells (Tager et al., 2008). A recent study showed that bleomycin challenge increased LPA_2 expression in lung, and down-regulation of LPA_2 in lung epithelial cells inhibited LPA-mediated activation of TGFβ (Xu et al., 2009). The activation of TGFβ by LPA in isolated embryonic fibroblasts from LPA_2 knockout mice was much less than isolated cells from wild-type mice (Xu et al., 2009). The role of LPA_2 in bleomycin-induced fibrosis will be further investigated by using LPA_2 knockout mice.

11.6.5   LPA1 Antagonist in Bleomycin-Induced Lung Fibrosis

The proinflammatory role of LPA_1 in lung fibrosis was recently confirmed with an antagonist selective for the LPA_1 in treating bleomycin-induced lung fibrosis. LPA_1 antagonist 4′-{4-[(R)-1-(2-chloro-phenyl)-ethoxycarbonylamino]-3-methyl-isooxazol-5-yl}-biphenyl-4-yl)-acetic acid (AM966) reduced lung injury, vascular leak, inflammation, and fibrosis following intratracheal instillation of bleomycin to female C57BL/6 mice (Swaney et al., 2010). These studies suggest that this orally bioavailable LPA_1 receptor antagonist may be beneficial in treating human pulmonary fibrosis, a finding that will require clinical trials to confirm.

11.7   LPA AND LPA RECEPTORS IN ACUTE LUNG INJURY

Acute lung injury (ALI) is characterized by increased lung vascular and epithelial permeability, inflammatory cell infiltration, and cytokine storm. Primary ALI is caused by direct injury to lung, and secondary ALI is caused by other diseases, such as prancreatitis. The role of lysophospholipids in the pathogenesis of ALI has been investigated. For example, sphingosine kinase-1 knockout mice show decreases in S1P levels in lung and BAL fluids and elevate LPS-induced inflammation and injury (Wadgaonkar et al., 2009). S1P lyase knockout mice show increases in S1P levels in lung and BAL fluids and reduce LPS-induced inflammation and injury (Zhao et al., 2009). Here, we will discuss the role of LPA in ALI.

11.7.1   LPA Levels in BAL Fluids and Plasma from ALI

LPA levels in BAL fluids and plasma were detected by LC-MS/MS in an LPS-induced murine model of ALI. LPS, a glycolipid present in the outer membrane of Gram-negative bacteria, is composed of a polar lipid head group (lipid A) and a chain of repeating disaccharides. Intratracheal administration of LPS induced lung injury and inflammation. LPA levels in BAL fluids and plasma were measured by LC-MS/MS after 24 h of intratracheal LPS challenge. LPA levels increased in both BAL fluids and plasma ~2–3-fold compared to LPA levels in control mice (Zhao, Y. et al. unpublished data).

11.7.2   Effects of LPA on Cytokine Release in Epithelial Cells

Airway epithelial cells release various cytokines and lipid mediators in response to stimuli. For example, particulate matter or influenza virus A induces IL-6 and IL-8 secretion in human bronchial epithelial cells. LPA treatment of human bronchial epithelial cells induces IL-8 secretion via $G\alpha_i$-coupled LPA receptors (Cummings et al., 2004; Saatian et al., 2006; Zhao et al., 2005, 2006). IL-8 plays a central role in neutrophil chemotaxis. LPA activates transcriptional factors, NF-κB, AP-1, and C/EBPβ, as evidenced by inducing phosphorylation of I-κB, JNK, and C/EBPβ, and nuclear translocation of NF-κB and AP-1 (Zhao and Natarajan, 2009). Inhibition or down-regulation of these transcriptional factors reduces LPA-induced IL-8 gene expression. Cross-talk between GPCR and receptor tyrosine kinases (RTKs) regulates GPCR-mediated signaling and biological function. LPA receptors cross-link with various RTKs such as PDGF-R, EGF-R, and c-Met (Wang et al., 2003; Zhao et al., 2006, 2007a) (Figure 11.4). EGF-R transactivation is regulated by PKCδ and Lyn kinase-mediated pro-HB-EGF shedding in human bronchial epithelial cells (Zhao and Natarajan, 2009). Unlike EGF-R, cross-talk with PDGF-R or c-Met is not involved in LPA-induced IL-8 secretion. Recent study also suggests that intracellular LPA, which is regulated by AGK, mediates extracellular LPA-mediated IL-8 secretion via regulation of EGF-R transactivation (Kalari et al., 2009). Furthermore, several other signal proteins and molecules, such as PKCδ, p38 MAPK, and intracellular Ca^{2+}, regulates LPA-induced activation of transcriptional factors and IL-8 secretion (Zhao and Natarajan, 2009).

11.7.3   Effects of LPA on Epithelial Barrier Function

Airway epithelial barrier integrity is maintained by tight and adherens junctions. Adherens junctions regulate whole integrity, since loss of adherens junctions

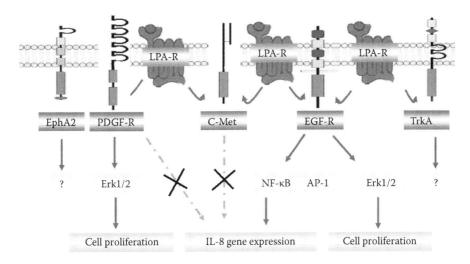

FIGURE 11.4 Cross-talk between LPA receptors and receptor tyrosine kinases (RTKs). Transactivation of RTKs by LPA receptors regulates LPA-mediated signal transduction and biological functions.

subsequently leads to impaired tight junctions and increases permeability. E-cadherin is the central protein in adherens junction. LPA treatment of airway epithelial cells stimulates E-cadherin accumulation at cell–cell contacts and increases transepithelial resistance that reflects barrier integrity (He et al., 2008; Zhao et al., 2007a). Inhibition of PKC isoforms PKCδ and PKCζ attenuates the effect of LPA on E-cadherin accumulation at cell–cell contacts (He et al., 2008; Zhao et al., 2007a) (Figure 11.5). LPS induces airway epithelial barrier disruption and E-cadherin translocation from cell–cell contacts in the cytoplasm. Post-treatment with LPA reverses

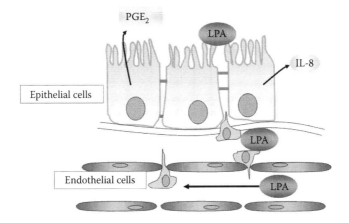

FIGURE 11.5 LPA in lung functions. LPA modulates different types of cells in the lung and mediates normal and pathophysiological responses.

LPS-induced airway epithelial barrier disruption and permeability (He et al., 2008). These results suggest that LPA plays an airway epithelial barrier enhancement effect.

As reported earlier, LPA receptors cross-talk with c-Met in airway epithelial cells. Unlike EGF-R or PDGF-R transactivation by LPA, LPA induces serine phosphorylation of c-Met and reduces HGF-induced activation of c-Met, suggesting that LPA induces c-Met transinactivation. However, LPA treatment induces c-Met accumulation at cell–cell contacts (Zhao et al., 2007a). Further, E-cadherin and c-Met interaction was determined by coimmunoprecipitation and coimmunostaining, indicating a role of c-Met in LPA-induced airway epithelial barrier integrity.

11.7.4 EFFECTS OF LPA ON NEUTROPHILS AND MACROPHAGES

Neutrophils are important contributors to acute lung information and injury. Neutrophils rapidly infiltrate into alveolar spaces in response to proinflammatory cytokines, such as IL-8, IL-1β, and TNF-α. Neutrophils play a key role in the first line of defense against invading pathogens by phagocytes and deregulation and also release proinflammatory cytokines, elastase, and oxygen metabolites, thus enhancing inflammation, increasing epithelial and endothelial permeability, and hurting host tissues. Mouse neutrophils express LPA_2 and LPA_5 (Tager et al., 2008). Human neutrophils express LPA_2, while LPA_1 is increased in neutrophils isolated from pneumonia patients (Rahaman et al., 2006). We have discussed that LPA treatment of epithelial cells stimulates IL-8 secretion, which induces chemotaxis of neutrophils. LPA itself stimulates polarization and motility of neutrophils; however, there is an argument for the role of LPA in chemotaxis of neutrophils. The higher concentration of LPA (~1 mM) significantly induces chemotaxis of neutrophils (Fischer et al., 2001). Chemotaxis of neutrophils from patients with pneumonia compared with control subjects have higher expression of LPA_1 and levels are significantly increased in response to LPA (Rahaman et al., 2006). However, several studies show that LPA has no effect on chemotaxis of neutrophils (Itagaki et al., 2005). The analysis of different expressions of LPA receptors in inactivated and activated neutrophils may provide a mechanism for this argument. LPA also induces the activation of PLD activity, increases in intracellular Ca^{2+}, neutrophil degranulation, and respiratory burst (Tou and Gill, 2005). The opposite effect of LPA on respiratory burst was revealed. Chettibi et al. (1994) show that LPA inhibits PMA-stimulated metabolic burst in neutrophils.

Alveolar macrophages are the first line of contact to inhaled stimuli. Macrophages play a central role in various lung diseases via phagocytosis. Macrophages clear bacteria as well as recruited cells, such as neutrophils. Human and rat alveolar macrophages express LPA_{1-3} (Hornuss et al., 2001) and mouse alveolar macrophages express LPA_2, LPA_4, and LPA_5 (Tager et al., 2008). LPA stimulates $O_2^{-\bullet}$ generation in human and rat alveolar macrophages (Hornuss et al., 2001). LPA upregulates oxidized low-density lipoprotein (Ox-LDL) uptake in mouse J774A.1 macrophages via ligation to $G\alpha_i$-coupled LPA_3 (Chang et al., 2008a). LPA also enhances IL-1β and TNF-α expression in mouse peritoneal macrophages (Chang et al., 2008b). In J774A.1 macrophages, ROS production mediates LPA-induced IL-1β expression (Chang et al., 2008b). LPA activates PI3K and protects macrophages from apoptosis induced by serum deprivation (Koh et al., 1998). Inhibition of PI3K and downstream kinase activated by PI3K siRNA blocks enhancement of macrophage survival by LPA. Recent study shows that LPA significantly inhibited LPS-induced TNF-α production, but not IL-6, in peritoneal macrophages (Fan et al., 2008). PI3K and Erk1/2 pathways contribute to the anti-inflammatory effect of LPA in macrophages.

11.7.5 LPA RECEPTOR KNOCKOUT MURINE MODELS OF ALI

Consistent with data from LPA receptor knockout mice in asthma and pulmonary fibrosis, LPA receptor knockout mice show decreases in inflammation in LPS-induced murine model of ALI. In wild-type mice, intratracheal administration of LPS for 24 h increased IL-6 and protein levels in BAL fluids (Zhao et al., 2010). LPA$_1$ and LPA$_2$ knockout mice show a reduction in inflammatory cell infiltration and IL-6 levels in BAL fluids and have no change in protein levels in BAL fluids by LPS, compared to LPS-challenged wild-type mice. This suggests that LPA receptor knockout mice reduce inflammation, but not injury, by LPS challenge (Zhao et al., 2010). The effect of LPA receptor 1 and 3 antagonist Ki16425 was determined by intratracheal administration before LPS challenge. Ki16425 reduces LPS-induced inflammatory cell infiltration and IL-6 levels in BAL fluids, but had no effect on protein level in BAL fluids (Zhao et al., 2010). Inhibition of LPS-induced inflammation, but not injury, by LPA receptor antagonist confirms the data from LPA receptor knockout mice.

11.7.6 LPA ADMINISTRATION INDUCES LUNG INFLAMMATION *IN VIVO*

Effects of LPA administration into lung in pulmonary inflammation were investigated. After 4 h of inhalation of LPA, increases in eosinophils and neutrophils in BAL fluids were detected in guinea pigs (Hashimoto et al., 2003). RhoA-associated protein kinase (ROCK) regulates this effect by LPA. Inhalation of LPA also enhanced the airway response induced by intravenous administration of acetylcholine in guinea pigs (Hashimoto et al., 2003). Intratracheal administration of LPA into mouse lung increased MIP-2 levels at 3 h and neutrophil infiltration in alveolar spaces at 6 h. After 12 h, the levels of MIP-2 and neutrophils returned to normal basal levels (Cummings et al., 2004; Zhao et al., 2009). The fact that LPA rapidly returns to original levels may be due to the short half-life of LPA. LPA can be dephosphorylated by LPPs on various cell surfaces. Administration of LPA has no effect on protein levels in BAL fluids (Cummings et al., 2004). LPA receptor knockout mice block the effect of LPA in neutrophil infiltration (Zhao et al., 2009). This result suggests that the administration of LPA increases pulmonary innate inflammation at early time points (<6 h), but not at later time periods (>12 h). LPA does not induce lung leakage. The effects of the administration of LPA are through LPA receptors, and not due to lipid nonspecific binding. Recent studies show that the administration of LPA by intravenous or intratracheal injection attenuates LPS challenge-induced acute lung inflammation and injury. Intravenous injection with LPA followed by intraperitoneal injection of LPS significantly increases mice survival, reduces TNF-α level in BAL fluids, and decreases MPO activity in lung by LPS (Fan et al.,

2008). Intratracheal administration of LPA at 1 h after LPS challenge reduces LPS-induced inflammatory cell infiltration and IL-6 and protein levels in BAL fluids (He et al., 2008).

11.8 LPA AND LPA RECEPTOR AGONISTS AND ANTAGONISTS

LPA and other lipid mediators such as S1P, FTY720, FTY720 phosphate, and 2-arachidonylglycerol are attractive targets for therapy, including immunosuppressant, antiobesity, lung injury, asthma, pulmonary fibrosis, and multiple sclerosis (Brown et al., 2007; Calne, 2004). In this context, there has been considerable interest and progress in developing inhibitors of LPA-producing enzymes, including ATX and antagonists and agonists for LPA receptors and LPA signaling. A recent review by Kano et al. (2008) summarizes the recent advances in the development of LPA receptor-targeted agonists and antagonists, LPA analogs, and ATX inhibitors as tools for the elucidation of LPA biology and drug development. None of the LPA analogs or LPA receptor agonists/antagonists has undergone Phase I or Phase II clinical trials. Future development of small-molecule inhibitors and second- and third-generation agonists/antagonists for LPA receptors will facilitate the prevention and treatment of various lung diseases. Table 11.2 summarizes some of the selective LPA analogs, LPA receptor agonists and antagonists, and ATX inhibitors. Several of the LPA receptor agonists and antagonists exhibit different specificity and potency, and majority of the small-molecular-weight agents have been designed to modulate LPA_{1-3} with a few recent studies focusing on other non-EDG-related LPA receptors (LPA_{4-7}) (Tables 11.3 and 11.4). The reported effects of LPA receptor agonists and antagonists have utilized *in vitro* assays such as receptor binding or intracellular calcium changes for validation of the specificity; however, only a few studies have

---

TABLE 11.2
Selective LPA Analogs, LPA Receptor Agonists and Antagonists, and Autotaxin Inhibitors

Target	Agonists	Antagonists
LPA_1	*N*-acyl ethanolamide phosphate; VPC12086; VPC31143	DGPP (C8:0); phosphatidic acid (C8:0); VPC12249; AM966
LPA_2	*N*-acyl ethanolamide phosphate	Compound 35: ki16425
LPA_3	OMPT; XY-17; T13; VPC12086; VPC31143	DGPP (C8:0); phosphatidic acid (C8:0); VPC12249; K916425
LPA_4	None	α-Bromomethylene LPA analogs
LPA	Difluoromethylene phosphate analogs (XX); 2-arachidonyl LPA	—

Target	Inhibitor	Activator
Autotaxin (lysophospholipase D; ENPP2)	Cyclic phosphatidic acid (cPA); carba cPA: oleoyl thiophosphate; Darmstoff compounds; VPC8a202	Phospholipase C gamma; integrinα6β4; Epstein–Barr virus

TABLE 11.3
Effective Concentrations (EC$_{50}$) of LPA Receptor Agonists

Agonists	Target(s)	EC$_{50}$ (nM)				
		LPA$_1$	LPA$_2$	LPA$_3$	LPA$_4$	LPA$_5$
LPA (18:1)	LPA$_{1-3}$	~1–10	~10–100	~100–250	~10–100	~10–100
NAEPA	LPA$_{1,2}$	~200–500	~10–100	>5000	NA	NA
2S-OMPT	LPA$_3$	>1000	>1000	~0.1–1	NA	NA
VPC31143	None	~1–10	~100–500	~100–500	NA	NA
VPC12086	None	~1–10	~10–100	~100–500	NA	NA
XY-17	LPA$_3$	>1000	>1000	<1	NA	NA
LPA analog T10	LPA$_3$	>1000	>1000	~1–10	NA	NA
LPA analog T13	LPA$_3$	~100–500	>1000	<1	NA	NA
Difluoromethylene phosphonade cyclic carba-LPA	None	>5000	>5000	>5000	NA	NA

Note: LPA, lysophosphatidic acid; NAEPA, *N*-acyl ethanolamide phosphoric acid; OMPT, 1-oleoyl-2-*o*-methyl-*rac*-glycerophosphothioate.

demonstrated the efficacy *in vivo* with animal models. The recently reported LPA$_1$ receptor antagonist, AM966, showed good pharmacokinetic profile following oral dosing in a mouse bleomycin model of lung fibrosis (Swaney et al., 2010). Additionally, AM966 reduced lung injury, vascular leakage, and inflammation at various time intervals after intratracheal instillation of bleomycin. Many of the LPA receptor agonists and antagonists (Figures 11.6 and 11.7) require *in vivo* validation in animal models to demonstrate the efficacy, specificity, and tolerance prior to clinical trials.

TABLE 11.4
Effective Concentrations (IC$_{50}$) of LPA Receptor Antagonists

Antagonists	Target(s)	IC$_{50}$ (nM)				
		LPA$_1$	LPA$_2$	LPA$_3$	LPA$_4$	LPA$_5$
DGPP (8:0)	LPA$_{1-3}$	>5000	NA	>100	NA	NA
VPC12249	LPA$_{1,3}$	>5000	NA	>5000	NA	NA
K:16425	LPA$_{1,2,3}$	~100–1000	>10,000	NA	NA	NA
Compound 35	LPA$_2$	>10,000	10–100	>10,000	NA	NA
1-Bromo-(3*S*)-hydroxy-4-(palmitoyl)butyl phosphate	LPA$_{1,4}$	>1000	>1000	>1000	~100–500	NA
AM966	LPA$_1$	17	1700	1600	7700	8600

Note: AM966, (4′-{4-[(*R*)-1-(2-chloro-phenyl)-ethoxycarbonylamino]-3-methyl-isooxazol-5-yl}-biphenyl-4-yl)-acetic acid; DGPP, diacylglycerol pyrophosphate.

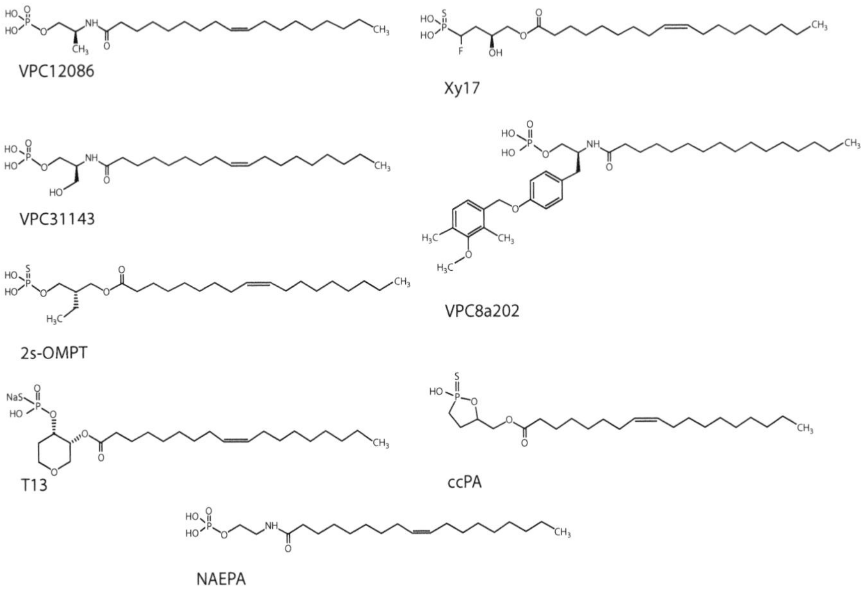

FIGURE 11.6   Structures of LPA and LPA receptor agonists.

11.9   EXPERT OPINION

LPA, a simple bioactive lipid mediator, has been implicated in normal cellular functions through LPA receptors and abnormal signal transduction in a number of human diseases, including atherosclerosis, cancer, idiopathic pulmonary fibrosis, pain, and inflammation. Therefore, LPA receptors that are G-protein-coupled and LPA

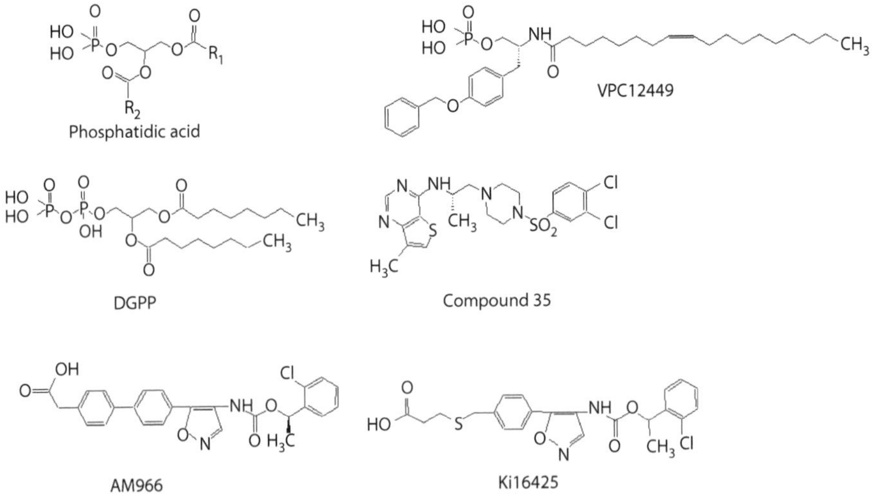

FIGURE 11.7   Structures of LPA and LPA antagonists.

metabolizing enzymes have been targets for drug development. Several novel inhibitors specific for LPA receptors and ATX/lyso PLD (LPA-producing enzyme) have been reported to be effective in ameliorating or minimizing fibrosis, inflammation, airway hyperresponsiveness, and tumor metastasis in cell culture and animal models. These novel inhibitors of LPA receptors and ATX need to be tested rigorously for cytotoxicity and specificity before being evaluated for human clinical Phase I and II trials. Most of the currently available LPA receptor agonists/antagonists and autotoxin inhibitors are lipid-based and the development of small-molecular-weight and water-soluble inhibitors will be ideal for better targeting and drug delivery. In this context, the recent development of LPA$_1$ receptor antagonist, AM966, may prove to be beneficial in treating lung injury in fibrosis as well as other conditions characterized by inflammation and edema. With an increasing role of LPA and LPA receptors in the pathobiology of human diseases, the development of novel antagonists and inhibitors should translate to new drugs for clinical trails and FDA approval.

ACKNOWLEDGMENTS

Part of this work was supported by grants from National Institutes of Health/Heart, Lung and Blood Institute RO1 HL079396 to Viswanathan Natarajan and RO1 HL091916 to Yutong Zhao. We thank Ms Donghong He for her technical assistance and Dr. Evgeny Berdyshev for LPA measurements by LC/MS/MS. We wish to thank Drs. P. V. Subbaiah and Ting Wang, University of Illinois at Chicago for drawing the chemical structures.

REFERENCES

Ali, G., M. S. Chishti, S. I. Raza et al. 2007. A mutation in the lipase H (LIPH) gene underlie autosomal recessive hypotrichosis. *Hum Genet.* 121: 319–325.

Ammit, A. J., A. T. Hastie, L. C. Edsall et al. 2001. Sphingosine 1-phosphate modulates human airway smooth muscle cell functions that promote inflammation and airway remodeling in asthma. *FASEB J.* 15: 1212–1214.

An, S., T. Bleu, O. G. Hallmark, and E. J. Goetzl. 1998. Characterization of a novel subtype of human G protein-coupled receptor for lysophosphatidic acid. *J Biol Chem.* 273: 7906–7910.

Aoki, J. 2004. Mechanisms of lysophosphatidic acid production. *Semin Cell Develop Biol.* 15: 477–489.

Aoki, J., A. Inoue, and S. Okudaira. 2008. Two pathways for lysophosphatidic acid production. *Biochim Biophys Acta.* 1781: 513–518.

Aoki, J., A. Taira, Y. Takanezawa et al. 2002. Serum lysophosphatidic acid is produced through diverse phospholipase pathways. *J. Biol. Chem.* 277: 48737–48744.

Baker, D. L., D. M. Desiderio, D. D. Miller et al. 2001. Direct quantitative analysis of lysophosphatidic acid molecular species by stable isotope dilution electrospray ionization liquid chromatography-mass spectrometry. *Anal Biochem.* 292: 287–295.

Baker, D. L., Y. Fujiwara, K. R. Pigg et al. 2006. Carba analogs of cyclic phosphatidic acid are selective inhibitors of autotaxin and cancer cell invasion and metastasis. *J Biol Chem.* 281: 22786–22793.

Baker, R. R. and H. Y. Chang. 1999. Evidence for two distinct lysophospholipase activities that degrade lysophosphatidylcholine and lysophosphatidic acid in neuronal nuclei of cerebral cortex. *Biochim Biophys Acta.* 1438: 253–263.

Bandoh, K., J. Aoki, H. Hosono et al. 1999. Molecular cloning and characterization of a novel human G-protein-coupled receptor, EDG7, for lysophosphatidic acid. *J Biol Chem.* 274: 27776–27785.

Bektas, M., S. G. Payne, H. Liu et al. 2005. A novel acylglycerol kinase that produces lyso-phosphatidic acid modulates cross talk with EGFR in prostate cancer cells. *J Cell Biol.* 169: 801–811.

Bligh, E. G. and W. J. Dyer. 1959. A rapid method of total lipid extraction and purification. *Can J Biochem Physiol.* 37: 911–917.

Brindley, D. N. 2004. Lipid phosphate phosphatases and related proteins: Signaling functions in development, cell division, and cancer. *J Cell Biochem.* 92: 900–912.

Brown, B. A., P. P. Kantesaria, and L. M. McDevitt. 2007. Fingolimod: A novel immunosup-pressant for multiple sclerosis. *Ann Pharmacother.* 41: 1660–1668.

Calne, S. R. 2004. FTY: Not just a homely drug. *Transplantation.* 77: 1327.

Cerutis, D. R., M. Nogami, J. L. Anderson et al. 1997. Lysophosphatidic acid and EGF stimulate mitogenesis in human airway smooth muscle cells. *Am J Physiol.* 273: L10–L15.

Chan, L. C., W. Peters, Y. Xu et al. 2007. LPA3 receptor mediates chemotaxis of immature murine dendritic cells to unsaturated lysophosphatidic acid (LPA). *J Leukoc Biol.* 82: 1193–1200.

Chang, C. L., H. Y. Hsu, H. Y. Lin et al. 2008a. Lysophosphatidic acid-induced oxidized low-density lipoprotein uptake is class A scavenger receptor-dependent in macrophages. *Prostaglandins Other Lipid Mediat.* 87: 20–25.

Chang, C. L., M. E. Lin, H. Y. Hsu et al. 2008b. Lysophosphatidic acid-induced interleukin-1 beta expression is mediated through Gi/Rho and the generation of reactive oxygen species in macrophages. *J Biomed Sci.* 15: 357–363.

Chen, J. H., F. Zou, N. D. Wang et al. 2000. Production and application of LPA polyclonal antibody. *Bioorg Med Chem Lett.* 10: 1691–1693.

Chen, R. J., J. Roman, J. Guo et al. 2006. Lysophosphatidic acid modulates the activation of human monocyte-derived dendritic cells. *Stem Cells Dev.* 15: 797–794.

Chettibi, S., A. J. Lawrence, R. D. Stevenson et al. 1994. Effect of lysophosphatidic acid on motility, polarisation and metabolic burst of human neutrophils. *FEMS Immunol Med Microbiol.* 8: 271–281.

Cortijo, J., M. Mata, J. Milara et al. 2011. Aclidinium inhibits cholinergic and tobacco smoke-induced MUC5AC in human airways. *Eur Respir J.* 37: 244–254.

Cummings, R., Y. Zhao, D. Jacoby et al. 2004. Protein kinase Cdelta mediates lysophospha-tidic acid-induced NF-kappaB activation and interleukin-8 secretion in human bronchial epithelial cells. *J Biol Chem.* 279: 41085–41094.

Duff, R. F., R. Block, A. Friedman et al. 2010. Semi-quantitative analysis of plasma lyso-phosphatidic acid levels in patients with and without asthma. *Am J Respir Crit Care Med.* 181: A5626.

Ediger, T. L., N. A. Schulte, T. J. Murphy et al. 2003. Transcription factor activation and mito-genic synergism in airway smooth muscle cells. *Eur Respir J.* 21: 759–769.

Eichholtz, T., K. Jalink, I. Fahrenfort et al. 1993. The bioactive phospholipid lysophosphatidic acid is released from activated platelets. *Biochem J.* 291 677–680.

Emmert, D. A., J. A. Fee, Z. M. Goeckeler et al. 2004. Rho-kinase-mediated Ca^{2+}-independent contraction in rat embryo fibroblasts. *Am J Physiol Cell Physiol.* 286: C8–C21.

Fan, H., B. Zingarelli, V. Harris et al. 2008. Lysophosphatidic acid inhibits bacterial endotoxin-induced pro-inflammatory response: Potential anti-inflammatory signaling pathways. *Mol Med (Cambridge).* 14: 422–428.

Fischer, L. G., M. Bremer, E. J. Coleman, et al. 2001. Local anesthetics attenuate lysophospha-tidic acid-induced priming in human neutrophils. *Anesth Analg.* 92: 1041–1047.

Folch, J., M. Lees, and G. H. Sloane Stanley. 1957. A simple method for the isolation and purification of total lipids from animal tissues. *J Biol Chem.* 226: 497–509.

Fujiwara, Y. 2008. Cyclic phosphatidic acid—A unique bioactive phospholipid. *Biochim Biophys Acta.* 1781: 519–524.

Fujiwara, Y., A. Sebok, S. Meakin et al. 2003. Cyclic phosphatidic acid elicits neurotrophin-like actions in embryonic hippocampal neurons. *J Neurochem.* 87: 1272–1283.

Fukushima, N. and J. Chun. 2001. The LPA receptors. *Prostaglandins.* 64: 21–32.

Furneisen, J. M. and G. M. Carman. 2000. Enzymological properties of the LPP1-encoded lipid phosphatase from *Saccharomyces cerevisiae. Biochim Biophys Acta.* 1484: 71.

Gendaszewska-Darmach, E. 2008. Lysophosphatidic acids, cyclic phosphatidic acids and autotaxin as promising targets in therapies of cancer and other diseases. *Act Biochim Pol.* 55: 227–240.

Georas, S. N., E. Berdyshev, W. Hubbard et al. 2007. Lysophosphatidic acid is detectable in human bronchoalveolar lavage fluids at baseline and increased after segmental allergen challenge. *Clin Exp Allergy.* 37: 311–322.

Hao, F., M. Tan, D. D. Wu et al. 2010. LPA induces IL-6 secretion from aortic smooth muscle cells via an LPA1-regulated, PKC-dependent, and p38alpha-mediated pathway. *Am J Physiol Heart Circ Physiol.* 298: H974–H983.

Hashimoto, T., M. Yamashita, H. Ohata et al. 2003. Lysophosphatidic acid enhances *in vivo* infiltration and activation of guinea pig eosinophils and neutrophils via a Rho/Rho-associated protein kinase-mediated pathway. *J. Pharmacol Sci.* 91: 8–14.

He D., V. Natarajan, R. Stern et al. 2008. Lysophosphatidic acid-induced transactivation of epidermal growth factor receptor regulates cyclo-oxygenase-2 expression and prosta-glandin E(2) release via C/EBPbeta in human bronchial epithelial cells. *Biochem J.* 412: 153–162.

Hecht, J. H., J. A. Weiner, S. R. Post et al. 1996. Ventricular zone gene-1 (vzg-1) encodes a lysophosphatidic acid receptor expressed in neurogenic regions of the developing cerebral cortex. *J Cell Biol.* 135: 1071–1083.

Hiroyama, M. and T. Takenawa 1998. Purification and characterization of a lysophosphatidic acid-specific phosphatase. *Biochem J.* 336: 483–489.

Hornuss, C., R. Hammermann, M. Fuhrmann et al. 2001. Human and rat alveolar macro-phages express multiple EDG receptors. *Eur J Pharmacol.* 429: 303–308.

Hotta, H., F. Kagitani, and K. Murakami-Murofushi. 2006. Cyclic phosphatidic acid stimu-lates respiration without producing vasopressor or tachycardiac effects in rats. *Eur J Pharmacol.* 543: 27–31.

Idzko, M., M. Laut, E. Panther et al. 2004. Lysophosphatidic acid induces chemotaxis, oxygen radical production, CD11b up-regulation, Ca^{2+} mobilization, and actin reorganization in human eosinophils via pertussis toxin-sensitive G proteins. *J Immunol.* 172: 4480–4485.

Itagaki, K., K. B. Kannan, and C. J. Hauser. 2005. Lysophosphatidic acid triggers calcium entry through a non-store-operated pathway in human neutrophils. *J Leukoc Biol.* 77: 181–189.

Ito, Y., U. Ponnappan, and D. A. Lipschitz. 1996. Excess formation of lysophosphatidic acid with age inhibits myristic acid-induced superoxide anion generation in intact human neutrophils. *FEBS Lett.* 394: 149–152.

Kai, M., I. Wada, S. Imai et al. 1997. Cloning and characterization of two human isozymes of Mg^{2+}-independent phosphatidic acid phosphatase. *J Biol Chem.* 272: 24572–24578.

Kalari, S., Y. Zhao, E. W. Spannhake et al. 2009. Role of acylglycerol kinase in LPA-induced IL-8 secretion and transactivation of epidermal growth factor-receptor in human bron-chial epithelial cells. *Am J Physiol.* 296: L328–L336.

Kano, K., N. Arima, M. Ohgami et al. 2008. LPA and its analogs—Attractive tools for elucidation of LPA biology and drug development. *Curr Med Chem.* 15: 2122–2131.

Kassel, K. M., T. A. Wyatt, Jr. R. A. Panettieri et al. 2008. Inhibition of human airway smooth muscle cell proliferation by beta 2-adrenergic receptors and cAMP is PKA independent: Evidence for EPAC involvement. *Am J Physiol.* 294: L131–L138.

Kazantseva, A., A. Goltsov, R. Zinchenko et al. 2006. Human hair growth deficiency is linked to a genetic defect in the phospholipase gene LIPH. *Science.* 314: 982–985.

Khan, E. M., R. Lanir, A. R. Danielson et al. 2008. Epidermal growth factor receptor exposed to cigarette smoke is aberrantly activated and undergoes perinuclear trafficking. *FASEB J.* 22: 910–917.

Kobayashi, T., R. Tanaka-Ishii, R. Taguchi et al. 1999. Existence of a bioactive lipid, cyclic phosphatidic acid, bound to human serum albumin. *Life Sci.* 65: 2185–2191.

Koh, J. S., W. Lieberthal, S. Heydrick et al. 1998. Lysophosphatidic acid is a major serum noncytokine survival factor for murine macrophages which acts via the phosphatidylinositol 3-kinase signaling pathway. *J Clin Invest.* 102: 716–727.

Leung, D. W., C. K. Tompkins, and T. White. 1998. Molecular cloning of two alternatively spliced forms of human phosphatidic acid phosphatase cDNAs that are differentially expressed in normal and tumor cells. *DNA Cell Biol.* 17: 377–385.

Lin, M. E., D. R. Herr, and J. Chun. 2010. Lysophosphatidic acid (LPA) receptors: Signaling properties and disease relevance. *Prostaglandins Other Lipid Mediat.* 91: 130–138.

Luquain, C., A. Singh, L. Wang et al. 2003. Role of phospholipase D in agonist-stimulated lysophosphatidic acid synthesis by ovarian cancer cells. *J Lipid Res.* 44: 1963–1975.

Malcolm, K. C., C. M. Elliott, and J. H. Exton. 1996. Evidence for Rho-mediated agonist stimulation of phospholipase D in rat1 fibroblasts. Effects of *Clostridium botulinum* C3 exoenzyme. *J Biol Chem.* 271: 13135–13139.

McIntyre, T. M., A. V. Pontsler, A. R. Silva et al. 2003. Identification of an intracellular receptor for lysophosphatidic acid (LPA): LPA is a transcellular PPARgamma agonist. *Proc Natl Acad Sci USA.* 100: 131–136.

Medoff, B. D., A. L. Landry, K. A. Wittbold et al. 2009. CARMA3 mediates lysophosphatidic acid-stimulated cytokine secretion by bronchial epithelial cells. *Am J Respir Cell Mol Biol.* 40: 286–294.

Mio, T., X. Liu, M. L. Toews et al. 2002. Lysophosphatidic acid augments fibroblast-mediated contraction of released collagen gels. *J Lab Clin Med.* 139: 20–27.

Murakami, M., A. Shiraishi, K. Tabata et al. 2008. Identification of the orphan GPCR, P2Y(10) receptor as the sphingosine-1-phosphate and lysophosphatidic acid receptor. *Biochem Biophys Res Commun.* 371: 707–712.

Okudaira, S., H. Yukiura, and J. Aoki. 2010. Biological roles of lysophosphatidic acid signaling through its production by autotaxin. *Biochimie.* 92: 698–696.

Panther, E., M. Idzko, S. Corinti et al. 2002. The influence of lysophosphatidic acid on the functions of human dendritic cells. *J Immunol.* 169: 4129–4135.

Petrache, I., V. Natarajan, L. Zhen L et al. 2005. Ceramide upregulation causes pulmonary cell apoptosis and emphysema-like disease in mice. *Nat Med.* 11: 491–498.

Pilquil, C., J. Dewald, A. Cherney et al. 2006. Lipid phosphate phosphatase-1 regulates lysophosphatidate-induced fibroblast migration by controlling phospholipase D2-dependent phosphatidate generation. *J Biol Chem.* 281: 38418–38429.

Pilquil, C., I. Singh, Q. X. Zhang et al. 2001. Lipid phosphate phosphatase-1 dephosphorylates exogenous lysophosphatidate and thereby attenuates its effects on cell signalling. *Prostaglandins Other Lipid Mediat.* 64: 83–92.

Rahaman, M., R. W. Costello, K. E. Belmonte et al. 2006. Neutrophil sphingosine 1-phosphate and lysophosphatidic acid receptors in pneumonia. *Am J Respir Cell Mol Biol.* 34: 233–241.

Roberts, R., V. A. Sciorra, and A. J. Morris. 1998. Human type 2 phosphatidic acid phospho-hydrolases. Substrate specificity of the type 2a, 2b, and 2c enzymes and cell surface activity of the 2a isoform. *J Biol Chem.* 273: 22059–22067.

Rubenfeld, J., J. Guo, N. Sookrung et al. 2006. Lysophosphatidic acid enhances interleukin-13 gene expression and promoter activity in T cells. *Am J Physio Lung Cell Mol Physiol.* 290: L66–L74.

Saatian, B., Y. Zhao, D. He et al. 2006. Transcriptional regulation of lysophosphatidic acid-induced interleukin-8 expression and secretion by p38 MAPK and JNK in human bronchial epithelial cells. *Biochem J.* 393: 657–668.

Saulnier-Blache, J. S., A. Girard, M. F. Simon et al. 2000. A simple and highly sensitive radioenzymatic assay for lysophosphatidic acid quantification. *J Lipid Res.* 41: 1947–1951.

Scherer, M., G. Schmitz, and G. Liebisch. 2009. High-throughput analysis of sphingosine 1-phosphate, sphinganine 1-phosphate, and lysophosphatidic acid in plasma samples by liquid chromatography-tandem mass spectrometry. *Clin Chem.* 55: 1218–1222.

Shan, L., K. Jaffe, S. Li, and L. Davis. 2008. Quantitative determination of lysophosphatidic acid by LC/ESI/MS/MS employing a reversed phase HPLC column. *J Chromatogr B Analyt Technol Biomed Life Sci.* 862: 22–28.

Shao, M. X., T. Nakanaga, and J. A. Nadel. 2004. Cigarette smoke induces MUC5AC mucin overproduction via tumor necrosis factor-alpha-converting enzyme in human airway epithelial (NCI-H292) cells. *Am J Physiol Lung Cell Mol Physiol.* 287: L420–L427.

Spiegel, S. and S. Milstien. 2005. Critical role of acylglycerol kinase in epidermal growth factor-induced mitogenesis of prostate cancer cells. *Biochem Soc Trans.* 33: 1362–1365.

Sugiura, H., X. Liu, S. Togo et al. 2007. Prostaglandin E(2) protects human lung fibroblasts from cigarette smoke extract-induced apoptosis via EP(2) receptor activation. *J Cell Physiol.* 210: 99–110.

Suman, O. E., J. D. Morrow, K. A. O'Malley et al. 2000. Airway function after cyclooxygenase inhibition during hyperpnea-induced bronchoconstriction in guinea pigs. *J Appl Physiol.* 89: 1971–1978.

Swaney, J. S., C. Chapman, L. D. Correa et al. 2010. A novel, orally active LPA(1) receptor antagonist inhibits lung fibrosis in the mouse bleomycin model. *Br J Pharmacol.* 160: 1699–1713.

Tager, A. M., P. LaCamera, B. S. Shea et al. 2008. The lysophosphatidic acid receptor LPA1 links pulmonary fibrosis to lung injury by mediating fibroblast recruitment and vascular leak. *Nat Med.* 14: 45–54.

Tang, W., J. Yuan, X. Chen et al. 2006. Identification of a novel human lysophosphatidic acid acyltransferase, LPAAT-theta, which activates mTOR pathway. *J Biochem Mol Biol.* 39: 626–635.

Tatler, A. L., A. E. John, L. Jolly, et al. 2010. Activation of transforming growth factor-2 (Tgf-2) in response to lysophosphatidic acid is increased in asthma. *Am J Respir Crit Care Med.* 181: A2299.

Thompson, F. J. and M. A. Clark. 1994. Purification of a lysophosphatidic acid-hydrolysing lysophospholipase from rat brain. *Biochem J.* 300: 457–461.

Tigyi, G. and R. Miledi. 1992. Lysophosphatidates bound to serum albumin activate membrane currents in *Xenopus* oocytes and neurite retraction in PC12 pheochromocytoma cells. *J Biol Chem.* 267: 21360–21367.

Toews, M. L., E. E. Ustinova, and H. D. Schultz. 1997. Lysophosphatidic acid enhances contractility of isolated airway smooth muscle. *J Appl Physiol.* 83: 1216–1222.

Tokumura, A. 2002. Physiological and pathophysiological roles of lysophosphatidic acids produced by secretory lysophospholipase D in body fluids. *Biochim Biophys Acta.* 1582: 18–25.

Tokumura, A., L. D. Carbone, Y. Yoshioka et al. 2009. Elevated serum levels of arachidonoyl-lysophosphatidic acid and sphingosine 1-phosphate in systemic sclerosis. *Int J Med Sci.* 6: 168–176.

Tokumura, A., K. Harada, K. Fukuzawa et al. 1986. Involvement of lysophospholipase D in the production of lysophosphatidic acid in rat plasma. *Biochim Biophys Acta.* 875: 31–38.

Tokumura, A., E. Majima, Y. Kariya et al. 2002. Identification of human plasma lysophospholipase D, a lysophosphatidic acid-producing enzyme, as autotaxin, a multifunctional phosphodiesterase. *J Biol Chem.* 277: 39436–39442.

Tokumura, A., T. Yotsumoto, Y. Masuda et al. 1995. Vasopressor effect of lysophosphatidic acid on spontaneously hypertensive rats and Wistar Kyoto rats. *Res Commun Mol Pathol Pharmacol.* 90: 96–102.

Tou, J. S. and J. S. Gill. 2005. Lysophosphatidic acid increases phosphatidic acid formation, phospholipase D activity and degranulation by human neutrophils. *Cell Signal.* 17: 77–82.

Umezu-Goto, M., Y. Kishi, A. Taira et al. 2002. Autotaxin has lysophospholipase D activity leading to tumor cell growth and motility by lysophosphatidic acid production. *J Cell Biol.* 158: 227–233.

Wadgaonkar, R., V. Patel, N. Grinkina et al. 2009. Differential regulation of sphingosine kinases 1 and 2 in lung injury. *American J Physiol Lung Cell Mol Physiol.* 296: L603–L613.

Wang, A. and E. A. Dennis. 1999. Mammalian lysophospholipases. *Biochim Biophys Acta.* 1439: 1–16.

Wang, L., R. Cummings, Y. Zhao et al. 2003. Involvement of phospholipase D2 in lysophosphatidate-induced transactivation of platelet-derived growth factor receptor-beta in human bronchial epithelial cells. *J Biol Chem.* 278: 39931–39940.

Wang, L., E. Knudsen, Y. Jin et al. 2004. Lysophospholipids and chemokines activate distinct signal transduction pathways in T helper 1 and T helper 2 cells. *Cell Signal.* 16: 991–1000.

West, J., C. K. Tompkins, N. Balantac et al. 1997. Cloning and expression of two human lyso-phosphatidic acid acyltransferase cDNAs that enhance cytokine-induced signaling responses in cells. *DNA Cell Biol.* 16: 691–701.

Xu, M. Y., J. Porte, A. J. Knox et al. 2009. Lysophosphatidic acid induces alphavbeta6 integrin-mediated TGF-beta activation via the LPA2 receptor and the small G protein G alpha(q). *Am J Pathol.* 174: 1264–1279.

Zhao, Y., I. Gorshkova, D. He et al. 2009. Sphingosine-1-phosphate lyase is a novel target of LPS-induced lung inflammation and injury. *Am J Respir Crit Care Med.* 179: A5556.

Zhao, Y., D. He, B. Saatian et al. 2006. Regulation of lysophosphatidic acid-induced epidermal growth factor receptor transactivation and interleukin-8 secretion in human bronchial epithelial cells by protein kinase Cdelta, Lyn kinase, and matrix metalloproteinases. *J Biol Chem.* 281: 19501–19511.

Zhao, Y., D. He, R. Stern et al. 2007a. Lysophosphatidic acid modulates c-Met redistribution and hepatocyte growth factor/c-Met signaling in human bronchial epithelial cells through PKC delta and E-cadherin. *Cell Signal.* 19: 2329–2338.

Zhao, Y., D. He, J. Zhao et al. 2007b. Lysophosphatidic acid induces interleukin-13 (IL-13) receptor alpha2 expression and inhibits IL-13 signaling in primary human bronchial epithelial cells. *J Biol Chem.* 282: 10172–10179.

Zhao, Y., D. He, S. Pendyala et al. 2010. Deletion of lysophosphatidic acid receptors 1 and 2 protects against lipopolysaccharide-induced acute lung injury in mice. *FASEB J.* 24: 111.1.

Zhao, Y. and V. Natarajan. 2009. Lysophosphatidic acid signaling in airway epithelium: Role in airway inflammation and remodeling. *Cell Signal.* 21: 367–377.

Zhao, Y., J. Tong, D. He et al. 2009. Role of lysophosphatidic acid receptor LPA2 in the development of allergic airway inflammation in a murine model of asthma. *Respir Res.* 10: 114.

Zhao, Y., P. V. Usatyuk, R. Cummings et al. 2005. Lipid phosphate phosphatase-1 regulates lysophosphatidic acid-induced calcium release, NF-kappaB activation and interleukin-8 secretion in human bronchial epithelial cells. *Biochem J.* 385: 493–502.

12 Sphingolipids in Airway Inflammation

Sameer Sharma and Carole A. Oskeritzian

CONTENTS

12.1 INTRODUCTION

Sphingolipids are the principal constituents of the plasma membrane as well as a source of important signaling molecules such as ceramide, ceramide-1-phosphate (C1P), and sphingosine-1-phosphate (S1P). Advances in our understanding of sphingolipid biology have made it clear that the intracellular mechanism for the generation and activation of sphingolipid-derived mediators is dynamic inside living cells. Distinct cell surface receptors have been identified and characterized that mediate the responses of these lipid mediators. Evidence is emerging that supports the idea that ceramide, C1P, sphingosine, and S1P play important roles in different inflammatory diseases, including bronchial asthma. In this chapter, we shall look at the evidence supporting a role of sphingolipid mediators in airway inflammation. We shall also recapitulate sphingolipid metabolism, the proinflammatory role of sphingolipid-derived products, and the receptors that mediate sphingolipid responses. Among the different sphingolipids, S1P receptors have been key targets for drug discovery. On

September 22, 2010, FTY720 (fingolimod), a molecule that acts as an S1P receptor
agonist, became the first oral disease-modifying drug approved by the Food and
Drug Administration as Gilenya to reduce relapses and to delay disability progression in patients with relapsing forms of multiple sclerosis. In this section, we shall
restrict our discussion to S1P receptors and drug discovery efforts related to them.

12.2 SPHINGOLIPID BIOSYNTHESIS

Sphingomyelin (SM), one of the major membrane sphingolipids, has been recognized
to be the precursor of several bioactive molecules. A brief description of SM metabolism is given in Figure 12.1. Ceramide is formed when SM is hydrolyzed by sphingo-
myelinases (SMase). Ceramide can also be glycosylated through the action of
glucosylceramide synthase (GCS), which is the first step in glycosphingolipid synthesis. Ceramide is phosphorylated by ceramide kinase to form ceramide-1-phosphate.
Both ceramide and C1P are biologically active. Deacylation of ceramide by ceramidase with the loss of the fatty acid moiety from the amide leads to the formation of
sphingosine. Sphingosine is further phosphorylated by sphingosine kinases (SphKs),
to generate S1P. Two sphingosine kinase isoforms have been cloned: sphingosine
kinase 1 (SphK1) and sphingosine kinase 2 (SphK2). Sphks are highly conserved

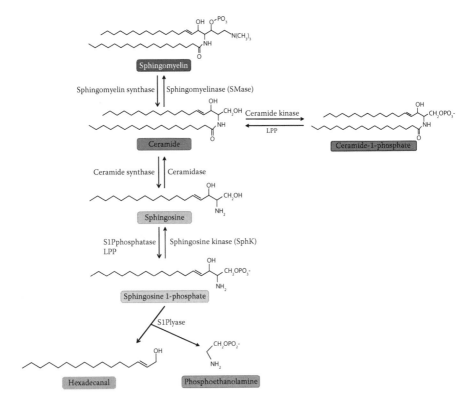

FIGURE 12.1   The sphingolipid metabolic pathway. LPP, lipid phosphate phosphatase.

ubiquitous lipid kinases that can be activated by numerous stimuli (Spiegel and Milstien, 2003; Melendez et al., 2007, Pushparaj et al., 2009). Given their contrasting kinetics and tissue expression patterns, SphK1 and SphK2 are likely to regulate distinct cellular functions with apparent differential subcellular localization, biochemical properties, and regulation (Spiegel and Milstien, 2003; Taha et al., 2006; Aarthi et al, 2011). Both SphKs need access to their lipid substrate, sphingosine. At physiological pH, it is estimated that 70% of cellular sphingosine is located within membranes and 30% is soluble, which suggests that both SphK1 and SphK2 are localized to membranes (Wattenberg et al., 2006; Hannun and Obeid, 2008; Pyne et al., 2009). The cellular levels of S1P are tightly regulated by its degradation through two S1P-specific phosphatases and S1P lyase, which respectively dephosphorylate S1P into sphingosine or irreversibly degrade S1P to phosphoethanolamine and hexadecenal (Melendez, 2008a). The constitutive activity of S1P lyase results in low intratissular concentrations of S1P, conversely to circulating levels of S1P in plasma and lymph where they reach sub- to low-micromolar ranges (Rivera et al., 2008).

12.3 BIOLOGY OF SPHINGOLIPIDS

Products of sphingomyelin metabolism have their distinct, often antagonistic, biological activities and are emerging as a network of interrelated bioactive lipids (Figure 12.2). While exploring the role of sphingomyelin metabolic products in cellular function, it is important to bear in mind that the endogenous levels of these

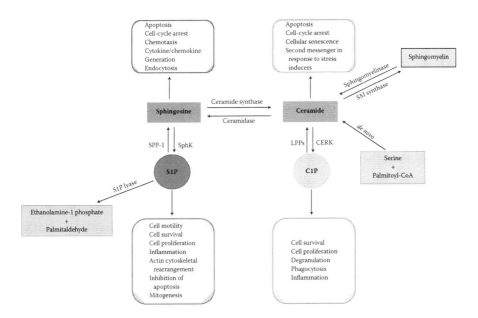

FIGURE 12.2 Biological functions of sphingosine, sphingosine-1-phosphate, ceramide, and ceramide-1-phosphate. CERK, ceramide kinase; SM, sphingomyelin; SphK, sphingosine kinase; SPP-1, S1P phosphatase-1.

lipids are tightly regulated. Usually, the highest level of a sphingolipid in the cell is ceramide and the lowest is S1P. A small change in ceramide can increase the levels of sphingosine or S1P (Bartke and Hannun, 2009). Exogenous addition of any given lipid readily allows for intracellular metabolic conversion, making it difficult to conclusively point to a particular lipid mediator as responsible for a specific biological process (Ogretmen et al., 2002).

Ceramide signaling has been intimately involved in the regulation of cell growth, differentiation, senescence, necrosis, proliferation, apoptosis, and airway smooth muscle contraction. The exact molecular receptor for ceramide and C1P remains unidentified. It has also been shown that ceramide may regulate protein kinase C (PKC), raf-1, and the kinase-suppressor of Ras, which significantly changes the level of phosphorylation of various key substrates. Another binding target for ceramide is the cellular protease cathepsin D, which may mediate the actions of lysosomally generated ceramide (Heinrich et al., 1999). Ceramide is a crucial mediator of alveolar destruction in emphysema and is increased in individuals displaying smoking-induced emphysema. Thus, inhibition of enzymes controlling *de novo* ceramide synthesis prevents alveolar cell apoptosis, oxidative stress, and emphysema caused by blockade of the vascular endothelial growth factor (VEGF) receptors in both rats and mice (Petrache et al., 2005). Emphysema can be triggered by intratracheal instillation of ceramide in naive mice. In contrast, S1P could functionally antagonize lung apoptosis, implying that a balance between ceramide and S1P is required for the maintenance of alveolar septal integrity. Ceramide can also act as a second messenger in response to various stress inducers such as TNF-α, interferon-γ (IFNγ), and interleukin 1β (IL-1β) and in response to UV light. Ceramide-1-phosphate has been implicated to play roles in inflammation and vesicular trafficking (Chalfant and Spiegel, 2005). It also mediates the activation of cytosolic phospholipase A$_2$ (cPLA$_2$) and the release of arachidonic acid in response to IL-1β. The demonstration that C1P generation is upstream of cPLA$_2$ activation suggests that therapeutics that could inhibit C1P or ceramide kinase could constitute a new generation of therapeutics for inflammatory disorders such as asthma, atherosclerosis, and Alzheimer's disease. Glycated ceramide has also been shown to be involved in post-Golgi trafficking and in drug resistance (Radin et al., 1993).

Sphingosine has been connected with cellular processes such as inducing cell cycle arrest and apoptosis by the modulation of protein kinases and other signaling pathways. It has roles in regulating the actin cytoskeleton and endocytosis and has been shown to inhibit PKC. S1P regulates numerous cellular functions, including proliferation and differentiation, chemotaxis, and cytokine/chemokine generation (Spiegel and Milstien, 2002).

S1P is a bioactive sphingolipid metabolite that plays important roles in allergic responses, including asthma and anaphylaxis. Mast cells are tissue-dwelling cells that are strategically located at the interfaces of host and environment and are key effectors of IgE-mediated allergic disorders, including anaphylaxis, hay fever, eczema, and asthma (Oskeritzian et al., 2007; Price et al., 2008). Upon the activation of the high-affinity receptors for IgE, which binds antigen (Ag)-specific IgE antibodies, and the subsequent exposure to Ag, the activation of SphK results in the production and secretion of S1P, as an additional lipid mediator to a plethora of mediators generated by mouse and human mast cells (Jolly et al., 2004; Mitra et al., 2006; Olivera et al., 2006). Although SphK activation resulting in S1P production and secretion is required for IgE- and mast cell-dependent allergic responses, conflicting results have been reported pertaining to the relative importance of each SphK isoenzyme in studies using mice deficient in SphK1 or SphK2 (Choi et al., 1996; Melendez and Khaw, 2002; Olivera et al., 2007; Pushparaj et al., 2009). SphK1, but not SphK2, was shown to be important for Ag-induced calcium mobilization, degranulation, and migration of rodent mast cells (Melendez and Khaw, 2002; Jolly et al., 2004; Pushparaj et al., 2009) and of human mast cells (Oskeritzian et al., 2008). However, using fetal liver-derived mouse mast cells from mice deficient in SphK1 or SphK2, it was suggested that SphK2, but not SphK1, modulates calcium influx and downstream signaling, leading to degranulation and eicosanoid and cytokine production (Olivera et al., 2007). Similarly, SphK2 was also necessary for the secretion of both TNF and IL-6 by mature human mast cells (Oskeritzian et al., 2008). Using siRNA gene silencing strategies, SphK1 was demonstrated as the dominant isoenzyme governing S1P generation in mast cells (Pushparaj et al., 2009; Oskeritzian et al., 2008). Since both SphK isoenzymes evenly utilize sphingosine as their lipid substrate, they might both contribute to mast cell functions and the phenotypes displayed by genetically deficient mice, a reflection of their overlapping functions and compensatory mechanisms aiming to supplement the deficiency. However, it is noteworthy that mice deficient in SphK1 display reduced levels of circulating S1P and are resistant to anaphylaxis, whereas SphK2-deficient mice undergo normal anaphylactic reaction (Olivera et al., 2007). Interestingly, an intriguing finding of this study is the association between circulating concentrations of S1P and susceptibility to anaphylaxis, suggesting that S1P is a key factor that determines anaphylactic responses.

S1P enhances inflammation by triggering cytokine and prostaglandin production, as well as promoting mitotic and chemotactic responses of immune cells (Kee et al., 2005). S1P binds to cell surface receptors, which are coupled with different G-proteins and have distinct constitutive and inducible patterns of expression. They trigger various intracellular signaling pathways, leading to diverse and specific functional outcomes (Hla et al., 2001; Taha et al., 2004).

12.4   SPHINGOLIPIDS AND AIRWAY INFLAMMATION

Evidence suggests the involvement of different components of the sphingolipid metabolic pathway in airway inflammation.

12.4.1   CERAMIDE

In experimental animal models, ceramide has been implicated in the development of allergen-induced eosinophilia, chemokine production, airway hyperresponsiveness (AHR), increase in bronchoconstriction and change in airway histopathology. Emphysema can be triggered by intratracheal instillation of ceramide in naive mice or by inhibition of VEGFR, which increases ceramides in lung structural cells of the alveolus, triggering lung tissue destruction (Petrache et al., 2005). The importance of ceramide activation for inflammation derives from the activation of various protein kinases and phosphatases in diverse downstream pathways and from its generating additional second messengers such as S1P. Administration of the S1P precursor sphingosine, S1P-mimetic FTY720 or the S1P receptor 1 agonist SEW2871 significantly counteracts ceramide-induced lung apoptosis (Diab et al., 2010). Furthermore, ceramide contributes to the lung inflammatory response by degranulating mast cells, recruiting neutrophils/eosinophils, and increasing the formation of cytokines (Bartke and Hannun, 2009).

It has been shown that aerosol administration of ovalbumin (commonly used as an antigen in experimental models of airway inflammation) in guinea pigs increases ceramide levels and ceramide synthase activity in the airway epithelium associated with respiratory abnormalities, such as cough, dyspnea, and severe bronchoconstriction. Inhibition of *de novo* ceramide synthesis with the competitive and reversible inhibitor of ceramide synthase, fumonisin B1, before allergen challenge attenuates oxidative/nitrosative stress, epithelial cell apoptosis, and airway inflammation while improving the respiratory and histopathological abnormalities (Masini et al., 2008). Ceramide kinase (CERK) generates C1P, which triggers histamine and prostaglandin D_2 (PGD_2) release from rat and human mast cell lines (Hewson et al., 2011). Moreover, the C1P effect is additive to Ag-mediated activation. Inhibition of CERK abrogated both histamine and PGD_2 release. Interestingly, CERK positioned itself differentially in the signaling cascade leading to histamine and PGD_2 release.

The possibility of NKT cells playing a role in ceramide response has also been reported. α-Galactosylceramide (α-GalCer) is a specific ligand of natural killer T cells (NKT cells) that regulates the immune responses such as tumor rejection and autoimmunity by producing IFN-γ and IL-4. Invariant NKT cells (iNKT cells) play a pivotal role in the development of allergen-induced AHR and inflammation. It has been observed that a single administration of α-GalCer at the time of ovalbumin challenge completely prevented eosinophilic infiltration in wild-type mice. This inhibitory effect of α-GalCer was associated with a decrease in airway hyperresponsiveness, an increase in IFN-γ, and decreases in IL-4, IL-5, and IL-13 levels in the bronchoalveolar lavage (BAL) fluids. It is suggested that ceramide and ceramide analogs act by tipping the T-helper cell type 1 (T_H1) and T-helper cell type 2 (T_H2) balance in favor of T_H1 (Morishima et al., 2005). In IFN-γ knockout mice α-GalCer failed to exert an inhibitory effect in the asthma model, indicating that α-GalCer prevents allergic airway inflammation possibly through IFN-γ production by ligand-activated NKT cells (Matsuda et al., 2005).

12.4.2 SPHINGOSINE-1-PHOSPHATE AND SPHINGOSINE KINASES

S1P is a proinflammatory sphingolipid (Baumruker and Prieschel, 2002). An elevated level of S1P in the airway has been reported following allergen challenge as well as in BAL fluid of patients suffering from asthma (Ammit et al., 2001; Rosen et al., 2008). S1P can induce bronchoconstriction (Rosenfeldt et al., 2003; Chiba et al., 2010a) as well as promote the migration of inflammatory cells (Prieschl et al., 1999; Roviezzo et al., 2004). Inflammation promotes S1P generation in the vicinity of its receptor in bronchial tissue, leading to bronchoconstriction. Intranasal application of

S1P aggravates Ag-induced airway inflammation in a mouse model of allergic bronchial asthma (Chiba et al., 2010a). But S1P also exerts anti-inflammatory effects by enhancing vascular barrier function in models of acute lung injuries (Ding et al., 2011). It has been reported that S1P at levels well below the circulating plasma level can induce a bronchial response favoring juxtacrine effects (Xia and Wadham, 2011). Moreover, because both SphK1 and SphK2 enzymatic activities can be activated by a wide range of stimuli, it is tempting to speculate that S1P could also be produced in very high levels, locally and transiently, in the tissues (Oskeritzian, personal observations; Xia and Wadham, 2011). Of particular interest in the field of inflammation, the TNF receptor complex has been shown to directly interact with SphK1 signaling unit through S1P binding to TNFR-associated factor 2 (TRAF2), leading to the activation of nuclear factor kappa-light-chain-enhancer of activated B cells (NF-κB), a protein complex that controls the transcription of DNA. (Xia et al., 2002; Alvarez et al., 2010). In addition to TRAF2, histone deacetylases (HDAC1 and HDAC2) constitute important intracellular targets of S1P (Hait et al., 2009). Nuclear SphK2 was found to be associated with histone H3 and produces S1P, which acts as the first characterized endogenous HDAC inhibitor.

Cyclooxygenase (COX) is the enzyme that constitutes the rate-limiting step in the formation of prostaglandins. Its inducible form, COX-2, mediates TNF-induced inflammatory responses. S1P induces COX-2 and PGE2 production (Pettus et al., 2003). Similarly, lipopolysaccharide (LPS)-induced increases of COX-2 and PGE2 in macrophages are also dependent on the SphK1/S1P pathway (Snider et al., 2010). Apart from its cytosolic and nuclear functions, S1P has extracellular signaling effects acting as a ligand for specific receptors. Both S1P and its dihydro analog bind to them with a high affinity.

The emerging role of S1P in regulating smooth muscle contraction has led to exploration of the possibility that this sphingolipid metabolite could represent a potential therapeutic target. Ingested antigen binding to FcεRI causes mast cell migration and degranulation with the release of histamine and other preformed mediators. It thus activates many downstream signaling molecules, including Src family of tyrosine kinases, calcium flux, PKC, and the Ras-mitogen-activated protein kinase (MAPK) pathway, which is dependent on transactivation of $S1P_2$ (Jolly et al., 2004). This complex thus leads to degranulation and secretion of cytokines and chemokines as well as the arachidonic acid metabolites, leukotrienes, and prostaglandins (Ryan and Spiegel, 2008; Price et al., 2008). The finding that S1P is an autocrine mediator of activated mast cells therefore establishes its role in the pathophysiology of asthma and/or airway hyperresponsiveness (Baumruker and Prieschel, 2002).

S1P level is also regulated by SphKs, which translocate to the plasma membrane in response to stimuli and convert sphingosine to S1P in the vicinity of its receptor (James, 2005). It has also been shown that SphK activity is increased in inflammatory cells such as neutrophil and macrophages in response to stimuli and regulates their functions such as chemotaxis and cytokine release. In an allergic mouse model, an inhibitor of SphK can lower the level of S1P, peroxidase activity, and eosinophil count in BAL fluid. Interestingly, SphK1$^{-/-}$ mice have increased inflammation-induced vascular leak in the lungs. SphK1$^{-/-}$ mice treated with either LPS or thrombin also had increased vascular permeability in lung tissue (Tauseef et al., 2008).

12.5   SPHINGOSINE-1-PHOSPHATE RECEPTORS

S1P acts as a signaling molecule that regulates fundamental cellular responses, including cell death, proliferation and differentiation, chemotaxis, and cytokine/chemokine and inflammatory mediator production through extracellular ligation to S1P receptors, in addition to its function as an intracellular second messenger (Spiegel and Milstien, 2002). The discovery of S1P as a ligand of five closely related G-protein-coupled receptors provided a fundamental clarification on how such a simple molecule can regulate so many diverse physiological and immunological processes. The marked difference in S1P concentrations between the circulation (high) and tissues (low) generates a naturally occurring S1P gradient that drives the trafficking of various immune cells (Schwab et al., 2005). In 2002, FTY720, a novel immunomodulatory molecule and synthetic analog of myriocin, was reported to induce lymphopenia by T-cell sequestration within lymphoid organs, acting through four out of the five S1PRs, with the exclusion of $S1P_2$ (Mandala et al., 2002; Brinkmann et al., 2002). In addition to its role in cell trafficking (Sanna et al., 2004), S1P has been assigned various functions such as cardiac rhythm regulation (Forrest et al., 2004) and being the critical linkage between inflammation and coagulation pathways downstream of the thrombin cascade (Niessen et al., 2008). Further important insights are emerging from the ability to combine genetic and chemical approaches both *in vitro* and *in vivo*. Gene deletion studies and reverse pharmacology have provided evidence that many of the biological effects of S1P are mediated via five specific G-protein-coupled receptors, now designated $S1P_{1-5}$, which are positioned between the cannabinoid receptors and the lysophosphatidic acid receptors and share some key structural and functional similarities with each of those families. Recent phylogenetic analysis of transmembrane regions of G protein-coupled receptors (GPCRs) indicates that the orphan receptor P2Y(10) is classified under the cluster of nucleotide and lipid receptors (Murakami et al., 2008). It is the first receptor recognized as a dual lysophospholipid receptor for S1P and LPA (Murakami et al., 2008).

The detailed features of the S1P receptors, including coupling, cellular distribution, and functions, are shown in Table 12.1.

12.5.1 $S1P_1$ RECEPTOR

$S1P_1$ receptor is found on chromosome location 1p21 and is highly expressed by endothelial cells and most immune cells mainly localized in the brain, heart, spleen, kidney, and skeletal muscles. One of its main functions is to govern astrocyte migration and chemotaxis, to name a few out of its various functionalities (Table 12.1). Of different receptors, deletion of $S1P_1$ alone is embryonic lethal at E13.5 (Liu et al., 2000) due to the abnormality in arterial wall formation, disturbed vascular maturation, and hemorrhage. S1P regulates vascular permeability at both basal and inflammatory states in an $S1PR_1$-dependent manner (Garcia et al., 2001; Tauseef et al., 2008). Thus, a short-term administration of S1P or $S1P_1$ agonists enhances endothelial barrier integrity and attenuates injury-induced vascular leakage in the lung and other organs *in vivo*. However, prolonged exposure to these agents abrogates the

TABLE 12.1

S1P Receptors, Their Subtypes, Coupling, Knockout Phenotypes, mRNA Distribution, and Specific Actions/Outcomes in Distinct Cell Type and Tissues

Receptor/Subtype	Coupling	Knockout Phenotypes	Distribution (mRNA)	Cellular Functional Expression and Consequences
S1P$_1$ (edg1) 1p21[a] 381[b] 0.47–0.67[c]	Gi/o	Embryonic lethal (E12.5–E14.5) Vascular maturation defects Failure of null thymocyte to egress in irradiated chimeras	Brain, heart, spleen, liver, lung, thymus, kidney, skeletal muscle, lymphoid[e]	Astrocyte: migration B cell: blockade of egress, chemotaxis Cardiomyocyte: increased β-AR positive inotropy Endothelial cell: early vascular system development, adherens junction assembly, APC-mediated increased barrier integrity Neural stem cell: increased migration Pericyte: early vascular system development T cell: blockade of egress, chemotaxis, decreased late-stage maturation VSMC (early vascular system development)
S1P$_2$ (edg5) 19p13.2[a] 353[b] 0.30–0.35[c]	Gi/o Gq G12/13	Slightly reduced viability[d] Seizures in certain genetic backgrounds Deafness	Brain, heart, spleen, liver, lung, thymus, kidney, skeletal muscle[e]	Cardiomyocyte: survival to ischemia–reperfusion Epithelial cell (stria vascularis): integrity/development Epithelial hair cells (cochlea): integrity/development Endothelial cell (retina): pathological angiogenesis, adherens junction disruption Hepatocyte: proliferation/matrix remodeling Fibroblast (MEF) Mast cell: degranulation VSMC: decreased PDGF-induced migration
S1P$_3$ (edg3) 9q22.2[a] 378[b] 0.17–0.26[c]	Gi/o Gq G12/13	Enhances sepsis outcome Slightly reduced viability	Brain, heart, spleen, liver, lung, thymus, kidney, testis, skeletal muscle[e]	Cardiomyocyte: survival to ischemia–reperfusion Dendritic cell (hematopoietic): worsening experimental sepsis lethality/ inflammation/coagulation

continued

TABLE 12.1 (continued)
S1P Receptors, Their Subtypes, Coupling, Knockout Phenotypes, mRNA Distribution, and Specific Actions/Outcomes in Distinct Cell Type and Tissues

Receptor/Subtype	Coupling	Knockout Phenotypes	Distribution (mRNA)	Cellular Functional Expression and Consequences
S1P₄ (edg6) 19q22.1[a] 384[b] 34–95[c]	Gi/o G12/13	N/A	Lymphoid, lung[f]	T cell: migration/cytokine secretion
S1P₅ (edg8) 19p13.2[a] 398[b] 0.50–0.61[c]	Gi/o G12/13	No obvious phenotype	Brain, skin, spleen[f]	NK cell: trafficking Oligodendrocyte: survival OPC: glial process retraction; inhibition of migration

Note: The results are derived from S1PR-null mice approaches, chemical approaches, or combination of both. APC, activated protein C; MEF, mouse embryo fibroblast; NK, national killer; OPC, oligodendrocyte progenitor cell; PDGF, platelet derived growth factor; VSMC, vascular smooth muscle cell.

[a] Chromosome location.
[b] Protein length.
[c] Measured by inhibition of [^{33}P]-S1P binding to stably expressed human S1P$_{1-5}$ in CHO-K1 cell membranes.
[d] Null S1P$_2$–S1P$_3$ mice have marked perinatal lethality.
[e] Widespread.
[f] Restricted.

barrier-protective effects of S1P, triggering pulmonary fibrosis and capillary barrier disruption as a result of an overzealous fibrogenic response (Shea et al., 2010). This underscores the therapeutic potential of developing highly selective $S1P_1$ agonists to reduce inflammatory acute lung (or other organs) injury after a careful consideration of this particular receptor expression in target tissues and the critical role of the S1P delivery route, in addition to agent concentrations (Sammani et al., 2010). Similarly, the activation of lung $S1P_1$ receptors reduces allergen-induced plasma leakage in mice (Blé et al., 2009), even inhibiting airway release of chemokines when locally delivered and thus suppressing eosinophilic and T-cell-triggered allergic airway inflammation (Marsolais et al., 2011). T_H2 are critical in allergy and asthma and a very recent report elucidated that their lymph node egress occurs through the reexpression of $S1P_1$ receptor, which is itself controlled by extracellular matrix protein (ECM)-1 (Li et al., 2011) through transcriptional regulation driven by KLF2 transcription factor in T cells (Carlson et al., 2006). It has been found that the administration of a new selective $S1P_1$ competitive antagonist (3-amino-4-(3-hexylphenylamino)-4-oxobutylphosphonic acid) induces disruption of barrier integrity in pulmonary endothelium (Rosen et al., 2007). Intriguingly, a single intravenous injection of S1P showed protective effects against lung injury caused by high-volume mechanical ventilation and intratracheal endotoxin instillation in an animal model that was attributed to expression of $S1P_1$ by endothelial cells (Rosen and Goetzl, 2005). The role of $S1P_1$ in T-cell trafficking was originally unraveled by the use of FTY720. However, whether *in vivo* phosphorylated FTY720 acts as an agonist or a functional antagonist to regulate lymphocyte trafficking remains unclear. Further studies are also required to define its direct cellular targets (i.e., lymphocytes or endothelial cells) (Matloubian et al., 2004; Wei et al., 2005). Intriguingly, FTY720 triggers rapid internalization of $S1P_1$ but $S1P_1$-associated signaling remains persistent hours after treatment (Mullershausen et al., 2009). To complicate the picture even further, FTY720 exerts receptor-independent immunomodulatory activities (Payne et al., 2007).

The highest levels of $S1P_1$ transcripts are found in the CD19[+] B cells and the cerebellum (Aarthi et al., 2011). S1P1 has also been reported as being responsible for the appearance of NK cells in the peripheral tissues (Allende et al., 2008) and regulatory T-cell proliferation (Wolf et al., 2009). $S1P_1$ can also form complexes with the activation of Ag CD69 to facilitate its internalization and degradation (Bankovich et al., 2010).

12.5.2 $S1P_2$ RECEPTOR

$S1P_2$ is found on chromosome location 19 (19p13.2.) and plays a key role in the amplification of mast cell responses. Mast cells express both $S1P_1$ and $S1P_2$ and, upon immunological activation via their high-affinity receptor for IgE, export S1P, which results in rapid transactivation of these receptors, another illustration of S1P's ability to signal "inside-out" (Rivera et al., 2008). $S1P_2$ expression is upregulated by FcεRI cross-linking (Oskeritzian, personal observation) and is required for *in vitro* degranulation of mouse and human mast cells (Rivera et al., 2008; Oskeritzian et al., 2010). In a mouse model of mast cell-dependent anaphylaxis, an exacerbated systemic form of allergic reaction, JTE-013, an antagonist for $S1P_2$ or $S1P_2$ deficiency, significantly attenuated the circulating levels of histamine and chemokines and mitigated the associated pulmonary edema (Oskeritzian et al., 2010). However, another report studying a mast cell-independent model of anaphylaxis established that $S1P_2$ was necessary for recovery from anaphylaxis (Olivera et al., 2010). Therefore, S1P mediates histamine degranulation through $S1P_2$ with both pharmacological and genetic evidences supporting a crucial role for this receptor. Both SphKs and $S1P_2$ regulate mast cell functions; however, mast cell- or tissue-specific or inducible knockout strategies will need to be developed to establish their functions *in vivo* (Chi, 2011). Nevertheless, these reports demonstrate that in addition to differential expression patterns for S1PRs, they can exert a variety of functions, reinforcing the wide range of functional outcomes of S1P depending upon its cellular target.

$S1P_2$ plays a pivotal role in the proper functioning of the auditory and vestibular systems (Herr et al., 2007; Kono et al., 2007). Vascular disturbance within the stria vascularis, a barrier epithelium containing the primary vasculature of the inner ear, is one of the components leading to deafness in $S1P_2$-deficient mice. $S1P_1$ and $S1P_2$ double-null mice do not survive because of a poorly developed vascular system (Kono et al., 2004). Hypoxia, a possible consequence of the narrowing of inflamed airways, has been shown to trigger the upregulation of $S1P_2$ in hypoxic endothelial cells of the mouse retina, resulting in pathological angiogenesis reversible upon treatment with $S1P_2$ antagonism using JTE-013 (Skoura et al., 2007).

12.5.3 $S1P_3$ RECEPTOR

$S1P_3$ receptor is found on chromosome location 9q22.2 and is mainly localized in brain, heart, lung, spleen, intestine, skeletal muscles, and kidney and plays a critical role in the regulation of the cardiac rhythm and the remodeling, proliferation, and differentiation of cardiac fibroblasts. Together with $S1P_2$, it mediates cardioprotection subsequent to ischemia/reperfusion injury *in vivo* (Means and Brown, 2009).

Dendritic cell (DC) thrombin signaling is coupled to migration, which is promoted by $S1P_3$ receptors. Thrombin generated in the lymphatic compartments perturbs DCs to promote systemic inflammation and intravascular coagulation in severe sepsis. Selective modulators of the $S1P_3$ receptor system have been employed to attenuate sepsis lethality, which suggests that such novel therapeutic approaches can be employed to correct such alterations (Ruf et al., 2009). Deletion of $S1P_2$ and $S1P_3$ receptors results in viable offspring with variable phenotypes and suggests that some compensation in receptor expression and signaling may occur. This means that coexpression of multiple S1P receptors subtypes in the same cells results in receptor responsiveness alterations, which could be further altered by gene deletion of one subtype.

12.5.4 $S1P_4$ AND $S1P_5$ RECEPTORS

$S1P_4$ is specifically expressed in lymphoid tissues such as spleen and lung and mainly functions in T-cell migration and cytokine secretion. Interestingly, $S1P_4$ seems to mediate the immunosuppressive effects of S1P through inhibition of proliferation and proinflammatory cytokine secretion, while promoting anti-inflammatory cytokine secretion such as IL-10 (Wang et al., 2005). S1P$_5$ receptors are mainly localized in the brain, skin, and spleen and function mainly in NK cell trafficking, oligodendrocyte migration, and survival. Studies have shown that S1P$_5$ is expressed in NK cells in mice and humans and that S1P$_5$-deficient mice had aberrant NK cell homing during steady-state conditions. In addition, S1P$_5$ is also required for the mobilization of NK cells to the inflamed organs. Jenne et al. (2009) identified a mouse strain designated Duane, in which NK cells are reduced in blood and spleen but increased in bone marrow and lymph nodes, reflecting their decreased ability to exit into lymphatic circulation. Duane-derived NK cells are defective in S1P$_5$ transcription. Interestingly, this strain carries a point mutation within a gene coding for the transcription factor T-bet that binds to the S1P$_5$ locus. Both S1P$_4$ and S1P$_5$ receptors are found on chromosome 19 with S1P$_4$ in close proximity on 19p13.2 chromosomal location (Walzer et al., 2007).

12.6 DRUG DISCOVERY EFFORTS RELATED TO SPHINGOSINE-1-PHOSPHATE

Chemical structures of select S1P receptor ligands are shown in Figure 12.3.

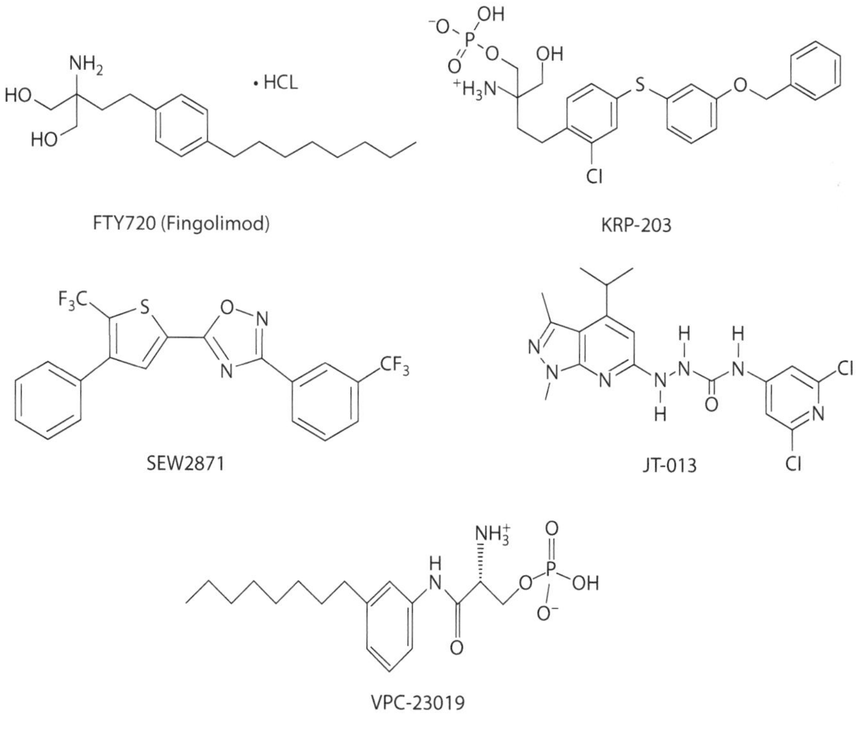

FIGURE 12.3 Structures of sphingosine-1-phosphate receptor agonists/antagonists.

12.6.1 RECEPTOR AGONISTS

The regulation of sphingolipids, in particular attempting to manipulate S1P signaling and its receptors, is becoming a topic of intensive research aimed to design novel therapeutic strategies for immune and inflammatory disorders.

As described in the previous section, five distinct receptors for S1P exist. Many of these receptors have widespread distribution in the body. The administration of a receptor agonist or antagonist may trigger off-target effects unless a specific receptor is targeted on a specific cell type or tissue. It also remains a dilemma how to interfere with sphingolipid function—using an agonist, which could also be a functional antagonist, or an antagonist.

Nevertheless, at this time the most advanced molecule that has reached Phase III clinical trial is an agonist of S1P receptor. FTY720, also known as Fingolimod (Novartis, Basel, Switzerland), was first synthesized in 1992 by the structural modification of myriocin. FTY720 is a fungal metabolite with immunosuppressive properties isolated from *Isaria sinclairii* culture broth (Fujita et al., 1994). It is now known that FTY720 is a prodrug that is phosphorylated both *in vitro* and *in vivo* by SphK2 (Paugh et al., 2003; Zemann et al., 2006) to biologically active FTY720-phosphate, a structural analog of S1P (Figure 12.3). FTY720-P binds to four of the five known S1P receptors, but not to $S1P_2$. FTY720-P induces internalization and degradation of the $S1P_1$ receptor that results in prolonged receptor downregulation (Matloubian et al., 2004; Yanagawa et al., 1998), thereby depriving thymocytes and lymphocytes of an S1P signal necessary for their egress from secondary lymphoid tissues (Cyster, 2005). Lymphocytes therefore remain in spleen and lymph nodes and do not migrate to the site of inflammation.

Airway DCs play a crucial role in the pathogenesis of allergic asthma, and interfering with their function could constitute a novel form of therapy. It has been shown that in a murine asthma model, administration of FTY720 via inhalation route prior to or during ongoing allergen challenge suppresses T_H2-dependent eosinophilic airway inflammation and bronchial hyperresponsiveness without causing lymphopenia and T cell retention in the lymph nodes. Therefore, the effectiveness of the local treatment can be achieved by inhibition of the migration of lung DCs to the mediastinal lymph nodes, which in turn inhibits the formation of allergen-specific T_H2 cells in lymph nodes (Sawicka et al., 2003; Idzko et al., 2006).

At present, no clinical data are available about the efficacy of FTY720 in human asthma or COPD. It has been speculated that lymphopenia can be a serious side effect. However, this has not hindered progression of FTY720 in advanced clinical development for multiple sclerosis and its approval for relapsing forms of multiple sclerosis. However, a more serious issue of its use in respiratory disease could be raised by the possible effect of FTY720 on diminished forced expiratory volume. This may be related to S1P-mediated modulation of bronchial smooth muscle contraction (Ammit et al., 2001). Serious adverse reactions have been associated with FTY720, including bradycardia, a consequence of its action on cardiac $S1P_3$ since FTY720 failed to induce bradycardia in $S1P_3$-deficient mice (Sanna et al., 2004). Other adverse events included atrioventricular block, certain infections, increased liver enzyme levels, hypertension, and macular edema (Chi, 2011).

Moreover, FTY720 has been recently shown to interfere with *de novo* sphingolipid synthesis. Thus, FTY720 inhibited ceramide synthetases, resulting in decreased levels of ceramides and sphingosine, but increased levels of dihydro-S1P mediated by SphK1, not SphK2 (Berdyshev et al., 2009). Therefore, FTY720 can also modulate the intracellular balance of signaling sphingolipids, another demonstration of its off-target effects.

A second generation of S1P receptor modulator has been designed by Novartis. BAF-312 exhibits a shorter half-life and higher selectivity for S1P receptors compared to fingolimod.

Unlike FTY720, a recently discovered compound, KRP-203, was found to be a very specific agonist for S1P1 (Shimizu et al., 2005). In experimental models, KRP-203 prolongs graft survival and attenuates chronic rejection in rat heart allograft models, suggesting that it not only regulates T-cell responses but also regulates those of B cells (Niessen et al., 2008). Furthermore, KRP-203 inhibits T_H1-type proinflammatory cytokine release in the lamina of the bowel, and regulates colitis in IL-10-deficient mice (Song et al., 2008), thus ameliorating the ongoing lesions of colitis. The use of KRP-203 with its apparent selectivity for $S1P_1$ would help avoid the clinical adverse bradycardia associated with the use of FTY720 (Shimizu et al., 2005). Thus, with the given results, KRP-203 is expected to be useful in immunointervention in inflammatory and/or autoimmune diseases and graft rejection and potentially more advantageous for long-term usage with improved side-effect profile compared to FTY720. The results of human clinical trials are awaited.

A selective agonist for $S1P_1$ receptor that was identified by high-throughput screening of commercial libraries is SEW2871 (Maybridge, UK). This compound is structurally unrelated to S1P. SEW2871 does not need to be phosphorylated to promote internalization of SIP_1 receptor. The receptor is recycled and not degraded (Wei et al., 2005). SEW2871 is known to induce lymphopenia in a mouse model (Sanna et al., 2004). As SEW2871 does not activate $S1P_3$, it also does not cause bradycardia (Brinkmann et al., 2004). Preliminary data suggest that agonism of $S1P_1$ alone using SEW2871 is sufficient to inhibit allergic airway inflammation and reduces neutrophil and macrophage infiltrates (Idzko et al., 2006).

Recently, a biospecific monoclonal antibody to S1P termed "sphingomab" was developed (Visentin et al., 2006). Sphingomab acts as a molecular sponge to selectively absorb and neutralize S1P. The murine anti-S1P, Sphingomab, was humanized for clinical development and was then optimized to retain the specificity and affinity characteristics of the murine mAb (O'Brien et al., 2009). This humanized, optimized antibody is referred to as sonepcizumab or LT1009 (Sabbadini, 2011). Sonepcizumab was formulated for use in oncology, which was investigated in a recently completed Phase I trial in cancer patients and is being considered for Phase II clinical trials. This antibody has also been formulated for use in age-related macular degeneration (AMD), which was also investigated in a recently completed Phase I trial for wet-AMD patients and will also likely be considered for Phase II clinical trials.

Many recent studies have investigated the effects of SphK inhibition and silencing using genetic deletion in mice or gene-specific knockdown with short interfering RNAs (siRNA) targeting SphK. The first competitive inhibitors of SphK activity were D,L-*threo*-dihydrosphingosine (DHS; Saphingol) and *N,N*-

dimethylsphingosine (DMS) (reviewed in Takabe et al., 2008). They were not only anti-inflammatory in many models, but also found to exert off-target effects in addition to insolubility and potential toxicity issues. Non-lipid-selective inhibitors were developed that block the ATP-binding site of SphKs, with high selectivity for SphKs but without discriminating SphK1 from isoenzyme SphK2. SKI-II was initially demonstrated to inhibit asthma (Nishiuma et al., 2008), but more recently to not ameliorate airway inflammation (Chiba et al., 2010b). More selective inhibitors are in development by many laboratories, including ours. ABC294640, a specific SphK2 inhibitor, has been developed, but has only been tested for its antitumor activity (French et al., 2010).

It was found that SphK1 inhibition suppresses inflammation in many preclinical models, including allergen-induced airway inflammation (Haberberger et al., 2009; Pushparaj et al., 2009; Melendez, 2008b; Lai et al., 2008). We have also found that modulating SphKs in human mast cells severely impairs their ability to degranulate and release exocytotic and secretory mediators (Oskeritzian et al., 2008).

Some of these beneficial therapeutic effects on airway inflammation have been recapitulated by siRNA-mediated silencing of SphK1 *in vivo*, which establishes SphK1 as a promoter of inflammation.

12.6.2 MISCELLANEOUS

Efforts are on to discover novel, more specific antagonists of S1P receptors. JTE-013, a specific $S1P_2$ receptor antagonist, is a pyrazolopyridine derivative discovered by Japan Tobacco Corporation (Osaka, Japan). JTE-013 is reported to reverse mast cell motility induced by S1P. We reported its efficacy in attenuating signs of anaphylaxis in a mast cell-dependent preclinical evaluation in a mouse model (Oskeritzian et al., 2010).

VPC23019, an aryl amide-containing S1P analog discovered by Avanti Polar Lipids (Alabaster, AL), acts as a nonselective antagonist of both $S1P_1$ and $S1P_3$ receptors (Davis et al., 2005). VPC23019 has been shown to inhibit S1P-induced migration of thyroid cancer cells (Balthasar et al., 2006) and human mast cell migration *in vitro* (Oskeritzian et al., 2008).

Effort has been made to interfere with the synthesis and degradation of S1P. Lexicon Pharmaceuticals is working on the development of LX 2931, an orally active small-molecule inhibitor of S1P lyase that induces lymphopenia and ablates S1P gradient, similar to FTY-720, for the treatment of autoimmune disorders such as rheumatoid arthritis.

12.7 SUMMARY AND PERSPECTIVES

The understanding of sphingolipid biology in physiology and pathophysiology is showing improvement. A testimony to this effect is the FDA approval of FTY720 for the treatment of relapsing multiple sclerosis. However, at the moment, a similar clarity has not emerged in our understanding of sphingolipid biology for inflammatory diseases of the airways. Sphingolipids such as ceramide, C1P, sphingosine, and S1P regulate the functions of many immune and inflammatory cells as well as of structural cells. A great deal of effort is under way to design and discover novel modulators to target sphingolipid signaling, given the specific signaling pathways activated by each subtype of S1P receptor. Currently, drugs targeting ligand S1P production and receptor function are in use for preclinical studies, in conjunction with genetic approaches. Several key issues remain regarding designing agonists versus antagonists, on the one hand, and designing selective versus nonselective molecules, on the other. It is likely that in the near future we may witness the discovery of molecules for airway inflammation that act by modulating sphingolipid signaling pathways. Through its pivotal role in cytokine/chemokine production and secretion, S1P is a key factor in inflammation and immune responses (reviewed in Xia and Wadham, 2011). Impetus exists to promote further investigations and discoveries of new therapeutic strategies for immune and inflammatory disorders, based on the valuable drug specificity and spectrum of action.

ACKNOWLEDGMENTS

The authors wish to thank all "sphingolipid enthusiasts" scientists for their contribution to the field of S1P receptor signaling and inflammation and express their regret for any omission due to space limitations. Carole A. Oskeritzian is supported by a National Institutes of Health grant from the National Institute of Arthritis and Musculoskeletal and Skin Diseases (K01AR053186).

REFERENCES

Aarthi, J. J., M. A. Darendeliler, and P. N. Pushparaj. 2011. Dissecting the role of S1P/S1PR axis in health and disease. *J Dent Res.* 90: 841–54.

Allende, M. L., D. Zhou, D. N. Kalkofen, S. Benhamed, G. Tuymetova, C. Borowski, A. Bendelac, and R. L. Proia. 2008. S1P1 receptor expression regulates emergence of NKT cells in peripheral tissues. *FASEB J.* 22(1): 307–15.

Ammit, A. J., A. T. Hastie., L. C. Edsall, R. K. Hoffman, Y. Amrani, V. P. Krymskaya, S. A. Kane et al. 2001. Sphingosine 1-phosphate modulates human airway smooth muscle cell functions that promote inflammation and airway remodeling in asthma. *FASEB J.* 15: 1212–4.

Alvarez, S. E., K. B. Harikumar, N. C. Hait, J. Allegood, G. M. Strub, E. Y. Kim, M. Maceyka et al. 2010. Sphingosine-1-phosphate is a missing cofactor for the E3 ubiquitin ligase TRAF2. *Nature* 465(7301): 1084–8.

Balthasar, S., J. Samulin, H. Ahlgren, N. Bergelin, M. Lundqvist, E. C. Toescu, M. C. Eggo, and K. Tornquist. 2006. Sphingosine 1-phosphate receptor expression profile and regulation of migration in human thyroid cancer cells. *Biochem J.* 398: 547–56.

Bartke, N. and Y. A. Hannun. 2009. Bioactive sphingolipids: Metabolism and function. *J Lipid Res.* 4: S91–6.

Bankovich, A. J., L. R. Shiow, and J. G. Cyster. 2010. CD69 suppresses sphingosine 1-phosophate receptor-1 (S1P1) function through interaction with membrane helix 4. *J Biol Chem.* 285: 22328–37.

Baumruker, T. and E. E. Prieschl. 2002. Sphingolipids and the regulation of the immune response. *Semin Immunol.* 14: 57–3.

Berdyshev, E. V., I. Gorshkova, A. Skobeleva, R. Bittman, X. Lu, S. M. Dudek, T. Mirzapoiazova, J. G. Garcia, and V. Natarajan. 2009. FTY720 inhibits ceramide synthases and up-regulates dihydrosphingosine 1-phosphate formation in human lung endothelial cells. *J Biol Chem.* 284(9): 5467–77.

Blé, F. X., C. Cannet, S. Zurbruegg, C. Gérard, N. Frossard, N. Beckmann, and A. Trifilieff. 2009. Activation of the lung S1P(1) receptor reduces allergen-induced plasma leakage in mice. *Br J Pharmacol.* 58(5): 1295–301.

Brinkmann, V., J. G. Cyster, and T. Hla. 2004. FTY720: Sphingosine 1-phosphate receptor-1 in the control of lymphocyte egress and endothelial barrier function. *Am J Transplant.* 4:1019–25.

Brinkmann, V., M. D. Davis, C. E. Heise, R. Albert, S. Cottens, R. Hof, C. Bruns et al. 2002. The immune modulator FTY720 targets sphingosine 1-phosphate receptors. *J Biol Chem.* 277(24): 21453–7.

Chalfant, C. E. and S. Spiegel. 2005. Sphingosine 1-phosphate and ceramide-1-phosphate: Expanding roles in cell signaling. *J Cell Sci.* 118: 4605–12.

Carlson, C. M., B. T. Endrizzi, J. Wu, X. Ding, M. A. Weinreich, E. R. Walsh, M. A. Wani, J. B. Lingrel, K. A. Hogquist, and S. C. Jameson. 2006. Kruppel-like factor 2 regulates thymocyte and T-cell migration. *Nature* 442(7100): 299–302.

Chi, H. 2011. Sphingosine-1-phosphate and immune regulation: Trafficking and beyond. *Trends Pharmacol Sci.* 32: 16–24.

Chiba, Y., K. Suzuki, E. Kurihara et al. 2010a. Sphingosine-1-phosphate aggravates antigen-induced airway inflammation in mice. *Open Respir Med J.* 4: 82–5.

Chiba, Y., H. Takeuchi, H. Sakai et al. 2010b. SKI-II, an inhibitor of sphingosine kinase, ameliorates antigen-induced bronchial smooth muscle hyperresponsiveness, but not airway inflammation, in mice. *J Pharmacol Sci.* 114: 304–10.

Choi, O. H., J. H. Kim, and P. Kinet. 1996. Calcium mobilization via sphingosine kinase in signaling by the FcεRI antigen receptor. *Nature* 380: 634–6.

Cyster, J. G. 2005. Chemokines, sphingosine-1-phosphate, and cell migration in secondary lymphoid organs. *Annu Rev Immunol.* 23: 127–59.

Davis, M. D., J. J. Clemens, T. L. Macdonald, and K. R. Lynch. 2005. Sphingosine 1-phosphate analogs as receptor antagonists. *J Biol Chem.* 280(11): 9833–41.

Diab, K. J., J. J. Adamowicz, K. Kamocki, N. I. Rush, J. Garrison, Y. Gu, K. S. Schweitzer et al. 2010 Stimulation of sphingosine 1-phosphate signaling as an alveolar cell survival strategy in emphysema. *Am J Respir Crit Care Med.* 181(4): 344–52.

Ding, R., J. Han, Y. Tian, R. Guo, and X. Ma. 2011. Sphingosine-1-phosphate attenuates lung injury induced by intestinal ischemia/reperfusion in mice: Role of inducible nitric-oxide synthase. *Inflammation.* In press.

Forrest, M., S. Y. Sun, R. Hajdu, J. Bergstrom, D. Card, G. Doherty, J. Hale et al. 2004. Immune cell regulation and cardiovascular effects of sphingosine 1-phosphate receptor agonists in rodents are mediated via distinct receptor subtypes. *J Pharmacol Exp Ther.* 309(2): 758–68.

French, K. J., Y. Zhuang, L. W. Maines, P. Gao, W. Wang, V. Beljanski, J. J. Upson, C. L. Green, S. N. Keller, and C. D. Smith. 2010. Pharmacology and antitumor activity of ABC294640, a selective inhibitor of sphingosine kinase-2. *J Pharmacol Exp Ther.* 333(1): 129–39.

Fujita, T., K. Inoue, S. Yamamoto, T. Ikumoto, S. Sasaki, R. Toyama, K. Chiba, Y. Hoshino, and T. Okumoto. 1994. Fungal metabolites: A potent immunosuppressive activity found in *Isaria sinclairii* metabolite. *J Antibiot (Tokyo).* 47(2): 208–15.

Garcia, J. G., F. Liu, A. D. Verin, A. Birukova, M. A. Dechert, W. T. Gerthoffer, J. R. Bamberg, and D. English. 2001. Sphingosine 1-phosphate promotes endothelial cell barrier integrity by Edg-dependent cytoskeletal rearrangement. *J Clin Invest.* 108(5): 689–701.

Haberberger, R. V., C. Tabeling, S. Runciman, B. Gutbier, P. König, M. Andratsch, H. Schütte, N. Suttorp, I. Gibbins, and M. Witzenrath. 2009. Role of sphingosine kinase 1 in allergen-induced pulmonary vascular remodeling and hyperresponsiveness. *J Allergy Clin Immunol.* 124(5): 933–41.

Hait, N. C., J. Allegood, M. Maceyka, G. M Strub, K. B. Harikumar, S. K. Singh, C. Luo et al. 2009. Regulation of histone acetylation in the nucleus by sphingosine-1-phosphate. *Science* 325(5945): 1254–7.

Hannun, Y. A. and L. M. Obeid. 2008. Principles of bioactive lipid signalling: Lessons from sphingolipids. *Nat Rev Mol Cell Biol.* 9: 139–50.

Heinrich, M. M., W. Wickel, C. Schneider-Brachert, C. Sandberg, J. Gahr, R. Schwandner, T. Weber et al. 1999. Cathepsin D targeted by acid sphingomyelinase-derived ceramide. *EMBO J.* 18(19): 5252–63.

Herr, D. R., N. Grillet, M. Schwander R. Rivera, U. Müller, and J. Chun. 2007. Sphingosine 1-phosphate (S1P) signaling is required for maintenance of hair cells mainly via activation of S1P2. *J Neurosci.* 27(6): 1474–8.

Hewson, C. A., J. R. Watson, W. L. Liu, and M. D. Fidock. 2011. A differential role for ceramide kinase in antigen/FcvarepsilonRI-mediated mast cell activation and function. *Clin Exp Allergy.* 41(3): 389–98.

Hla, T., M. J. Lee, N. Ancellin J. H. Paik, and M. J. Kluk. 2001. Lysophospholipids—Receptor revelations. *Science* 294(5548): 1875–8.

Idzko, M., H. Hammad, M. van Nimwegen, M. Kool, T. Müller, T. Soullié, M.A. Willart, D. Hijdra, H.C. Hoogsteden, and B.N. Lambrecht. 2006. Local application of FTY720 to the lung abrogates experimental asthma by altering dendritic cell function. *J Clin Invest.* 116(11): 2935–44.

James, A. 2005. Airway remodeling in asthma. *Curr Opin Pulm Med.* 11: 1–21.

Jenne, C. N., A. Enders, R. Rivera, S. R. Watson, A. J. Bankovich, J. P. Pereira, Y. Xu et al. 2009. T-bet-dependent S1P5 expression in NK cells promotes egress from lymph nodes and bone marrow. *J Exp Med.* 206(11): 2469–81.

Jolly, P. S., M. Bektas, A. Olivera, C. Gonzalez-Espinosa, R. L. Proia, J. Rivera, S. Milstien, and S. Spiegel. 2004. Transactivation of sphingosine-1-phosphate receptors by FcepsilonRI triggering is required for normal mast cell degranulation and chemotaxis. *J Exp Med.* 199(7): 959–70.

Kee, T. H., P. Vit, and A. J. Melendez. 2005. Sphingosine kinase signaling in immune cells. *Clin Exp Pharmcol Physiol.* 32: 153–61.

Kono, M., I. A. Belyantseva, A. Skoura, G. I. Frolenkov, M. F. Starost, J. L. Dreier, D. Lidington et al. 2007. Deafness and stria vascularis defects in S1P2 receptor-null mice. *J Biol Chem.* 282(14): 10690–6.

Kono, M., Y. Mi, Y. Liu, T. Sasaki, M. L. Allende, Y. P. Wu, T. Yamashita, and R. L. Proia. 2004. The sphingosine-1-phosphate receptors S1P1, S1P2, and S1P3 function coordinately during embryonic angiogenesis. *J Biol Chem.* 279: 29367–73.

Lai, W. Q., H. H. Goh, Z. Bao, W. S. Wong, A. J. Melendez, and B. P. Leung. 2008. The role of sphingosine kinase in a murine model of allergic asthma. *J Immunol.* 180(6): 4323–9.

Li, Z., Y. Zhang, Z. Liu X. Wu, Y. Zheng, Z. Tao, K. Mao et al. 2011. ECM1 controls T(H)2 cell egress from lymph nodes through re-expression of S1P(1). *Nat Immunol.* 12(2): 178–85.

Liu, H., M. Sugiura, V. E. Nava, L. C. Edsall, K. Kono, S. Poulton, S. Milstien, T. Kohama, and S. Spiegel. 2000. Molecular cloning and functional characterization of a novel mammalian sphingosine kinase type 2 isoform. *J Biol Chem.* 275(26): 19513–20.

Mandala, S., R. Hajdu, J. Bergstrom, E. Quackenbush, J. Xie, J. Milligan, R. Thornton et al. 2002. Alteration of lymphocyte trafficking by sphingosine-1-phosphate receptor agonists. *Science* 296(5566): 346–9.

Marsolais, D., S. Yagi, T. Kago, N. Leaf, and H. Rosen. 2011. Modulation of chemokines and allergic airway inflammation by selective local sphingosine-1-phosphate receptor 1 agonism in lungs. *Mol Pharmacol.* 79(1): 61–8.

Masini, E., L. Giannini, S. Nistri, L. Cinci, R. Mastroianni, W. Xu, S. A. Comhair, D. Li, S. Cuzzocrea, G. M. Matuschak, and D. Salvemini. 2008. Ceramide: A key signaling molecule in a Guinea pig model of allergic asthmatic response and airway inflammation. *J Pharmacol Exp Ther.* 324(2): 548–57.

Matloubian, M., C. G. Lo, G. Cinamon, M. J. Lesneski, Y. Xu, V. Brinkmann, M. L. Allende, R. L. Proia, and J. G. Cyster. 2004. Lymphocyte egress from thymus and peripheral lymphoid organs is dependent on S1P receptor 1. *Nature* 427(6972): 355–60.

Matsuda, H., T. Suda, J. Sato, T. Nagata, Y. Koide, K. Chida, and H. Nakamura. 2005. Alpha-galactosylceramide, a ligand of natural killer T cells, inhibits allergic airway inflammation. *Am J Respir Cell Mol Biol.* 33(1): 22–31.

Means, C. K. and J. H. Brown. 2009. Sphingosine-1-phosphate receptor signalling in the heart. *Cardiovasc Res.* 82: 193–200.

Melendez, A. J., M. M. Harnett, P. N. Pushparaj et al. 2007. Inhibition of Fc epsilon RI-mediated mast cell responses by ES-62, a product of parasitic filarial nematodes. *Nat Med.* 13: 1375–81.

Melendez, A. J. and A. K. Khaw. 2002. Dichotomy of Ca^{2+} signals triggered by different phospholipid pathways in antigen stimulation of human mast cells. *J Biol Chem.* 277: 17255–62.

Melendez, A. J. 2008a. Sphingosine kinase signaling in immune cells: Potential as novel therapeutic targets. *Biochim Biophys Acta.* 1784: 66–75.

Melendez, A. J. 2008b. Allergy therapy: The therapeutic potential of targeting sphingosine kinase signalling in mast cells. *Eur J Immunol.* 38: 2969–74.

Mitra, P., C. A. Oskeritzian, S. G. Payne, M. A. Beaven, S. Milstien, and S. Spiegel. 2006. Role of ABCC1 in export of sphingosine-1-phosphate from mast cells. *Proc Natl Acad Sci USA.* 103(44): 16394–9.

Morishima, Y., Y. Ishii, T. Kimura, A. Shibuya, K. Shibuya, A. E. Hegab, T. Iizuka et al. 2005. Suppression of eosinophilic airway inflammation by treatment with alpha-galactosylceramide. *Eur J Immunol.* 35(10): 2803–14.

Mullershausen, F., F. Zecri, C. Cetin, A. Billich, D. Guerini, and K. Seuwen. 2009. Persistent signaling induced by FTY720-phosphate is mediated by internalized S1P1 receptors. *Nat Chem Biol.* 5(6): 428–34.

Murakami, M., A. Shiraishi, K. Tabata, and N Fujita. 2008. Identification of the orphan GPCR, P2Y(10) receptor as the sphingosine-1-phosphate and lysophosphatidic acid receptor. *Biochem Biophys Res Commun.* 371(4): 707–12.

Niessen, F., F. Schaffner, C. Furlan-Freguia, R. Pawlinski, G. Bhattacharjee, J. Chun, C. K. Derian, P. Andrade-Gordon, H. Rosen, and W. Ruf. 2008. Dendritic cell PAR1-S1P3 signalling couples coagulation and inflammation. Nature. 2008 Apr 3; 452(7187): 654–8.

Nishiuma, T., Y. Nishimura, T. Okada, E. Kuramoto, Y. Kotani, S. Jahangeer, and S.-I. Nakamura. 2008. Inhalation of sphingosine kinase inhibitor attenuates airway inflammation in asthmatic mouse model. *Am J Physiol Lung Cell Mol Physiol.* 294(6): L1085–93.

O'Brien, N., S. T. Jones, D. G. Williams, H.B. Cunningham, K. Moreno, B. Visentin, A. Gentile et al. 2009. Production and characterization of monoclonal anti-sphingosine-1-phosphate antibodies. *J Lipid Res.* 50(11): 2245–57.

Ogretmen, B., B. J. Pettus, M. J. Rossi, R. Wood, J. Usta, Z. Szulc, A. Bielawska, L. M. Obeid, and Y. A. Hannun. 2002. Biochemical mechanisms of the generation of endogenous long chain ceramide in response to exogenous short chain ceramide in the A549 human lung adenocarcinoma cell line. Role for endogenous ceramide in mediating the action of exogenous ceramide. *J Biol Chem.* 277(15): 12960–9.

Olivera, A., C. Eisner, Y. Kitamura, S. Dillahunt, L. Allende, G. Tuymetova, W. Watford et al. 2010. Sphingosine kinase 1 and sphingosine-1-phosphate receptor 2 are vital to recovery from anaphylactic shock in mice. *J Clin Invest.* 120(5): 1429–40.

Olivera, A., K. Mizugishi, A. Tikhonova, L. Ciaccia, S. Odom, R. L. Proia, and J. Rivera. 2007. The sphingosine kinase-sphingosine-1-phosphate axis is a determinant of mast cell function and anaphylaxis. *Immunity* 26(3): 287–97.

Olivera, A., N. Urtz, K. Mizugishi, Y. Yamashita, A. M. Gilfillan, Y. Furumoto, H. Gu, R. L. Proia, T. Baumruker, and J. Rivera. 2006. IgE-dependent activation of sphingosine kinases 1 and 2 and secretion of sphingosine 1-phosphate requires Fyn kinase and contributes to mast cell responses. *J Biol Chem.* 281(5): 2515–25.

Oskeritzian, C. A., M. M. Price, N. C. Hait, D. Kapitonov, Y. T. Falanga, J. K. Morales, J. J. Ryan, S. Milstien, and S. Spiegel. 2010. Essential roles of sphingosine-1-phosphate receptor 2 in human mast cell activation, anaphylaxis, and pulmonary edema. *J Exp Med.* 207(3): 465–74.

Oskeritzian, C. A., S. E. Alvarez, N. C. Hait, M. M. Price, S. Milstien, and S. Spiegel. 2008. Distinct roles of sphingosine kinases 1 and 2 in human mast-cell functions. *Blood* 111(8): 4193–200.

Oskeritzian, C. A., S. Milstien, and S. Spiegel. 2007. Sphingosine-1-phosphate in allergic responses, asthma and anaphylaxis. *Pharmacol Ther.* 115: 390–9.

Paugh, S. W., B. S. Paugh, M. Rahmani, D. Kapitonov, J. A. Almenara, T. Kordula, S. Milstien et al. 2008. A selective sphingosine kinase 1 inhibitor integrates multiple molecular therapeutic targets in human leukemia. *Blood* 112(4): 1382–91.

Paugh, S. W., S. G. Payne, S. E. Barbour, S. Milstien, and S. Spiegel. 2003. The immunosuppressant FTY720 is phosphorylated by sphingosine kinase type 2. *FEBS Lett.* 554(1–2): 189–93.

Payne, S. G., C. A. Oskeritzian, R. Griffiths, P. Subramanian, S. E. Barbour, C. E. Chalfant, S. Milstien, and S. Spiegel. 2007. The immunosuppressant drug FTY720 inhibits cytosolic phospholipase A2 independently of sphingosine-1-phosphate receptors. *Blood* 109(3): 1077–85.

Petrache, I., V. Natarajan, L. Zhen, T. R. Medler, A. T. Richter, C. Cho, W. C. Hubbard, E. V. Berdyshev, and R. M. Tuder. 2005. Ceramide upregulation causes pulmonary cell apoptosis and emphysema-like disease in mice. *Nat Med.* 11(5): 491–8.

Pettus, B. J., J. Bielawski, A. M. Porcelli, D. L. Reames, K. R. Johnson, J. Morrow, C. E. Chalfant, L. M. Obeid, and Y. A. Hannun. 2003. The sphingosine kinase 1/sphingosine-1-phosphate pathway mediates COX-2 induction and PGE2 production in response to TNF-alpha. *FASEB J.* 17(11): 1411–21.

Price, M. M., C. A. Oskeritzian, S. Milstien, and S. Spiegel. 2008. Sphingosine-1-phosphate synthesis and functions in mast cells. *Future Lipidol.* 3(6): 665–74.

Prieschl, E. E., R. Csonga, V. Novotny, G. E. Kikuchi, and T. Baumruker. 1999. The balance between sphingosine and sphingosine-1-phosphate is decisive for mast cell activation after Fc epsilon receptor I triggering. *J Exp Med.* 190(1): 1–8.

Pushparaj, P. N., J. Manikandan, H. K. Tay, S. C. H'ng, S. D. Kumar, J. Pfeilschifterr, A. Huwile, and A. J. Melendez. 2009. Sphingosine kinase 1 is pivotal for Fc epsilon RI-mediated mast cell signaling and functional responses *in vitro* and *in vivo*. *J Immunol.* 183(1): 221–7.

Pyne, S., S. C. Lee, J. Long, and N. J. Pyne. 2009. Role of sphingosine kinases and lipid phosphate phosphatases in regulating spatial sphingosine 1-phosphate signalling in health and disease. *Cell Signal.* 21(1): 14–21.

Radin, N. S., J. A. Shayman, and J. Inokuchi. 1993. Metabolic effects of inhibiting glucosylceramide synthesis with PDMP and other substances. *Adv Lipid Res.* 26: 183–213.

Rivera, J., R. L. Proia, and A. Olivera. 2008. The alliance of sphingosine-1-phosphate and its receptors in immunity. *Nat Rev Immunol.* 8: 753–63.

Rosen, H. and E. J. Goetzl. 2005. Sphingosine 1-phosphate and its receptors: An autocrine and paracrine network. *Nat Rev Immunol.* 5: 560–70.

Rosen, H., P. Gonzalez-Cabrera, D. Marsolais, S. Cahalan, A. S. Don, and M. G. Sanna. 2008. Modulating tone: The overture of S1P receptor immunotherapeutics. *Immunol Rev.* 223: 221–35.

Rosen, H., M. G. Sanna, S. M. Cahalan, and P. J. Gonzalez-Cabrera. 2007. Tipping the gate-keeper: S1P regulation of endothelial barrier function. *Trends Immunol.* 28(3): 102–7.

Rosenfeldt, H. M., Y. Amrani, K. R. Watterson K.S. Murthy, R.A. Panettieri Jr., and S. Spiegel. 2003. Sphingosine-1-phosphate stimulates contraction of human airway smooth muscle cells. *FASEB J.* 17(13): 1789–99.

Roviezzo, F., F. Del Galdo, G. Abbate, M. Bucci, B. D'Agostino, E. Antunes, G. De Dominicis et al. 2004. Human eosinophil chemotaxis and selective *in vivo* recruitment by sphingosine 1-phosphate. *Proc Natl Acad Sci USA.* 101(30): 11170–5.

Ruf, W., C. Furlan-Freguia, and F. Niessen. 2009. Vascular and dendritic cell coagulation sig-naling in sepsis progression. *J Thromb Haemost.* 1: 118–21.

Ryan, J. J. and S. Spiegel. 2008. The role of sphingosine-1-phosphate and its receptors in asthma. *Drug News Perspect.* 21: 89–6.

Sabbadini, R. A. 2011. Sphingosine-1-phosphate antibodies as potential agents in the treatment of cancer and age-related macular degeneration. *Br J Pharmacol.* 162: 1225–38.

Sanna, M. G., J. Liao, E. Jo, C. Alfonso, M. Y. Ahn, M. S. Peterson, B. Webb et al. 2004. Sphingosine 1-phosphate (S1P) receptor subtypes S1P1 and S1P3, respectively, regulate lymphocyte recirculation and heart rate. *J Biol Chem.* 279(14): 13839–48.

Sammani, S., L. Moreno-Vinasco, T. Mirzapoiazova, P. A. Singleton, E. T. Chiang, C. L. Evenoski, T. Wang et al. 2010. Differential effects of sphingosine 1-phosphate receptors on airway and vascular barrier function in the murine lung. *Am J Respir Cell Mol Biol.* 43(4): 394–402.

Sawicka, E., C. Zuany-Amorim, C. Manlius, A. Trifilieff, V. Brinkmann, D. M. Kemeny, and C. Walker. 2003. Inhibition of T_H1- and T_H2-mediated airway inflammation by the sphingosine 1-phosphate receptor agonist FTY720. *J Immunol.* 171(11): 6206–14.

Schwab, S. R., J. P. Pereira, M. Matloubian,Y. Xu, Y. Huang, and J. G. Cyster. 2005. Lymphocyte sequestration through S1P lyase inhibition and disruption of S1P gradients. *Science* 309(5741): 1735–9.

Shea, B. S., S. F. Brooks, B. A. Fontaine, J. Chun, A. D. Luster, and A. M Tager. 2010. Prolonged exposure to sphingosine 1-phosphate receptor-1 agonists exacerbates vascular leak, fibro-sis, and mortality after lung injury. *Am J Respir Cell Mol Biol.* 43(6): 662–73.

Shimizu, H., M. Takahashi, T. Kaneko, T. Murakami, Y. Hakamata, S. Kudou, T. Kishi et al. 2005. KRP-203, a novel synthetic immunosuppressant, prolongs graft survival and attenuates chronic rejection in rat skin and heart allografts. *Circulation* 111(2): 222–9.

Skoura, A., T. Sanchez, K. Claffey, S. M. Mandala, R. L. Proia, and T. Hla. 2007. Essential role of sphingosine 1-phosphate receptor 2 in pathological angiogenesis of the mouse retina. *J Clin Invest.* 117(9): 2506–16.

Snider, A. J., K. A. Orr Gandy, and L. M. Obeid. 2010. Sphingosine kinase: Role in regulation of bioactive sphingolipid mediators in inflammation. *Biochimie* 92: 707–15.

Song, J., C. Matsuda, Y. Kai, T. Nishida, K. Nakajima, T. Mizushima , M. Kinoshita, T. Yasue, Y. Sawa, and T. Ito. 2008. A novel sphingosine 1-phosphate receptor agonist, 2-amino-2-propanediol hydrochloride (KRP-203), regulates chronic colitis in interleukin-10 gene-deficient mice. *J Pharmacol Exp Ther.* 324(1): 276–83.

Spiegel, S. and S. Milstien. 2002. Sphingosine 1-phosphate, a key cell signaling molecule. *J Biol Chem.* 277: 25851–4.

Spiegel, S. and S. Milstien. 2003. Sphingosine-1-phosphate: An enigmatic signalling lipid. *Nat Rev Mol Cell Biol.* 4: 397–407.

Taha, T. A., Y. A. Hannun, and L. M. Obeing. 2006. Sphingosine kinase: Biochemical and cel-lular regulation and role in disease. *J Biochem Mol Biol.* 39: 113–31.

Taha, T. A., K. M. Argraves, and L. M. Obeid. 2004. Sphingosine-1-phosphate receptors: Receptor specificity versus functional redundancy. *Biochim Biophys Acta.* 1682: 48–5.

Takabe, K., S. W. Paugh, S. Milstein, and S. Spiegel. 2008. Inside-out signaling of sphingo-sine-1-phosphate: Therapeutic targets. *Pharmacol Rev.* 60(2): 181–95.

Tauseef, M., V. Kini, N. Knezevic, M. Brannan, R. Ramchandaran, H. Fyrst, J. Saba, S. M. Vogel, A. B. Malik, and D. Mehta. 2008. Activation of sphingosine kinase-1 reverses the increase in lung vascular permeability through sphingosine-1-phosphate receptor signaling in endothelial cells. *Circ Res.* 103(10): 1164–72.

Visentin, B., J. A. Vekich, B. J. Sibbald, A. L. Cavalli, K. M. Moreno, R. G. Matteo, W. A. Garland et al. 2006. Validation of an anti-sphingosine-1-phosphate antibody as a potential therapeutic in reducing growth, invasion, and angiogenesis in multiple tumor lineages. *Cancer Cell* 9(3): 225–38.

Walzer, T., L. Chiossone, J. Chaix, A. Calver, C. Carozzo, L. Garrigue-Antar, E. Jacques, M. Baratin, E. Tomasello, and Y. Vivier. 2007. Natural killer cell trafficking *in vivo* requires a dedicated sphingosine 1-phosphate receptor. *Nat Immunol.* 8(12): 1337–44.

Wang, W., M. H. Graeler, and E. A. Goetzl. 2005. Type 4 sphingosine 1-phosphate G protein-coupled receptor (S1P4) transduces S1P effects on T cell proliferation and cytokine secretion without signaling migration. *FASEB J.* 19: 1731–3.

Wattenberg, B. W., S. M. Pitson, and D. M. Raben. 2006. The sphingosine and diacylglycerol kinase superfamily of signaling kinases: Localization as a key to signaling function. *J Lipid Res.* 47: 1128–39.

Wei, S. H., H. Rosen, M. P. Matheu, M. G. Sanna, S. K. Wang, E. Jo, C. H. Wong, I. Parker, and M. D. Cahalan. 2005. Sphingosine 1-phosphate type 1 receptor agonism inhibits transendothelial migration of medullary T cells to lymphatic sinuses. *Nat Immunol.* 6(12): 1228–35.

Wolf, A. M., K. Eller, R. Zeiser, C. Dürr, U.V. Gerlach, M. Sixt, L. Markut, G. Gastl, A. R. Rosenkranz, and D. Wolf. 2009. The sphingosine 1-phosphate receptor agonist FTY720 potently inhibits regulatory T cell proliferation *in vitro* and *in vivo*. *J Immunol.* 183(6): 3751–60.

Xia, P. and C. Wadham. 2011. Sphingosine 1-phosphate, a key mediator of the cytokine network: Juxtacrine signaling. *Cytokine Growth Factor Rev.* 22: 45–53.

Xia, P., L. Wang, P. A. Moretti, N. Albanese, F. Chai, S.M. Pitson, R. J. D'Andrea, J. R. Gamble, and M. A. Vadas. 2002. Sphingosine kinase interacts with TRAF2 and dissects tumor necrosis factor-alpha signaling. *J Biol Chem.* 277(10): 7996–8003.

Yanagawa, Y., K. Sugahara, H. Kataoka, T. Kawaguchi, Y. Masubuchi, and K. Chiba. 1998. FTY720, a novel immunosuppressant, induces sequestration of circulating mature lymphocytes by acceleration of lymphocyte homing in rats. II. FTY720 prolongs skin allograft survival by decreasing T cell infiltration into grafts but not cytokine production *in vivo*. *J Immunol.* 160(11): 5493–9.

Zemann, B., B. Kinzel, M. Muller, R. Reuschel, D. Mechtcheriakova, N. Urtz, F. Bornancin, T. Baumruker, and A. Billich. 2006. Sphingosine kinase type 2 is essential for lymphopenia induced by the immunomodulatory drug FTY720. *Blood* 107(4): 1454–8.

13 Protein Kinase C Isozymes and Airway Inflammation

*Puneet Chopra, V. Senthil,
and Manish Diwan*

CONTENTS

13.1 INTRODUCTION

Protein kinase C (PKC) includes a family of serine/threonine kinases. To date, 12 isoforms of PKC family have been identified. PKC family members exhibit broad tissue distribution and are known to regulate a wide variety of cellular processes. Using molecular tools, the contribution of different isoforms of PKC has been delineated in immune, antigen-presenting, inflammatory, and structural cells of the airway. Evidence is emerging in support of a role of specific PKC isoforms in experimental airway inflammation. Indeed, there has been excellent progress toward the discovery of highly isoform-selective inhibitors for PKC. A few PKC inhibitors have also been shown to be efficacious in preclinical animal models of airway diseases.

In this chapter, we will discuss the detailed biochemistry and molecular biology of the protein kinase family members. The role of PKC isoforms in airway inflammation and drug discovery attempts around PKC will also be discussed.

13.2   BIOLOGY OF PKC

Twenty years after the discovery of phosphorylase kinase, PKC was one of the very first identified protein kinases. PKC was first defined as a histone kinase activated by limited proteolysis (Inoue et al., 1977), but further studies revealed that it could also be activated by phosphatidylserine (PS) and diacylglycerol (DAG) in a Ca^{2+}-dependent manner and also by tumor-promoting phorbol esters such as phorbol-12-myristate-13-acetate (PMA) and 12-O-tetradecanoylphorbol-13-acetate (TPA) (Nishizuka, 1984). Remarkable heterogeneity exists within the PKC signal transduction pathway and it comprises of a large superfamily of AGC serine/threonine kinases, such as protein kinase A, protein kinase G, and protein kinase C. The PKC family includes 12 distinct family members in mammalian cells categorized into three subfamilies based on the presence or absence of motifs that dictate cofactor requirements for optimal catalytic activity, sequence homology, and activator and cofactor requirements (Figure 13.1).

The most studied and best understood of these groups is the conventional PKCs (cPKCs), which comprise the cPKC-α, βI, βII, and γ isotypes (the PKC-β gene is

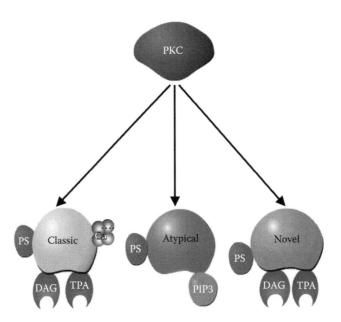

FIGURE 13.1   Classification of PKC. The PKC family consists of 12 isoforms that are categorized into three groups, the classic, novel, and atypical PKCs. The groups differ in their requirement for Ca^{2+} and their ability to be activated by various lipid species, including DAG and TPA. All the isoforms require phosphatidylserine for full enzyme activity.

alternatively spliced to produce two gene products, which differ only in their extreme C-terminal ends) (Coussens et al., 1987). These PKC isoforms are regulated by calcium, DAG, and phospholipids. The next well-characterized PKC subfamily is the novel PKC isoforms (nPKCs), which consist of the PKC-δ, ϵ, η, and θ isoforms. These kinases are Ca^{2+} insensitive, but are still activated by DAG or phorbol esters in the presence of PS (Ono et al., 1988). The least understood isoforms are the atypical PKCs (aPKCs), which comprise the aPKC-ζ and γ or ι (PKC-γ is the mouse ortholog of human PKC-ι). These kinases are insensitive to calcium and do not respond to PMA or DAG (Ono et al., 1989). An additional family of PKC was discovered independently by two laboratories and is also known as PKC-μ or PKD (Johannes et al., 1994; Valverde et al., 1994). Another more distantly related family includes the PKC-related kinases (PRK) comprising of three members PRK1–3, which are also known as PKN (Palmer et al., 1995).

In resting cells, PKCs are predominantly localized in the cytosol and are catalytically inactive due to autoinhibition by their pseudosubstrate domain. Upon cell activation, PKC isoform-specific signals trigger translocation from the cytosol to the membrane and induce conformational changes, which displace the pseudosubstrate moiety from the catalytic domain and enable PKC isoforms to phosphorylate specific protein substrates. PKC isoforms are activated by a variety of extracellular signals, and in turn modify the activities of cellular proteins, including receptors, enzymes, cytoskeletal proteins, and transcription factors. Accordingly, the PKC family members regulate a vast array of cellular processes such as mitogenesis, proliferation of cells, apoptosis, platelet activation, remodeling of the actin cytoskeleton, modulation of ion channels, and secretion (Figure 13.2).

PKC isoform-selective knockout and transgenic mice studies along with pharmacological and biochemical tools utilizing PKC isoform-specific cDNA, antisense oligonucleotides, RNA interference, and pharmacological inhibitors have highlighted that PKC-regulated signaling pathways also play a significant role in the immune intracellular signaling. These PKC-mediated signaling events vary from development, differentiation, activation, and survival of lymphocytes to macrophage activation. How these individual isoforms of PKC contribute to the regulation of diverse cell responses is an important area of signal transduction research. The specific cofactor requirements, tissue distribution, and cellular compartmentalization suggest differential functions and fine tuning of specific signaling cascades for each PKC isoform. Thus, specific stimuli can lead to differential responses via isoform-specific PKC signaling regulated by their expression, localization, and phosphorylation status in particular biological settings.

13.3   STRUCTURE OF PKC

The structure of each PKC isoforms (Figure 13.3) consists of a single polypeptide chain with two structurally well-defined domains: the amino-terminal regulatory domain (~20–40 kDa) and the carboxyl-terminal catalytic domain (~45 kDa).

The regulatory and the catalytic domains are connected by a hinge region that is highly sensitive to proteolytic cleavage by cellular proteases. The regulatory region possesses the motifs involved in the binding of the phospholipid cofactors and Ca^{2+}

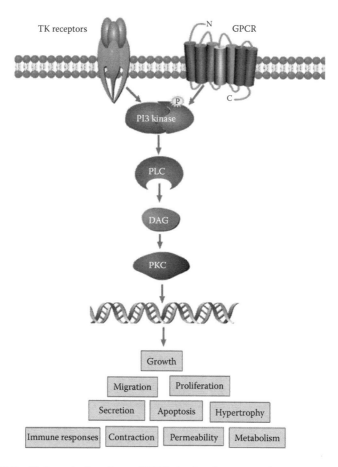

FIGURE 13.2 Pleiotropic functions of PKC. At the plasma membrane, receptor-mediated activation occurs through either G-protein-coupled receptors (GPCR) or tyrosine kinase (TK) receptors. PI3K generates phosphatidylinositol-3,4,5-trisphosphate (PtdIns(3,4,5)P3), which activates PLC. PLC generates the second messenger DAG that activates PKC. Activated PKC controls a vast array of cellular functions. The cellular effect of PKC activation is dependent on the isoform, the duration of activity, the cellular compartmentalization, and additional signaling events occurring at the cellular level.

and is involved in protein–protein interactions that regulate PKC activity and localization. The carboxyl-terminal region is the kinase domain and includes motifs involved in ATP and substrate binding.

The family of PKC isoforms is grouped into three subclasses on the basis of the domain composition of the regulatory moiety. Cloning of the first isoforms in the mid-1980s revealed four highly conserved domains (C1–C4) (Coussens et al., 1986) and five variable domains (V1–V5). The C1 and C2 domains form a part of the regulatory region and C3 and C4 domains form the kinase region. The C1 region is present in all PKC isoforms. Extensive biochemical and mutational analysis studies have revealed the function of each of these domains. The C1 domain consists of two

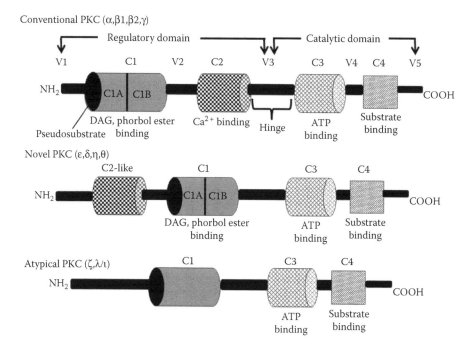

FIGURE 13.3   Structural architecture of different isoforms of PKC. Each PKC family member contains a highly homologous C-terminal catalytic domain and an N-terminal regulatory domain containing key motifs responsible for cofactor binding and substrate accessibility. The autoinhibitory pseudosubstrate sequence (PS) present in the regulatory domain of all PKC isoenzymes interacts directly with the substrate-binding cavity in the catalytic domain. The C1 domain that contains two repeated cysteine-rich zinc finger motifs (C1A and C1B) is functional in cPKCs and nPKCs and mediates the DAG/phorbol ester binding. Atypical PKCs contain only one Cys-rich motif. The C2 domain mediates Ca^{2+} binding in cPKCs but the C2 domain of novel PKCs lacks amino acids involved in calcium binding but has key conserved residues involved in maintaining the C2 fold (hence it is described as "C2-like"). C3 and C4 form the ATP- and substrate-binding lobes of the kinase core.

cysteine-rich zinc motifs referred to as C1a and C1b, conserved in most of the isoforms, and serves as a binding site for the second messenger DAG/PMA in cPKC and nPKC isoforms (Burns and Bell, 1991; Kazaneitz et al., 1995a, b; Ono et al., 1989). The aPKCs contain a C1 domain with only one zinc finger that does not bind to DAG/PMA, while the PRKs do not possess this structure at all (Hurley et al., 1997). The C1 domain of cPKCs, nPKCs, and aPKCs is preceded by an autoinhibitory domain or a so-called pseudosubstrate that binds to the substrate-binding site in the catalytic domain and keeps the enzyme in an inactive state in the absence of cofactors and activators (Orr et al., 1992). While the cPKCs and nPKCs have two copies of these motifs in tandem, only a single copy is found in the aPKCs.

The cPKCs possess a C2 domain that binds to phospholipids in a calcium-dependent way, which contributes to PKC activation. A C2-like domain is present close to the amino-terminal end in nPKCs, although this domain is Ca^{2+} independent (Sossin and

Schwartz, 1993). C2 domains can also interact with receptors for activated C kinase (RACKs) in addition to Ca^{2+} and PS-binding sites (Mochly-Rosen et al., 1991). RACKs along with variable region V5 contribute to the translocation of activated PKCs from the cytosol to the plasma membrane (Ron and Mochly-Rosen, 1995).

13.4 EXPRESSION OF PKC ISOFORMS

Members of the PKC family are expressed in many different cell types, where they are known to regulate a wide variety of cellular processes that eventually impact cell growth and differentiation, cytoskeletal remodeling, and gene expression in response to various stimuli. However, recent PKC isoform-selective knockout studies and transgenic mice have highlighted distinct functions of individual PKCs in the immune system (Table 13.1).

The expression pattern of different isoforms of PKC plays key roles in regulating the various signaling events of hematopoietic cells in response to various stimuli. In spite of the fact that PKC isoforms exhibit small differences in their dependencies on lipid cofactors and substrate specificities, they differ significantly in their tissue distribution and cellular localization (Table 13.1). Although most of the PKC isoforms are ubiquitous in mammalian tissues, some isoforms have a disproportionate distribution in specific cells, for example, PKC-α in T cells, PKC-β in B cells, and mast

TABLE 13.1
Properties of PKC Family Members

PKC Isoforms	Tissue Expression	PKC Knockout Phenotype	References
		cPKC subfamily	
PKC-α	Ubiquitous, high in T cells	Transgenic PKC-α display hyperproliferative T cells	Iwamoto et al. (1992)
PKC-β	Ubiquitous, high in B cells	BCR signaling and survival defects; mast cell defects	Leitges et al. (1996), Ivaska et al. (2003)
		nPKC subfamily	
PKC-δ	Ubiquitous	Hyperproliferative B and mast cells; B cell anergy defect	Miyamoto et al. (2002), Mecklenbrauker et al. (2002), Leitges et al. (2002)
PKC-ε	Ubiquitous	Macrophage activation defect	Castrilo et al. (2001)
PKC-θ	T lymphocytes, skeletal muscle	TCR signaling defect	Sun et al. (2000), Pfeifhofer et al. (2003)
		aPKC subfamily	
PKC-ζ	Ubiquitous	BCR signaling defect	Martin et al. (2002)

cells. Neutrophils possess at least two phospholipid-dependent forms of protein kinase C, a classical PKC-β isoform- and a Ca-independent but PS/DG-dependent novel protein kinase C (nPKC) (Majumdar et al., 1993). Using isoform-specific polyclonal antibodies, the presence of the PKC-α, βI, βII, and ζ isoforms and the low-level expression of the PKC-δ, ε, ι, and μ isoforms were detected in circulating eosinophils of both normal and asthmatic subjects. However, among these isoforms, the activation of only PKC-α and βII isoforms was detected in circulating eosinophils from asthmatic patients. In contrast, increase in PKC-ζ expression and its significant translocation to the membrane were observed after allergen challenge (Evans et al., 1999). Expression of different isoforms of PKCs has also been observed in the airway smooth muscle (ASM) cells and among these isoforms, PKC-α, β1, δ, ε, μ, γ, and ξ are found in the cytosol and PKC-βII in the membrane of ASM cells under basal conditions (Pang et al., 2002). Expression of PKC-θ is primarily restricted to T lymphocytes and is recruited to the immunological synapse upon T cell receptor signaling (Meller et al., 1999).

Diverse expression patterns of different isoforms of PKC suggest that most cell types express only a subset of isoforms, which might account for the functional diversity in different isoforms of PKC. In addition to difference in expression pattern, difference also exists in the subcellular localization of various isoforms of PKC. Thrombin causes rapid nuclear translocation of PKC-α in fibroblasts but not of PKC-ε and PKC-ζ, which are also expressed in these cells (Leach et al., 1992). Translocation of PKC to the nucleus also seems to be a cell-type-specific effect and depends on the activator used, as bryostatin (artificial activator of PKC) causes PKC nuclear association in HL60 cells whereas phorbol esters (another activator of PKC) do not (Hocevar and Fields, 1991). PKC-α has been found to be associated with the focal contact points in rat embryo fibroblasts whereas no such localization is reported for PKC-ε and PKC-ζ, which are present in the same cells (Jaken et al., 1989). Such differences might be expected to reflect specificity in activating molecules, which could be generated at different membrane locations.

13.5   RELEVANCE OF PKC IN RESPIRATORY AILMENTS

Among chronic respiratory diseases, asthma and chronic obstructive pulmonary disease (COPD) are the most common clinical entities (Barnes, 2008). The site and specific characteristics of the inflammatory response may vary from one respiratory disease condition to another but it mainly involves the recruitment and activation of immune cells and inflammatory cells, and changes in the structural cells of the lung. The pathology of asthma includes the release of inflammatory mediators and the activation and degranulation of tissue mast cells, eosinophils, granulocytes, B cells, and airway epithelial and smooth muscle cells (Barnes, 2008; Trivedi and Lloyd, 2007). Immunologically, COPD is distinct from asthma as it primarily involves innate/T_H1 immunity, including recruitment of neutrophils, macrophages, T helper 1 (T_H1), and Type 1 cytotoxic T (Tc1) cells (Barnes, 2008; Chung and Adcock, 2008) and it is orchestrated by a series of proinflammatory mediators, including chemokines such as CXCL1 and CXCL8, cytokines such as TNF-α, IL-1β, IL-6, and interferon-γ (IFN-γ), and proteases such as neutrophil elastase and metalloproteinase-9 (Barnes, 2008).

Recent advances in the understanding of the molecules and intracellular signaling events controlling these processes are now translating to new therapeutic entities. Therefore, the reduction in the recruitment and/or activation of immune cells, inflammatory cells, and structural cells may be beneficial for controlling the progression of various inflammatory processes. However, the identification and development of viable alternative anti-inflammatory therapies for these diseases is still a challenging endeavor. The search for new therapeutic targets has led to the consideration of a plethora of kinases as targets since many of these kinases such as PKC are important in regulating receptor-mediated signaling and cellular function.

Individual PKC isoforms have been implicated in many cellular responses that are important in both normal lung function and the pathogenesis of pulmonary disease. These responses include permeability, contraction, migration, hypertrophy, proliferation, apoptosis, and secretion (Dempsey et al., 2000). PKC family members exert their influence on the expression and activation of inflammatory mediators, inflammatory cell recruitment, and airway remodeling. PKC isoforms have been implicated in asthma and COPD as they can induce airway inflammation, bronchospasm, and mucus production, a hallmark of airway inflammation. (Dempsey et al., 2007). PKC family members contribute at multiple steps of both innate and adaptive immune responses and hence offer an attractive drug target for inflammatory and autoimmune diseases, including respiratory diseases.

Figure 13.4 shows that most immune, inflammatory, and structural cells relevant to asthma and COPD are controlled by different isoforms of PKC. Therefore, the inhibition of PKC isoforms would provide a novel therapeutic target for intervention in a broad spectrum of respiratory diseases, including asthma and COPD.

In the subsequent sections, we will discuss the evidence for the roles of specific PKC isoforms in the airway inflammatory response and evaluate their potential as therapeutic targets for respiratory diseases.

13.5.1 PKC IN IMMUNE CELL FUNCTIONS

Immune cells play a crucial role in a vast array of cellular functions by responding to several stimuli, such as microbial antigens, mitogens, and cytokines. However, the signal transduction mechanism involved in immune cell functions needs to be tightly regulated, as overreactions to self-antigens can lead to autoimmune disease, whereas insufficient response to foreign antigens can result in the increased susceptibility to infection. Recent gene knockout studies have shown that different isoforms of PKCs are important signaling mediators of various immune cells such as B cells, T cells, macrophages, and dendritic cells.

13.5.1.1 PKC in B Cell Functions

PKC plays a crucial role in B cell functions and the first direct evidence for the role of PKC came from the study of Leitges et al. (1996). They showed that PKC-β knockout mice have reduced splenic B cells, a significantly lower number of B-1 lymphocytes, and low levels of serum IgM and IgG3. These mutant mice also showed defective IgM-induced B cell proliferation, despite normal T cell activation in response to TCR stimulation (PKC-β is also expressed in T cells). In two independent studies, BCR-dependent

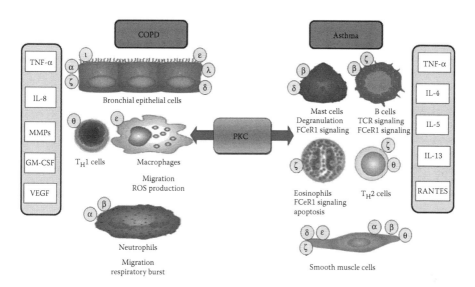

FIGURE 13.4 PKCs play a crucial role in the pathogenesis of asthma and COPD. PKCs play a central role in the regulation of many biological functions that eventually contribute to the physiological processes. PKCs are expressed in the majority of inflammatory cells, including macrophages, neutrophils, mast cells, eosinophils, B cells, and T cells. All these cells are key players in the pathogenesis of inflammatory diseases such as asthma and COPD. The PKCs regulate the production of key inflammatory mediators, including proinflammatory cytokines, chemokines, degradative enzymes, growth factors, and adhesion molecules. Therefore, inhibition of PKC pathway is expected to inhibit not only the production of various inflammatory mediators but also their actions, thereby interrupting the vicious cycle that often occurs in inflammatory diseases. IL, interleukin; MMP, matrix metalloproteases; RANTES, regulated on activation, normal T expressed and secreted; TNF-α, tumor necrosis factor-α; VEGF, vascular endothelial growth factor.

impaired cell proliferation and survival of PKC-β knockout mice were attributed to the defective induction of the nuclear factor κB (NF-κB)-regulated cell survival genes Bcl-2 and Bcl-xL (Su et al., 2002; Saijo et al., 2002). Moreover, IκB degradation was compromised in the PKC-β-deficient B cells activated by IgM cross-linking, but not when stimulated through CD40 ligation (Su et al., 2002; Saijo et al., 2002).

The targeted disruption of PKC-ζ gene in mice also suggests a potential role of PKC-ζ in B cell functions (Martin et al., 2002). B cells from PKC-ζ-deficient mice showed increased spontaneous apoptosis and impaired proliferation and survival in response to IgM cross-linking, whereas both peripheral T cells and thymocytes showed normal growth and proliferation. The defective survival of B cells in these mice was attributed to the defects in the activation of extracellular-signal-regulated kinase (ERK) and the transcription of NF-κB-dependent survival genes, including Bcl-xL, IκB, and IL-6. The involvement of a PKC cascade with both PKC-β and PKC-ζ is also a viable therapeutic option (Duran et al., 2003; Savkovic et al., 2003).

Self-reactive B cells normally exhibit tolerance to self-antigens (B cell anergy), which is essential for the prevention of autoimmune disease. *In vivo* studies using

PKC-δ knockout mice defined the physiological role of PKC-δ in the control of B cell tolerance. Mice deficient in PKC-δ showed significant splenomegaly and lymphadenopathy due to an increased number of peripheral B cells with no abnormality observed in T cells. Moreover, PKC-δ deficiency prevents B cell anergy and the mice die prematurely due to severe autoimmune disease, indicating that PKC-δ is essential for the prevention of autoimmune disease. The exact molecular basis for the opposing function of the two PKC isoforms in B cell signaling is still not well understood. In addition to the role of PKC isoforms (PKC-β and PKC-ζ) in B cell survival and proliferation, a recent study suggested that PKC-λ may have a role in the NF-κB activation during early B cell development in the bone marrow (Saijo et al., 2003).

13.5.1.2   PKC in T Cell Functions

T cells play critical roles in initiating and controlling immune response; inappropriate or excessive stimulation of T cells results in various chronic inflammatory diseases (Boschelli, 2009a). The pathology of asthma involves aberrant activation of CD4$^+$ lymphocytes differentiated along the T helper (T$_H$2) lineage (Luster and Tager, 2004). T$_H$2 cells produce a set of crucial cytokines, including IL-4, IL-5, IL-10, and IL-13, making T$_H$2 cells a key player in the orchestration of the networks activated during allergic airway inflammation (Shuai and Lu, 2003). Isoform-specific genetically inactivated mice divulge the selective involvement of various isoforms of PKC in cell-specific aspects of the immune response. PKC-ζ$^{-/-}$ knockout mice are unable to mount an optimal immune response and this results in a defective humoral response to both T-independent and T-dependent antigen, specifically in the levels of IgG1, IgG2α, and IgG2b (Martin et al., 2002). PKC-ζ-mediated T cell (possibly in the T$_H$2 lineage) alteration was also confirmed by the dramatic reduction of basal IgE levels in PKC-ζ$^{-/-}$ mice compared to wild-type controls (Martin et al., 2002). In addition, loss of PKC-ζ results in the impaired secretion of T$_H$2 cytokines in *ex vivo* and *in vivo* experiments due to the inability of the PKC-ζ$^{-/-}$ CD4$^+$ T cells to differentiate adequately along the T$_H$2 lineage (Martin et al., 2005). This result suggests that loss of PKC-ζ leads to the generation of T$_H$2 cells that poorly activate GATA3, c-Maf, Stat6, and NFATc1 during the T$_H$2 differentiation.

PKC-θ is another isoform of PKC that plays an important role in T cell signaling as it is the only isoform that is selectively translocated to the T cell/antigen-presenting cell contact site immediately after cell–cell interaction (Srivastava et al., 2004). Overexpression and inhibition studies of PKC-θ showed that it mediates the activation of the transcription factors, activator protein-1 (AP-1) and NF-κB in response to TCR/CD28 costimulation in a cell-type-dependent manner (Baier-Bitterlich et al., 1996; Lin et al., 2000; Bauer et al., 2000). Furthermore, PKC-θ is crucial for IL-2 production, a prerequisite for the proliferation of T cells. PKC-θ-deficient mice are defective in NF-κB activation and are resistant to experimental autoimmune encephalomyelitis, probably due to impaired production of IFN-γ and IL-17 (Tan et al., 2006). In contrast to PKC-ζ$^{-/-}$ mice, which have a role restricted to only the T$_H$2 polarization mechanism, PKC-θ$^{-/-}$ mice has a broader impact in T cell function as in these mice defects were also observed in NF-κB activation and proliferation of naive T cells (Pfeifhofer et al., 2003; Sun et al., 2000). These reports indicate that both

PKC-ζ and PKC-θ are critical modulators of the T cell functions and these isoforms are potentially relevant target for pathological alterations of the immune system.

In addition to PKC-ζ and PKC-θ, PKC-βI, and PKC-δ also play a role in the T cell signaling as these isoforms control the intercellular cell adhesion molecule-1 (ICAM-1)-mediated lymphocyte function-associated antigen-1 (LFA-1) signaling, which is crucial for the T cell recruitment (Volkov et al., 1998). These studies collectively suggest that blocking T cell activation with PKC-θ and PKC-ζ inhibitors may attenuate airway inflammation in established allergic disease and may be beneficial in treating chronic allergic asthma.

13.5.1.3   PKC in Macrophage Functions

Inflammatory cells, such as macrophages, are the main orchestrators of inflammatory and injurious responses in lungs of smokers and in COPD patients (Barnes, 2004; Tetley, 2002). The numbers of alveolar macrophages are markedly increased in the lung and alveolar space of patients with COPD (Jeffery, 1998; Finkelstein et al., 1995). Macrophages from COPD patients can stimulate the secretion of all major inflammatory mediators of COPD, including certain cytokines, chemokines, reactive oxygen species, and elastolytic enzymes (Barnes, 2004). Therefore, any anti-inflammatory therapies that inhibit macrophage functioning in chronic inflammation of COPD would be a therapy of choice.

In a recent study, enhanced activation of PKC-ζ was reported in the BAL of wild-type mice exposed to cigarette smoke (CS) and lipopolysaccharide (LPS). A similar increase in the phosphorylation of PKC-ζ was also observed in the cigarette smoke extract (CSE) and LPS-treated MonoMac6 cells and primary mouse alveolar macrophages. Inhibition of PKC-ζ by PS-PKC (a myristoylated PKC-ζ pseudosubstrate peptide inhibitor) reduced the CSE- and LPS-mediated release of proinflammatory cytokines from both MonoMac6 cells and mouse alveolar macrophages, indicating a crucial role of PKC-ζ in CS- and LPS-induced lung inflammation. This study was further corroborated by the observation that PS-PKC significantly decreased the acrolein and acetaldehyde (the reactive components of CS)-mediated release of proinflammatory mediator in primary mouse alveolar macrophages and MonoMac6 cells. Collectively, this study suggests that the PKC-ζ signaling pathway plays a crucial role in the CS-mediated sustained macrophage activation and lung inflammation (Yao et al., 2010).

PKC-ε is an important mediator of macrophage function as macrophage-specific inhibition of PKC-ε led to the inhibition of IL-4-induced NO production (Sands et al., 1994). The role of PKC-ε in macrophage biology was corroborated by knockout mice, as the loss of PKC-ε in peritoneal macrophages resulted in dramatically reduced levels of NO, prostaglandin E_2 (PGE$_2$), TNF-α, and IL-1β in response to LPS and IFN-γ costimulation. PKC-ε-deficient mice also showed low survival rate due to their inability to clear bacterial infections as a result of reduced NO synthase expression and attenuated IKK and NF-κB activation and partial inhibition of ERK and p38 MAPK. Interestingly, the loss of PKC-ε did not affect the differentiation of monocytes and macrophages from bone marrow precursors (Castrillo et al., 2001). In addition, PKC-ε-deficient mouse embryonic fibroblasts display a reduced response to IFN-γ as a function of integrin engagement (Ivaska et al., 2003). Collectively, these results suggest that

PKC-ε and PKC-ζ play a critical role in integrating different signaling cascades of macrophage that impact the establishment of an effective innate immune response.

13.5.1.4   PKC in Dendritic Cell Functions

Dendritic cells (DC) are specialized in antigen processing and presentation; therefore, they play a very important role in the induction of immune responses within the airways (Banchereau et al., 2000). In one study, the number of dendritic cells was shown to be greater in the airways of asthmatics patients compared to control subjects. Moreover, the number of dendritic cells is altered postexposure to topical and systemic corticosteroids (Bocchino et al., 1997). These studies indicate that dendritic cells may play some role in the pathogenesis of asthma/COPD; however, the exact role is still not known. Since dendritic cells are the most potent antigen-presenting cells, the manipulation of DC maturation provides a strategy for the treatment of allergic and inflammatory diseases.

LPS results in the phosphorylation of conventional PKC-α/β and novel PKC-ε isoforms in DC. In one study, the treatment of DC derived from human monocytes with LPS resulted in the phosphorylation of PKC. Bisindolylmaleimide (Bis), a pan-PKC inhibitor, dose dependently inhibited the PKC activation and LPS-induced IL-12 production. DC stimulated with LPS in the presence of Bis was deficient in the induction of IFN-γ production by allogeneic CD4+ T cells. Moreover, Bis also impaired the LPS-induced IκB-α degradation and subsequent NF-κB activation in DC. Since Bis is a pan-PKC inhibitor, the contribution of individual isoforms of PKC in the LPS signaling of DC cells was not clear. In the same study, using pseudosubstrate peptides specific for PKC isoforms, PKC-ε was identified as a PKC isoform involved in the production of IL-12 and TNF-α. Collectively, this study suggests that PKC-ε inhibition impairs LPS signaling in DC (Aksoy et al., 2002).

13.5.2   PKC in Inflammatory Cell Functions

Chronic obstructive airway diseases such as asthma and COPD are caused by a myriad of mediators from a variety of cellular sources. The more chronic inflammatory response is due to perennial infiltration and activation, at the local site, by inflammatory cells such as eosinophils, mast cells, and neutrophils. Recent studies have confirmed that different isoforms of PKC control a vast array of signaling events related to these inflammatory cells.

13.5.2.1   PKC in Mast Cell Functions

Mast cells are a key source of inflammatory mediators and they regulate ASM functions (Siddiqui et al., 2007). Degranulation of mast cells is largely responsible for sustaining the allergic reaction through the release of several proinflammatory and bronchoconstrictive mediators (including histamine, proteoglycans, and serine proteases) and thereafter the synthesis of other inflammatory mediators (including prostaglandin D2, cysteinyl leukotrienes, and eosinophil chemotactic factor) (Hakim-Rad et al., 2009; Boyce et al., 2009). In this context, mast cells derived from PKC-β-deficient mice showed substantial decrease in mast cell degranulation and also produced less IL-6 in response to IgE antigen or Ca^{2+} ionophore stimulation (Nechushtan et al., 2000; Leitges et al., 2002; Kawakami et al., 2003). This data corroborated the results from a previous study in which an increased level of IL-6 and IL-2 mRNA was observed in PKC-β-overexpressing mast cells. Leitges and coworkers, utilizing bone marrow-derived mast cells from knockout mice, have shown that PKC-δ, but not PKC-ε, is a negative regulator of antigen-induced mast cell degranulation (Leitges et al., 2002; Lessmann et al., 2006; Kawakami et al., 2003), as PKC-δ-deficient mast cells exhibited a more sustained Ca^{2+} mobilization and a significantly higher level of mast cell degranulation. The exact molecular basis for the opposing function of the two PKC isoforms in mast cell degranulation is still not well understood. Figure 13.5 depicts the proposed mechanism of PKC-mediated mast cell degranulation.

13.5.2.2 PKC in Eosinophil Function

Eosinophils are considered to be the prime regulators of the late phase of asthmatic reaction. These cells infiltrate the asthmatic airway and release cytokines and chemokines that promote their adhesion to the activated endothelium, and generate inflammatory response at the sites of allergen exposure (Sampson, 2000). Eosinophil

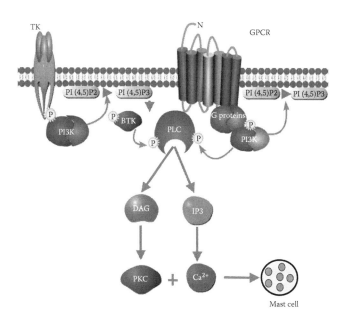

FIGURE 13.5 PKC-mediated mast cell degranulation. The activation of PKC is controlled by a sequence of posttranslational phosphorylation and receptor-mediated activation by lipid second messengers. At the plasma membrane, receptor-mediated activation occurs through either G-protein-coupled or tyrosine kinase (TK) receptors. PI3K generates phosphatidylinositol-3,4,5-trisphosphate (PtdIns(3,4,5)P3), which activates Bruton's tyrosine kinase (BTK), which in turn amplifies the activation of PLC. PLC generates the second messengers diacylglycerol (DAG) and inositol-1,4,5-trisphosphate (IP3) by the hydrolysis of phosphatidylinositol-4,5-bisphosphate (PtdIns(4,5)P2), and IP3 initiates the release of calcium from internal stores. DAG activates PKC isoforms that, together with free calcium, trigger mast cell degranulation.

degranulation is tightly associated with the structural changes that lead to airway remodeling.

In recent years, using selective activators and inhibitors of individual isoforms of PKC, several studies have convincingly proved that PKCs are involved in a variety of eosinophil functions such as cell adhesion (Sepulveda et al., 2005), chemotaxis (Gilbert et al., 1994), superoxide anion generation (Sedgwick et al., 1990), degranulation (Egesten and Malm, 1998), and release of other mediators. Cellular adhesion is the most critical step for eosinophil activation. It has been proposed that β2 integrin-dependent cellular adhesion, particularly αMβ2 (CD11b/CD18), which is induced by a lipid mediator [e.g., platelet-activating factor (PAF)], a cytokine (e.g., IL-5), or an immunoglobulin (e.g., IgG), is critical for the effector functions of the eosinophils (Kaneko et al., 1995; Kato et al., 1998). In a study, the effects of several PKC inhibitors, including a broad-spectrum PKC inhibitor such as bisindolylamleimide I (Bis I; inhibitor of PKC-α, βI, βII, γ, δ, and ε), peptide 20–28 (inhibitor of PKC-α and PKC-β), and a specific PKC-δ inhibitor (rottlerin) or PKC-ζ inhibitor (myristoylated PKC-ζ inhibitor), on CD11b expression on the surface of eosinophils was evaluated. In this study, PKC-ζ inhibitor attenuated the PAF- or C5a-induced cell adhesion. In another study, rottlerin blocked the IL-5-induced β2 integrin-dependent adhesion of human eosinophils (Takizawa et al., 2002; Kato et al., 2005). These results indicate that the PKC-δ or PKC-ζ pathway mediates β2 integrin-dependent adhesion. Once eosinophils are activated, the next step is their stimulation on contact with chemoattractants, and this process also results in a change in their shape, which is a prerequisite for chemotaxis. In this context, treatment with broad-spectrum PKC inhibitors such as Bis I or staurosporine resulted in a striking inhibition of eosinophil shape change induced by IL-5 but not by eotaxin. A previous study has reported that Bis I had no effect on the PAF-induced eosinophil chemotaxis (Choi et al., 2003). These studies suggest that the involvement of PKCs is limited only to the specific induction such as IL-5 signaling pathway in this study.

In inflammatory conditions such as asthma, eosinophils migrate into the airways and generate highly toxic ROS, including superoxide anion and hydrogen peroxide. The nonspecific PKC inhibitor Bis I at a concentration of 3 μM significantly inhibited PAF-induced superoxide anion generation (Takizawa et al., 2002). Another study demonstrated that three PKC inhibitors—staurosporine, Ro318220, and Go 6983—inhibited the PMA-induced superoxide anion generation in human eosinophils (Woschnagg et al., 2001). These data suggest that PKCs are involved in the superoxide anion generation from eosinophils. The contribution of individual isoforms of PKC in the superoxide generation came from the study of Bankers-Fulbright et al. (2001) who showed that PKC-α, βI, βII, γ, δ, and ζ are constitutively expressed in human eosinophils and a PKC-δ inhibitor blocked the IL-5- or LTB4-mediated superoxide anion generation, indicating that PKC-δ regulates superoxide anion generation in eosinophils. Moreover, another PKC-δ inhibitor, rottlerin, also inhibited the superoxide generation from PAF-stimulated eosinophils (Takizawa et al., 2003). In addition to PKC-δ, PKC-ζ inhibitor also suppressed the PAF- or C5a-induced superoxide anion generation in a dose-dependent manner (Takizawa et al., 2002). These evidences suggest that PKC-δ and PKC-ζ play a crucial role in the superoxide anion generation in eosinophils.

Degranulation of eosinophil accompanied by the release of toxic granule proteins is a major eosinophil effector function that is directly related to the pathogenesis of bronchial asthma. Several reports have validated the crucial role of PKCs in eosinophil degranulation. In one such study, a specific PKC-ζ inhibitor attenuated the PAF- and C5a-induced eosinophil-derived neurotoxin (EDN) release from eosinophils (Kato et al., 2005). Furthermore, the inhibition of PKC by 3 μM of Bis I led to a reduction in the PAF-induced degranulation from human eosinophils; however, lower concentrations of Bis I enhanced the degranulation (Kato et al., 2005). In contrast, another report demonstrated that the activation of PKC-δ by PMA inhibited degranulation in human eosinophils. Similarly, a high concentration of rottlerin enhanced the PAF- or C5a-induced degranulation in human eosinophils. These results confirm that PKCs have dual modes of regulation of the PAF-induced signaling pathway in human eosinophils.

In conclusion, these data collectively confirm that PKCs play a crucial role in regulating diverse signaling events of eosinophils. The identification of the precise role of individual isoforms of PKCs in human eosinophils would provide a novel strategy for the treatment of allergic diseases such as bronchial asthma.

13.5.2.3 PKC in Neutrophil Functions

Neutrophils not only represent the first line of defense against bacterial infection but are also involved in many diseases in which the functions of adaptive immune cells are important, such as COPD and chronic severe asthma (Nathan, 2006). Neutrophils contribute to acute asthma exacerbations and are also present in high numbers in the airways of patients with chronic severe asthma. Neutrophil produces several inflammatory mediators such as reactive oxygen intermediates (ROI), lipid mediators, and proteases such as elastase. These mediators contribute to airflow obstruction, epithelial damage, and remodeling, a hallmark of pulmonary inflammatory diseases such as COPD (Sampson, 2000). PKCs are involved in a variety of neutrophil functions such as membrane alteration and motility, oxidative phosphorylation, and apoptosis modulation of neutrophils.

Neutrophils possess a superoxide-generating oxidase system that is essential for the efficient killing of microbes, and PKCs have been shown to be crucial mediators for the activation of superoxide-generating oxidase system (Heyworth and Badwey, 1990). The activation of human neutrophils triggers an assembly of NADPH oxidase in the membrane, and the generation of O_2^-. PKC-β has been proposed to play a role in the stimulus-induced phosphorylation and association of a cytosolic 47-kDa protein (p47-phox) with the membrane NADPH oxidase. In one study, PKC-β phosphorylated both endogenous and recombinant p47-phox and this phosphorylation was inhibited with a pseudosubstrate, derived from the C terminus of Ca-dependent PKC isoforms (Majumdar et al., 1993). Pseudosubstrate also inhibited the O_2^- generation in activated neutrophils. This study indicates that PKC-β may play a role in the assembly or maintenance of an active NADPH oxidase system (Majumdar et al., 1993).

In another study, the phosphorylation of p47-phox in fMLP-stimulated neutrophils was potently inhibited by 0.3 μM of RO 31-8220, a selective inhibitor of PKC (Bengis-Garber and Gruener, 1995). In a recent study, dehydroepiandrosterone sulfate (DHEAS), the most abundant steroid in the human circulation, has been shown to increase superoxide generation in primed human neutrophils in a dose-dependent manner. In this study, DHEAS directly activated recombinant protein PKC-β in a cell-free assay and this enhanced PKC-β activation led to the increased phosphorylation of p47-phox, a central component of the active reduced nicotinamide adenine dinucleotide phosphate complex responsible for the neutrophil superoxide generation (Radford et al., 2010). This study demonstrates that PKC-β acts as an intracellular receptor for DHEAS in human neutrophils, which includes reduced neutrophil superoxide generation in response to pathogens. In another study, in formyl–Met–Leu–Phe (fMLP)-stimulated rat neutrophils, 2-benzyl-3-(4-hydroxymethylphenyl) indazole (CHS-111) inhibited superoxide anion generation through the blockade of PAK, Akt, and PKC signaling pathways (Chang et al., 2009). An antisense approach also corroborated the important role of PKC-β in superoxide generation. In this study, selective depletion of PKC-β in HL60 cells differentiated to a neutrophil-like phenotype (dHL60 cells) elicited selective inhibition of O_2^- generation in response to fMLP trigger, immune complexes, or phorbol myristate acetate, an activator of PKC (Korchak et al., 1998).

13.5.3 PKC IN STRUCTURAL CELL FUNCTIONS

The pathophysiology of both asthma and COPD involves a complex process of airway hyperresponsiveness (AHR) and airway remodeling and it involves all the tissues of the airway from the epithelium to the adventitia (Mcvicker et al., 2007; Palmans et al., 2000). In asthma, ASM hypertrophy and hyperplasia occur in the bronchial wall at the level of large and medium airways, while in COPD these changes occur at the level of the small airways (Zhou and Hershenson, 2003). ASM cells are the central component of airway remodeling in asthma (Janssen and Killian, 2006). Smooth muscle mass is usually increased in large and/or small airways in asthma and COPD. ASM cells also participate in the chronic airway inflammation by interacting with both T_H1- and T_H2-derived cytokines to modulate chemoattractant activity for eosinophils, activated T lymphocytes, and monocytes/macrophages. Smooth muscle cells also have the potential to alter the composition of the extracellular matrix environment and orchestrate key events in the process of chronic airway remodeling (Hirst, 1996). Therefore, the inhibition of mediators of these structural changes would provide us new avenues for the treatment of respiratory diseases. Recent studies have indicated that PKCs play a crucial role in the airway remodeling. The treatment of ASM cells with proinflammatory neuropeptide bradykinin (BK) activates PKC-α, β1, δ, and ε. In addition to this, BK, in a PKC-ε-dependent mechanism, induces COX-2 expression and PGE_2 accumulation in human ASM cells (Zhang et al., 2004). PKC-ξ activity is also increased in proliferating human ASM cells. Similarly, increased activation of PKC-β1 and decreased activation of PKC-δ have been documented in the actively proliferating ASM cells from hyperresponsive rats, suggesting that PKC-β1 and PKC-δ are a positive and a negative regulator of ASM cell growth, respectively (Zhou and Hershenson, 2003). PKC-α is shown to be involved in the proliferation of passively sensitized human airway smooth muscle cells (HASMCs). In this study, the activation of PKC-α with PMA resulted in the concomitant upregulation of cyclin D1 expression and also increased the proliferation of passively sensitized HASMCs. The

PKC-mediated effect on cell proliferation and cyclin D expression was significantly attenuated by specific inhibition of PKC-α with Go6976 (Du et al., 2010).

Occlusion of bronchioles with mucus is one of the most striking morphological changes associated with both asthma and COPD. Hypersecretion of mucus results from the hypertrophy of submucosal mucus glands and increased numbers of goblet cells in bronchiolar epithelium. In one study, PMA-induced expression of MUC5AC (one of the prominent airway mucins) was shown to be attenuated with the pan-PKC inhibitor, calphostin, in human respiratory epithelial cell line (Hewson et al., 2004). This study suggests that PKC inhibition is a viable option to treat asthma and COPD. Collectively, these studies demonstrate that multiple PKCs (α, β1, δ, ε, θ, and ξ) are important in the characteristic airway constriction, inflammation, mucus production, and cellular growth observed in asthma and COPD.

13.6 PKC AS A TARGET FOR ASTHMA AND COPD

As discussed above, the pathogenesis of respiratory diseases such as asthma and COPD involves the migration and aberrant activation of various immune, structural, and inflammatory cells such as macrophages, neutrophils, B cells, T cells, eosinophils, mast cells, and ASM cells in the lung or airway. These cells release several proinflammatory and bronchioconstrictive mediators that eventually help in the manifestation and potentiation of inflammatory responses. As discussed in the previous sections, PKCs regulate signaling events of immune, inflammatory, and structural cells relevant to the establishment and propagation of inflammation. In the context of airway inflammation, PKCs have been implicated in bronchospasm, and mucus production, a hallmark of airway diseases. Therefore, the modulation of PKCs as a therapeutic target may help in the regulation of airway inflammation. The literature suggests the validation of PKCs as a potential target for inflammatory respiratory diseases using gene knockouts, *in vitro* cultured cells from animal models and humans, tissue samples from patients, and in certain cases, small-molecule inhibitors as a tool compound as described below.

In one such example, a significant increase in total, cytosolic, and membrane PKC activity was observed in the peripheral blood lymphocytes of asthmatic patients (27 patients classified as mild, moderate to severe, and cases in remission) as compared to 14 healthy volunteers (Bansal et al., 1997). These data were further corroborated by another study in which the percentage of eosinophils, lymphocytes, and PKC-α-positive cells, and the concentration of IL-5 in the induced sputum of asthmatics patients were higher than the controls ($p < 0.01$). In the same study, glucocorticosteroid suppressed the expression of PKC-α in the airway inflammatory cells and there was a concomitant decrease in the production of IL-5 in asthmatic airways (Tang et al., 2004). Similarly, significantly higher expression of PKC-α and NF-κB was observed in the lung tissues of patients with COPD. These studies suggest that increased expression of PKC isoforms may have some role in disease progression in asthma and COPD patients (Zhang et al., 2004). These reports also indicate that strategies to antagonize PKC may act in a selective manner during inflammatory conditions.

The discovery of small-molecule inhibitors of PKCs has also been reported in the literature. Roche's PKC inhibitors, Ro-318220 and Ro-317549, have been frequently used as a tool compound to validate PKC as a plausible therapeutic target for respiratory disorders. Wyeth has also reported the identification of certain *in vitro* active nanomolar hits and their preclinical evaluation validating their role in asthma. Ro-318220 is a competitive, selective inhibitor for protein kinase C (IC$_{50}$: 10 nM).

Ro-318220   Compound A

Ro-317549   Compound B

Severe chronic asthma leads to ASM hyperplasia and hence structural remodeling of the airway. In one of the reports, ASM isolated from rabbit trachealis were stimulated with fetal calf serum (FCS) to induce mitogenesis. PKC inhibitors Ro-318220 and Ro-317549 showed concentration-dependent inhibition of FCS-stimulated proliferation of ASM (Hirst et al., 1995). Similar observations have been made in the human ASM cells as well. Cigarette smoke extract (CSE) has been shown to passively sensitize human ASM cells (HASMCs) to proliferate *in vitro* when cultured in the presence of serum from asthma patients. Lin et al. (2005) reported that the proliferation was accompanied by the upregulation of PKC-α expression in HASMCs. Treatment with 5 μM Ro-318220 resulted in the significant decrease in the proliferation of HASMCs, suggesting the involvement of PKC signaling pathway in the proliferation of these cells (Lin et al., 2005). Similarly, IL-17F involvement has been implicated in the airway inflammation and remodeling through the induction of IL-11. Kawaguchi et al. (2009) showed that bronchial cells when cultured in the presence of IL-17F expressed IL-11, which could be blocked using Ro-318220 (Kawaguchi et al., 2009).

Bronchoalveolar lavage from asthmatic patients has been shown to carry elevated levels of macrophage migration inhibitory factor (MIF), a proinflammatory cytokine.

This cytokine is secreted by the circulating eosinophils. Rossi et al. reported the presence of higher MIF concentration (up to 797 pg/mL) in asthmatic patients compared to normal volunteers (274 pg/mL). Similarly, when human eosinophils were stimulated *in vitro* with PMA (100 nM), about 10-fold increase in the MIF secretion (1539 pg as compared to 142 pg basal level) was observed that could be significantly blocked by the PKC inhibitor Ro-318220, indicating the potential of PKC inhibitors in the treatment of asthma (Rossi et al., 1998).

Different irritant agents, including capsaicin, resinferatoxin, and citric acid, elicit cough response in humans and experimental animals through the activation of transient receptor potential vanilloid 1 (TRPV-1). Gatti and colleagues investigated the role of protease-activated receptor-2 (PAR2) via PKC signaling in the lowering of the threshold to the cough response to citric acid and resinferatoxin as irritants in guinea pigs. It was observed that inhalation of aerosolized PAR2 agonist did not produce cough in normal animals. However, the cough response was potentiated in the induced animals, which could be completely prevented by capsazepine, a TRPV1 receptor antagonist. The administration of a PKC inhibitor GF-109203X also showed an effect similar to capsazepine and could prevent the aggravation of cough response. These observations suggest that the PAR2 stimulation of TRPV-1-dependent cough involves the PKC pathway. Further, the role of PAR2 and hence PKC was also implicated, by sensitizing TRPV1 in primary sensory neurons, in the exaggerated cough response in chronic airway inflammation conditions of asthma and COPD (Gatti et al., 2006).

Chromatin remodeling on proinflammatory genes plays a pivotal role in cigarette smoke (CS) and LPS-induced abnormal lung inflammation in diseases such as COPD. However, the underlying mechanism is not clear. In a recent study, PKC-ζ has been implicated in CS/aldehyde- and LPS-induced lung inflammation, as the lung inflammatory response was decreased in PKC-$\zeta^{-/-}$ mice compared with wild-type mice exposed to CS and LPS. Moreover, the inhibition of PKC-ζ by a specific pharmacological PKC-ζ inhibitor (a myristoylated PKC-ζ pseudosubstrate peptide inhibitor) attenuated CS extract and LPS-mediated proinflammatory mediator release from macrophages. These data in addition to deciphering the molecular mechanisms underlying the pathogenesis of chronic inflammatory lung diseases, also suggest PKC-ζ as a novel target for the treatment of respiratory diseases such as COPD (Yao et al., 2010).

PKC-ζ has also been suggested to have a role in the antigen-specific T_H2 immune responses. PKC-ζ was validated as a therapeutic target for asthma using a mouse model. In this report, mouse allergic asthma was induced by repeated sensitization followed by intranasal challenge with OVA. AHR was measured by betamethacoline-induced airflow obstruction. When PKC-ζ pseudosubstrate inhibitor (PPI) was intratracheally instilled before each OVA challenge, asthmatic manifestations were significantly reduced, including AHR and reduction in eosinophil presence in bronchoalveolar lavage fluid (BALF). PPI instillation could also decrease IL-5 and IL-13 levels in BALF, TNF-α, and OVA-specific IgE in serum. In one similar study, PKC-ζ mutant mice showed a dramatically reduced response to OVA-induced airway inflammation (Do et al., 2009). Collectively, these data suggest that blockade of PKC-ζ is a promising target for the treatment of asthma.

PKC-θ is primarily expressed in T cells and is recruited to the immunological synapse upon T cell receptor signaling. It is also required for the mature T cell activation. The selective role of PKC-θ in T cell effector function and not in T cell development makes it a potential target for T cell-mediated diseases (Chaudhary and Kasaian, 2006). PKC-θ-deficient mice display a diminished pulmonary inflammatory response following allergen challenge. DeClercq et al. (2009) investigated the allergic airway response in mice when PKC-θ was specifically blocked with a small-molecule inhibitor (compound A). Mice were sensitized with ovalbumin (OVA) on days 0 and 14, and then challenged with OVA on days 26, 27, and 28. Four weeks later, mice were rechallenged with aerosolized OVA. Airway inflammation was monitored after both primary challenges and rechallenge. Compound A was administered intraperitoneally twice daily at 30 mg/kg dose for 3–7 days during and after antigen challenge in three dosing regimens, that is, for 4 days concurrent with and after the primary challenges only; for 3 days starting before the OVA rechallenge only; and both the above regimens. The administration of compound A significantly inhibited the influx of eosinophils, lymphocytes, and neutrophils into the airways. *Ex vivo* antigen-induced T_H2 cytokine production by splenocytes was reduced in cells obtained from mice treated with compound A in either the primary challenge phase or both primary and rechallenge phases, but not in the rechallenge phase alone (DeClercq et al., 2009). These observations indicate that blocking of T cell activation with a PKC-θ inhibitor may attenuate airway inflammation in established allergic disease and may potentially benefit in treating chronic allergic asthma. However, the identification of PKC-θ selective inhibitors is still in its infancy. Wyeth has reported the identification of a 70-nM PKC-θ inhibitor, 5-(3,4-dimethoxyphenyl)-4-(1*H*-indol-5-ylamino)-3-pyridinecarbonitrile, optimized from an HTS initial hit of a low-micromolar ATP-competitive PKC-θ inhibitor. This compound was selective against a panel of 21 serine/threonine, tyrosine, and phosphoinositol kinases. In cellular assays, this compound inhibited IL-2 production in anti-CD3/anti-CD28-activated T cells enriched from splenocytes (Cole et al., 2008). Further optimization of this molecule resulted in a 10-fold improvement in the enzyme inhibition activity for PKC-θ to an IC_{50} value of 7.4 nM as reported for the compound A substitution with a 4-methylindol-5-ylamino group at C-4 and a 4-(2-(4-methylpiperazin-1-yl) ethoxy) phenyl group at C-5. Compound B has an IC_{50} of 3.8 nM against PKC-θ and also showed more than 300-fold selectivity over other isoforms of PKCs (Boschelli et al., 2009b).

Matrix metalloproteinase-9 (MMP-9, a major proteolytic enzyme that induces airway remodeling in asthma) represents one of the potential drug targets for airway diseases such as asthma and COPD, as its expression is associated with airway remodeling and tissue injury. However, the exact mechanism responsible for the MMP-9 upregulation in airway epithelial cells is still poorly understood. In one study, PMA has been shown to induce MMP-9 expression in a protein kinase C alpha-dependent signaling cascade in BEAS-2B human lung epithelial cells. In the same study, pretreatment with GF109203X, a general PKC inhibitor, or Go6976, a PKC-alpha/beta isoform-specific inhibitor, inhibited the PMA-induced activation of MMP-9 (Shin et al., 2007). Similar results were also observed with the transient transfection with PKC-α antisense oligonucleotides. These data suggest that PKC inhibitor-mediated down-regulation of MMP expression might help in the control of airway inflammatory diseases (Shin et al., 2007). In another study, PKC inhibitors, calphostin C and chelerythrine chloride, reduced airway inflammation and hyper-responsiveness by reducing the MMP-9 expression (Park and Lee, 2004).

In summary, the preclinical data using PKC inhibitors further confirm the role of PKC in the pathophysiology of respiratory diseases such as asthma and COPD. At present there is no PKC inhibitor that has reportedly reached the development stage for respiratory disorders as a therapeutic indication. Further investigations are warranted to identify clinical candidates.

13.7 EXPERT OPINION

Despite the advances in treatment, pulmonary disease such as asthma and COPD continue to result in significant mortality. All available therapies come with associated side effects and long onset of beneficial effects, and also show restricted efficacy in few patients. All these information suggests that there is an utmost need of finding a therapeutic solution to these inflammatory diseases. Understanding the intracellular machinery, including signal transduction pathways that regulate gene expression in inflammatory conditions, has led to the exploration of novel therapeutic interventions. Among the various signaling molecules that regulate the inflammatory pathway, kinases have an important role to play, as their aberrant or upregulated expression is the key feature in the regulation of inflammatory disease. In this respect, PKC offers a target-based approach that will affect the multiple pathways involved in inflammation. PKC activation is crucial for a number of cellular immune responses that underlie acute and chronic respiratory diseases, including asthma and COPD; therefore, selective inhibition of PKC isoform might be effective in treating a wide range of human diseases associated with aberrant immune functions, including airway diseases.

Recent studies have provided convincing evidence that different PKC isoforms play a crucial role in the signaling of various immune, inflammatory, and structural cells. Among the different isoforms of PKCs, PKC-ζ has emerged as a central player for the regulation of critical intracellular signaling pathways involved in various respiratory diseases. The loss of PKC-ζ impairs signaling through the B cell receptor and also fails to mount an optimal T-cell-dependent immune response (Martin et al., 2002). In concordance with this, in the lungs of PKC-$\zeta^{-/-}$ mice, inflammatory cell influx and proinflammatory mediator release were significantly attenuated as compared to wild-type mice (Martin et al., 2005). Intratracheal instillation of PKC-ζ pseudosubstrate inhibitor has been shown to significantly reduce the asthmatic manifestations (Do et al., 2009). Similarly, PKC-ζ mutant mice also showed a dramatically reduced response to OVA-induced airway inflammation (Martin et al., 2005). Collectively, these observations suggest that blockade of PKC-ζ is a promising target for the treatment of both asthma and COPD.

The first-generation inhibitors of PKC provided tools to aid research and development of a new generation of compounds with a range of selectivity. In the recent past, few isoform-selective inhibitors of PKC have been identified, supporting that isoform-selective inhibitors can be generated despite the similarities in the ATP-binding sites (Way et al., 2000). In spite of the fact that PKC-ζ represents an excellent target for anti-inflammatory therapy, research in the field of specific PKC-ζ inhibitors is still in its infancy. Therefore, more efforts should be made to identify novel, selective, and efficacious PKC-ζ inhibitors for airway diseases.

Since PKC regulates diverse crucial cellular events, achievement of an acceptable therapeutic index by targeting the PKC pathway is still a matter of concern. Therefore, the suitability of a PKC isoform or combinations thereof, as a target for inhibition, depends mainly on a delicate balance between protein expression and function, disease relevance, potential toxicity profile, and tolerance within the proposed disease indication. The achievement of desirable selectivity is another hurdle, as the inhibition of multiple PKC isoforms may be expected to result in additive, or even synergistic, adverse effects, which might restrict the use of these inhibitors in chronic diseases. Rational synthesis of molecules that target specific sites within the PKC isoforms is in progress. Peptide fragments that act as inhibitors or activators of translocation and antisense approaches are also being pursued (Way et al., 2000). In spite of the strong rationale for PKC inhibitors in human disease, direct proof of concept in clinic is yet to be demonstrated and efforts are continuing to identify newer, more selective inhibitors for PKCs.

PKC is ubiquitous in all cell types and is involved in several key cellular adaptive responses, such as apoptosis, cell cycle regulation, and proliferation. As a consequence of the wide-ranging regulatory role of PKC in diverse cellular processes, PKC inhibition is expected to result in undesired pharmacological activity. However, the phenotype of various PKC knockout mice and the recent development of PKC-specific inhibitors suggest that the toxicity of PKC inhibition is anticipated to be rather limited due to probable compensatory or complementary functions of PKC family members. One of the strategies for safe PKC inhibitors for the treatment of respiratory disease includes the delivery of PKC inhibitors locally to minimize systemic exposure of the drugs. In the case of the inhalation mode of drug delivery, as the drug is delivered directly to its required site of action, only a small quantity is required for an adequate therapeutic response; consequently, there is a lower probability of systemic side effects compared to oral or intravenous administration. In addition, the onset of action of inhaled drugs is generally faster than that achieved by oral administration. The particle size and solubility of compounds are the key concerns that need to be taken care of in addition to major challenges to optimize the delivery. Selective compounds for a single PKC isoform will at least restrict potential adverse effects to those with a known association.

In summary, the PKC family undoubtedly represents an interesting and challenging target for the development of novel therapeutic agents for various diseases, including respiratory disease. Small-molecule inhibitors for PKC provide a significant opportunity as they can be used earlier in the course of the disease to prevent the disease progression prior to major damage. However, very limited *in vivo* data (including side effect and toxicological assessment) have been reported for the PKC inhibitors in airway inflammatory disease. Consequently, a number of questions regarding the merits of PKC inhibition remain unanswered. The outcome of drug discovery efforts focused toward the development of PKC inhibitors is eagerly awaited. In spite of safety concerns and less success so far, PKC inhibitors show an attractive profile and demand further studies for a successful drug in an appropriate therapeutic area.

REFERENCES

Aksoy, E., Z. Amraoui, S. Goriely et al. 2002. Critical role of protein kinase C epsilon for lipopolysaccharide-induced IL-12 synthesis in monocyte-derived dendritic cells. *Eur J Immunol* 32: 3040–49.

Baier-Bitterlich, G., F. Uberall, B. Bauer et al. 1996. Protein kinase C-theta isoenzyme selective stimulation of the transcription factor complex AP-1 in T lymphocytes. *Mol Cell Biol* 16: 1842–50.

Banchereau, J., F. Briere, C. Caux et al. 2000. Immunobiology of dendritic cells. *Annu Rev Immunol* 18: 767–811.

Bankers-Fulbright, J.L., H. Kita, G.J. Gleich et al. 2001. Regulation of human eosinophil NADPH oxidase activity: A central role for PKCdelta. *J Cell Physiol* 189: 306–15.

Bansal, S.K., A. Jha, A.S. Jaiswal et al. 1997. Increased levels of protein kinase C in lymphocytes in asthma: Possible mechanism of regulation. *Eur Respir J* 10: 308–13.

Barnes, P.J. 2004. Alveolar macrophages as orchestrators of COPD. *COPD* 1: 59–70.

Barnes, P.J. 2008. Immunology of asthma and chronic obstructive pulmonary disease. *Nat Rev Immunol* 8: 183–92.

Bauer, B., N. Krumbock, N. Ghaffari-Tabrizi et al. 2000. T cell expressed PKCtheta demonstrates cell-type selective function. *Eur J Immunol* 30: 3645–54.

Bengis-Garber, C. and N. Gruener. 1995. Involvement of protein kinase C and of protein phosphatases 1 and/or 2A in p47 phox phosphorylation in formylmet-Leu-Phe stimulated neutrophils: Studies with selective inhibitors RO 31-8220 and calyculin A. *Cell Signal* 7: 721–32.

Bocchino, V., G. Bertorelli, X. Zhuo et al. 1997. Short-term treatment with a low dose of inhaled fluticasone propionate decreases the number of CD1a+ dendritic cells in asthmatic airways. *Pulm Pharmacol Ther* 10: 253–59.

Boschelli, D.H. 2009a. Small molecule inhibitors of PKCtheta as potential antiinflammatory therapeutics. *Curr Top Med Chem* 9: 640–54.

Boschelli, D.H., D. Wang, A.S. Prashad et al. 2009b. Optimization of 5-phenyl-3-pyridinecarbonitriles as PKCtheta inhibitors. *Bioorg Med Chem Lett* 19: 3623–26.

Boyce, J.A., D. Broide, K. Matsumoto et al. 2009. Advances in mechanisms of asthma, allergy, and immunology in 2008. *J Allergy Clin Immunol* 123: 569–74.

Burns, D.J. and R.M. Bell. 1991. Protein kinase C contains two phorbol ester binding domains. *J Biol Chem* 266: 18330–38.

Castrillo, A., D.J. Pennington, F. Otto et al. 2001. Protein kinase Cepsilon is required for macrophage activation and defense against bacterial infection. *J Exp Med* 194: 1231–42.

Chang, L.C., R.H. Lin, L.J. Huang et al. 2009. Inhibition of superoxide anion generation by CHS-111 via blockade of the p21-activated kinase, protein kinase B/Akt and protein kinase C signaling pathways in rat neutrophils. *Eur J Pharmacol* 615: 207–17.

Chaudhary, D. and M. Kasaian. 2006. PKCtheta: A potential therapeutic target for T-cell-mediated diseases. *Curr Opin Investig Drugs* 7: 432–37.

Choi, E.N., M.K. Choi, C.S. Park et al. 2003. A parallel signal-transduction pathway for eotaxin- and interleukin-5-induced eosinophil shape change. *Immunology* 108: 245–56.

Chung, K.F. and I.M. Adcock. 2008. Multifaceted mechanisms in COPD: Inflammation, immunity, and tissue repair and destruction. *Eur Respir J* 31: 1334–56.

Cole, D.C., M. Asselin, A. Brennan et al. 2008. Identification, characterization and initial hit-to-lead optimization of a series of 4-arylamino-3-pyridinecarbonitrile as protein kinase C theta (PKCtheta) inhibitors. *J Med Chem* 51: 5958–63.

Coussens, L., P.J. Parker, L. Rhee et al. 1986. Multiple, distinct forms of bovine and human protein kinase C suggest diversity in cellular signaling pathways. *Science* 233: 859–66.

Coussens, L., L. Rhee, P.J. Parker et al. 1987. Alternative splicing increases the diversity of the human protein kinase C family. *DNA* 6: 389–94.

Dempsey, E.C., A.C. Newton, D. Mochly-Rosen et al. 2000. Protein kinase C isozymes and the regulation of diverse cell responses. *Am J Physiol Lung Cell Mol Physiol* 279: L429–38.

Dempsey, E.C., C.D. Cool, and C.M. Littler. 2007. Lung disease and PKCs. *Pharmacol Res* 55: 545–59.

DeClercq, C., D. Demers, A. Bree et al. 2009. Specific blockade of protein kinase C (PKC)-theta with a small molecule antagonist attenuates lung inflammation in a mouse model of established allergic airway disease. *J Immunol* 182: 140–21.

Do, J.S., K.S. Park, H.J. Seo et al. 2009. Therapeutic target validation of protein kinase C(PKC)-zeta for asthma using a mouse model. *Int J Mol Med* 23: 561–66.

Du, C.L., Y.J. Xu, X.S. Liu et al. 2010. Up-regulation of cyclin D1 expression in asthma serum-sensitized human airway smooth muscle promotes proliferation via protein kinase C alpha. *Exp Lung Res* 36: 201–10.

Duran, A., M.T. Diaz-Meco, and J. Moscat. 2003. Essential role of RelA Ser311 phosphorylation by zetaPKC in NF-kappaB transcriptional activation. *Embo J* 22: 3910–18.

Egesten, A. and J. Malm. 1998. Eosinophil leukocyte degranulation in response to serum-opsonized beads: C5a and platelet-activating factor enhance ECP release, with roles for protein kinases A and C. *Allergy* 53: 1066–73.

Evans, D.J., M.A. Lindsay, B.L. Webb et al. 1999. Expression and activation of protein kinase C-zeta in eosinophils after allergen challenge. *Am J Physiol* 277: L233–39.

Finkelstein, R., R.S. Fraser, H. Ghezzo et al. 1995. Alveolar inflammation and its relation to emphysema in smokers. *Am J Respir Crit Care Med* 152: 1666–72.

Gatti, R., E. Andre, S. Amadesi et al. 2006. Protease-activated receptor-2 activation exaggerates TRPV1-mediated cough in guinea pigs. *J Appl Physiol* 101: 506–11.

Gilbert, S.H., K. Perry, and F.S. Fay. 1994. Mediation of chemoattractant-induced changes in [Ca^{2+}]i and cell shape, polarity, and locomotion by InsP3, DAG, and protein kinase C in newt eosinophils. *J Cell Biol* 127: 489–503.

Hakim-Rad, K., M. Metz, and M. Maurer. 2009. Mast cells: Makers and breakers of allergic inflammation. *Curr Opin Allergy Clin Immunol* 9: 427–30.

Hewson, C.A., M.R. Edbrooke, and S.L. Johnston. 2004. PMA induces the MUC5AC respiratory mucin in human bronchial epithelial cells, via PKC, EGF/TGF-alpha, Ras/Raf, MEK, ERK and Sp1-dependent mechanisms. *J Mol Biol* 344: 683–95.

Heyworth, P.G. and J.A. Badwey. 1990. Protein phosphorylation associated with the stimulation of neutrophils. Modulation of superoxide production by protein kinase C and calcium. *J Bioenerg Biomembr* 22: 1–26.

Hirst, S.J., B.L. Webb, M.A. Giembycz et al. 1995. Inhibition of fetal calf serum-stimulated proliferation of rabbit cultured tracheal smooth muscle cells by selective inhibitors of protein kinase C and protein tyrosine kinase. *Am J Respir Cell Mol Biol* 12: 149–61.

Hirst, S.J. 1996. Airway smooth muscle cell culture: Application to studies of airway wall remodelling and phenotype plasticity in asthma. *Eur Respir J* 9: 808–20.

Hocevar, B.A. and A.P. Fields. 1991. Selective translocation of beta II-protein kinase C to the nucleus of human promyelocytic (HL60) leukemia cells. *J Biol Chem* 266: 28–33.

Hurley, J.H., A.C. Newton, P.J. Parker et al. 1997. Taxonomy and function of C1 protein kinase C homology domains. *Protein Sci* 6: 477–80.

Inoue, M., A. Kishimoto, Y. Takai et al. 1977. Studies on a cyclic nucleotide-independent protein kinase and its proenzyme in mammalian tissues. II. Proenzyme and its activation by calcium-dependent protease from rat brain. *J Biol Chem* 252: 7610–16.

Ivaska, J., L. Bosca, and P.J. Parker. 2003. PKCepsilon is a permissive link in integrin-dependent IFN-gamma signalling that facilitates JAK phosphorylation of STAT1. *Nat Cell Biol* 5: 363–69.

Iwamoto, T., M. Hagiwara, H. Hidaka et al. 1992. Accelerated proliferation and interleukin-2 production of thymocytes by stimulation of soluble anti-CD3 monoclonal antibody in transgenic mice carrying a rabbit protein kinase C alpha. *J Biol Chem* 267: 18644–48.

Jaken, S., K. Leach, and T. Klauck. 1989. Association of type 3 protein kinase C with focal contacts in rat embryo fibroblasts. *J Cell Biol* 109: 697–704.

Janssen, L.J. and K. Killian. 2006. Airway smooth muscle as a target of asthma therapy: History and new directions. *Respir Res* 7: 123.

Jeffery, P.K. 1998. Structural and inflammatory changes in COPD: A comparison with asthma. *Thorax* 53: 129–36.

Johannes, F.J., J. Prestle, S. Eis et al. 1994. PKCu is a novel, atypical member of the protein kinase C family. *J Biol Chem* 269: 6140–48.

Kaneko, M., S. Horie, M. Kato et al. 1995. A crucial role for beta 2 integrin in the activation of eosinophils stimulated by IgG. *J Immunol* 155: 2631–41.

Kato, M., R.T. Abraham, S. Okada et al. 1998. Ligation of the beta2 integrin triggers activation and degranulation of human eosinophils. *Am J Respir Cell Mol Biol* 18: 675–86.

Kato, M., T. Yamaguchi, A. Tachibana et al. 2005. An atypical protein kinase C, PKC zeta, regulates human eosinophil effector functions. *Immunology* 116: 193–202.

Kawaguchi, M., J. Fujita, F. Kokubu et al. 2009. IL-17F-induced IL-11 release in bronchial epithelial cells via MSK1-CREB pathway. *Am J Physiol Lung Cell Mol Physiol* 296: L804–10.

Kawakami, Y., J. Kitaura, L. Yao et al. 2003. A Ras activation pathway dependent on Syk phosphorylation of protein kinase C. *Proc Natl Acad Sci USA* 100: 9470–75.

Kazaneitz, M.G., J.J. Barchi, Jr., J.G. Omichinski et al. 1995a. Low affinity binding of phorbol esters to protein kinase C and its recombinant cysteine-rich region in the absence of phospholipids. *J Biol Chem* 270: 14679–84.

Kazaneitz, M.G., S. Wang, G.W. Milne et al. 1995b. Residues in the second cysteine-rich region of protein kinase C delta relevant to phorbol ester binding as revealed by site-directed mutagenesis. *J Biol Chem* 270: 21852–59.

Korchak, H.M., M.W. Rossi, and L.E. Kilpatrick. 1998. Selective role for beta-protein kinase C in signaling for O-2 generation but not degranulation or adherence in differentiated HL60 cells. *J Biol Chem* 273: 27292–99.

Leach, K.L., V.A. Ruff, M.B. Jarpe et al. 1992. Alpha-thrombin stimulates nuclear diglyceride levels and differential nuclear localization of protein kinase C isozymes in IIC9 cells. *J Biol Chem* 267: 21816–22.

Leitges, M., C. Schmedt, R. Guinamard et al. 1996. Immunodeficiency in protein kinase cbeta-deficient mice. *Science* 273: 788–91.

Leitges, M., K. Gimborn, W. Elis et al. 2002. Protein kinase C-delta is a negative regulator of antigen-induced mast cell degranulation. *Mol Cell Biol* 22: 3970–80.

Lessmann, E., M. Leitges, and M. Huber. 2006. A redundant role for PKC-epsilon in mast cell signaling and effector function. *Int Immunol* 18: 767–73.

Lin, J., Y. Xu, Z. Zhang et al. 2005. Effect of cigarette smoke extract on the role of protein kinase C in the proliferation of passively sensitized human airway smooth muscle cells. *J Huazhong Univ Sci Technol Med Sci* 25: 269–73.

Lin, X., A. O'Mahony, Y. Mu et al. 2000. Protein kinase C-theta participates in NF-kappaB activation induced by CD3-CD28 costimulation through selective activation of IkappaB kinase beta. *Mol Cell Biol* 20: 2933–40.

Luster, A.D. and A.M. Tager. 2004. T-cell trafficking in asthma: Lipid mediators grease the way. *Nat Rev Immunol* 4: 711–24.

Majumdar, S., L.H. Kane, M.W. Rossi et al. 1993. Protein kinase C isotypes and signal-transduction in human neutrophils: Selective substrate specificity of calcium-dependent beta-PKC and novel calcium-independent nPKC. *Biochim Biophys Acta* 1176: 276–86.

Martin, P., A. Duran, S. Minguet et al. 2002. Role of zeta PKC in B-cell signaling and function. *Embo J* 21: 4049–57.

Martin, P., R. Villares, S. Rodriguez-Mascarenhas et al. 2005. Control of T helper 2 cell function and allergic airway inflammation by PKCzeta. *Proc Natl Acad Sci USA* 102: 9866–71.

McVicker, C.G., S.Y. Leung, V. Kanabar et al. 2007. Repeated allergen inhalation induces cytoskeletal remodeling in smooth muscle from rat bronchioles. *Am J Respir Cell Mol Biol* 36: 721–27.

Mecklenbrauker, I., K. Saijo, N.Y. Zheng et al. 2002. Protein kinase Cdelta controls self-antigen-induced B-cell tolerance. *Nature* 416: 860–65.

Meller, N., Y. Elitzur, and N. Isakov. 1999. Protein kinase C-theta (PKCtheta) distribution analysis in hematopoietic cells: Proliferating T cells exhibit high proportions of PKCtheta in the particulate fraction. *Cell Immunol* 193: 185–93.

Miyamoto, A., K. Nakayama, H. Imaki et al. 2002. Increased proliferation of B cells and auto-immunity in mice lacking protein kinase Cdelta. *Nature* 416: 865–69.

Mochly-Rosen, D., H. Khaner, and J. Lopez. 1991. Identification of intracellular receptor proteins for activated protein kinase C. *Proc Natl Acad Sci USA* 88: 3997–4000.

Nathan, C. 2006. Neutrophils and immunity: Challenges and opportunities. *Nat Rev Immunol* 6: 173–82.

Nechushtan, H., M. Leitges, C. Cohen et al. 2000. Inhibition of degranulation and interleukin-6 production in mast cells derived from mice deficient in protein kinase Cbeta. *Blood* 95: 1752–57.

Nishizuka, Y. 1984. The role of protein kinase C in cell surface signal transduction and tumour promotion. *Nature* 308: 693–98.

Ono, Y., T. Fujii, K. Ogita et al. 1988. The structure, expression, and properties of additional members of the protein kinase C family. *J Biol Chem* 263: 6927–32.

Ono, Y., T. Fujii, K. Igarashi et al. 1989. Phorbol ester binding to protein kinase C requires a cysteine-rich zinc-finger-like sequence. *Proc Natl Acad Sci USA* 86: 4868–71.

Orr, J.W., L.M. Keranen, and A.C. Newton. 1992. Reversible exposure of the pseudosubstrate domain of protein kinase C by phosphatidylserine and diacylglycerol. *J Biol Chem* 267: 15263–66.

Palmans, E., J.C. Kips, and R.A. Pauwels. 2000. Prolonged allergen exposure induces structural airway changes in sensitized rats. *Am J Respir Crit Care Med* 161: 627–35.

Palmer, R.H., J. Ridden, and P.J. Parker. 1995. Cloning and expression patterns of two members of a novel protein-kinase-C-related kinase family. *Eur J Biochem* 227: 344–51.

Pang, L., M. Nie, L. Corbett et al. 2002. Protein kinase C-epsilon mediates bradykinin-induced cyclooxygenase-2 expression in human airway smooth muscle cells. *FASEB J* 16: 1435–37.

Park, K.S. and Y.C. Lee. 2004. PKC inhibitor reduces airway inflammation and hyperresponsiveness by reducing MMP-9 expression. *J Allergy Clin Immunol* 113: S326.

Pfeifhofer, C., K. Kofler, T. Gruber et al. 2003. Protein kinase C theta affects Ca^{2+} mobilization and NFAT cell activation in primary mouse T cells. *J Exp Med* 197: 1525–35.

Radford, D.J., K. Wang, J.C. McNelis et al. 2010. Dehdyroepiandrosterone sulfate directly activates protein kinase C-beta to increase human neutrophil superoxide generation. *Mol Endocrinol* 24: 813–21.

Ron, D. and D. Mochly-Rosen. 1995. An autoregulatory region in protein kinase C: The pseudoanchoring site. *Proc Natl Acad Sci USA* 92: 492–96.

Rossi, A.G., C. Haslett, N. Hirani et al. 1998. Human circulating eosinophils secrete macrophage migration inhibitory factor (MIF). Potential role in asthma. *J Clin Invest* 101: 2869–74.

Saijo, K., I. Mecklenbrauker, A. Santana et al. 2002. Protein kinase C beta controls nuclear factor kappaB activation in B cells through selective regulation of the IkappaB kinase alpha. *J Exp Med* 195: 1647–52.

Saijo, K., C. Schmedt, I.H. Su et al. 2003. Essential role of Src-family protein tyrosine kinases in NF-kappaB activation during B cell development. *Nat Immunol* 4: 274–79.

Sampson, A.P. 2000. The role of eosinophils and neutrophils in inflammation. *Clin Exp Allergy* 30 Suppl 1: 22–27.

Sands, W.A., V. Bulut, A. Severn et al. 1994. Inhibition of nitric oxide synthesis by interleukin-4 may involve inhibiting the activation of protein kinase C epsilon. *Eur J Immunol* 24: 2345–50.

Savkovic, S.D., A. Koutsouris, and G. Hecht. 2003. PKC zeta participates in activation of inflammatory response induced by enteropathogenic *E. coli*. *Am J Physiol Cell Physiol* 285: C512–21.

Sedgwick, J.B., K.M. Geiger, and W.W. Busse. 1990. Superoxide generation by hypodense eosinophils from patients with asthma. *Am Rev Respir Dis* 142: 120–25.

Sepulveda, M.F., E.C. Greenaway, M. Avella et al. 2005. The role of protein kinase C in regulating equine eosinophil adherence and superoxide production. *Inflamm Res* 54: 97–105.

Shin, Y., S.H. Yoon, E.Y. Choe et al. 2007. PMA-induced up-regulation of MMP-9 is regulated by a PKCalpha-NF-kappaB cascade in human lung epithelial cells. *Exp Mol Med* 39: 97–105.

Shuai, K. and B. Liu. 2003. Regulation of JAK-STAT signalling in the immune system. *Nat Rev Immunol* 3: 900–911.

Siddiqui, S., F. Hollins, S. Saha et al. 2007. Inflammatory cell microlocalisation and airway dysfunction: Cause and effect? *Eur Respir J* 30: 1043–56.

Sossin, W.S. and J.H. Schwartz. 1993. Ca (2+)-independent protein kinase Cs contain an amino-terminal domain similar to the C2 consensus sequence. *Trends Biochem Sci* 18: 207–08.

Srivastava, K.K., S. Batra, A. Sassano et al. 2004. Engagement of protein kinase C-theta in interferon signaling in T-cells. *J Biol Chem* 279: 29911–20.

Su, T.T., B. Guo, Y. Kawakami et al. 2002. PKC-beta controls I kappa B kinase lipid raft recruitment and activation in response to BCR signaling. *Nat Immunol* 3: 780–86.

Sun, Z., C.W. Arendt, W. Ellmeier et al. 2000. PKC-theta is required for TCR-induced NF-kappaB activation in mature but not immature T lymphocytes. *Nature* 404: 402–07.

Takizawa, T., M. Kato, H. Kimura et al. 2002. Inhibition of protein kinases A and C demonstrates dual modes of response in human eosinophils stimulated with platelet-activating factor. *J Allergy Clin Immunol* 110: 241–48.

Takizawa, T., M. Kato, M. Suzuki et al. 2003. Distinct isoforms of protein kinase C are involved in human eosinophil functions induced by platelet-activating factor. *Int Arch Allergy Immunol* 131 Suppl 1: 15–19.

Tan, S.L., J. Zhao, C. Bi et al. 2006. Resistance to experimental autoimmune encephalomyelitis and impaired IL-17 production in protein kinase C theta-deficient mice. *J Immunol* 176: 2872–79.

Tang, Y.J., Y.J. Xu, S.D. Xiong et al. 2004. [The effect of inhaled glucocorticosteroid on protein kinase C alpha expression and interleukin-5 production in induced sputum inflammatory cells of asthma patients]. *Zhonghua Nei Ke Za Zhi* 43: 849–52.

Tetley, T.D. 2002. Macrophages and the pathogenesis of COPD. *Chest* 121: 156S–9.

Trivedi, S.G. and C.M. Lloyd. 2007. Eosinophils in the pathogenesis of allergic airways disease. *Cell Mol Life Sci* 64: 1269–89.

Valverde, A.M., J. Sinnett-Smith, J. Van Lint et al. 1994. Molecular cloning and characterization of protein kinase D: A target for diacylglycerol and phorbol esters with a distinctive catalytic domain. *Proc Natl Acad Sci USA* 91: 8572–76.

Volkov, Y., A. Long, and D. Kelleher. 1998. Inside the crawling T cell: Leukocyte function-associated antigen-1 cross-linking is associated with microtubule-directed translocation of protein kinase C isoenzymes beta(I) and delta. *J Immunol* 161: 6487–95.

Way, K.J., E. Chou, and G.L. King. 2000. Identification of PKC-isoform-specific biological actions using pharmacological approaches. *Trends Pharmacol Sci* 21: 181–87.

Woschnagg, C., R. Garcia, S. Rak et al. 2001. IL-5 priming of the PMA-induced oxidative metabolism of human eosinophils from allergic and normal subjects during a pollen season. *Clin Exp Allergy* 31: 555–64.

Yao, H., J.W. Hwang, J. Moscat et al. 2010. Protein kinase C zeta mediates cigarette smoke/
    aldehyde- and lipopolysaccharide-induced lung inflammation and histone modifica-
    tions. *J Biol Chem* 285: 5405–16.
Zhang, H.P., Y.J. Xu, Z.X. Zhang et al. 2004. [Expression of protein kinase C and nuclear fac-
    tor kappa B in lung tissue of patients with chronic obstructive pulmonary disease].
    Zhonghua Nei Ke Za Zhi 43: 756–59.
Zhou, L. and M.B. Hershenson. 2003. Mitogenic signaling pathways in airway smooth muscle.
    Respir Physiol Neurobiol 137: 295–308.

14 Prospects for PI3 Kinase Signaling Inhibition in Obstructive Airway Diseases

Kazuhiro Ito and Nicolas Mercado

CONTENTS

14.1 INTRODUCTION

Mild to moderate asthma are now mostly controllable by current treatment; however, in contrast, several respiratory diseases, such as chronic obstructive pulmonary disease (COPD), severe asthma, and smoking asthma, have little clinical benefit for the corticosteroid treatment (Vestbo et al., 1999; Barnes, 2010). In addition, cystic fibrosis is also a therapy-resistant inflammatory disease with high mortality, which shows increased neutrophilia, a wide range of proinflammatory gene expression, and enhanced activation of several transcription factors (Nichols et al., 2008). Therefore, these diseases represent a profound and growing public health problem. There is still a fundamental lack of knowledge about the cellular, molecular, and genetic causes of these diseases, and current therapies are inadequate. Recently, many researches suggest that a number of kinases are involved in chronic airway inflammation and its exacerbation (Adcock et al., 2006). Especially, we demonstrated that the phosphoinositide 3-kinase (PI3K) family is expected to be involved in the corticosteroid-insensitive airway inflammatory response, particularly under conditions of oxidative stress (Marwick et al., 2009; To et al., 2010). In this chapter, we discuss the potential of PI3K inhibitors for the treatment of chronic respiratory diseases.

14.2 PI3K FAMILY

PI3Ks have been divided into three classes based on their structure and lipid substrate specificity (Vanhaesebroeck et al., 2005; Ward and Finan, 2003; Ghigo et al., 2010) (Table 14.1). The most extensively documented PI3Ks are class I PI3Ks. Class I PI3Ks are activated by cell surface receptors, such as growth factors, insulin, and G-protein-coupled receptors (GPCRs). Upon activation, class I PI3Ks convert phosphatidylinositol 4,5-bisphosphate (PI(4,5)P2) to phosphatidylinositol 3,4,5-trisphosphate (PI(3,4,5)P_3) (Figure 14.1). PI(3,4,5)P_3 then acts as a ubiquitous second messenger and binds the pleckstrin homology (PH) domains containing proteins, such as protein serine/threonine kinases (protein kinase B/Akt) and phosphoinositide-dependent kinase 1 (PDK1). Akt is also activated by PDK and mTOR complex 2 (mTORC2) (Kong and Yamori, 2009) (Figure 14.1). Akt stabilizes cyclin D1 by glycogen synthe-

TABLE 14.1
Characteristics of the PI3K Family

PI3K Family		Catalytic Subunits	Regulatory Subunits	Upstream Regulator	Distribution
Class IA	PI3Kα	p110α	p85α, β	Receptor tyrosine kinase	Ubiquitous
	PI3Kβ	p110β	p55γ		Ubiquitous
	PI3Kδ	p110δ	p85α, β		Leukocytes, thymus
Class IB	PI3Kγ	p110γ	p101 84/ p87[FIKAP]	GPCR	Leukocytes, thymus
Class II '		C2α	Clathrin?		Widely expressed
		C2β	Clathrin?		Widely expressed
		C2β	Clathrin?		Prostate, liver, breast
Class III		Vps34p	Vps15p (p150) Beclin 1		Ubiquitous, constitutive
Class IV	mTOR			Insulin, growth factor, mitogen	Ubiquitous
	DNA-PK			Oxidative stress	Ubiquitous
	ATM				Ubiquitous

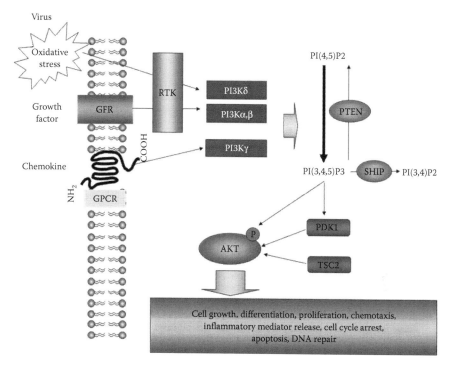

FIGURE 14.1   Scheme of the signal transduction pathways involving PI3Ks. Upon activation of PI3K through phosphotyrosine kinase (PTK) or G-protein-coupled receptor, PI(4,5)P2 converts to PI(3,4,5)P3. Akt is then phosphorylated directly or via PDK1 activation or TSC2 activation. The phosphorylated Akt has multiple targets involved in cell growth, differentiation, apoptosis, and cytokine production. PTEN dephosphorylates PI(3,4,5)P3 (PIP3) and returns it to PI(4,5)P2. SHIP can also dephosphorylate PI(3,4,5)P3 (PIP3) to produce PI(3,4)P2.

sis kinase 3 (GSK3) inhibitions and blocks the gene expression of the Cdk inhibitor, p27, by the FOXO (forkhead) transcription factor, resulting in promoting cell cycle progression. Furthermore, Akt is involved in cell growth, cell differentiation, and inflammation (Sale and Sale, 2008). Thus, these signaling proteins are actively involved in the modulation of cell growth, proliferation and shape, apoptosis (prevent/enhance), cell movement, and activation of cells.

Class I PI3Ks are further divided into subclass IA and subclass IB as shown in Table 14.1. PI3Ks IA enzymes are heterodimers consisting of a catalytic p110 subunit (designated as α, β, or δ) and a regulatory subunit (designated p85, p55, p50). There are five regulatory isoforms that are responsible for protein–protein interactions via the SH2 domain and phosphotyrosine residues of proteins (p85α, p85β, and p55γ, which are encoded by specific genes, and alternate splicing of the p85α gene (p55α and p50α)) (Table 14.1). P110α and p110β isoforms are ubiquitously expressed and genetic knockout leads to early embryonic death, although conditional knockout in mice of PI3Kβ are not lethal (Jia et al., 2008). PI3Kα is involved in glucose metabolism and insulin signaling and PI3Kβ is involved in platelet activation in thrombotic diseases. In contrast, expressions of the p110δ and p110γ isoforms are largely restricted to the hematopoietic systems and mice lacking expression of PI3Kδ and PI3Kγ are still viable, although they show an impaired immune system (Kong and Yamori, 2009). PI3K IA family members were stimulated by signals downstream of receptor tyrosine kinase (RTK) and Ras in response to growth factor, cytokines, and oxidative stress (Figure 14.1).

The single class PI3K IB consists of the p110γ catalytic subunit and the p101 regulatory subunit (or p84). This isoform was regulated by Ras, which is activated by βγ subunits from GPCRs in response to GPCR ligands, such as chemokines (Figure 14.1).

Class II PI3Ks are comprised of α, β, and γ isoforms, which are characterized by the presence of a C2 domain at the C terminus (Table 14.1). They predominantly utilize phosphatidylinositol (PI) and phosphatidylinositol 4-phosphate (PI(4)P) as substrates and are mainly involved in membrane trafficking and receptor internalization (Falasca and Maffucci, 2007; Maffucci et al., 2005).

The class III PI3Ks utilize only PI as a substrate. This class of PI3K contains only one member (Vps34: vacuolar protein sorting 34) and is reported to be involved in macroautophagy, which is a multistep process responsible for the degradation of long-lived proteins and organelle renewal and starts with the formation of an autophagosome, which ultimately fuses with the endosomal/lysosomal compartment (Backer, 2008). This pathway is known to be important in the maintenance of cell function during periods of nutrient deprivation as well as endocytosis of pathogens, such as virus and bacteria. In addition, other PI3K-related kinases are termed as class IV PI3Ks. This class includes mammalian target of rapamycin (mTOR), DNA-dependent protein kinase (DNA-PK), and ataxia telangiectasis mutated gene product (ATM) (Foster and Fingar, 2010; Smith et al., 1999; Hoekstra, 1997). These kinases play an important role in protein synthesis and response to DNA damage (Table 14.1).

Furthermore, phosphatidylinositol 4-kinases (PI4K) are also identified as another group of lipid kinase, which phosphorylate PI to PIP at 4-OH position (D'Angelo et al., 2008; Balla and Balla, 2006). PI4Ks are divided into two classes, type II and III. Type III PI4Ks are classified as PIK4IIIα and PIK4IIIβ. PI4Ks play important roles in vesicular trafficking and endocytosis. These kinases are also inhibited by the classical PI3K inhibitors, wortmannin and LY-294002, although the potency is 10-fold less than those in PI3K (Balla and Balla, 2006).

14.3 LIPID PHOSPHATASE

The termination of PI3K signaling by degradation of PI(3,4,5)P$_3$ can be mediated by at least two different types of phosphatases, namely SH2-containing inositol 5-phosphatase (SHIP) and phosphatase and tensin homologue deleted on chromosome 10 protein (PTEN) (Koyasu, 2003). PTEN removes the 3-phosphate of PI(3,4,5)P3 and thus directly counteracts all types of PI3K by catalyzing the opposite reaction. PTEN knockout in mice is embryonic lethal, but PTEN[+/−] mice showed tumorigenesis, T lymphocyte activation, and increased T cell chemotaxis. PTEN has been implicated in cancer, where mutated PTEN with lost function are frequently seen. PTEN is also reported to play a pivotal role in T$_H$2-mediated airway inflammation and airway responsiveness (Kwak et al., 2003) and PTEN overexpression also reduced airway hyperresponsiveness and vascular endothelial growth factor (VEGF) Expression in a murine model of asthma (Lee et al., 2006a). Preliminarily, we also found a reduction of PTEN expression in the peripheral lung in COPD. In contrast, SHIP removes the 5-phosphate from the inositol ring of PI(3,4,5)P3 to generate PI(3,4)P2 and this dephosphorylation of PI(3,4,5)P$_3$ by SHIP impairs downstream effects of PI3K, although PI(3,4)P2, the metabolic product of SHIP, can also mediate PI3K-dependent responses. SHIP KO mice showed increased mast cell degradation, B cell activation, and increased chemotaxis (Brauweiler et al., 2000). These mice also suffer from a lethal accumulation of macrophages and neutrophils in the lungs and therefore persistent high levels of PI(3,4,5)P$_3$ and subsequent activation of its downstream effectors might lead to excessive inflammation (Helgason et al., 1998). In addition, SHIP is shown to be involved in macrophage and dendritic cell (DC) function, and cytokine and growth factor signaling (Sly et al., 2007; Kalesnikoff et al., 2003; Huber et al., 1999; Neill et al., 2007; Antignano et al., 2010).

14.4 PI3K AND CELL FUNCTION

The role of PI3Kδ and PI3Kγ in cell function has been listed in Figure 14.2.

14.4.1 GRANULOCYTES (NEUTROPHILS AND EOSINOPHILS)

A hallmark of inflammation is the migration of leukocytes (e.g., eosinophils, neutrophils, macrophages, and T cells) to the inflammatory lesion in response to chemokines

PI3Kδ inhibition	Cell	PI3Kγ inhibition
ROS production↓	 Macrophage	Migration↓ ROS production↓
T$_H$2 differentiation↓ T$_H$2 cytokin↓	 T cell	Thymocyte maturation↓ Cytokine production↓
IgM/IgG production↓ IgE production↑ Defective maturation	 B cell	
Chemotaxis/migration↓	 Eosinophil	Chemotaxis/migration↓
Chemotaxis/migration↓ ROS production↓	 Neutrophil	Chemotaxis/migration↓
Proliferation↓	 Smooth muscle	Proliferation↓
Mast cell degranulation↓ Goblet? Fibloblast?	Others	Mast cell deglanulation↓

FIGURE 14.2 Role of PI3Kδ and PI3Kγ in functions of inflammatory cells. Contributions of PI3Kδ and PI3Kγ to cellular function are listed.

and other chemoattractants (Curnock et al., 2002). This migration has been shown to be dependent on the activation of PI3K. Control of cell polarity is essential for neutrophil chemotaxis. That is dependent on GPCR-mediated myosin assembly at the tailing edge of the cell, F-action polymerization, and phospho-PKB/Akt colocalization at the leading edge of the cell. Studies in mice lacking PI3Kγ have shown that this isoform is essential for PIP_3 production, Akt/PKB activation, and superoxide production in neutrophils exposed to chemoattractants such as N-formyl–Met–Leu–Phe (fMLP), C5a, and IL-8 as well as neutrophil chemotactic events (rather than chemokinetic events) (Thomas et al., 2005). The chemotaxis of cells involved in mounting an effective immune response to a pathogen or a foreign body (e.g., neutrophils, macrophages, and T lymphocytes) was also impaired in the absence of PI3Kγ, both *in vitro* and *in vivo* (Gruen et al., 2010; Liu et al., 2007; Reutershan et al., 2010).

There is evidence to suggest that other PI3K isoforms are also activated by chemokines in PI3Kγ$^{-/-}$ mice. There is incomplete (e.g., 50–70%) reduction in the capacity of neutrophils to migrate to a range of chemoattractants, and PI3Kγ knockout does not prevent chemoattractant-induced actin polymerization (Liu et al., 2007). Thus, GPCR stimulation can activate p85/p110 PI3K as well as PI3Kγ through G-protein βγ subunits and/or $Gα_i$ subunits (Figure 14.2).

Along with chemotaxis, reactive oxygen species (ROS) production is also a key factor for the pathogenesis of chronic inflammation although this is an antimicrobial function at the site of infection. Genetic ablation of PI3K inhibited fMLP-induced ROS production by neutrophils (Sasaki et al., 2000; Hirsch et al., 2000). The ROS production is known to be a biphasic reaction with initial PI3Kδ-dependent phase and subsequent amplification of the signal by PI3Kδ. Thus, IA and IB coordinate on ROS production (Condliffe et al., 2005).

Eosinophils are abundant cells in allergic disease and asthma, and ROS and myeloperoxidase (MPO) as well as cytokines produced from eosinophils are believed to contribute to the pathogenesis of asthma and atopic dermatitis. There is no study to test the effects of the PI3K inhibitor on eosinophil function *in vitro*; a lot of *in vivo* studies demonstrated the involvement of PI3K in eosinophil accumulation. In OVA-sensitized and OVA-challenged model, LY294002 and PI3Kγ knockout inhibited eosinophil accumulation in airways (Duan et al., 2005; Takeda et al., 2009). Pinho-V and colleagues demonstrated that only PI3Kγ regulates the maintenance of eosinophilic inflammation, and other isoforms are actually involved in eosinophil recruitment (Pinho et al., 2005). On the other hand, recently, the double-mutant of PI3Kγ and PI3Kδ (PI3Kγ$^{-/-}$δ$^{D910A/D910A}$) mice showed increased eosinophil accumulation as well as increased levels of IL-4/IL-5 and IgE (Ji et al., 2007). As knockout mice do not always show the same results with pharmacological intervention, it is not clear whether these potential side effects seen in double kinases-deficient mice happen by pharmacological intervention of both kinases (PI3Kγ/δ inhibitor) or not.

14.4.2 MACROPHAGES AND DENDRITIC CELLS

DCs are also a target of PI3Kγ and DC obtained from PI3Kγ$^{-/-}$ mice showed a reduced ability to respond to chemokines *in vitro* and *ex vivo*, and to travel to draining lymph nodes under inflammatory conditions (Del et al., 2004). In addition, PI3Kγ$^{-/-}$ mice had a selective defect in the number of skin Langerhans cells and a reduced capacity to mount contact hypersensitivity and delayed-type hypersensitivity reactions (Del et al., 2004). Thus, PI3Kγ plays a nonredundant role in DC trafficking and in the activation of specific immunity.

The contribution of PI3Kγ to macrophage responses to chemoattractants has also been investigated. They observed that early membrane ruffling induced by MCP-1, which activates a GPCR, or by CSF-1, which activates a tyrosine kinase receptor, is unaltered in PI3Kγ$^{-/-}$ mice compared to wild-type macrophages (Jones et al., 2003). Furthermore, macrophages from PI3K γ$^{-/-}$ mice showed reduced migration speed and translocation, and no chemotaxis to MCP-1. This study also indicated that the initial actin reorganization induced by either a GPCR or a tyrosine kinase receptor agonist is not dependent on PI3Kγ, whereas PI3Kγ is needed for optimal migration of macrophages to either agonist.

Along with type I PI3K, PI3K-C2α activation by MCP-1 is also reported to be involved in cell (THP-1: Human acute monocytic leukemia cell line) migration, and this activation exhibits the same resistance to wortmannin and sensitivity to pertussis toxin as MCP-1-stimulated increases in 3'-phosphoinositide lipid generation. Recent work using RNA interference (RNAi) suggested that a class II PI3K (PI3K-C2β) regulates the lysophosphatidic acid (LPA)-stimulated migration of HeLa cell (Maffucci et al., 2005) and HEK293 cell (Domin et al., 2005). Thus, type II PI3K might be involved in growth-factor-mediated cell migration, but its role in inflammatory cell migration has not been established.

14.4.3 LYMPHOCYTES

In asthma, T helper (T$_H$)2-type cytokine (IL-4, IL-5, IL-13) production is dominant; however, in COPD and perhaps in severe asthma, T$_H$1 cytokines (e.g., Interferon (IFN) gγ) are reported to be produced more than T$_H$2 cytokines, suggesting that altered T$_H$1/T$_H$2 balance might establish the phenotype of these diseases. Several papers using both pharmacological tools and gene-modified mice have implicated PI3K in T and B lymphocyte activation/function (Ward and Finan, 2003; Vanhaesebroeck et al., 2005; Koyasu, 2003). Recent studies on the role of PI3Ks in innate immunity have also highlighted their involvement in the control of cellular responses to pathogens. In this regard, the amount of IL-12 produced by stimulation through Toll-like receptors (TLRs) is critical in the balance between T$_H$1 and T$_H$2 responses. Interestingly, mice lacking the p85α regulatory subunit (single knockout) show impaired immunity against the intestinal nematode *Strongyloides venezuelensis* as a result of impaired T$_H$2 responses. In contrast, p85α$^{-/-}$ mice and PI3Kδ KO (PI3Kγ$^{-/-}$δ$^{D910A/D910A}$) demonstrate enhanced T$_H$1 responses and, unlike wild-type mice, are resistant to *Leishmania major* infection (Fukao et al., 2002; Liu et al., 2009). These observations indicate that class IA PI3Ks are important in the T$_H$1 versus T$_H$2 balance *in vivo*, and that they control the induction of the T$_H$2 response and/or the suppression of the T$_H$1 response. p85α$^{-/-}$ splenic and bone-marrow-derived DCs produce more IL-12 than wild-type DCs, suggesting that PI3K is one of the key regulators in the T$_H$1 versus T$_H$2 balance through the control of IL-12 production. Overproduction of IL-12 by DCs might cause the skewed T$_H$1 response in p85α$^{-/-}$ mice. These observations indicate that PI3K plays a critical negative regulatory role during the induction of the T_H1 immune response by suppressing the production of IL-12 from DCs, although it is unclear which of the three class 1A catalytic isoforms is involved (Fukao and Koyasu, 2003). Along with TLR, CD40L and RANKL (ligand to receptor activator of NF-κB-ligand) also induce IL-12 via activation of class 1A PI3K.

As well as T lymphocyte, PI3Kδ also regulates B cell function. PI3Kδ-deficient mice showed an increase of the pro-B/pre-B ratio, indicating a maturation block between these two stages. In addition, PI3Kδ[−/−] mice and PI3Kδ[D910A/D910]A mice showed reduced antibody production by T-cell-dependent and T-cell-independent stimulation (Clayton et al., 2002; Jou et al., 2002). Very interestingly and also paradoxically, both IC87114 and PI3Kδ mutant increased IgE production despite reduced T_H2-type cytokines (Zhang et al., 2008). The clinical trial of IC87114 is now ongoing and this is a key clinical (safety) point that needs to be confirmed at the earliest.

14.4.4 OTHER CELLS

Mast cells are tissue-resident cells and contain histamine, cytokines, and heparin. Although the cells play an important role in innate immunity (defense against pathogens) and wound healing, they are known to be involved in allergy/anaphylaxis or pathogenesis of asthma. It has been reported that the pan-isoform PI3K inhibitors, wortmannin and LY-294002, or the PI3Kδ-selective inhibitor, IC87114, inhibits mast cell degranulation (Ali et al., 2004, 2008). The mast cells expressing catalytically inactive PI3Kδ (PI3Kδ[D910A]) showed impaired mast cell degranulation (Lam et al., 2008). In the *in vivo* study, passive cutaneous anaphylaxis reaction induced by IgE and allergen was reduced in mice having inactive PI3Kδ (PI3Kδ[D910A/D910A]) (Ali et al., 2004, 2008). As in PI3Kδ-mutated mice, PI3Kγ[−/−] mice are also resistant to passive systemic anaphylaxis reaction (Wymann et al., 2003; Laffargue et al., 2002). Thus, both PI3Kδ and PI3Kγ are crucial for mast cell degranulation, but the regulations are complicated. PI3Kδ seems to be involved in the early phase in IgE-allergen response and PI3Kγ will work in the later phase to maximize degranulation (Ghigo et al., 2010).

In terms of neutrophil accumulation, several studies showed the importance of PI3Kγ function in endothelial cells. PI3Kγ inhibition in endothelial cells reduced selectin-mediated attachment and increased neutrophil rolling velocity (Puri et al., 2004, 2005). In contrast, PI3Kδ inhibition in endothelial cells increased rolling velocity (Puri et al., 2004). Double knockout mice showed a higher reduction of neutrophil attachment to TNFα-activated endothelial cells as compared with the single mutant of each kinase (Puri et al., 2005). Chemotaxis of airway epithelial cells is also controlled by PI3K. Shahabuddin et al. (2006) showed that wortmannin concentration-dependently inhibited the chemotactic response of epithelial cells to I-TAC (CXCR3 ligand).

14.5 PI3K AND OXIDATIVE STRESS

The inflammation in COPD is an amplification of the normal inflammatory response to inhaled noxious agents (cigarette smoke or other irritants) (Rahman and Adcock,

2006). ROS such as H_2O_2 (hydrogen peroxide) and O_2^- (superoxide free radical) have emerged as key mediators of intracellular signaling, which is elevated by various types of extracellular stimuli, including growth factors, cytokines, and environmental stresses. In COPD or smoking asthma, a high level of ROS from exposure to cigarette smoke or other irritants, or endogenously produced from inflammatory cells such as neutrophils and macrophages is thought to be an important component of amplification of inflammation in the lung (Rahman and Adcock, 2006). It is widely accepted that ROS can modulate cell functions by activating MAPKs, phospholipase C, protein kinase C, and various other types of signaling components.

ROS induction is often accompanied by the activation of PI3K. For example, LY294002, a specific inhibitor for PI3K, was shown to abolish chemokine-induced ROS generation in phagocytes, which was further confirmed by studies using PI3K knockout mice. It was similarly reported that the ROS accumulation induced by TNFα, platelet-derived growth factor (PDGF), or VEGF in various other cell types was suppressed when PI3K activity/activation was blocked by pharmacological or molecular tools. Therefore, PI3K appears to be commonly involved in the ROS accumulation induced by cytokines, growth factors, and viruses (Qin and Chock, 2003). It was also reported that serum withdrawal (SW) killed human U937 blood cells by elevating cellular ROS levels, which occurred through PI3K activation (Lee et al., 2005).

Along with the role of PI3K in ROS induction, evidence exists that supports the opposite hierarchical relationship between ROS and PI3K. PI3K in various cell types was activated in response to the exogenous application of H_2O_2. Consistent with the ability of H_2O_2 to activate PI3K, the PI3K activation induced by ultraviolet irradiation or Zn^{2+} treatment was blocked by the addition of antioxidants. Exogenous H_2O_2 can activate an array of nonreceptor-type protein tyrosine kinases (PTKs). H_2O_2 stimulation leads to the initiation of downstream signaling events, such as stimulation of PLCγ2, mitogen-activated protein kinase (MAPK) and activation of PI3K. This activation of PI3K is selective as H_2O_2-induced tyrosine phosphorylation of the p110 but not the p85 subunit of PI3K in DT40 cells (chicken B cell line) (Qin and Chock, 2003). In addition, hydrogen peroxide treatment caused an increase in the amount of p85 PI3K associated with the particulate fraction. Along with H_2O_2, cigarette smoke also activates PI3K signaling, which was confirmed by Akt phosphorylation in cigarette smoke condition medium-treated cells and also cigarette smoke-exposed mice (To et al., 2010). Thus, PI3K is involved in ROS-induced cell signaling as well as cell signaling for ROS induction.

14.6 PI3K AND AIRWAY INFLAMMATION

14.6.1 INFLAMMATORY COMPONENT OF ASTHMA AND COPD

Lower airway inflammation is a central feature of many lung diseases, including asthma and COPD. This involves the recruitment and activation of inflammatory cells and changes in the structural cells of the lung. Inflammation in asthma is associated with increased airway hyperresponsiveness, leading to recurrent episodes of wheezing, breathlessness, chest tightness, and coughing, particularly at night or in the early morning. These conditions are characterized by an increased expression of chemokines, cytokines, growth factors, enzymes, noxious gas, reactive oxygen, receptors, and adhesion molecules.

The inflammation of asthma is characterized by the infiltration of T lymphocytes of the T_H2 phenotype, eosinophils, macrophages/monocytes, and mast cells into the airway wall. Airway inflammation is also amplified during exacerbation, with an increase in eosinophils and sometimes neutrophils. Chronic inflammation may also lead to structural changes in the airway, including increased thickness of airway smooth muscle, increased number of mucus-secreting cells, subepithelial fibrosis, and increased number of blood vessels (angiogenesis), which are referred to as airway remodeling. These changes may not be fully reversible with current treatments.

COPD is a chronic inflammatory disease of the lower airways and lung, which is enhanced during exacerbations (Barnes, 2003). The pathological characteristics of COPD are destruction of the lung parenchyma (emphysema) and inflammation of the peripheral airways and the central airways. There is a marked increase in macrophages and neutrophils in bronchoalveolar lavage fluid and induced sputum, and T lymphocytes and B lymphocytes in lung parenchyma. Our recent study suggested that the activity and expression of histone deacetylase (HDAC), a transcriptional corepressor, and sirtuin 1, an anti-aging molecule, were decreased in COPD due to oxidative stress, and consequently, cytokine transcription was increased (Ito et al., 2001, 2005; Nakamaru et al., 2009).

Another important feature of severe asthma and COPD is corticosteroid insensitivity (Ito, 2005). Several large studies suggest that long-term treatment with corticosteroids did not stop the decline of lung function in COPD patients. This is consistent with the demonstration that inhaled or oral corticosteroids fail to reduce inflammatory cell numbers, cytokines, chemokines, or proteases in induced sputum or bronchial biopsies of patients with COPD. The molecular mechanisms for corticosteroid insensitivity are not fully elucidated, but may include overexpression of transcriptional factors to trap glucocorticoid receptor (GR), GR degradation by oxidative stress, and/or decoy GR (GRβ) overexpression. We found that HDAC2 reduction in COPD and severe asthma is involved in corticosteroid resistance possibly via hyperacetylation of GR (Ito et al., 2005, 2006) although kinases such as MAPK may also be important in corticosteroid insensitivity in severe asthma (Irusen et al., 2002).

14.6.2 PI3K AND AIRWAY INFLAMMATION

Despite limitations in selectivity, the two commercially available PI3K inhibitors, wortmannin and LY294002, have contributed greatly to our understanding of the biological role of PI3K in lung inflammation.

Intratracheal administration of LY294002 is reported to significantly inhibit ovalbumin (OVA)-induced increases in total cell counts, eosinophil counts, and interleukin (IL)-5, IL-13, and CCL11(eotaxin) levels in bronchoalveolar lavage fluid and dramatically inhibit OVA-induced tissue eosinophilia and airway mucus production (Duan et al., 2005). This was associated with a significant suppression of OVA-induced airway hyperresponsiveness to inhaled methacholine. Importantly, this study confirmed that LY294002 markedly attenuated OVA-induced serine phosphorylation of Akt, a direct downstream substrate of PI3K. In addition, the antigen-induced anaphylactic contraction of bronchial rings, and the release of histamine and peptide leukotrienes (LTs) from chopped lung preparations from sensitized guinea pigs were blocked by LY294002. LY294002 and wortmannin also attenuated eosinophilic airway inflammation and airway hyperresponsiveness in a murine asthma model (Lee et al., 2006b; Kwak et al., 2003; Ezeamuzie et al., 2001). Although human rhinovirus (HRV) is known to be involved in the exacerbation of asthma (and COPD), HRV type B infection-induced IL-8 release was inhibited by LY294002 in mice and also *in vitro* bronchial epithelial cell (Newcomb et al., 2008). Furthermore, PI3Kβ is reported to be involved in human rhinovirus infection (Bentley et al., 2007). Thus, PI3K inhibition was indicated to have a therapeutic potential for the treatment of asthmatic airway inflammation.

Recent development of novel PI3K inhibitors overcomes poor selectivity of PI3K inhibition. IC87114, a selective p110δ inhibitor, has been used to investigate the role of p110δ in allergic airway inflammation and hyperresponsiveness using a mouse asthma model (Lee et al., 2006b). IC87114 significantly reduced the serum levels of total immunoglobulin (IgE) and OVA-specific IgE and LTC_4 release into the airspace, OVA-induced lung tissue eosinophilia, airway mucus production and inflammation score, and, importantly, OVA-induced increase in expression of IL-4, IL-5, IL-13, intercellular adhesion molecule (ICAM)-1, vascular cell adhesion molecule (VCAM)-1, CCL5 (RANTES), and CCL11. Furthermore, IC87114 significantly suppressed OVA-induced airway hyperresponsiveness to inhaled methacholine and this corresponded to a reduction in OVA-induced Akt serine phosphorylation. These results were supported by the *in vitro* findings that p110δ is involved in B and T cell antigen receptor signaling and activation and allergen-IgE-induced mast cell degranulation. Mutation of p110δ also leads to defects in mast cells and possibly neutrophils (Puri et al., 2004; Ali et al., 2004). However, recent evidence showed that IC87114 treatment actually increased IgE expression (Zhang et al., 2008). The molecular mechanism has not been fully elucidated.

Along with PI3Kδ, the studies using knockout mice revealed that PI3Kγ is an important component in the pathogenesis of asthma. Wymann et al. (2003) have demonstrated that murine mast cell responses are exacerbated *in vitro* and *in vivo* by autocrine signals, and require functional PI3Kγ. Adenosine, acting through the A_3 adenosine receptor, as well as other agonists of $G\alpha_i$-coupled receptors, transiently increased PI(3,4,5)P3 exclusively via PI3Kγ. Additionally, mice that lacked PI3Kγ did not form edema when challenged by passive systemic anaphylaxis. PI3Kγ thus relays inflammatory signals through various GPCRs, and is thus central to mast cell function. Eosinophil accumulation was also reported to be inhibited at 48 h in these PI3K γ-deficient mice, as compared with wild-type mice but not at earlier time points (6 and 24 h), suggesting that PI3Kγ plays a role in the maintenance of eosinophilic inflammation *in vivo*.

There are some published reports on the effect of PI3K inhibitors on experimental models of COPD, and data suggesting a potential role of PI3K in the pathogenesis of COPD are now accumulating. In mice exposed to cigarette smoke subchronically, neither IC87114 nor steroid inhibits airway neutrophilia, but a combination of IC87114 and steroid significantly inhibited neutrophil accumulation in the airway (To et al., 2010). The PI3Kγ/δ inhibitor, TG100-115, also inhibited subacute cigarette smoke exposure-induced neutrophilia (Doukas et al., 2009). Matrix metalloproteinase (MMP)-9 degrades extracellular matrix components (particularly elastin) and is related to the pathogenesis of pulmonary emphysema. MMP9 is present in low quantities in the healthy adult lung but is much more abundant in COPD, and the inappropriate expression of MMP-9 is thought to contribute to the pathogenesis of COPD (Barnes et al., 2003). MMP-9 expression, whether stimulated by PAF or fibronectin, probably through an action on NF-κB, is also regulated by PI3K signaling pathways (Ko et al., 2005). Furthermore, several lines of evidence point to PI3K being important in the activation of macrophage and neutrophils, which are key players in COPD inflammation (Thomas et al., 2005) as shown in Figure 14.3.

14.6.3 PI3K and Corticosteroid Insensitivity

As discussed above, patients with COPD and severe asthma do not respond well to corticosteroids although corticosteroids are very effective in controlling mild to moderate asthma. This is consistent with the demonstration that inhaled or oral steroids fail to reduce inflammatory cell numbers, cytokines, chemokines, or proteases in induced sputum or airway biopsies of patients with COPD and severe asthma. Previously, we have reported a reduction in corepressor, HDAC2 expression, and total HDAC activity

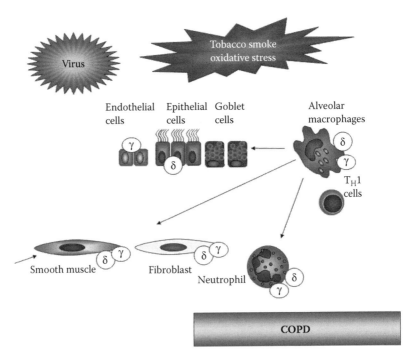

FIGURE 14.3 Targets for PI3Kδ and PI3Kγ in airway inflammation in COPD. PI3K has been shown to be involved in almost all target cells shown in the figure using nonspecific PI3K inhibitors. Targets of PI3Kδ and PI3Kγ are specified as "δ" and "γ" based on reports using selective knockout mice or selective inhibitors and shown in this figure.

in COPD patients (Ito et al., 2005). By overexpression and knockdown of HDAC2, we have shown that HDAC2 is a prerequisite molecule for corticosteroid action in airway macrophages and reduction of HDAC2 is one of the causes of corticosteroid insensitivity in COPD (Ito et al., 2006). *In vitro* experiments have also shown that oxidative stress raised by hydrogen peroxide reduced HDAC2 expression, and LY294002 and an Akt inhibitor (SH-5) restored defective HDAC2 expression and activity in these cells. Marwick et al. (2009) also demonstrated that PI3Kδ knockout mice did not establish steroid insensitivity in cigarette smoke mice although cigarette smoke exposure induced steroid insensitivity on neutrophilia in wild-type mice and PI3Kγ knockout mice. This has been confirmed by To et al., demonstrating the superior efficacy of combination of IC87114 (PI3Kδ inhibitor) and steroid as well as the combination of theophylline (weak PI3Kδ inhibitor) and steroid (To et al., 2010) than the efficacies by each component alone. This is also confirmed by RNAi of PI3Kδ *in vitro* (To et al., 2010). This suggests that PI3Kδ may be involved in corticosteroid sensitivity by reducing HDAC activity. Further studies are in progress to elucidate this aspect.

14.7 PI3K INHIBITION

14.7.1 PI3K INHIBITORS (ISOFORM-SPECIFIC AND FUNCTIONAL PI3K INHIBITOR)

The availability of two PI3K inhibitors, wortmannin, an irreversible inhibitor (isolated from *Penicillium wortmanni*), and the morpholino derivative, LY294002, a reversible inhibitor that is derived from the broad-spectrum kinase inhibitor, quercetin, has contributed greatly to our understanding of the biological role of PI3K and its effector proteins (Ward and Finan, 2003). However, wortmannin and LY294002 have no selectivity for individual PI3K isoforms, and have poor stability, solubility, and absorption as well as their toxicity. In addition, wortmannin and LY294002 possess off-target effects (wortmannin: myosin light-chain kinase inhibition; LY294002: casein kinase-2 inhibition). Developments of pharmacokinetics-improved nonspecific inhibitors [such as ZSTK474 (Ward et al., 2003), PEG-wortmannin, LY294002 stable derivative, SF1126 (Semafore)] are underway or these have already been tested in the clinical trial of cancer (Table 14.2).

As discussed above, selective PI3Kδ inhibitors have the potential to treat steroid-insensitive chronic airway disease. A number of patent specifications have been published that describe isoform-specific inhibitors of PI3K (Powis et al., 2006). ICOS Corporation has described several p110δ inhibitors, including IC87114, which contains a quinazoline core structure. Pomel et al. (2006) also reported furan-2-ylmethylene thiazolidinediones as novel, potent, and selective inhibitors of PI3Kγ by structure-based design and x-ray crystallography of complexes formed by inhibitors bound to PI3Kγ. AS-604850 and related compounds are selective PI3K γ inhibitors that show efficacy in a murine model of rheumatoid arthritis (Camps et al., 2005). In addition, a number of companies have declared active programs in PI3Kγ inhibitor development (Novartis, Boehringer, Pfizer, Bayer, etc.) (Pomel et al., 2006) for cancer and chronic inflammatory disease, but no published results are available for these compounds indicating anti-inflammatory efficacy in respiratory disease models.

TABLE 14.2

Potency of Published PI3K Inhibitors

Compound (Originator)	Specificity	Activity (Type I PI3K)	Other Targets
Wortmannin	Nonspecific	0.001 μM (α); 0.014 μM (β); 0.009 μM (δ); 0.005 μM (γ)	MLCK (0.004 μM); PLK (0.009 μM); mTOR/DNA-PK (0.001 μM)
LY29002 (Eli Lilly)	Nonspecific	0.72 μM (α); 0.31 μM (β); 1.3 μM (δ); 7.3 μM (γ)	Casein kinase 2 (2–10 μM); mTOR/DNA-PK (11 μM)
PX-866 (ProIX)	Nonspecific	0.0055 μM (α); >0.3 μM (β); 0.0027 μM (δ); 0.009 μM (γ)	
PI-103 (Piramed)	Nonspecific	0.002 μM (α); 0.003 μM (β); 0.003 μM (δ); 0.015 μM (γ)	mTOR/DNA-PK
NVP-BEZ235 (Novartis)	Nonspecific	0.004 μM (α); 0.076 μM (β); 0.005 μM (δ); 0.007 μM (γ)	mTOR/DNA-PK
ZSTK474 (Zenyaku Kogyo)	Nonspecific	0.016 μM (α); 0.017 μM (β); 0.006 μM (δ); 0.053 μM (γ)	mTOR/DNA-PK
YM-024 (Yamanouchi)	p110α ?	0.3 μM (α); 2.65 μM (β); 0.33 μM (δ); 9.1 μM (γ)	
TGX-221 (Kinacia Pty Ltd.)	p110β	5 μM (α); 0.007 μM (β); 0.1 μM (δ); 3.5 μM (γ)	
IC-87114 (Calistoga)	p110δ	>100 μM (α); 75 μM (β); 0.5 μM (δ); 29 μM (γ)	
Theophylline	p110δ	400 μM (α); 400 μM (β); 75 μM (δ); 1000 μM (γ)	PDE inhibition; adenosine antagonism
AS-604850 (Serono)	p110γ	4.5 μM (α); >20 μM (β); >20 μM (δ); 0.25 μM (γ)	
TG100-115 (TargeGen)	p110δ/γ	1.3 μM (α); 1.2 μM (β); 0.24 μM (δ); 0.083 μM (γ)	

We recently found that old drugs that are currently used in clinics showed functional PI3Kδ inhibition and restored corticosteroid sensitivity. The first drug discovered as PI3K inhibitor is methylxanthines, such as caffeine and theophylline, which are used as bronchodilator, but rediscovered as anti-inflammatory agents. Although the use has been declined due to side effects, recently, (safe) lower-dose theophylline restores corticosteroid sensitivity *in vitro* in COPD disease cells, *in vivo* in smoking mice model, and also in COPD patients clinically. These compounds were reported as selective inhibitors for p110 δ isoforms, although their activity is rather low (Foukas et al., 2002) and also possess several off-target effects, such as phosphodiesterase inhibition and adenosine A receptor antagonism. Very importantly, the PI3Kδ inhibiting activity of theophylline increased in oxidative stress-exposed PI3Kδ enzymes (To et al., 2010). The mechanisms are not elucidated yet, but conformational change by oxidation of cysteine or tyrosine nitration will be involved in altered sensitivity to theophylline. Furthermore, resveratrol is also reported as a functional PI3K pathway inhibitor (Frojdo et al., 2007).

14.8 CONCLUSIONS AND EXPERT OPINIONS

Pharmacological and genetic intervention revealed that PI3Kδ and PI3Kγ play prominent roles in various inflammatory cells by controlling cell growth, differentiation, survival, proliferation, migration, and mediator production (such as cytokines), which are involved in the pathogenesis of severe asthma and COPD. Several PI3K inhibitors are under development for the treatment of asthma and COPD as well as cancer, diabetes, thrombosis, cardiac contractility during heart failure, hypertension, rheumatoid arthritis, and inflammatory bowel disease. It is most likely that PI3K inhibitors, maybe the combination of PI3Kδ and PI3Kγ, will be more efficacious in more severe steroid-insensitive asthma and in COPD where corticosteroids are of limited effectiveness and no alternative therapy is available. In addition, it is possible that specific PI3Kδ inhibitors will be more efficient in augmenting current therapies, particularly corticosteroids, rather than as a monotherapy because PI3Kδ inhibitors have the potential to restore corticosteroid sensitivity *in vitro* and *in vivo* as discussed above. The success of selective PI3K inhibitors in reaching the clinic will depend upon the specific isoform activated in each disease as a group and importantly in each individual patient.

ACKNOWLEDGMENTS

This research is funded by the Wellcome Trust, the Medical Research Council, GlaxoSmithKline, AstraZeneca, Mitsubishi, and Pfizer.

We regret that we were unable to cite all important original works due to space constraints and apologize to those authors whose work we have not cited.

REFERENCES

Adcock, I. M., K. F. Chung, G. Caramori, and K. Ito. 2006. Kinase inhibitors and airway inflammation. *Eur J Pharmacol* 533:118–32.

Ali, K., A. Bilancio, M. Thomas, W. Pearce, A. M. Gilfillan, C. Tkaczyk, N. Kuehn et al. 2004. Essential role for the p110delta phosphoinositide 3-kinase in the allergic response. *Nature* 431:1007–11.

Ali, K., M. Camps, W. P. Pearce, H. Ji, T. Ruckle, N. Kuehn, C. Pasquali et al. 2008. Isoform-specific functions of phosphoinositide 3-kinases: p110 delta but not p110 gamma promotes optimal allergic responses *in vivo. J Immunol* 180:2538–44.

Antignano, F., M. Ibaraki, C. Kim, J. Ruschmann, A. Zhang, C. D. Helgason, and G. Krystal. 2010. SHIP is required for dendritic cell maturation. *J Immunol* 184:2805–13.

Backer, J. M. 2008. The regulation and function of Class III PI3Ks: Novel roles for Vps34. *Biochem J* 410:1–17.

Balla, A. and T. Balla. 2006. Phosphatidylinositol 4-kinases: Old enzymes with emerging functions. *Trends Cell Biol* 16:351–61.

Barnes, P. J. 2003. New concepts in chronic obstructive pulmonary disease. *Annu Rev Med* 54:113–29.

Barnes, P. J. 2010. Inhaled corticosteroids in COPD: A controversy. *Respiration* 80:89–95.

Barnes, P. J., S. D. Shapiro, and R. A. Pauwels. 2003. Chronic obstructive pulmonary disease: Molecular and cellular mechanisms. *Eur Respir J* 22:672–88.

Bentley, J. K., D. C. Newcomb, A. M. Goldsmith, Y. Jia, U. S.Sajjan, and M. B. Hershenson. 2007. Rhinovirus activates interleukin-8 expression via a Src/p110beta phosphatidylino-sitol 3-kinase/Akt pathway in human airway epithelial cells. *J Virol* 81:1186–94.

Brauweiler, A. M., I. Tamir, and J. C. Cambier. 2000. Bilevel control of B-cell activation by the inositol 5-phosphatase SHIP. *Immunol Rev* 176:69–74.

Camps, M., T. Ruckle, H. Ji, V. Ardissone, F. Rintelen, J. Shaw, C. Ferrandi et al. 2005. Blockade of PI3Kgamma suppresses joint inflammation and damage in mouse models of rheumatoid arthritis. *Nat Med* 11:936–43.

Clayton, E., G. Bardi, S. E. Bell, D. Chantry, C. P. Downes, A. Gray, L. A. Humphries et al. 2002. A crucial role for the p110delta subunit of phosphatidylinositol 3-kinase in B cell development and activation. *J Exp Med* 196:753–63.

Condliffe, A. M., K. Davidson, K. E. Anderson, C. D. Ellson, T. Crabbe, K. Okkenhaug, B. Vanhaesebroeck et al. 2005. Sequential activation of class IB and class IA PI3K is important for the primed respiratory burst of human but not murine neutrophils. *Blood* 106:1432–40.

Curnock, A. P., M. K. Logan, and S. G. Ward. 2002. Chemokine signalling: Pivoting around multiple phosphoinositide 3-kinases. *Immunology* 105:125–36.

D'Angelo, G., M. Vicinanza, C. A. Di et al. 2008. The multiple roles of PtdIns(4)P—not just the precursor of PtdIns(4,5)P2. *J Cell Sci* 121:1955–63.

Del, P. A., W. Vermi, E. Dander, K. Otero, L. Barberis, W. Luini, S. Bernasconi et al. 2004. Defective dendritic cell migration and activation of adaptive immunity in PI3Kgamma-deficient mice. *EMBO J* 23:3505–15.

Domin, J., L. Harper, D. Aubyn, M. Wheeler, O. Florey, D. Haskard, M. Yuan, and D. Zicha. 2005. The class II phosphoinositide 3-kinase PI3K-C2beta regulates cell migration by a PtdIns3P dependent mechanism. *J Cell Physiol* 205:452–62.

Doukas, J., L. Eide, K. Stebbins, A. Racanelli-Layton, L. Dellamary, M. Martin, E. Dneprovskaia et al. 2009. Aerosolized phosphoinositide 3-kinase gamma/delta inhibitor TG100-115 [3-[2,4-diamino-6-(3-hydroxyphenyl)pteridin-7-yl]phenol] as a therapeutic candidate for asthma and chronic obstructive pulmonary disease. *J Pharmacol Exp Ther* 328:758–65.

Duan, W., A. M. guinaldo Datiles, B. P. Leung, C. J. Vlahos, and W. S. Wong. 2005. An anti-inflammatory role for a phosphoinositide 3-kinase inhibitor LY294002 in a mouse asthma model. *Int Immunopharmacol* 5:495–502.

Ezeamuzie, C. I., J. Sukumaran, and E. Philips. 2001. Effect of wortmannin on human eosinophil responses in vitro and on bronchial inflammation and airway hyperresponsiveness in Guinea pigs in vivo. *Am J Respir Crit Care Med* 164:1633–39.

Falasca, M. and T. Maffucci. 2007. Role of class II phosphoinositide 3-kinase in cell signalling. *Biochem Soc Trans* 35:211–14.

Foster, K. G. and D. C. Fingar. 2010. Mammalian target of rapamycin (mTOR): Conducting the cellular signaling symphony. *J Biol Chem* 285:14071–77.

Foukas, L. C., N. Daniele, C. Ktori, K. E. Anderson, J. Jensen, and P. R. Shepherd. 2002. Direct effects of caffeine and theophylline on p110 delta and other phosphoinositide 3-kinases. Differential effects on lipid kinase and protein kinase activities. *J Biol Chem* 277:37124–30.

Frojdo, S., D. Cozzone, H. Vidal, and L. Pirola. 2007. Resveratrol is a class IA phosphoinositide 3-kinase inhibitor. *Biochem J* 406:511–18.

Fukao, T. and S. Koyasu S. 2003. PI3K and negative regulation of TLR signaling. *Trends Immunol* 24:358–63.

Fukao, T., M. Tanabe, Y. Terauchi, T. Ota, S. Matsuda, T. Asano, T. Kadowaki, T. Takeuchi, and S. Koyasu. 2002. PI3K-mediated negative feedback regulation of IL-12 production in DCs. *Nat Immunol* 3:875–81.

Ghigo, A., F. Damilano, L. Braccini, and E. Hirsch. 2010. PI3K inhibition in inflammation: Toward tailored therapies for specific diseases. *Bioessays* 32:185–96.

Gruen, M., C. Rose, C. Konig, M. Gajda, R. Wetzker, and R. Brauer. 2010. Loss of phosphoinositide 3-kinase gamma decreases migration and activation of phagocytes but not T cell activation in antigen-induced arthritis. *BMC Musculoskelet Disord* 11:63.

Helgason, C. D., J. E. Damen, P. Rosten, R. Grewal, P. Sorensen, S. M. Chappel, A. Borowski, F. Jirik, G. Krystal, and R. K. Humphries. 1998. Targeted disruption of SHIP leads to hemopoietic perturbations, lung pathology, and a shortened life span. *Genes Dev* 12:1610–20.

Hirsch, E., V. L. Katanaev, C. Garlanda, O. Azzolino, L. Pirola, L. Silengo, S. Sozzani, A. Mantovani, F. Altruda, and M. P. Wymann. 2000. Central role for G protein-coupled phosphoinositide 3-kinase gamma in inflammation. *Science* 287:1049–53.

Hoekstra, M. F. 1997. Responses to DNA damage and regulation of cell cycle checkpoints by the ATM protein kinase family. *Curr Opin Genet Dev* 7:170–75.

Huber, M., C. D. Helgason, J. E. Damen, M. Scheid, V. Duronio, L. Liu, M. D. Ware, R. K. Humphries, and G. Krystal. 1999. The role of SHIP in growth factor induced signalling. *Prog Biophys Mol Biol* 71:423–34.

Irusen, E., J. G. Matthews, A. Takahashi, P. J. Barnes, K. F. Chung, and I. M. Adcock. 2002. p38 Mitogen-activated protein kinase-induced glucocorticoid receptor phosphorylation reduces its activity: Role in steroid-insensitive asthma. *J Allergy Clin Immunol* 109:649–57.

Ito, K. 2005. Corticosteroid resistance in COPD. In: *Chronic Obstructive Pulmonary Disease* (Barnes, P. J. ed) pp. 367–89, Taylor & Francis: Boca Raton, FL.

Ito, K., M. Ito, W. M. Elliott, B. Cosio, G. Caramori, O. M. Kon, A. Barczyk et al. 2005. Decreased histone deacetylase activity in chronic obstructive pulmonary disease. *N Engl J Med* 352:1967–76.

Ito, K., S. Lim, G. Caramori, K. F. Chung, P. J. Barnes, and I. M. Adcock. 2001. Cigarette smoking reduces histone deacetylase 2 expression, enhances cytokine expression, and inhibits glucocorticoid actions in alveolar macrophages. *FASEB J* 15:1110–12.

Ito, K., S. Yamamura, S. Essilfie-Quaye, B. Cosio, M. Ito, P. J. Barnes, and I. M. Adcock. 2006. Histone deacetylase 2-mediated deacetylation of the glucocorticoid receptor enables NF-{kappa}B suppression. *J Exp Med* 203:7–13.

Ji, H., F. Rintelen, C. Waltzinger, M. D. Bertschy, A. Bilancio, Pearce W, Hirsch E et al. 2007. Inactivation of PI3Kgamma and PI3Kdelta distorts T-cell development and causes multiple organ inflammation. *Blood* 110:2940–47.

Jia, S., Z. Liu, S. Zhang, M. D. Bertschy, A. Bilancio, W. Pearce, E. Hirsch et al. 2008. Essential roles of PI(3)K-p110beta in cell growth, metabolism and tumorigenesis. *Nature* 454:776–79.

Jones, G. E., E. Prigmore, R. Calvez, C. Hogan, G. A. Dunn, E. Hirsch, M. P. Wymann, and A. J. Ridley. 2003. Requirement for PI 3-kinase gamma in macrophage migration to MCP-1 and CSF-1. *Exp Cell Res* 290:120–31.

Jou, S. T., N. Carpino, Y. Takahashi, R. Piekorz, J. R. Chao, N. Carpino, D. Wang, and J. N. Ihle. 2002. Essential, nonredundant role for the phosphoinositide 3-kinase p110delta in signaling by the B-cell receptor complex. *Mol Cell Biol* 22:8580–91.

Kalesnikoff, J., L. M. Sly, M. R. Hughes, T. Buchse, M. J. Rauh, L. P. Cao, V. Lam, A. Mui, M. Huber, and G. Krystal. 2003. The role of SHIP in cytokine-induced signaling. *Rev Physiol Biochem Pharmacol* 149:87–103.

Ko, H. M., J. H. Kang, J. H. Choi, S. J. Park, S. Bai, and S. Y. Im. 2005. Platelet-activating factor induces matrix metalloproteinase-9 expression through Ca(2+)- or PI3K-dependent signaling pathway in a human vascular endothelial cell line. *FEBS Lett* 579:6451–58.

Kong, D. and T. Yamori. 2009. Advances in development of phosphatidylinositol 3-kinase inhibitors. *Curr Med Chem* 16:2839–54.

Koyasu, S. 2003. The role of PI3K in immune cells. *Nat Immunol* 4:313–19.

Kwak, Y. G., C. H. Song, H. K. Yi, P. H. Hwang, J. S. Kim, K. S. Lee, and Y. C. Lee. 2003. Involvement of PTEN in airway hyperresponsiveness and inflammation in bronchial asthma. *J Clin Invest* 111:1083–92.

Laffargue, M., R. Calvez, P. Finan, A. Trifilieff, M. Barbier, F. Altruda, E. Hirsch, and M. P. Wymann. 2002. Phosphoinositide 3-kinase gamma is an essential amplifier of mast cell function. *Immunity* 16:441–51.

Lam, R. S., E. Shumilina, N. Matzner, I. M. Zemtsova, M. Sobiesiak, C. Lang, E. Felder, P. Dietl, S. M. Huber, and F. Lang. 2008. Phosphatidylinositol-3-kinase regulates mast cell ion channel activity. *Cell Physiol Biochem* 22:169–76.

Lee, K. S., S. R. Kim, S. J. Park, H. K. Lee, H. S. Park, K. H. Min, S. M. Jin, and Y. C. Lee. 2006a. Phosphatase and tensin homolog deleted on chromosome 10 (PTEN) reduces vascular endothelial growth factor expression in allergen-induced airway inflammation. *Mol Pharmacol* 69:1829–39.

Lee, K. S., H. K. Lee, J. S. Hayflick, Y. C. Lee, and K. D. Puri. 2006b. Inhibition of phosphoinositide 3-kinase delta attenuates allergic airway inflammation and hyperresponsiveness in murine asthma model. *FASEB J* 20:455–65.

Lee, S. B., E. S. Cho, H. S. Yang, H. Kim, and H. D. Um. 2005. Serum withdrawal kills U937 cells by inducing a positive mutual interaction between reactive oxygen species and phosphoinositide 3-kinase. *Cell Signal* 17:197–204.

Liu, D., T. Zhang, A. J. Marshall, K. Okkenhaug, B. Vanhaesebroeck, and J. E. Uzonna. 2009. The p110delta isoform of phosphatidylinositol 3-kinase controls susceptibility to Leishmania major by regulating expansion and tissue homing of regulatory T cells. *J Immunol* 183:1921–33.

Liu, L., K. D. Puri, J. M. Penninger, and P. Kubes. 2007. Leukocyte PI3Kgamma and PI3Kdelta have temporally distinct roles for leukocyte recruitment in vivo. *Blood* 110:1191–98.

Maffucci, T., F. T. Cooke, F. M. Foster, C. J. Traer, M. J. Fry, and M. Falasca. 2005. Class II phosphoinositide 3-kinase defines a novel signaling pathway in cell migration. *J Cell Biol* 169:789–99.

Marwick, J. A., G. Caramori, C. S. Stevenson, P. Casolari, E. Jazrawi, P. J. Barnes, K. Ito, I. M. Adcock, P. A. Kirkham, and A. Papi. 2009. Inhibition of PI3Kdelta restores glucocorticoid function in smoking-induced airway inflammation in mice. *Am J Respir Crit Care Med* 179:542–48.

Nakamaru, Y., C. Vuppusetty, H. Wada, J. C. Milne, M. Ito, C. Rossios, M. Elliot et al. 2009. A protein deacetylase SIRT1 is a negative regulator of metalloproteinase-9. *FASEB J* 23:2810–19.

Neill, L., A. H. Tien, J. Rey-Ladino, and C. D. Helgason. 2007. SHIP-deficient mice provide insights into the regulation of dendritic cell development and function. *Exp Hematol* 35:627–39.

Newcomb, D. C., U. S. Sajjan, D. R. Nagarkar, Q. Wang, S. Nanua, Y. Zhou, C. L. McHenry et al. 2008. Human rhinovirus 1B exposure induces phosphatidylinositol 3-kinase-dependent airway inflammation in mice. *Am J Respir Crit Care Med* 177:1111–21.

Nichols, D., J. Chmiel, and M. Berger. 2008. Chronic inflammation in the cystic fibrosis lung: Alterations in inter- and intracellular signaling. *Clin Rev Allergy Immunol* 34:146–62.

Pinho, V., D. G. Souza, M. M. Barsante, F. P. Hamer, M. S. De Freitas, A. G. Rossi, and M. M. Teixeira. 2005. Phosphoinositide-3 kinases critically regulate the recruitment and survival of eosinophils in vivo: Importance for the resolution of allergic inflammation. *J Leukoc Biol* 77:800–10.

Pomel, V., J. Klicic, D. Covini, D. D. Church, J. P. Shaw, K. Roulin, F. Burgat-Charvillon et al. 2006. Furan-2-ylmethylene thiazolidinediones as novel, potent, and selective inhibitors of phosphoinositide 3-kinase gamma. *J Med Chem* 49:3857–71.

Powis, G., N. Ihle, and D. L. Kirkpatrick. 2006. Practicalities of drugging the phosphatidylinositol-3-kinase/Akt cell survival signaling pathway. *Clin Cancer Res* 12:2964–66.

Puri, K. D., T. A. Doggett, J. Douangpanya, Y. Hou, W. T. Tino, T. Wilson, T. Graf et al. 2004. Mechanisms and implications of phosphoinositide 3-kinase delta in promoting neutrophil trafficking into inflamed tissue. *Blood* 103:3448–56.

Puri, K. D., T. A. Doggett, C. Y. Huang, J. Douangpanya, J. S. Hayflick, M. Turner, J. Penninger, and T. G. Diacovo. 2005. The role of endothelial PI3Kgamma activity in neutrophil trafficking. *Blood* 106:150–57.

Qin, S. and P. B. Chock. 2003. Implication of phosphatidylinositol 3-kinase membrane recruitment in hydrogen peroxide-induced activation of PI3K and Akt. *Biochemistry* 42:2995–3003.

Rahman, I. and I. M. Adcock 2006. Oxidative stress and redox regulation of lung inflammation in COPD. *Eur Respir J* 28:219–42.

Reutershan, J., M. S. Saprito, D. Wu, T. Ruckle, and K. Ley. 2010. Phosphoinositide 3-kinase gamma required for lipopolysaccharide-induced transepithelial neutrophil trafficking in the lung. *Eur Respir J* 35:1137–47.

Sale, E. M. and G. J. Sale 2008. Protein kinase B: Signalling roles and therapeutic targeting. *Cell Mol Life Sci* 65:113–27.

Sasaki, T., J. Irie-Sasaki, R. G. Jones, A. J. Oliveira-dos-Santos, W. L. Stanford, B. Bolon, A. Wakeham et al. 2000. Function of PI3Kgamma in thymocyte development, T cell activation, and neutrophil migration. *Science* 287:1040–46.

Shahabuddin, S., R. Ji, P. Wang, E. Brailoiu, N. Dun, Y. Yang, M. O. Aksoy, and S. G. Kelsen. 2006. CXCR3 chemokine receptor-induced chemotaxis in human airway epithelial cells: Role of p38 MAPK and PI3K signaling pathways. *Am J Physiol Cell Physiol* 291:C34–39.

Sly, L. M., V. Ho, F. Antignano, J. Ruschmann, M. Hamilton, V. Lam, M. J. Rauh, and G. Krystal. 2007. The role of SHIP in macrophages. *Front Biosci* 12:2836–48.

Smith, G. C., N. Divecha, N. D. Lakin, and S. P. Jackson. 1999. DNA-dependent protein kinase and related proteins. *Biochem Soc Symp* 64:91–104.

Takeda, M., W. Ito, M. Tanabe, S. Ueki, H. Kato, J. Kihara, T. Tanigai et al. 2009. Allergic airway hyperresponsiveness, inflammation, and remodeling do not develop in phosphoinositide 3-kinase gamma-deficient mice. *J Allergy Clin Immunol* 123:805–12.

Thomas, M. J., A. Smith, D. H. Head, L. Milne, A. Nicholls, W. Pearce, B. Vanhaesebroeck et al. 2005. Airway inflammation: Chemokine-induced neutrophilia and the class I phosphoinositide 3-kinases. *Eur J Immunol* 35:1283–91.

To, Y., K. Ito, Y. Kizawa, M. Failla, M. Ito, T. Kusama, W. M. Elliott, J. C. Hogg, I. M. Adcock, and P. J. Barnes. 2010. Targeting phosphoinositide-3-kinase-{delta} with theophylline reverses corticosteroid insensitivity COPD. *Am J Respir Crit Care Med* 182:897–904.

Vanhaesebroeck, B., K. Ali, A. Bilancio, B. Geering, and L. C. Foukas. 2005. Signalling by PI3K isoforms: Insights from gene-targeted mice. *Trends Biochem Sci* 30:194–204.

Vestbo, J., T. Sorensen, P. Lange, A. Brix, P. Torre, and K. Viskum. 1999. Long-term effect of inhaled budesonide in mild and moderate chronic obstructive pulmonary disease: A randomised controlled trial. *Lancet* 353:1819–23.

Ward, S., Y. Sotsios, J. Dowden, I. Bruce, and P. Finan. 2003. Therapeutic potential of phosphoinositide 3-kinase inhibitors. *Chem Biol* 10:207–13.

Ward, S. G. and P. Finan. 2003. Isoform-specific phosphoinositide 3-kinase inhibitors as therapeutic agents. *Curr Opin Pharmacol* 3:426–34.

Wymann, M. P., K. Bjorklof, R. Calvez, P. Finan, M. Thomast, A. Trifilieff, M. Barbier, F. Altruda, E. Hirsch, and M. Laffargue. 2003. Phosphoinositide 3-kinase gamma: A key modulator in inflammation and allergy. *Biochem Soc Trans* 31:275–80.

Zhang, T. T., K. Okkenhaug, B. F. Nashed, K. D. Puri, Z. A. Knight, K. M. Shokat, B. Vanhaesebroeck, and A. J. Marshall. 2008. Genetic or pharmaceutical blockade of p110delta phosphoinositide 3-kinase enhances IgE production. *J Allergy Clin Immunol* 122:811–19.

Index

A

AA. *See* Arachidonic acid (AA)
AA-derived lipoxin A_4 (LXA_4), 46–47
Acetyl-glyceryl-ether-phosphoryl-choline
    (AGEPC). *See* Platelet-activating
    factor (PAF)
Acetylsalicylic acid (ASA), 215
Acetyltransferase, 240
1-(1-Acetypiperidin-4-yl)-3-adamantanylurea
    (APAU), 136
Acrolein, 190
Acute asthma, 5
Acute bronchoconstriction, 106
Acute chest syndrome, 58–59
Acute lung injury (ALI), 51, 269–274
Acute pneumonia, rat model of, 134
Acute respiratory distress syndrome (ARDS),
    48–49, 51–53
12-(3-Adamantan-1-yl-ureido)-dodecanoic acid
    (AUDA), 132, 136
Adenosine, 347
Adenoviral overexpression, of $sPLA_2$, 49
β-Adrenoceptor agonist, 13–15
Adverse effect
    FTY720 trials, 34
    LABA, 14
    in roflumilast, 18
Age-related macular degeneration (AMD), 299
AHR. *See* Airway hyperreactivity (AHR);
    Airway hyperresponsiveness (AHR)
Airway disease
    clinical evidence, 225–228
    COX inhibitors in, 188–191
    drug discovery effort, 222
    preclinical evidence, 224–225
    prostaglandin receptor antagonists in, 191
    in vitro biology, 222–224
Airway eosinophilia, 83
Airway epithelial cells, 266, 270
Airway hyperreactivity (AHR), 194
    mast cells and, 50
Airway hyperresponsiveness (AHR), 7, 243–244,
    324
    LTB_4, 109
Airway inflammation, 30
    ceramide in, 33
    chronic, 338
    of COPD, 10
    $cPLA_2α$ in, 48

CysLTs and, 107–108
evidence for role in, 219–222
lipid-mediated gene expression in, 34–35
lipid mediators in, 31, 46–47
5-LO and, 82–84
LPA in, 31
lysosomal PLA_2s in, 55
mast cells and, 50
mediators in, 244–245
PI3K and
    asthma and COPD, 345–346
    corticosteroid insensitivity, 348–349
    HRV, 347
    inhibition, 347
role of 5-Oxo-ETE in, 166–167
sphingolipids and, 289–291
$sPLA_2$s in, 48–55
in steroid-resistant asthma, 9
Airway inflammation mediator, 220
Airway neutrophilia, 84
Airway smooth muscle (ASM), 131, 166, 190
Airway smooth muscle cell (ASMC), 243, 264
Akt, 338–339
ALI. *See* Acute lung injury (ALI)
Alveolar phagocytes, 3, 55
Alveolar sac, damage to, 10
Alveoli, 3
    loss of elastic recoil, 10
ALX/FPR2 receptor, 216
AMD. *See* Age-related macular degeneration
    (AMD)
AM-103 inhibitor, 119
Antagonists, 27, 31
Antigens, 8
Anti-IgE monoclonal antibody, 18
Anti-inflammatory activity
    in rats, 90
    sEH inhibitors, 137
Anti-inflammatory agents, 228
Anti-inflammatory effects, by EETs, 129
Anti-inflammatory lipid mediators, 31–33
Antileukotriene drugs, 84–86
    direct 5-LO inhibitors
        compounds inhibition, 89–91
        iron-ligand inhibitors, 87
        nonredox-type 5-LO inhibitors, 87–89
        redox-active 5-LO inhibitors, 86–87
    therapy, 83
Apoptotic inflammatory cells, 214
Apoptotic neutrophils, 156

Printed and bound by CPI Group (UK) Ltd, Croydon, CR0 4YY

21/10/2024

01777089-0014